# From Cardinals to Chaos

*Reflections on the Life and Legacy of Stanislaw Ulam*

Editor

NECIA GRANT COOPER

Associate Editors

Roger Eckhardt

Nancy Shera

Design

Gloria Sharp

Katherine Norskog

*Los Alamos National Laboratory*



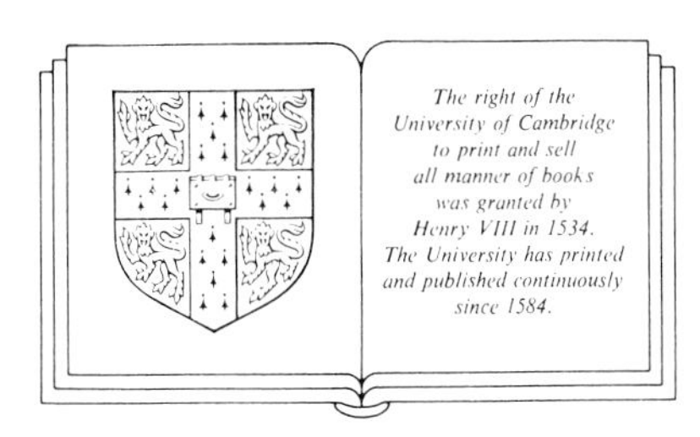


CAMBRIDGE UNIVERSITY PRESS

CAMBRIDGE

NEW YORK NEW ROCHELLE MELBOURNE SYDNEY

# Contents

*The staff of* Los Alamos Science *is deeply indebted to Françoise Ulam for her help on this issue. She generously opened Stan's files to us and gave advice and moral support whenever she was called upon. Her unfailing kindness and patience added a special element to the privilege and challenge of working on this volume.*

PART I

# Stan Ulam

THE MAN ♠ HIS LIFE ♠ HIS STYLE

*For forty years on and off Stan Ulam was a catalyst at Los Alamos, influencing the thinking of those around him. He may be less remembered than some great minds who wrote more and focused on fewer things. But his was the kind of genius that is unforgettable to those who knew him. In this issue we have tried to capture some of what he was and what he started. We hope the contents will resonate with the playful, expansive, thinking part all of us have inside us.*

א
Ulam
Everett
SEGLER

# Esquisse

*by Françoise Ulam*

Stan was larger than life; his person defies description. Not any one, I think (myself included, of course), ever viewed the whole of him. I hope that from this special issue a composite picture will emerge. All I am able to contribute, at this time, is the following quick thumbnail sketch. Stan never passed unnoticed; he came on fast and strong, as the expression goes. (I think that trait is genetic, for it exists in varying degrees among members of his family I have known.)

He was a loner, a maverick, a very complicated man, a Pole, and, above all, a study in contrasts and contradictions, which often aroused mixed and conflicting emotions in people. He moved only to the beat of his own drum and never kowtowed to anyone or stooped to promote himself. Given to bragging in jest about inconsequential prowesses, he was singularly modest about his scientific accomplishments. "Posterity will decide!" was his way of putting it to me.

He electrified the air around all who came in touch with him, for his wit, culture, and erudition were dazzling. It was a sport for him to beat people at their own game. With a humanist he would display his classical education and "one-up" him with Latin and Greek. With historians he loved to debate obscure points of their specialties. And when he met a chemist, he would expound on his own views of chemistry—a subject he himself admitted he knew very little about.

Characteristically, he measured himself against the great of the past and not against his contemporaries, from whom he had a sort of Olympian detachment. But it's lonely up there, so to test his own thoughts and opinions he constantly craved company, though often he felt excruciatingly bored. "People never tell me anything interesting," he would complain.

He lived mainly in the confines of his mind, in a world of abstract cogitations. This made him shun most other forms of activity except chess and tennis, which he enjoyed. Outside science what attracted him the most were history, the classics, antiquity, the Renaissance.

Another characteristic of his was an absolute self-confidence and unflinching optimism. This served him in good stead in the dark moments of his life, for it enabled him to block them off.

But he paid a price. This urbane, gentle man who always appeared smiling, affable, and at ease suffered from what he called a "nervous stomach." When alone he tended to dwell on such discomforts, except at the moment of death, which by contrast he took with extraordinary detachment. I believe that in a way he sensed it coming, though at the time, neither he nor I fully gauged the gravity of his final malaise.

We lived in New Mexico the better part of our lives, and from the first he loved its vistas and the quality of its air. Despite a deep-seated longing for his old world, Stan thrived in this country, and he loved its openness, dynamism, and scientific audacity.

In his youth he had the reputation of a Don Juan and remained very attractive to women, whose attentions he enjoyed. "Women seem to like me. I wonder why?" he would ask. As for me, over the years I became a partner on whom he leaned increasingly for most practical matters, his "Home Secretary," in the British sense of the term, and managed most of his time and his affairs, which left him free to indulge his mind.

A more extraordinary person I have never known. I came alive when we first met at Harvard in 1939 and consider myself most fortunate to have basked in his aura, at the frontiers of twentieth–century science, for nearly half a century. ■

Santa Fe
October 1984

*The text of "Vita" consists solely of quotations from Stan's autobiography, *Adventures of a Mathematician*, which are reproduced with permission from Charles Scribner's Sons, an imprint of Macmillan Publishing Company. *Adventures* was assembled by Françoise Ulam from hours of recorded reminiscences. The selections reprinted here not only chronicle Stan's life but also highlight those areas of his work that are featured in Part II of this volume.

# Vita*

## POLISH YEARS

*1909 Born April 13 in Lwów, Poland, then part of Austro-Hungarian Empire*

*1916 Russian troops occupy Lwów. Family moves temporarily to Vienna*

*1918 Family returns to Lwów, now part of Republic of Poland. Ukrainians besiege the city*

*1919 Enters gymnasium*

*1927 Matriculates from gymnasium. Enters Lwów Polytechnic Institute*

*Father (left) and uncle Szymon seeing Stan and his young brother, Adam, off for the last time at Gdynia, Poland, 1939*

*Passport photo, 1935*

My father, Jozef Ulam, was a lawyer. He was born in Lwów, Poland, in 1877. At the time of his birth the city was the capital of the province of Galicia, part of the Austro-Hungarian Empire. When I was born in 1909 this was still true ... My mother, Anna Auerbach, was born in Stryj, a small town some sixty miles south of Lwów, near the Carpathian Mountains. Her father was an industrialist who dealt in steel and represented factories in Galicia and Hungary.

In November of [1918] the Ukrainians besieged the city ... Our house was in a relatively safe part of town, even though occasional artillery shells struck nearby ... Many of our relatives came to stay with us ... some thirty of them, half being children. There were not nearly enough beds, of course, and I remember people sleeping everywhere on rolled rugs on the floor ... Strangely enough, my memories of these days are of the fun I had playing, hiding, learning card games with the children for the two weeks before the siege was lifted ... For children wartime memories are not always traumatic.

At the age of ten in 1919 I passed the entrance examination to the gymnasium. This was a secondary school patterned after the German gymnasia and the French lycées. Instruction usually took eight years. I was an A student, except in penmanship and drawing, but did not study much.

Around [that time] so much was written in newspapers and magazines about the theory of relativity that I decided to find out what it was all about ... This interest became known among friends of my father, who remarked that I "understood" the theory of relativity ... This gave me a reputation I felt I had to maintain, even though I knew that I did not genuinely understand any of the details. Nevertheless, this was the beginning of my reputation as a "bright child."

I had mathematical curiosity very early. My father had in his library a wonderful series of German paperback books—*Reklam*, they were called. One was Euler's *Algebra*. I looked at it when I was perhaps ten or eleven, and it gave me a mysterious feeling. The symbols looked like magic signs; I wondered whether one day I could understand them.

In high school, I was stimulated by ... the problem of the existence of odd perfect numbers. An integer is perfect if it is equal to the sum of all its divisors including one but not itself. For instance: $6 = 1 + 2 + 3$ is perfect. So is $28 = 1 + 2 + 4 + 7 + 14$. You may ask: does there exist a perfect number that is odd? The answer is unknown to this day.

Poincaré molded portions of my scientific thinking. Reading one of his books today demonstrates how many wonderful truths [remain], although everything in mathematics has changed almost beyond recognition and in physics perhaps even more so. I admired Steinhaus's book almost as much, for it gave many examples of actual mathematical problems.

In 1927 I passed my three-day matriculation examinations and a period of indecision began. The choice of a future career was not easy. My father, who had wanted me to become a lawyer so I could take over his large practice, now recognized that my inclinations lay in other directions ... My parents urged me to become an engineer, and so I applied for admission at the Lwów Polytechnic Institute as a student of either mechanical or electrical engineering.

# Vita

## POLISH YEARS

*1928 Writes his first paper, published in* Fundamenta Mathematicae *in 1929*

*1931 Attends mathematical congress in Wilno*

*1932 M.A. from Polytechnic Institute*

*1933 D.Sc. from Polytechnic Institute*

In the fall of 1927 I began attending lectures at the Polytechnic Institute in the Department of General Studies, because the quota of Electrical Engineering already was full. The level of the instruction was obviously higher than that at high school, but having read Poincaré and some special mathematical treatises, I naively expected every lecture to be a masterpiece of style and exposition. Of course, I was disappointed.

Soon I could answer some of the more difficult questions in [Kuratowski's] set theory course, and I began to pose other problems. Right from the start I appreciated Kuratowski's patience and generosity in spending so much time with a novice. Several times a week I would accompany him to his apartment at lunch time, a walk of about twenty minutes, during which I asked innumerable mathematical questions . . . Between classes, I would sit in the offices of some of the mathematics instructors. At that time I was perhaps more eager than at any other time in my life to do mathematics to the exclusion of almost any other activity.

At the beginning of the second semester of my freshman year, Kuratowski told me about a problem in set theory that involved transformations of sets. It was connected with a well-known theorem of Bernstein: if $2A = 2B$, then $A = B$, in the arithmetic sense of infinite cardinals. This was the first problem on which I really spent arduous hours of thinking. I thought about it in a way which now seems mysterious to me, not consciously or explicitly knowing what I was aiming at. So immersed in some aspects was I, that I did not have a conscious overall view. Nevertheless, I managed to show by means of a construction how to solve the problem, devising a method of representing by graphs the decomposition of sets and the corresponding transformations. Unbelievably, at the time I thought I had invented the very idea of graphs.

*Joint mathematics-physics meeting, Lwów, 1930. Stan is number 10*

*Panorama of Lwów in the 1970s*

*Kuratowski and a photo of Banach, circa 1968*

It was Mazur (along with Kuratowski and Banach) who introduced me to certain large phases of mathematical thinking and approaches. From him I learned much about the attitudes and psychology of research. Sometimes we would sit for hours in a coffee house. He would write just one symbol or a line like y = f(x) on a piece of paper, or on the marble table top. We would both stare at it as various thoughts were suggested and discussed. These symbols in front of us were like a crystal ball to help us focus our concentration.

Beginning with the third year of studies, most of my mathematical work was really started in conversations with Mazur and Banach. And according to Banach some of my own contributions were characterized by a certain "strangeness" in the formulation of problems and in the outline of possible proofs. As he told me once some years later, he was surprised how often these "strange" approaches really worked.

He [Banach] enjoyed long mathematical discussions with friends and students. I recall a session with Mazur and Banach at the Scottish Café which lasted seventeen hours without interruption except for meals.

These long sessions in the cafés were probably unique. Collaboration was on a scale and with an intensity I have never seen surpassed, equaled or approximated anywhere—except perhaps at Los Alamos during the war years . . . Needless to say such mathematical discussions were interspersed with a great deal of talk about science in general (especially physics and astronomy), university gossip, politics, the state of affairs in Poland; or to use one of John von Neumann's favorite expressions, the "rest of the universe." The shadow of coming events, of Hitler's rise in Germany and the premonition of a world war loomed ominously.

The second big congress I attended [of mathematicians from the Slavic countries] was held in Wilno in 1931 . . . At the congress I gave a talk about the results obtained with Mazur on geometrical isometric transformations of Banach spaces, demonstrating that they are linear. Some of the additional remarks we made at the time are still unpublished. In general, the Lwów mathematicians were on the whole somewhat reluctant to publish. Was it a sort of pose or a psychological block?

If I had to name one quality which characterized the development of this school, made up of the mathematicians from the University [of Lwów] and the Polytechnic Institute, I would say that it was their preoccupation with the heart of the matter that forms mathematics. On a set theoretical and axiomatic basis we examined the nature of a general space, the general meaning of continuity, general sets of points in Euclidean space, general functions of real variables, a general study of the spaces of functions, a general idea of the notions of length, area and volume, that is to say, the concept of measure and the formulation of what should be called probability.

In 1932 I was invited to give a short communication at the International Mathematical Congress in Zürich. This was the first big international meeting I attended, and I felt very proud to have been invited. In contrast to some of the Polish mathematicians I knew, who were terribly impressed by western science, I had confidence in the equal value of Polish mathematics. Actually this confidence extended to my own work. Von Neumann once told my wife, Françoise, that he had never met anyone with as much self-confidence—adding that perhaps it was somewhat justified.

By 1934 I had become a mathematician rather than an electrical engineer. It was not so much that I was doing mathematics, but rather that mathematics had taken possession of me . . . At twenty-five, I had established some results in measure theory which soon became well known. These solved certain set theoretical problems attacked earlier by Hausdorff, Banach, Kuratowski, and others. These measure problems again became significant years later in connection with the work of Gödel and more recently with that of Paul Cohen. I was also working in topology, group theory, and probability theory. From the beginning I did not become too specialized. Although I was doing a lot of mathematics, I never really considered myself as only a mathematician. This may be one reason why in later life I became involved in other sciences.

[Nevertheless] ever since I started learning mathematics I would say that I have spent—regardless of any other activity—on the average two to three hours a day thinking and two to three hours reading or conversing about mathematics.

# Vita

## PRINCETON
## HARVARD
## WISCONSIN

Stan in Warsaw, 1972

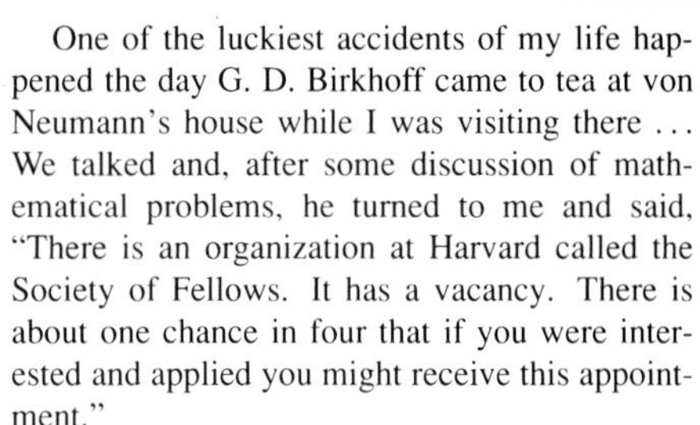

Oxtoby and Stan at Harvard, circa 1936

*1934* *Postdoctoral travels and studies in Vienna, Zürich, Paris, and Cambridge (England)*

*1935* *Scottish Book originates*

*Returns to Poland. Receives letter of invitation to Institute for Advanced Study in Princeton*

*December: Sails to America*

*1936–39* *Academic years with Harvard Society of Fellows. Summers at home in Poland*

*1939* *Leaves home for the last time in the fall of 1939, accompanied by his young brother, Adam*

*1939–40* *Lecturer at Harvard*

*1940–41* *Instructor at University of Wisconsin. Meets C. J. Everett. Works with him on ordered groups and projective algebras*

*1941* *Becomes American citizen. Tries to volunteer in the U.S. Air Force*

*1941–43* *Assistant Professor at University of Wisconsin*

In 1934, the international situation was becoming ominous. Hitler had come to power in Germany. His influence was felt indirectly in Poland. There were increasing displays of inflamed nationalism ... and anti-Semitic demonstrations ... For years my uncle Karol Auerbach had been telling me: "Learn foreign languages!" Another uncle, Michael Ulam, an architect, urged me to try a career abroad. For myself, unconscious as I was of the realities of the situation in Europe, I was prompted to arrange a longish trip abroad ... to meet other mathematicians ... and in my extreme self-confidence, try to impress the world with some new results. My parents were willing to finance the trip.

It was only toward the end of 1934 that I entered into correspondence with von Neumann. He was then in the United States, a very young professor at the Institute for Advanced Study in Princeton. I wrote him about some problems in measure theory. He had heard about me from Bochner, and in his reply he invited me to come to Princeton for a few months, saying that the Insitiue could offer me a $300 stipend. I met him [in Warsaw] shortly after my return from England ... Von Neumann appeared quite young to me, although he was ... some five or six years older than I ... At once I found him congenial. His habit of intermingling funny remarks, jokes, and paradoxical anecdotes or observations of people into his conversation, made him far from remote or forbidding.

[At the Institute] I went to lectures and seminars, heard Morse, Veblen, Alexander, Einstein, and others, but was surprised how little people talked to each other compared to the endless hours in the coffee houses in Lwów ... There was another way in which the Princeton atmosphere was entirely different from what I expected: it was fast becoming a way station for displaced European scientists. In addition, these were still depression days and the situation in universities in general and in mathematics in particular was very bad.

One of the luckiest accidents of my life happened the day G. D. Birkhoff came to tea at von Neumann's house while I was visiting there ... We talked and, after some discussion of mathematical problems, he turned to me and said, "There is an organization at Harvard called the Society of Fellows. It has a vacancy. There is about one chance in four that if you were interested and applied you might receive this appointment."

I came to the Society of Fellows during its first few years of existence ... I was given a two-room suite in Adams House, next door to another new fellow in mathematics by the name of John Oxtoby ... He was interested in some of the same mathematics I was: in set theoretical topology, analysis, and real function theory. Right off, we started to discuss problems concerning the idea of "category" of sets. "Category" is a notion in a way parallel to but less quantitative than the measure of sets ... We quickly established some new results, and the fruits of our conversations ... were published as two notes in *Fundamenta*. We followed this with an ambitious attack on the problem of the existence of ergodic transformations. The ideas and definitions connected with this had been initiated in the nineteenth century by Boltzmann.

Birkhoff, in his trail-breaking papers and in his book on dynamical systems, had defined the notion of "transitivity." Oxtoby and I worked on the completion to the existence of limits in the ergodic theorem itself ... We wanted to show that on every manifold (a space representing the possible states of a dynamical system)—the kind used in statistical mechanics—such ergodic behavior is the rule ... It took us more than two years to break through and to finish a long paper, which appeared in *The Annals of Mathematics* in 1941 and which I consider one of the more important results that I had a part in.

While I was at Harvard, Johnny came to see me a few times, and I invited him to dinner at the Society of Fellows. We would also take automobile drives and trips together during which we discussed everything from mathematics to literature and talked without interruption while still paying attention to our surroundings. Johnny liked this kind of travel very much.

Each summer between 1936 and 1939, I returned to Poland for a full three months. The first time, after only a few months' stay in America, I was surprised that street cars ran, electricity and telephones worked. I had become imbued with the idea of America's absolute technological superiority and unique "know-how." My main emotional reactions were, of course, related to reunion with my family and friends, and the familiar scenes of Lwów, followed by a longing to return to the free and hopeful "open-ended" conditions of life in America.

I had to go to the American consulate in Warsaw each summer I was in Poland to apply for a new visitor's visa in order to return to the United States. Finally, the consul said to me, "Instead of coming here every summer for a new visa, why don't you get an immigration visa?" It was lucky that I did, for just a few months later these became almost impossible to obtain.

*Claire, at 14 months, and Françoise, Los Angeles, 1945*

[My brother] Adam and I were staying in a hotel on Columbus Circle [in New York] ... It must have been around one or two in the morning when the telephone rang ... my friend the topologist Witold Hurewicz began ... "Warsaw has been bombed, the war has begun." That is how I learned about the beginning of World War II ... Adam was asleep; I did not wake him. There would be time to tell him the news in the morning. Our father and sister were in Poland, so were many other relatives. At that moment, I suddenly felt as if a curtain had fallen on my past life ... There has been a different color and meaning to everything ever since.

Birkhoff helped me to secure the job [at the University of Wisconsin] ... Almost at once I met congenial, intelligent people not only in mathematics and science, but also in the humanities and arts ... So I found Madison not at all the intellectual desert I had feared it would be ... I was given a light teaching load ... But the very expression ... implied physical effort and fatigue—two things I have always been afraid of, lest they interfere with my own thinking and research.

Something else happened to make Madison most important to me. It was there that I married a French girl, who was an exchange student at Mount Holyoke College and whom I had met in Cambridge, Françoise Aron. Marriage, of course, changed my way of life, greatly influencing my daily mode of work, my outlook on the world, and my plans for the future.

It was in Madison that I met C. J. Everett ... [He] and I hit it off immediately. As a young man he was already eccentric, original, with an exquisite sense of humor, wry, concise, and caustic in his observations. He was totally devoted to mathematics ... I found in him much that resembled my friend Mazur in Poland, the same kind of epigrammatic comments and jokes ... We collaborated on difficult problems of "order"—the idea of order for elements in a group. In our mathematical conversations, as always, I was the optimist, and had some general, sometimes only vague ideas. He supplied the rigor, the ingenuities in the details of the proof, and the final constructions. Everett exhibited a trait of mind whose effects are, so to speak, non-additive: persistence in thinking. Thinking continuously ... for an hour, is at least for me—and I think for many mathematicians—more effective than doing it in two half-hour periods. It is like climbing a slippery slope. If one stops, one tends to slide back. Both Everett and Erdös have this characteristic of long-distance stamina.

I was asked to run the mathematics colloquium, which took place every two weeks ... The colloquium was run differently from what I had known in Poland, where speakers gave ten- or twenty-minute informal talks. At Madison they were one-hour lectures. There is quite a difference between short seminar talks like those at our math society in Lwów, and the type of lecture which necessitates talking about major efforts. The latter were better prepared, of course, but their greater formality removed some of the spontaneity and stimulation of the shorter exchanges.

*Harvard Society of Fellows, 1938*

# Vita

## LOS ALAMOS

*1943* *Receives invitation to join the Manhattan Project*

*1944* *February 4: Arrives in Los Alamos*

*Joins Edward Teller's group in the Theoretical Division under Hans Bethe*

*Teller's group is transferred to Fermi's Division*

Von Neumann wrote this letter to Stan in longhand. It is reproduced here in typewritten form to decrease its size and increase its legibility.

COSMOS CLUB
WASHINGTON 5, D. C

November 9, 1943

Dear Stan,

I am very glad that Mr. Hughes ''and all he stands for'' have come through. I told them about you, because you wrote me several times in the past that you definitely wanted a war job, and because this is a very real possibility, where you could do very effective and useful work.

The project in question is exceedingly important, probably beyond all adjectives I could affix to it. It is very interesting, too, and the theoretical (and other) physicists connected with it are probably the best group existing anywhere at this moment. It does require some computational work, but there is no doubt, that everybody will be most glad and give you all the encouragement you can wish in doing original research on the subject, for which there is ample opportunity. I can also assure you of my own cooperation in this respect.

The secrecy requirements of this project are rather extreme, and it will probably necessitate your and your families essentially staying on the premises (except for vacations) as long as you choose to be associated with it.

To repeat: If you want war work, this is probably a quite exceptional opportunity.

I may be able to give you a better idea orally, and I would be glad to do so, if we can meet somewhere before you answer---but I suppose that there is no time. (I will be in Princeton on 13-15 and in Washington on 16-17 of this month.)

So you really count on a quite short war? I don't see that from a purely technical standpoint  Germany need be broken before next fall. Of course a collapse may come any day from now on for moral and political reasons, but I can't [know] how to judge that, without knowing much more about the present state and efficiency of the Nazi political machine.

And there is still a year's worth of Asiatic war after that. Anyhow, qui vivra verra ...

It seems that Morgenstern's and my book on ''games'' will be out in 3 months or so.

Best regards, and looking forward to seeing you soon---here or ''there''---

as ever

*Von Neumann, circa 1950*

During the late spring of 1943, I wrote to von Neumann about the possibility of war work . . . I received an official invitation to join an unidentified project that was doing important work, the physics having something to do with the interior of stars. The letter inviting me was signed by the famous physicist Hans Bethe.

*On the terrace of Fuller Lodge. Clockwise from lower left: Davis, Metropolis, Ulam, an unidentified person, McMillan, and de Hoffman*

*Left to right: Metropolis, Ulam, Morgenstern, Carol Stein, Brillouin, Paul Stein*

*Enrico Fermi*

*Left to right: von Neumann, Feynman, Ulam at Bandelier, 1949*

Finally I learned that we were going to New Mexico, to a place not far from Santa Fe. Never having heard about New Mexico, I went to the library and borrowed the Federal Writers' Project Guide to New Mexico. At the back of the book, on the slip of paper on which borrowers signed their names, I read the names of Joan Hinton, David Frisch, Joseph McKibben, and all the other people who had been mysteriously disappearing [from Madison] to hush-hush war jobs without saying where. I had uncovered their destination in a simple and unexpected fashion. It is next to impossible to maintain absolute secrecy and security in war time.

[Upon my arrival at Los Alamos, Johnny] took me aside and . . . told me of all the possibilities which had been considered, of the problems relating to the assembling of fissionable materials, about plutonium (which did not yet physically exist even in the most microscopic quantities at Los Alamos). I remember very well, when a couple of months later I saw Robert Oppenheimer running excitedly down a corridor holding a small vial in his hand, with Victor Weisskopf trailing after him. He was showing some mysterious drops of something at the bottom of the vial. Doors opened, people were summoned, whispered conversations ensued, there was great excitement. The first quantity of plutonium had just arrived at the lab.

It is one thing to know about physics abstractly, and quite another to have a practical encounter with problems directly connected with experimental data . . . I found out that the main ability to have was a visual, and also an almost tactile, way to imagine the physical situations, rather than a merely logical picture of the problems . . . Very few mathematicians seem to possess [such an imagination] to any great degree.

Strangely enough, the actual working problems did not involve much of the mathematical apparatus of quantum theory although it lay at the base of the phenomena, but rather dynamics of a more classical kind—kinematics, statistical mechanics, large-scale motion problems, hydrodynamics, behavior of radiation . . . Compared to quantum theory the project work was like applied mathematics as compared with abstract mathematics.

[Edward] Teller, in whose group I was supposed to work, talked to me on that first day about a problem in mathematical physics that was part of the necessary theoretical work in preparation for developing the idea of a "super" bomb, as the proposed thermonuclear hydrogen bomb was then called . . . Teller's problem concerned the interaction of an electron gas with radiation . . . This was the first technical problem in theoretical physics I had ever tackled in my life . . . It was a messy little job. Edward was not satisfied with my rather elementary derivations.

After this first work on Edward's problem, I spread out my interests to other related questions, one being the problem of statistics of neutron multiplication. This was more tangible for me from the purely mathematical side. I discussed such problems of branching and multiplying patterns with David Hawkins.

A discussion with von Neumann . . . [in] early 1944 took several hours, and concerned ways to calculate the course of an implosion more realistically than the first attempts outlined by him and his collaborators. The hydrodynamical problem was simply stated, but very difficult to calculate—not only in detail, but even in order of magnitude . . . In this discussion I stressed pure pragmatism and the necessity for attempting to get a heuristic survey of the general problem by simpleminded brute force—that is, more realistic, massive numerical work . . . With the available computing facilities, the accuracy of the necessary numerical work could not be satisfactory. This was one of the first reasons for pressing for the development of electronic computers.

Fermi was short, sturdily built, strong in arms and legs, and rather fast moving. His eyes, darting at times, would be fixed reflectively when he was considering some questions . . . He would try to elucidate other persons' thoughts by asking questions in a Socratic manner, yet more concretely than in Plato's succession of problems.

I think he had a supreme sense of the important. He did not disdain work on the so-called smaller problems; at the same time, he kept in mind the order of importance of things in physics. This quality is more vital in physics than in mathematics, which is not so uniquely tied to "reality." Strangely enough, he started as a mathematician. Some of his first papers with very elegant results were devoted to the problem of ergodic motion. When he wanted to, he could do all kinds of mathematics. To my surprise, once on a walk he discussed a mathematical question arising from statistical mechanics which John Oxtoby and I had solved in 1941.

[Fermi] could be also quite a tease. I remember his Italian inflections when he taunted Teller with statements like "Edward-a how com-a the Hungarians have not-a invented anything?"

*Clockwise from lower left: An unidentified person, Mark, Matthias, Ulam, Evans, Cowan, Metropolis*

# Vita

## LOS ALAMOS

*1945 July 16: Trinity Test*

*September: Moves to Los Angeles as Associate Professor at University of Southern California*

*1946 January: Acute attack of encephalitis*

*April: Attends secret conference at Los Alamos*

*May: Returns to Los Alamos*

*1947 C. J. Everett joins the Laboratory*

*Seminars on the Monte Carlo Method and hydrodynamical calculations*

*Beginning of heuristic studies on electronic computing machines*

*Ulam and Everett develop theory of multiplicative processes*

*1949 Russian atomic bomb test*

*Truman directs AEC to proceed with work on the hydrogen bomb*

One thing that relieved the repetition and alternation of work, intellectual discussions, evening gatherings, social family visits and dinner parties, was when a group of us would play poker about once a week. The group included Metropolis, Davis, Calkin, Flanders, Langer, Long, Konopinski, von Neumann (when he was in town), Kistiakowski sometimes, Teller, and others. We played for small stakes; the naïveté of the game and the frivolous discussions laced with earthy exclamations and rough language provided a bath of refreshing foolishness from the very serious and important business that was the raison d'être of Los Alamos.

The Trinity test, Hiroshima, V-J Day, and the story of Los Alamos exploded over the world almost simultaneously with the A-Bomb. Publicity over the secret wartime Project filled the newspapers and its administrative heads were thrown in the limelight.

As I was reading [such items,] something else flashed through my mind, a story of a "pension" in Berlin before the war ... One man was taking most of the asparagus that was on the platter. Whereupon another man stood up shyly and said: "Excuse me, Mr. Goldberg, we also like asparagus!" And the expression "asparagus" became a code word in our private conversations for trying to obtain an unduly large share of credit for scientific work or any other accomplishment of a joint or group character. Johnny loved this story ... We would plan to write a twenty-volume treatise on "Asparagetics through the Ages" ... But levities like these could hardly alleviate the general feeling of foreboding upon entering into the era of history that would be called the Atomic Age.

In early September of 1945, I went to Los Angeles to look for housing and to prepare our move from Los Alamos.

It was a stormy day; on the walk from the bus to the house in Balboa the violent winds almost choked me. That same night I developed a fantastic headache ... The following night ... I noticed that my speech was confused, that I was barely able to form words. I tried to talk ... but it was mostly a meaningless mumble—a most frightening experience ... A severe attack of brain troubles began, which was to be one of the most shattering experiences of my life ... Many of the recollections of what preceded my operation are hazy. Thanks to what Françoise told me later I was able to put it together ... She feared I was dying and made a frantic telephone call to the surgeon, who decided the operation should be performed immediately. This probably saved my life; the emergency operation relieved the severe pressure on my brain which was causing all the trouble ... The illness was tentatively diagnosed as a kind of virus encephalitis. But the disquietude about the state of my mental faculties remained with me for a long time, even though I recovered speech completely.

Many friends came to visit me ... Metropolis came all the way from Los Alamos. His visit cheered me greatly. I found out that the security people in Los Alamos had been worried that in my unconscious or semi-conscious states I might have revealed some atomic secrets.

As I was preparing to leave [the hospital], ... Erdös appeared at the end of the hall ... In the car on the way home from the hospital, Erdös plunged immediately into a mathematical conversation. I made some remarks, he asked me about some problem, I made a comment, and he said: "Stan, you are just like before." These were reassuring words.

*Stan and Everett in Madison, 1942*

In the days that followed we had more and more mathematical discussions and longer and longer walks on the beach. Once he stopped to caress a sweet little child and said in his special language: "Look, Stan! What a nice epsilon." A very beautiful young woman, obviously the child's mother, sat nearby, so I replied, "but look at the capital epsilon." This made him blush with embarrassment.

Two seminar talks I gave shortly after my return [to Los Alamos] turned out to have good or lucky ideas and led to successful further developments. One was on what was later called the Monte Carlo method, and the other was about some new possible methods of hydrodynamical calculations. Both talks laid the groundwork for very substantial activity in the applications of probability theory and in the mechanics of continua. [Both ideas required extensive machine computation.]

Computing machines came about through the confluence of scientific and technological developments. On one side was the work in mathematical logic, in the foundations of mathematics, in the detailed study of formal systems, in which von Neumann played such an important role; on the other was the rapid progress of technological discoveries in electronics which made it possible to construct electronic computers.

Almost immediately after the war Johnny and I also began to discuss the possibilities of using computers heuristically to try to obtain insights into questions of pure mathematics. By producing examples and by observing the properties of special mathematical objects one could hope to obtain clues as to the behavior of general statements which have been tested on examples.

*The Gamow cartoon*

It was in 1949 . . . that George Gamow, whom I had met briefly in Princeton before the war, came to Los Alamos for a lengthy visit . . . There was nothing dry about him. A truly "three-dimensional" person, he was exuberant, full of life, interested in copious quantities of good food, fond of anecdotes, and inordinately given to practical jokes.

Banach once told me, "Good mathematicians see analogies between theorems or theories, the very best ones see analogies between analogies." Gamow possessed this ability to see analogies between models for physical theories to an almost uncanny degree . . . It was along the great lines of the foundations of physics, in cosmology, and in the recent discoveries in molecular biology that his ideas played an important role. His pioneering work in explaining the radioactive decay of atoms was followed by his theory of the explosive beginning of the universe, the "big bang" theory (he disliked the term by the way), and the subsequent formation of galaxies.

Shortly after President Truman's announcement directing the AEC to proceed with work on the H-Bomb, E. O. Lawrence and Luis Alvarez visited Los Alamos from Berkeley and started discussions with Bradbury and then with Gamow, Teller, and myself about the feasibility of constructing a "super." This visit played a part in the politics of this enterprise.

Several different proposals of ideas existed on how to initiate the thermonuclear reaction, using fission bombs as starter. One of Gamow's was called "the cat's tail." Another was Edward's original proposal. Gamow drew a humorous cartoon with symbolic representations of these various schemes. In it he squeezes a cat by the tail, I spit in a spittoon, and Teller wears an Indian fertility necklace, which according to Gamow is the symbol for the womb, a word he pronounced "vombb." This cartoon has appeared among the illustrations in his autobiography, *My World Line*, published by The Viking Press in 1970.

A first committee was formed to organize all work on the 'super' and investigate all possible schemes for constructing it. The committee's work was directed by Teller, as chairman, Gamow and myself . . . Both Gamow and I showed a lot of independence of thought in our meetings and Teller did not like this very much. Not too surprisingly, the original 'super' directing committee soon ceased to exist.

UNIVERSITÉ DE PARIS

**FACULTÉ DES SCIENCES**

Année scolaire 1953-1954

M. S. M. ULAM, de l'Université de Californie, fera une conférence *le Mardi 1er Juin 1954, à 18 heures*, à l'Institut Henri Poincaré, 11, rue Pierre-Curie (Amphithéâtre Darboux), sur le sujet suivant :

**Etudes heuristiques de divers problèmes mathématiques sur les machines à calculer électroniques**

Vu et approuvé : Le Recteur, Président du Conseil de l'Université, Jean SARRAILH.

Le Doyen de la Faculté des Sciences, Joseph PÉRÈS

# Vita

## LOS ALAMOS

*Hand calculations by Ulam and Everett suggest that ignition scheme for "super" won't work*

*Results of calculations confirmed by von Neumann and Evans on Princeton computer*

*Ulam suggests new approach to ignition*

*Teller suggests a related approach and presents it to the General Advisory Committee*

1951–52 *Fall semester: Visiting Professor at Harvard*

*Begins serious discussions of cellular automata with von Neumann*

1952 *Summer: Studies nonlinear systems with Fermi and Pasta*

1955 *Fermi dies*

1956–57 *Visiting Professor at Massachusetts Institute of Technology*

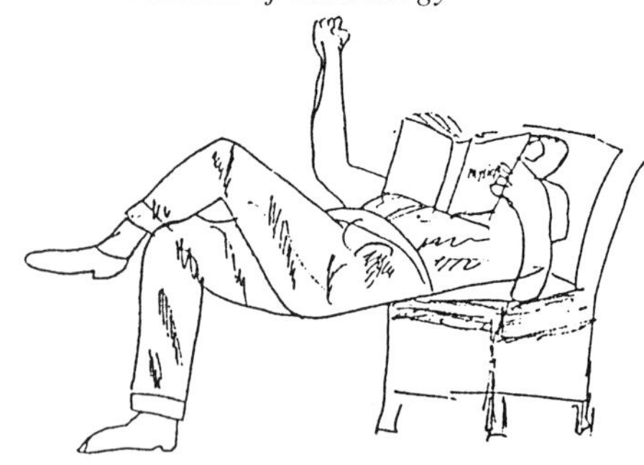

*A sketch by Françoise of Stan's favorite reading position*

Still Teller kept on hinting that not enough work was being done on his original scheme for the ignition of the "super" ... In collaboration with my friend Everett one day I decided to try a schematic pilot calculation which could give an order of magnitude, at least, a "ballpark" estimate of the promise of his scheme ... Before we started this calculation of the progress of a thermonuclear reaction (burning in a mass of deuterium or deuterium-tritium mixture), Everett and I had done a lot of work on probablility questions connected with the active assemblies of uranium and with neutron multiplications. We worked out a theory of multiplicative processes, as we called it. (Now the preferred name is "branching processes.") ... Our calculation ... threw grave doubts on the prospects of Edward's original approach to the initial ignition conditions of the "super."

As the results of the von Neumann-Evans calculation on the big electronic Princeton machine slowly started to come in, they confirmed broadly what we had shown. There, in the course of the calculation, in spite of an initial, hopeful-looking "flare up," the whole assembly started slowly to cool down. Every few days Johnny would call in some results. "Icicles are forming," he would say dejectedly.

Perhaps the change came with a proposal I contributed. I thought of a way to modify the whole approach by injecting a repetition of certain arrangements. Unfortunately, the idea or set of ideas involved is still classified and cannot be described here.

At once Edward took up my suggestions, hesitantly at first, but enthusiastically after a few hours. He had seen not only the novel elements, but had found a parallel version, an alternative to what I had said, perhaps more convenient and generalized. From then on pessimism gave way to hope ... Teller lost no time in presenting these ideas, perhaps with most of the emphasis on the second half of our paper, at a General Advisory Committee meeting in Princeton which was to become quite famous because it marked the turning point in the development of the H-bomb.

Contrary to those people who were violently against the bomb on political, moral or sociological grounds, I never had any questions about doing purely theoretical work ... I felt that one should not initiate projects leading to possibly horrible ends. But once such possibilities exist, is it not better to examine whether or not they are real? An even greater conceit is to assume that if you yourself won't work on it, it can't be done at all ... When I reflected on the end results, they did not seem so qualitatively different from those possible with existing fission bombs. After the war it was clear that A-bombs of enormous size could be made. The thermonuclear schemes were neither very original nor exceptional. Sooner or later the Russians or others would investigate and build them.

The Oppenheimer Affair, which grew out of the violent hydrogen-bomb debate—even though the animosity between Strauss and Oppenheimer had personal and perhaps petty origins—greatly affected the psychological and emotional role of scientists.

Oppenheimer's opposition to the development of the H-bomb were not exclusively on moral, philosophical, or humanitarian grounds. I might say cynically that he struck me as someone who, having been instrumental in starting a revolution (and the advent of nuclear energy does merit this appellation), does not contemplate with pleasure still bigger revolutions to come ...

It seems to me this was the tragedy of Oppenheimer. He was more intelligent, receptive, and brilliantly critical than deeply original. Also he was caught in his own web, a web not of politics but of phrasing. Perhaps he exaggerated his role when he saw himself as "Prince of Darkness, the destroyer of Universes." Johnny used to say, "Some people profess guilt to claim credit for the sin."

Computers were brand-new; in fact the Los Alamos MANIAC was barely finished. The Princeton von Neumann machine had met with technical and engineering difficulties that had prolonged its perfection.

As soon as the machines were finished, Fermi, with his great common sense and intuition, recognized immediately their importance for the study of problems in theoretical physics, astrophysics, and classical physics. We discussed this at length ... After deliberating about possible problems, we found a typical one requiring long-range prediction and long-time behavior of a dynamical system. It was the consideration of an elastic string with two fixed ends, subject not only to the usual elastic force but having, in addition, a physically correct small non-linear term.

Our problem turned out to have been felicitously chosen. The results were entirely different qualitatively from what even Fermi, with his great knowledge of wave motions, had expected. The original objective had been to see at what rate the energy of the string, initially put into a single sine wave (the note was struck as one tone), would gradually develop higher tones with the harmonics, and how the shape would finally become "a mess" both in the form of the string and in the way the energy was distributed among higher and higher modes. Nothing of the sort happened. To our surprise the string started playing a game of musical chairs, only between several low notes, and perhaps even more amazingly, after what would have been several hundred ordinary up and down vibrations, it came back almost exactly to its original ... shape.

I know that Fermi considered this to be, as he said, "a minor discovery." And when he was invited a year later to give the Gibbs Lecture, he intended to talk about this. He became ill before the meeting.

These were the days of defense research contracts. Even mathematicians frequently were recipients. Johnny and I commented on how in some of their proposals scientists sometimes described how useful their intended research was for the national interest, whereas in reality they were motivated by bonafide scientific curiosity and an urge to write a few papers. Sometimes the utilitarian goal was mainly a pretext. This reminded us of the story of the Jew who wanted to enter a synagogue on Yom Kippur. In order to sit in a pew he had to pay for his seat, so he tried to sneak in by telling the guard he only wanted to tell Mr. Blum inside that his grandfather was very ill. But the guard refused, telling him: "Ganev, Sie wollen beten" ["You thief! You really want to pray"]. This, we liked to think, was a nice abstract illustration of the point.

*Jim Tuck and Stan*

Just after Johnny was offered the post of AEC Commissioner and before he accepted and became one in 1954 we had a long conversation. He had profound reservations about his acceptance because of the ramifications of the Oppenheimer Affair ... In a two-hour visit to Frijoles Canyon one afternoon he bared his doubts and asked me how I felt about it. He joked, "I'll become a commissionnaire." (In French the term is used to mean errand boy.) But he was flattered and proud that although foreign born he would be entrusted with a high governmental position of great potential influence in directing large areas of technology and science.

Our usual conversations were either about mathematics or about his new interest in a theory of automata. These conversations had started in a sporadic and superficial way before the war at a time when such subjects hardly existed. After the war and before his illness we held many discussions on these problems. I proposed to him some of my own ideas about automata consisting of cells in a crystal-like arrangement.

It is evident that Johnny's ideas on a future theory of automata and organisms had roots that went back in time, but his more concrete ideas developed after his involvement with electronic machines. I think that one of his motives for pressing for the development of electronic computers was his facination with the working of the nervous system and the organization of the brain itself. After his death some of his collaborators collected his writings on the outlines of the theory of automata.

Von Neumann's reputation and fame as a mathematician and as a scientist have grown steadily since his death. More than his direct influence on mathematical research, the breadth of his interests and of his scientific undertakings, his personality and his fantastic brain are becoming almost legendary.

Now Banach, Fermi, von Neumann were dead —the three great men whose intellects had impressed me the most. These were sad times indeed.

# Vita

LOS ALAMOS

BOULDER

*1957 February 8: Von Neumann dies in Walter Reed Hospital*

*1957–67 Research advisor, with John Manley, to Norris Bradbury, Director of Los Alamos*

*1960 Publication of* A Collection of Mathematical Problems

*1961–62 Fall semester: Visiting Professor at University of Colorado*

*1963 Winter quarter: Visiting Professor at University of California, San Diego*

*1965–77 Visiting Professor of Mathematics, and later Chairman of Mathematics Department, at University of Colorado. Spends vacations in Santa Fe*

*1967 Retires from Los Alamos but maintains loose connection with the Laboratory as a dollar-a-year consultant*

*1968–75 Professor of Biomathematics at University of Colorado Medical School*

*1972–73 Sabbatical year at Massachusetts Institute of Technology, University of Paris, and Los Alamos*

As a result of my work on the hydrogen bomb, I became drawn into a maze of involvements. These had to do precisely with government science and with work as a member of various Space and Air Force committees. Also, in some circles I became regarded as Teller's opponent, and I suspect I was consulted as a sort of counterweight. Some of these political activities included my stand on the Test Ban Treaty ... The cartoonist Herblock drew in the *Washington Post* a picture of the respective positions of Teller and me in which I fortunately appeared as the "good guy."

The idea of nuclear propulsion of space vehicles was born as soon as nuclear energy became a reality ... I think Feynman was the first in Los Alamos during the war to talk about using an atomic reactor which would heat hydrogen and expel the gas at high velocity. A simple calculation shows that this would be more efficient than expelling the products of chemical reactions.

I became involved with two such projects ... The first was Project Rover, a nuclear-reactor rocket which was being designed in Los Alamos already quite a few years before the Russian Sputnik, but with very limited funds. The second was a space vehicle, later named Orion. Around 1955 Everett and I wrote a paper about a space vehicle propelled by successive explosions of small nuclear charges ...

When it was decided to do something in earnest about Project Rover, Wiesner named a Presidential Committee to look into the matter. I was one of its members ... The committee wrote a report which by faint praise, essentially condemned Project Rover to a de facto death by proposing to make it a purely theoretical study without funds for experimental work or any investment in construction. The physicist Bernd Matthias was the only member of the committee who joined me in writing a dissenting opinion.

*Morgenstern and Stan at Bandelier, 1949*

At the same time, I was continuing my own work. After Fermi's death Pasta and I decided to continue exploratory heuristic experimental work on electronic computers in mathematical and physical problems ...

The problem of clusters of stars was I think the first study of this nature using computers. We took a great number of mass points representing stars in a cluster. The idea was to see what would happen in the long-range time scale of thousands of years to the spherical-looking cluster whose initial conditions imitated the actual motions of such stars.

*Edward Lasker (International Chess Master) and Stan, late sixties*

While such astrophysical calculations were going on, I began in an amateurish way to work on some questions of biology. After reading about the new discoveries in molecular biology which were coming fast, I became curious about a conceptual role which mathematical ideas could play in biology.

*Rota and Stan, 1973*

*President Kennedy in Los Alamos*

In 1960 my book, [*A Collection of Mathematical Problems*], was published. Many years ago Françoise asked Steinhaus what it was that made me what people seemed to consider a fairly good mathematician. According to her, Steinhaus replied: "C'est l'homme du monde qui pose le mieux les problèmes." Apparently my reputation, such as it is, is founded on my ability to pose problems and to ask the right kind of questions.

[In 1964] I met Gian-Carlo Rota, a mathematician who is almost a quarter-century younger than I ... Our relationship is not built on our age difference. Rota claims that he is greatly influenced by me. So I coined the expression "influencer and influencee." Rota is one of my best influencees ...

Rota's personality is compatible with mine. His general education, active interest in philosophy (he is an expert on the work of Edmund Husserl and Martin Heidegger), and, above all, his knowledge of classical Latin and ancient history, have made him fill the gap left by the loss of von Neumann.

During the Los Alamos years I frequently took time off to return to academic life and around 1965 I started visiting the University of Colorado on a more regular basis ... In 1967 I decided to retire from Los Alamos and accept a professorship in Boulder ... The University of Colorado was flourishing ... and the mathematics department experienced an explosive growth in size and quality. Besides Boulder was sufficiently close to Los Alamos ... so I could continue as a consultant and visit frequently ... The mathematics department was acquiring excellent researchers ... [among them] a younger, brilliant Pole, Jan Mycielski, a student of Steinhaus, whom I invited to accept a professorship.

In 1967 ... Mark Kac and I were invited to write a long article [for *Britannica Perspectives*,] ... a semi-popular presentation of modern ideas and perspectives of ... the great concepts of mathematics ... Since then it has appeared separately under the title *Mathematics and Logic*.

Mark Kac had also studied in Lwów, but since he was several years younger than I (and I had left when only twenty-six myself), I knew him then only slightly ... After the war he visited Los Alamos, and we developed our scientific collaboration and friendship ... Mark is one of the very few mathematicians who possess a tremendous sense of what the real applications of pure mathematics are and can be ... He was one of Steinhaus's best students.

[After I retired from the University of Colorado, we] sold our Boulder house and bought another one in Santa Fe, which has become our base. From Santa Fe I commute three or four times a week to the Los Alamos Laboratory. Its superb scientific library and computing facilities allow me to continue working ... Dan Mauldin, a professor at North Texas State University [and I] are now collaborating on a collection of new unsolved problems. This book will have a different emphasis from that of my Collection of Mathematical Problems. The new collection will deal more with mathematical ideas connected to theoretical physics and biological schemata.

In the short span of my life great changes have taken place in the sciences ... Sometimes I feel that a more rational explanation for all that has happened during my lifetime is that I am still only thirteen years old, reading Jules Verne or H. G. Wells, and have fallen asleep.

*Mark Kac*

It is still an unending source of surprise for me to see how a few scribbles on a blackboard or on a sheet of paper could change the course of human affairs. I became involved in the work on the atomic bomb, then in the work on the hydrogen bomb, but most of my life has been spent in more theoretical realms.

# Vita

BOULDER

SANTA FE

*1974–84* *Winter trimesters: Graduate Research Professor, University of Florida*

*1975* *Retires from University of Colorado*

*1979–84* *Visiting Professor of Biomathematics, University of Colorado Medical School*

*1980–84* *Visiting Professor at Neurosciences Institute of Rockefeller University*

*1982–83* *Fall semester: Visiting Professor at University of California, Davis*

*1984* *Dies in Santa Fe on May 13*

*At the time of his death, Stanislaw M. Ulam was an elected* Fellow of the American Academy of Arts and Sciences *and an elected* member of the National Academy of Sciences and the American Philosophical Society. *He sat on* the Board of Governors and the Scientific Advisory Committee of the Weizmann Institute of Science (Rehovot, Israel) and the Board of the Jurzykowski Foundation (New York, New York). *He belonged to* the Polish Mathematical Society, the American Mathematical Society, the American Physical Society, and the American Association for the Advancement of Science.

*He held honorary degrees from* the University of New Mexico, the University of Pittsburgh, and the University of Wisconsin *and was recipient of* the Sierpinski Medal, the Heritage Award, and the Polish Millenium Prize.

*He had been a member and/or chairman of* the Committee on Innovations of the National Academy of Sciences, the Committee on Applications of Mathematics of the National Research Council, the Visiting Commmittee for Mathematics and the Visiting Committee for Applied Mathematics of Harvard University, the Gibbs Lecture Committee of the American Mathematical Society, and the Mathematics Research Committee of the Mathematical Association of America.

*He had served as consultant to* President Kennedy's Science Advisory Committee, Air Force General Twining's Space Research Committee, IBM Corporation, General Atomic Corporation, North American Aviation Corporation, Hycon Corporation, and other organizations.

*Building in Lwów in which the Ulam family resided, photographed after the war*

# The Lost Café

*by Gian-Carlo Rota*

**This is a chapter of a forthcoming autobiographical volume now being written by the author and to be published under the auspices of the Sloan Foundation**

One morning in 1946 in Los Angeles, Stan Ulam, a newly appointed professor at the University of Southern California, awoke to find himself unable to speak. A few hours later he underwent an emergency operation. His skull was sawed open and his brain tissue sprayed with newly discovered antibiotics. The diagnosis—encephalitis, an inflammation of the brain. After a short convalescence he managed to recover, apparently unscathed.

In time, however, some changes in his personality became obvious to those who knew him. Paul Stein, one of his collaborators at Los Alamos, remarked that, while before his operation Stan had been a meticulous dresser, a dandy of sorts, afterwards he became visibly careless in the details of his attire, even though his clothing was still expensively chosen.

When I met him, many years after the event, I could not help noticing that his trains of thought were unusual, even for a mathematician. In conversation he was livelier and wittier than anyone I had ever met, and his ideas, which he spouted out at odd intervals, were fascinating beyond anything I have witnessed before or since. However, he seemed to studiously avoid going into any details. He would dwell on a given subject no longer than a few minutes, then impatiently move on to something entirely unrelated.

Out of curiosity I asked Oxtoby, Stan's collaborator in the thirties, about their working habits before his operation. Surprisingly, Oxtoby described how at Harvard they would sit for hours on end, day after day, in front of the blackboard. Since I met him, Stan never did anything of the sort. He would perform a calculation, even the simplest, only when he had absolutely no other way out. I remember once watching him at the blackboard trying to solve a quadratic equation. He furrowed his brow in rapt absorption, while scribbling formulas in his tiny handwriting. When he finally got the answer, he turned around and said with relief, "I feel I have done my work for the day."

The Germans have aptly called *Sitzfleisch* the ability to spend endless hours at a desk doing gruesome work. *Sitzfleisch* is considered by mathematicians to be a better gauge of success than any of the attractive definitions of talent with which psychologists regale us from time to time. Stan Ulam was able to get by without any *Sitzfleisch* whatsoever. After his bout with encephalitis, he came to lean instead on his own unimpaired imagination for new ideas and on the *Sitzfleisch* of others for technical support. The beauty of his insights and the promise of his proposals kept him amply supplied with young collaborators always willing to lend (and sometimes risking to waste) their time.

A crippling technical weakness coupled with an extraordinarily creative imagination is the drama of Stan Ulam. Soon after I met him, I was made to understand that, as far as our conversations went, his drama would be a Forbidden Topic. Perhaps he discussed it with his daughter, Claire, the only person with whom he would occasionally have brutally frank discussions, but certainly not with anyone else. But he knew I knew, and I knew he knew I knew.

Stan Ulam was born into a family that stood as high on the social ladder as a Jewish family could at the time. He was the golden boy from one of the richest families of Lwów. In central Europe the Ulam name was then a synonym of banking wealth, not unlike the Rothschilds' in western Europe. He was educated by private tutors and in the best schools. As a child he already showed an unusual interest in astronomy ("I am star-struck," he would often tell me) and in physics. At the age of twelve he was reasonably familiar with the outlines of the special theory of relativity, a great novelty at the time. In high school he was a top student, far too bright for his age. His quick wit got him good grades with little effort but lent free rein to his laziness.

The two authors he read thoroughly in his teens were Karl May and Anatole France. They had a formative influ-

ence on his personality, and throughout his life he kept going back to them for comfort. From Karl May's numerous adventure novels (popular enough in the German-speaking world to be among the favorite books of both Einstein and Hitler) he derived the childlike and ever fresh feeling of wonder that is often found in great men. From Anatole France he took his man-of-the-world mannerisms, which in later life would endear him to young ladies.

He kept a complete set of Karl May's novels (in German, the other language of his childhood) behind his desk until he died. He regretted that a *Pléiade* edition of Anatole France had not been published, which he could keep by his bedside. He often gave me paperbacks of Anatole France, bought on his frequent trips to Paris and dedicated with inscriptions urging me to read them. I regret to admit I haven't.

There was never any doubt that he would study mathematics when, at age seventeen, he enrolled at Lwów Polytechnic Institute. Shortly after classes started he discovered with relief that the mathematics that really mattered was not taught in the classroom, but was instead to be found alive in one of the large cafés in town, the Scottish Café. There the Lwów mathematicians would congregate daily. Between a shot of brandy and a cup of coffee, they would pose (and often solve) what turned out to be some of the outstanding mathematical conjectures of their time, conjectures that would be dashed off on the marble of coffee tables in the late evenings, in loud and uninhibited brawls.

The Lwów school was made up of offbeat, undisciplined types. Stan's teacher Banach was an alcoholic, and his best friend Mazur was a Communist. They cultivated the new fields of measure theory, set theory, and functional analysis, which at the time required very little background. The rival Warsaw mathematicians, more conservative, looked down on the Lwów mathematicians as amateurish upstarts, but the results of the Lwów school soon came to be better known and appreciated the world over, largely after the publication of Banach's book on linear operators, in which Ulam's name is the most frequently mentioned.

One day the amateur Ulam went one up on the Warsaw mathematicans, who cultivated the equally new field of algebraic topology. While chatting at the Scottish Café with Borsuk, an outstanding Warsaw topologist, he saw in a flash the truth of what is now called the Borsuk-Ulam theorem. Borsuk had to commandeer all his technical resources to prove it. News of the result quickly swept across the ocean, and Ulam became an instant topologist.

Stan took to café-mathematics like a fish to water. He quickly became the most daring of the Lwów mathematicians in formulating bold new mathematical conjectures. Almost all his guesses of that time have been proved true and are now to be found as theorems scattered in graduate textbooks.

In the casual ambiance of the Scottish Café, Stan blossomed into one of the most promising mathematicians of his generation. He also began to display the contradictory traits in behavior that after his operation were to become dominant: deep intuition and impatience with detail, playful inventiveness and dislike of prolonged work. He began to view mathematics as a game, one that a well-bred gentleman should not take too seriously. His insights have opened whole new areas of mathematics, all of them still actively cultivated today, but he himself could not bear to give his discoveries more than a passing interest, and at times he would make merciless fun of those who did take them too seriously.

The papers in mathematics that he wrote by himself date back to this period. Most were written in one sitting, often in a night's work, probably in response to some colleague's challenge at the Scottish Café. Much of his present reputation as a mathematician rests on these short, brilliant notes. His measurable cardinals, the best idea he had in this period, are still the mainspring of much present work in set theory. More often, however, his flashes of originality, scattered as they are in unexpected contexts, have been appropriated by others with little acknowledgement, and have proved decisive in making more than one career in mathematics. For example, his paper with Łomnicki on the foundations of probability, which also dates back to his Polish period, contains a casual remark on the existence of prime ideals in Boolean algebras, later developed by Tarski and others in several formidable papers.

The Borsuk-Ulam theorem was striking enough to catch the attention of Solomon Lefschetz, the leading topologist of the time and the chairman of the Princeton mathematics department. Through Lefschetz and von Neumann, with whom he had started to correspond, Ulam was invited in 1936 to visit the Institute for Advanced Study in Princeton.

For four years he commuted between Poland and America where, first in Princeton and later at the Harvard Society of Fellows, he lived in luxury on his parents' monthly checks. In the summer of 1939, shortly after he returned to the United States with his brother from what would be the last visit to his family, World War II broke out. By accident he had been saved from almost certain extinction. He would never leave the United States again, except on short trips.

The *belle époque*, the period that runs between 1870 and the 1930s (though some claim that it ended with World War I), was one of the happiest times of our civilization. Vienna, Prague, Lwów, and Budapest were capitals of turn-of-the-century sophistication, though they lacked the staid traditions of Paris, Florence, or Aranjuez. Robert Musil, Gustav Mahler, Franz Kafka, Ludwig Wittgenstein, and the philosophers of the Vienna Circle have become for us symbols of *mitteleuropaische Kultur*. Most of those now legendary figures betrayed personality traits similar to Stan's: restlessness, intolerance, a dialectic of arrogance and contrition, and an unsatisfied need for affection, compounded by their society's failure to settle on a firm code for the expression of emotion. Perhaps the roots of the tragedy that befell central Europe should be looked for in those men's tragic lives and flawed personalities, rather than in the scurrilous outbursts of some demented housepainter.

When the catastrophe came, those among them who were still alive to watch their world go up in flames never recovered from the shock. They remained emotionally crippled for the rest of their lives.

Stan Ulam was one of them. Had he been able to remain in Poland and survive the war, as Steinhaus, Kuratowski, and a few others did, he would have gone on to become one of the leading international figures of pure mathematics, at least on a par with Banach. But after he bade farewell to his friends at the Scottish Café, something died forever within him, and his career as a pure mathematician went permanently adrift.

Like other immigrants from the European leisure class, Stan arrived in the United States ill-equipped for the rigors of puritan society.

The big open spaces of America, the demands for aloneness and self-reliance made him feel estranged. He wished to belong, and he loved this country, but he never came to feel fully at home in the United States, whether in Cambridge, Madison, or Los Alamos. He missed the lively street life of European cities, the culture, the rambling conversations (what the Spanish call *tertulias*) and viewed with alarm the decay of that art, which in our day has become all but extinct.

By now the effective American way of scientific exchange has imposed itself on the rest of the world. But fifty years ago life in American universities was incomparably duller than the café-science of Lwów. The atmosphere of Cambridge in the thirties was too cold, and, what was worse, there were no cafés. And then, in Europe, the war started.

In the fall of 1939, Stan would spend endless hours watching the Charles River from his room at Harvard, stupefied by the sudden turn of events that had changed his life and that of so many others. He learned of the fall of Poland, of the deportation of his family

to a concentration camp (his sister and uncles were killed in gas chambers), of the sacking of the great Ulam bank.

He was all alone now. His father's monthly checks had stopped, his Junior Fellowship would soon run out, and he would have to support his brother's college education at Brown. He pinned great hopes on his big paper on ergodic theory, which he had just finished writing with Oxtoby and which had been accepted for publication in *Annals of Mathematics*, the most prestigious mathematics journal. In the solitude of Adams House, he could not bring himself to finish a paper by himself, though his lectures on the theory of functions of several real variables were the most brilliant he ever delivered (some former students still cherish the notes they took of that course).

G. D. Birkhoff, the ranking Harvard mathematician and the absolute monarch of American mathematics, took a liking to Stan Ulam. Like other persons rumored to be anti-Semitic, he would occasionally feel the urge to shower his protective instincts on some good-looking young Jew. Ulam's sparkling manners were diametrically opposite to Birkhoff's hard-working, aggressive, touchy personality. Birkhoff tried to keep Ulam at Harvard, but his colleagues balked at the idea. After all, Ulam had only one long paper in course of publication, and it can be surmised that the Harvard mathematicians of the thirties turned up their noses at the abstract lucubrations of a student of Banach.

Birkhoff then began to write letters to his friends at several universities, suggesting Ulam's name for appointment. It didn't take long before Stan received an offer from the University of Wisconsin in Madison, an assistant professorship carrying a rather high stipend for the time, over two thousand dollars. He had no choice but to accept it.

For the first time in his life, Stan had to do "an honest day's work," and he didn't like the thought. The teaching load of some twelve hours a week of pre-calculus soon turned into a torture. Rumor had it that he had occasionally fallen asleep while lecturing. Madison, a friendly little Midwestern town, was the end of the world for a worldly young European. The ambiance was more non-existent than dismal. His colleagues, upright men and world-renowned mathematicians like Everett and Kleene, were not the garrulous Slavic types he was used to. Then after Stan's second year at Wisconsin, America entered the war.

Once more John von Neumann came to Stan's rescue.

Of all escapes from reality, mathematics is the most successful ever. It is a fantasy that becomes all the more addictive because it works back to improve the same reality we are trying to evade. All other escapes—love, drugs, hobbies, whatever—are ephemeral by comparison. The mathematician's feeling of triumph, as he forces the world to obey the laws his imagination has freely created, feeds on its own success. The world is permanently changed by the workings of his mind, and the certainty that his creations will endure renews his confidence as no other pursuit. The mathematician becomes totally committed, a monster like Nabokov's chess player, who eventually sees all life as subordinate to the game of chess.

Many of us remember the feeling of ecstasy we experienced when we first read von Neumann's series of papers on rings of operators in Hilbert space. It is a paradise from which no one will ever dislodge us (as Hilbert said of Cantor's set theory). But von Neumann's achievements went far beyond the reaches of pure mathematics. Together with Ulam he was the first to have a vision of the boundless possibilities of computing, and he had the resolve to gather the considerable intellectual and engineering resources that led to the construction of the first computer. No other mathematician of this century has had as deep and lasting an influence on the course of civilization.

Von Neumann was a lonely man with deep personal problems. He had two difficult marriages. He had trouble relating to others except on a strictly impersonal level. Whoever spoke to him noticed a certain aloofness, a distance that would never be bridged. He was always formally dressed in impeccable business suits, and he always kept his jacket on (even on horseback), as if to shield himself from the world.

Stan was probably the only close friend von Neumann ever had. A similar background and a common culture shock brought them together. They would spend hours on end gossiping and giggling, swapping Jewish jokes, and drifting in and out of mathematical talk.

Stan was the more original mathematician of the two, though he accomplished far less in mathematics than von Neumann did. Von Neumann had an incomparably stronger technique. From their free play of ideas came some of the great advances in applied mathematics of our day: the Monte Carlo method, mathematical experiments on the computer, cellular automata, simulated growth patterns.

Like everyone who works with abstractions, von Neumann needed constant reassurance against deep-seated

and recurring self-doubts. Following his uncanny instinct for doing the right thing at the right time, Stan soon found the way to cheer up his brooding friend. He began to make fun of von Neumann's accomplishments. He would mercilessly ridicule continuous geometries, Hilbert space, and rings of operators, cleverly picking on weaknesses in von Neumann's work that were obvious and expected. Stan's jibes were an indirect but firm expression of admiration. Rather than feel offended, von Neumann would burst out in a laughter of relief.

Much later, when Stan related to me these events, he affected to regret never having said a kind word to von Neumann about his work in pure mathematics. But I could feel he was not serious. Deep inside he knew he had been good to his friend.

Stan didn't fully realize how much von Neumann meant to him until his friend began to die of cancer, in 1955. Stan would make frequent trips to Walter Reed Hospital in Washington, where for months on end his honored friend was confined to a bed in the Presidential Suite. Stan came prepared with a bagful of the latest jokes and prurient Los Alamos gossip. The little hospital bed would shake with the vibrations of von Neumann's big belly as he laughed himself to tears, the very tears that Stan was fighting to control. Then weeks passed when von Neumann could no longer recognize anyone. When he finally died, Stan broke into tears. It was probably the only time in his life when he openly lost control of his emotions.

Back in 1941 shortly after the United States entered the war, Stan (then still at Wisconsin) began to notice that von Neumann's letters were becoming infrequent. Curious about his friend's mysterious unavailibility, Stan managed one day to corner him in Chicago. He implored von Neumann to drag him out of his Wisconsin rut and to get him a job related to the war effort. The request fit perfectly with von Neumann's plans. He had already made up his mind to bring Stan with him to the newly founded Los Alamos laboratory, where the atomic bomb project was being launched.

The choice of a set theorist for work in applied physics might seem eccentric, but in retrospect von Neumann made the right choice. Besides, as the token mathematician in a sea of physicists (though he was probably one of the finest minds among them, together with Fermi and Feynman), von Neumann was relieved to have his cohort join him.

The assembly of geniuses who roamed the corridors of the Los Alamos laboratory during World War II has not been matched in recorded history, with the possible exception of ancient Greece. In the hothouse of the Manhattan Project, Stan's mind opened up as it hadn't since the days of the Scottish Café. The joint efforts of the best scientists of the time, their talents stimulated and strained by the challenge of a difficult project, made what could have been a drab weapons laboratory into a cradle of new ideas. In welcome breaks between long stints at the bench, in a corner at some loud drinking party, the postwar revolutions in science were being hatched.

Los Alamos was a turning point in Stan Ulam's career. From that time on physics, not mathematics, became the center of his interest. After watching Fermi and Feynman at the blackboard, he discovered that he too had a knack for accurately estimating physical quantities by doing simple calculations with orders of magnitude. In fact, he turned out to be better at that game than just about anyone around him.

It is hard to overstate how rare such an ability is in a mathematician. The literalness of mathematics is as far removed from the practical needs of the physicist as might be the story of the Wizard of Oz. As Stan began to display his newly found talent, he came to rely less and less on standard mathematical techniques and to view ordinary mathematics with some contempt. He admired Fermi's genius for solving physical problems with no more than the minimum amount of math. Since that time Fermi remained for him the ideal of a scientist. In his old age he liked to repeat (perhaps with a touch of exaggeration) that Fermi had been the last physicist.

But the Magic Mountain lasted only as long as the war. In 1945 it seemed that the Los Alamos laboratory might close down, like many other wartime projects, and Stan began to look for a job elsewhere. Unfortunately, his list of publications was hardly longer now than it had been in 1939, and unpublished work gets no credit. To his chagrin he was ignored by the major universities. He finally had to accept the offer of a professorship at the University of Southern California, at the time a second-rate institution but one with great plans for the future.

Suddenly he found himself in the middle of an asphalt jungle, teaching calculus to morons. The memories of his friends in Los Alamos, of the endless discussions, of the all-night poker games, haunted him as he commuted daily among the tawdry streets of Los Angeles. The golden boy had lost the company of great minds, his audience of admirers. Like anguish that could no longer be contained, encephalitis struck.

We still tend to regard disease as a mere physical occurrence, as an unfore-

seen impairment of the body that also, mysteriously, affects the mind. But this is an oversimplification. After a man's death, at the time of the final reckoning, an event that might once have appeared accidental is viewed as inevitable. Stan Ulam's attack of encephalitis was the culmination of his despair.

After recovering from his operation, Stan resigned his position in a hurry and went back to Los Alamos.

The year was 1946, and the Los Alamos laboratory was now a different place. Gone were most of the luminaries (though many of them would make cameo appearances as consultants), and the federal government was lavishing limitless funds on the laboratory. For a few years Los Alamos scientists found themselves coddled, secure and able to do or not do whatever they pleased, free to roam around the world in red-carpeted MATS flights (that is, until Americans decided to give up the Empire they had won).

Ulam came back to Los Alamos haunted by the fear that his illness might have irreparably damaged his brain. He knew his way of thinking had never been that of an ordinary mathematician, and now less than ever. He also feared that whatever was left of his talents might quickly fade. He decided the time had come to engage in some substantial project that would be a fair test of his abilities, and one with which his name might perhaps remain associated.

While at Wisconsin, Stan had met Everett. They had jointly written the first paper on the subject that is now called algebraic logic (a beautiful paper that has been plundered without acknowledgement). Everett, a seclusive and taciturn man, was richly endowed with the ability to compute. He was a good listener, and he suffered from a paranoid fear of being fired for wasting Lab time on research in pure mathematics. He was a perfect complement to Stan. After he had accepted Stan's invitation to come to Los Alamos, they joined forces on a long and successful collaboration.

As their first project they chose the theory of branching processes. They believed they were the first to discover the probabilistic interpretation of functional composition. (They had ignored all previous work, all the way back to Galton and Watson in the nineteenth century! Stan never had the patience to leaf through published research papers. He hated to learn from others what he thought he could invent by himself and often did). They rediscovered all that had been already done, and added at least as much of their own. Their results were drafted by Everett in three lengthy lab reports, which found substantial applications in the theory of neutron diffusion, an essential step in the understanding of nuclear reactions. These reports were never published, but they nevertheless had a decisive influence on the development of what is still a thriving branch of probability theory. The authors have received little acknowledgement for their work, perhaps as a spiteful punishment for their own neglect of the work of others.

Their second project was the hydrogen bomb.

Stan Ulam and Edward Teller had disliked each other from the moment they had met. Since the days of the Manhattan Project, Teller had been somewhat of a loner. His behavior put him outside the main-line Bethe-Fermi-Oppenheimer group, and not even his fellow Hungarian von Neumann felt at ease with him. This despite the fact that he distinguished himself from the first days of Los Alamos as one of the most brilliant applied physicists there.

Teller related with difficulty and diffidence to other scientists of his age. He felt more at ease either with young people or with celebrities, highly placed politicians, generals and admirals. His group (what eventually became the Lawrence Livermore Laboratory after he left Los Alamos in a huff) was highly disciplined, rank-conscious, and loyal. He would sagely guide his students and assistants to doing the best research work they were capable of, and he would reward his followers with top-rank positions in academic administration or in government.

Since the success of the first bomb, Teller had been obsessed by the idea of the "Super." Because of disagreements between him and Oppenheimer, his project had more than once been on the verge of being cancelled. Now, Stan Ulam was out to get him by proving that his plans for the new bomb would not work.

For about two years Everett and Ulam worked frantically in competition with Teller's group. They met every morning for several hours in a little office out of the way. Ulam would generate an endless stream of ideas and guesses, and Everett would check each one of them with feverish computations. In a few months' time Everett wore out several slide rules. At last they proved Teller wrong. And then, adding insult to injury, Stan, in a sudden flash of inspiration, came upon a trick to make the first hydrogen weapon work.

The full extent of Stan's contribution to the design of the first hydrogen bomb will never be precisely established. It is

certain, however, that he was instrumental in demolishing misguided proposals that would have resulted in considerable waste of time and funds. It is all but certain that the seed idea that finally worked was his own. At any rate, the ensuing loud dispute with Teller over the priority of the invention brought him wide publicity. (The patent application for the device was jointly submitted by Teller and Ulam.) The Democrats soon saw their advantage in adopting Ulam as a bulwark against the Republicans, who had Teller on their side. He was invited to sit in on important Washington committees and later became a darling of the Kennedy era.

At last some of the glitter of his Polish youth had come back, if not in the form of tangible wealth, at least in the guise of public recognition.

The late forties and fifties were the high point of Stan Ulam's life. His personality thrived. His conversation, always lively, became all the more witty and engaging. The better part of his day was spent telling jokes and funny stories and inventing one interesting mathematical idea after another, like a wheel of fortune that never stopped. The joke was the literary form he most appreciated. He would come up with anecdotes, ideas, and stories on any subject of his acquaintance, however little his competence. He so liked to dominate a conversation that some of his colleagues began to take pains to avoid him. Now he had to win every argument. When he felt he was on the losing side, he would abruptly change the subject, but not before seeing the bottom of the other person's position and summarizing it with irritating accuracy. Considering how fast it all happened, it is remarkable how seldom he misunderstood. Mathematicians felt put down, and Ulam's ways alienated him from the guild. He retaliated by claiming not to be a "professional" mathematician and by going into rambling tirades against the myopia of much contemporary mathematics.

The free rein Ulam gave to his fantasy fed on one of his latent weaknesses—his wishful thinking. He became an artist at self-deception. He would go to great lengths to avoid facing the unpleasant realities of daily life. When anyone close to him became ill, he would seize on every straw to pretend that nothing was really wrong. When absolutely forced to face an unpleasant fact, he would drop into a chair and fall into a silent and wide-eyed panic.

His severest critics were those close to him who felt excluded from his private world, who stood outside the mighty fortress of mathematics. His daughter would browbeat him and cut him to pieces at regular intervals, incredulous of her father's achievements. He took her criticisms in silence, and was fond of quoting one of James Thurber's lovely generalizations: "Generals are afraid of their daughters."

Despite the comfort of the Los Alamos Laboratory (in the fifties and sixties Ulam was one of two research advisors to the Director of the laboratory), Stan could find no peace there. Since his return in 1946, he had, unbelievable as it may sound, lived out of a suitcase. He owned beautiful homes in Los Alamos and Boulder, but he thought of himself as permanently on the road. (Significantly, his ashes are now in Montparnasse Cemetery in Paris.) The Scottish Café was gone forever, and he was a passenger on an imaginary ship, who survived on momentary thrills designed to get him through the day. He surrounded himself with traveling companions who were fun to be with and to talk to. He went to great lengths to avoid being alone. When he was, only the lure of mathematics could draw his mind away from the clamor of his memories.

I will always treasure the image of Stan Ulam sitting in his study in Santa Fe early in the morning, rapt in thought, scribbling formulas in drafts that would probably fill a couple of postage stamps.

The traits of Stan Ulam's personality that became dominant in his later years were laziness, generosity, considerateness, and most of all, depth of thought.

Those who knew Stan and did not know what to make of him covered up the mixture of envy and resentment they felt toward him by pronouncing him lazy. He was in fact lazy, in the dictionary sense of the word. In the thirties he would take a taxi to Harvard from his apartment in Boston to avoid tackling the petty decisions that a ride on the subway required. In Los Alamos there is a spot on a pathway up the Jémez Mountains that is called Ulam's Landing. It is as far as Stan ever went on a hike before turning back. More often, he would watch the hikers with binoculars from the porch of his house, while sipping gin and tonics and talking to his friends.

Like all words denoting human conditions, laziness, taken by itself, is neutral. It is a catchall that conceals a tension of opposites. *Fata ducunt, non trahunt.* Ulam turned his laziness into elegance in mathematics and into *grand seigneur* behavior in his life. He had to give all of his thinking an epigrammatic twist of elegant definitiveness. His failing became an imperious demand to get to

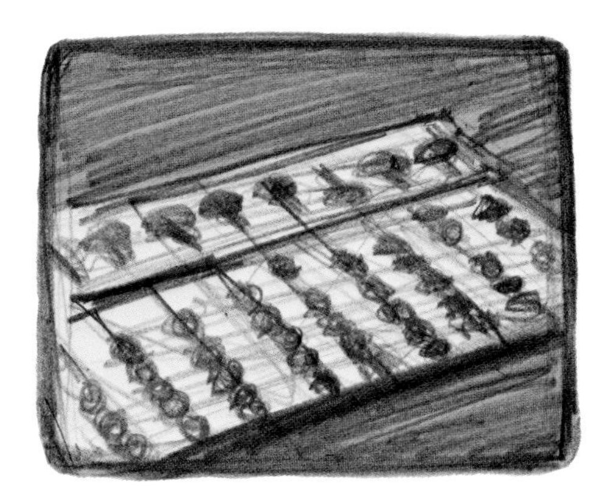

the heart of things with a minimum of jargon.

He had a number of abrupt conversation stoppers that he used to get rid of bores. One of them was a question designed to stop some long tirade: "What is this compared to $E = mc^2$?" When I first heard it (undoubtedly it was being used to stop me), I thought it a sign of conceit. But I was wrong. He would wake up in the middle of the night and compare his own work, too, to $E = mc^2$, and he developed ulcers from these worries. In truth, his apparent conceit was a way of concealing from others, and most of all from himself, the aging of his brain. On rare occasions he felt overwhelmed by guilt at his inability to concentrate, which he viewed as avoidance of "serious" work. He looked at me, his intense blue-green eyes popping and slightly twitching (they were the eyes of a prophet, like Madame Blavatsky's), his mask about to come down, and asked, "Isn't it true that I am a charlatan?" I proceeded to set his mind to rest by giving him, as a sedative, varied examples of flaming charlatans taken from scientists we both knew (both with and without Nobel Prizes). But soon his gnawing doubts would start all over again. He knew he would remain to the end a Yehudi Menuhin who never practiced.

His generosity was curiously linked to his laziness. A generous action is often impulsive and calls for little foresight. Its opposite requires the careful advance planning that Stan loathed. He fancied himself a *grand seigneur* of bottomless means, and in matters of money he was apt to practice the art of self-deception. In his penurious years he went to great lengths to conceal his shaky financial condition. He always lived as the spirit moved him, sometimes beyond his means. He carried on his person bundles of fifty and one-hundred dollar bills, partly from a remnant of the refugee mentality, partly to impress whomever he met during his travels.

He was also too much of a *grand seigneur* to insist on his priority for the many new ideas he contributed to science. His nonchalance as to the fate and success of his work has unjustly lowered his standing as a scientist. When he saw one of his ideas circulating without credit, he remarked, "Why should they remember me? No one quotes Newton or Einstein in the bibliographies of their papers."

His way of expressing himself lent itself to his being exploited. He would speak in sibylline pronouncements that seemed to make little sense. Those of his listeners who decided to pursue his proposals (and often ended up writing dozens of research papers on them) felt they had spent enough of an effort in figuring out what Stan really meant to reward themselves by claiming full credit.

A seed idea is the last thing we want to acknowledge, all the more so when it originates from a native intelligence seemingly blessed with inexhaustible luck. After we silently appropriate it, we will soon enough figure out a way to obliterate all memory of its source. In a last-ditch effort to salvage our pride, we will also manage to find fault with the person to whom we are indebted. Stan Ulam's weaknesses were all too apparent and made him more vulnerable than most. But the strength of his thinking more than made up for what he lost to the pettiness of others.

Stan once showed me in five minutes the central idea of the theory of continued fractions and thereby saved me much work. Once I bragged to him about some computations I had done on the speed of convergence in the central limit theorem, and he showed me how to derive the same result by an elegant argument with ordinary square roots.

Stan did his best work in fields where no one daréd to tread, where he would be sure of having the first shot, free from all fear of having been anticipated. He used to brag about being lucky. But the source of his luck was his boundless intellectual courage, which let him see an interesting possibility where everyone else could see only a blur.

He refused to write down some of his best ideas. He thought he would find some day the time and the help he needed to work them out. But he was misjudging the time he had left. His best problems will survive only if his students ever write them down.

Two of them have struck me. In the nineteenth century mathematicians could not conceive of a surface unless it was defined by specific equations. After a tortuous period of abstraction, the point-set topologists in this century arrived at the abstract notion of a topological space, which renders in precise terms our intuitive grasp of the notion of extension. Ulam proposed going through a similar process of refinement on Maxwell's equations to arrive at an abstract structure for electromagnetic theory free of algebraic irrelevancies.

The second problem bore on ergodic dynamical systems. Poincaré, and several others after him, taught us that in such a system every state is visited infinitely often, given a sufficiently long time. In practice, however, the recurrence times are so large that one cannot observe successive visits, and the practical import of ergodicity is nil. This paradox became strikingly evident after the Fermi-Pasta-Ulam computer sim-

ulations of coupled nonlinear oscillators. (These were written up in one of Fermi's last papers. It is rumored that Fermi considered this to have been his most important discovery.) In these nonlinear systems the initial state is visited several times before another set of available states is even approached. After observing this phenomenon, Ulam guessed that in some ergodic systems the phase space ought to be measure-theoretically represented by two or more big blobs connected by thin tubes. He wanted to express his guess in terms of ergodic theory. I wish we knew how.

Stan's fascination with physics led him to formulate mathematical thoughts that had a background of physics, but they invariably bore the unmistakable ring of mathematics. (He once started to draft a long paper that was to be titled "Physics for Mathematicians.") One of the most striking is his proposal for the reconstruction of the cgs system (distance, mass, and time) on the basis of a random walk. Another, which Dan Mauldin has recently proved true, is the existence of a limiting energy distribution for systems in which energy is redistributed through particle collisions.

Stan Ulam's best work is a game played in the farthest reaches of abstraction, where the cares of the world cannot intrude: in set theory, in measure theory, and in the foundations of mathematics. He used to refer to his volume of collected papers as a slim volume of poems. It is just that.

As a mathematician, his name is most likely to survive for his two problem books, which will remain bedside books for young mathematicians eager to make their mark by solving at least one of them. He also wanted to be remembered for those of his insights that found substantial practical applications, such as the Monte Carlo method, for which he will share the credit with Metropolis and von Neumann, and the bomb, for which he will be remembered alongside Teller.

Only in the last years of his life did his thinking take a decisively speculative turn. He always professed to dislike philosophical discussions, and he excoriated ponderous treatises in philosophy. He thought them in bad taste, "Germanic" (one of his words of reprobation). Nonetheless, he had an instinctive grasp of philosophical issues, which he refused to express in words. When forced to take a philosophical stand, he would claim to agree with the naive scientism of H. G. Wells and with the positivism of the Vienna Circle (the reigning philosophy of his time), but in his actual thinking he was closer to the phenomenology of Husserl and Heidegger. His knowledge of philosophy suffered from his habit of scanning without reading. He seldom read a book from top to bottom; more often he would handle it long enough to pick out the main point, sometimes after correcting a few misprints, and then literally toss it away. I once set up a little test of his understanding of existentialism, by way of teasing him. I gave him a collection of poems written by Trakl, the first existential poet in German. Stan read them all and was visibly moved. I will always regret not being able to hold his attention long enough for him to get the basic idea of Husserl's phenomenology. He would have liked it.

Those of us who were close to him at the end of his life (Bednarek, Beyer, Everett, Mauldin, Metropolis, Mycielski, Stein, and I, to name a few) were drawn to him by a fascination that went beyond the glitter of new ideas of arresting beauty, beyond the trenchant remarks that laid bare the hidden weakness of some well-known theory, beyond the endless repertoire of amusing anecdotes. The fascination of Stan Ulam's personality rested in his supreme self-confidence. His self-confidence was not the complacency of success. It rested on the realization that the outcome of all undertakings, no matter how exalted, will be ultimate failure. From this unshakeable conviction he drew his strength.

This conviction of his, of course, was kept silent. What we heard from him instead were rambling tirades against mathematicians and scientists who took themselves too seriously. He would tear to shreds some of the physics that goes on today, which is nothing but poor man's mathematics, poorly learned and poorly dressed up in a phoney physical language. But his faith in a few men whom he considered great remained unshaken: Einstein, Fermi, Brouwer, President Truman.

Thinking back and recalling the ideas, insights, analogies, nuances of style that I drew from my association with him for twenty-one years, I am at a loss to tell where Ulam ends and where I really begin. Perhaps this is one way he chose to survive.

He could not bear to see unhappiness among his friends, and he went to any lengths to cheer us up when we were down. One day, we were driving towards the Jémez Mountains, along the stretch of straight road that starts right after the last site of the laboratory. I felt depressed, and drove silently, looking straight ahead. I could feel his almost physical discomfort at my unhappiness. He tried telling some funny stories, but they didn't work. After a minute of silence, he deployed another tactic. He

knew I had been interested in finding out just how much physics he really knew, and that I had unsuccessfully tried to quiz him. Now he launched on a description of the Planck distribution (which he knew I didn't know) and its role in statistical mechanics. I turned around, surprised at the thoroughness of his knowledge, and he smiled. But a few minutes later he again fell silent, and the gloom started all over. After a pause that was undoubtedly longer than he could bear, he blurted out: "You are not the best mathematician I have ever met, because von Neumann was a better one. You are not the best Italian I have ever met, because Fermi was a better one. But you are the best psychologist I have ever met." This time I smiled. It was his way of acknowledging our friendship. He knew that I could see through his weaknesses, through his laziness, through his inability to do any prolonged stint of work. He knew that I discounted those weaknesses, and that I saw, beyond them, the best of his person. That he appreciated.

No other period of civilization has been so dependent on hypocrisy for survival as the *belle époque*, the Victorian Age. It has bequeathed us a heritage of lies that we are now charged with erasing, like a huge national debt: the image of the hero as the fair-haired boy, and the sharp partition of all people into "good guys" and "bad guys." These false illusions must now make way for biographies in which ambiguity, duplicity, and the tension of opposites are seen as the fundamental forces that drive every person.

The prejudice that the scientist, as a seeker of the truth, is immune from the passions of the world and is capable of doing no wrong, a prejudice propagated for over a century by bigoted biographers, has done harm. One shudders to guess how many talented young minds have been discouraged from a career in science by reading such unrealistic portrayals of the scientist as a saint. Moreover the presumption that "good" behavior (as interpreted by the biographer) is a prerequisite for success in science betrays a lack of faith in science. Lastly, one should tell the truth, even when such a truth belies our ideas of how things ought to be.

Stan Ulam was lazy, he talked too much, he was hopelessly self-centered (though not egotistical), he had an overpowering personality. But he bequeathed us a view that bears the imprint of depth and elegance, one that enriches our lives and will enrich the lives of those who come after us. For this he will always be remembered. ■

**Gian-Carlo Rota** is a Professor of Applied Mathematics and Philosophy at the Massachusetts Institute of Technology. He has served as a consultant to the Laboratory for over twenty years.

# From Above *the* Fray

*by Carson Mark*

Many people thought of Stan as lazy. In one way of speaking they were right: he never was observed cutting the lawn or washing dishes, and he engaged only extremely rarely in the tedious, straightforward—albeit demanding—business of actually carrying through long and complicated and step-wise dependent numerical calculations.

But in a more important way of speaking they were wrong. He was never asleep. He was constantly alert to the strange ways in which people applied and immersed themselves in immediate and detailed problems. He did not consider such efforts to be undeserving of credit—indeed he recognized them as essential to any practical outcome—but for the most part they were outside the range of his direct interest and personal style. He thought, and thought constantly, in a qualitative and unconstrained way. In this fashion he made a number of exceedingly important observations bearing on the weapons program.

He was among the first to realize how valuable the Monte Carlo method would be when electronic computers should make extensive applications of the method feasible. Even before that time, in connection with a particular aspect of the Super Program that was essentially impregnable against analytic efforts, he and C. J. Everett applied the method, by hand, in a highly schematic but still enormously time-demanding manner. Stan was not fired by any desire that the hydrogen bomb should add a new dimension to the already intolerable capabilities of the atomic bomb—indeed he hoped that it might be possible to show that a thermonuclear weapon was impractical. And actually his and Everett's work strongly indicated impracticality for the particular pattern envisaged at that time (1950). In the meantime, others had prepared a much more elaborate treatment of the same central problem to be handled on electronic computers. The point had been indicated, and was probably right, and further detailed examination was in proper hands. It was time for him to drop the problem.

Stan subsequently turned his attention to the unique conditions—temperatures, pressures, and energy densities—that are established in the near neighborhood of a fission explosion, and he asked himself about the possible ways one might apply these to realize unprecedented effects. In discussing these speculations with Edward Teller, a completely new approach to obtaining a thermonuclear explosion came in sight. Very quickly the whole program of the Los Alamos Laboratory was redirected to the problems involved in this new approach. From the theoretical point of view, the new effort immediately required intensive and extensive calculations that could be carried out only on the most capable computing machines then available. This was, of course, done, and Stan followed the results with interest—but again, the conduct of this work was not his style. He turned his own personal attention to more long-standing questions having to do, for example, with random processes, nuclear propulsion, mathematical models for biological processes, patterns of growth, and so on.

At the time he retired from the staff of Los Alamos (1967), and to a very large extent since then, the weapons-related activities of the Laboratory were directed mainly to realizing modifications, improvements, or refinements of devices of the sort presaged in the Ulam-Teller proposal of 1951. ■

**Carson Mark** first came to Los Alamos in May 1945 as a Canadian collaborator on the Manhattan Project. In 1946 he became a member of the permanent staff of the laboratory that arose from the wartime project and was head of its Theoretical Division from 1947 until he retired in 1973.

PART II

# The Ulam Legacy

TO INTERDISCIPLINARY SCIENCE

*"There is not a single mathematician who in the slightest degree reminds one of Stan ... Mathematicians are much more in a mold, because finally the only method they have is logic. There is imagination and all that, but there are no experiments, no external things like those that tickle a physicist's imagination. Most mathematicians don't experiment at all. Stan did. There is nobody like him, absolutely nobody." (Mark Kac, 1984) The contents that follow were inspired by that instinct to experiment, not only in mathematics, but in physics and biology as well.*

Banach
Borsuk
Fermi
Gamow
Oxtoby
S Ulam
J Von Neuman

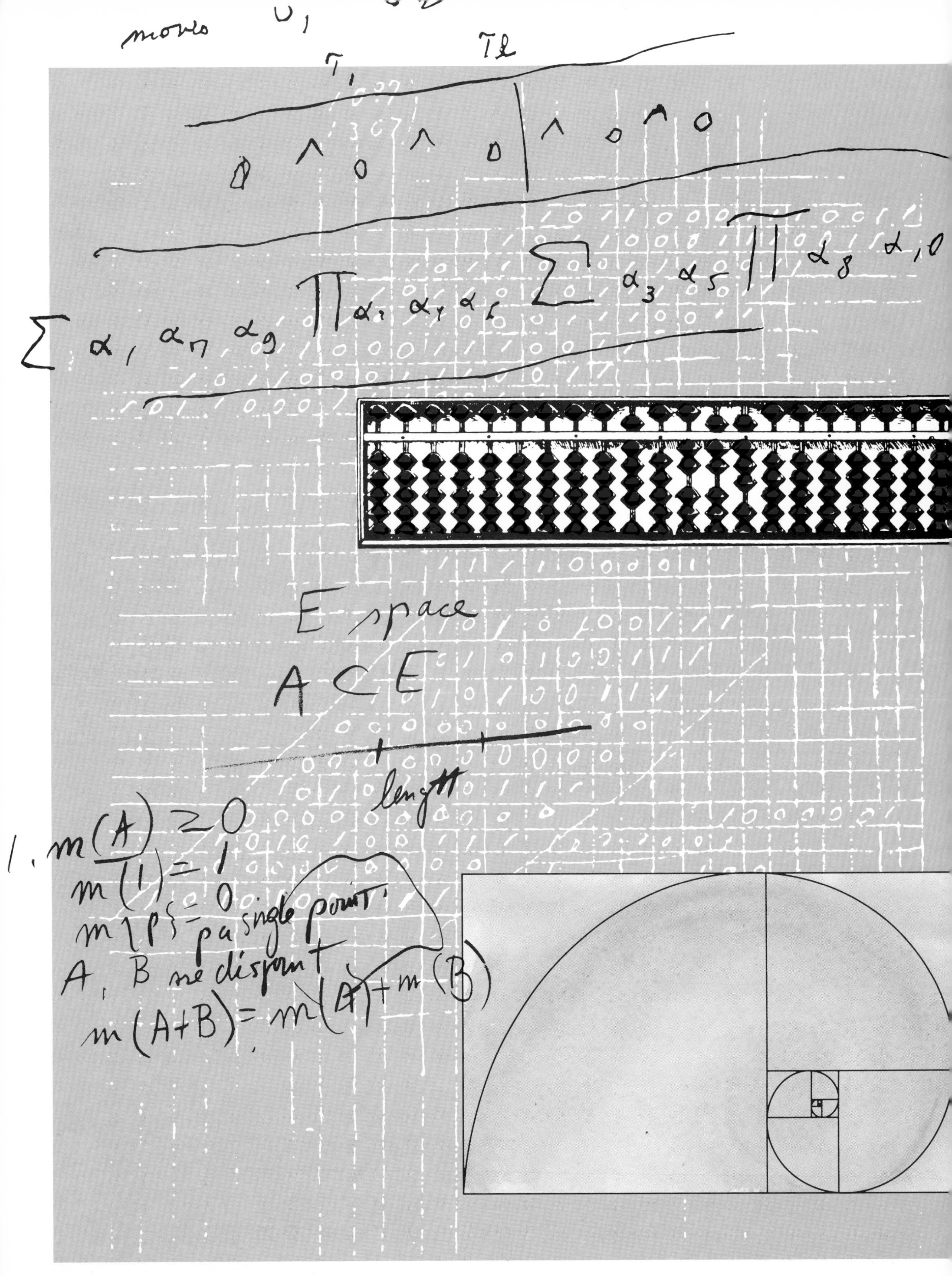
E space
A C E
length
1. m(A) ≥ 0
m(I) = 1
m{p} = 0
p a single point
m(A+B) = m(A) + m(B)

# MATHEMATICS

Stan Ulam was first and foremost a mathematician, but his nature was far too expansive to be contained within a single discipline. He saw even the most abstract realms of mathematics in relation to the natural sciences and looked to the natural sciences for inspiration in mathematics. His extraordinary facility for turning a simple question into a bona fide mathematical problem is evident in the mathematics articles presented here. Also evident is his pioneering interest in the use of computers for what he called heuristic studies in mathematics and nonlinear science—what we now call experimental mathematics. Following directly from Stan's example, experimental mathematics has become the primary tool for exploring the complex behavior of nonlinear systems and as such appears as a recurrent theme throughout "The Ulam Legacy."

Stan's great generosity and ingenuity attracted a long list of collaborators—among them the authors who so gladly contributed to this section. Each has a unique experience of the man to relate, and each introduces the reader to one or several areas of mathematics by discussing problems that Stan posed. Our authors—undoubtedly influenced by Ulam—go straight to the heart of the matter and thereby make mathematics a delight to learn.

We start with an article by David Hawkins, a philosopher from the University of Colorado, Boulder, who met Stan at Los Alamos during the Manhattan Project. In his autobiography Stan characterizes Hawkins as "the most talented amateur mathematician I know." During the war he and Hawkins, "teacher" and "student," developed a formalism to describe the multiplication of neutrons during a fission chain reaction. (So began the abstract theory of branching processes later developed with C. J. Everett.) Through Stan's vision the neutron-multiplication problem and other topics discussed by Hawkins (prime numbers, the Monte Carlo method, "sexual" reproduction) took shape as iterative processes of a nonstandard kind, perfectly suited for study on electronic computers. Hawkins, in "The Spirit of Play," has portrayed better than anyone else the fun of working with Stan.

Dan Mauldin, professor of mathematics at North Texas State University and the author of "Probability and Nonlinear Systems," was a major collaborator and close friend of Stan's during the last ten years of his life. Mark Kac, another Polish mathematician from the Lwów school, gave a clue to the source of this strong collaboration: "[Dan] is a first-rate mathematician and he has the Polish soul with regard to mathematics. ... He was on his way to becoming an all-American linebacker on the famous Longhorn team, and he gave it up for mathematics!" Dan's love for mathematics and gift for teaching shine through as he introduces the readers to the basic tools of measure and probability and then shows how to apply these tools to the fashionable and challenging problems in nonlinear science. His tutorial on probability and measure is intended to fill some gaps in our mathematical background as it reminds us of the beauty and precision of the mathematicians' world. Of the three research problems that follow the tutorial, the first and third demonstrate how intuition gained from computer experiments leads to strict mathematical proofs. The second, "Geometry, Invariant Measures, and Dynamical Systems," is most closely tied to the physicists' approach to nonlinear systems. We know now that many deterministic nonlinear systems live on strange attractors, delicate fractal structures that describe a never-repeating orbit confined to a finite region of space. These systems exhibit what appears to be chaotic behavior. How would one describe the long-term behavior of such a system? Dan shows us how to define a probability measure on strange attractors that could be used to calculate the average properties of complex nonlinear systems.

In the article that follows, Paul Stein takes us back a step in history to the computer studies done with Stan in the early sixties on iterations of nonlinear transformations. Their work, apart from leading to perhaps the first discovery of a strange attractor, also led to the 1973 paper by Metropolis, Stein, and Stein on the iteration of the famous one-dimensional logistic map. This paper was a source of inspiration for Feigenbaum's 1976 breakthrough on the universal nature of the transition to chaos by "period doubling." Stein describes the earlier results on period doubling as well as some recent work with Mauldin on an alternate route to chaos that has been observed in chemical experiments.

The last two authors in this section break off from the theme of experimental mathematics and nonlinear systems. Mycielski, a Polish mathematician well known in the field of logic and set theory, took this opportunity to introduce to the non-mathematician two of Ulam's formidable contributions to pure mathematics; first his measurable cardinals, which allow one to talk about orders of infinity well beyond what anyone had dreamed of, and, second, his proof with Oxtoby of the existence of ergodic transformations, one important step toward proving the ergodic hypothesis, the most controversial assumption in the foundations of statistical mechanics. (The significance of ergodicity or the lack thereof is discussed again in the physics section of "The Ulam Legacy").

Ronald Graham from Bell Laboratories closes our foray into mathematics with a little introduction to graph theory. Graham focuses on a problem that illustrates Stan's continuing fascination with quantifying exactly how alike (or different) mathematical objects or structures are. This theme will recur in another form when Walter Goad takes up the question of "alikeness" of DNA sequences in the biology section of "The Ulam Legacy."

# the Spirit of Play

*a memoir for Stan Ulam*

***by David Hawkins***

Sometime in early 1944 I passed the open door of a small office near my own: S. ULAM. He had arrived at Los Alamos only a few days before and seemed unoccupied. We introduced ourselves—he a young mathematician, I an even younger philosopher, one with mathematical leanings. My field of work was the philosophy of mathematics and science. I had listened in on the shop talk of the theoretical physicists at Berkeley and knew their style. They thought of me for managerial chores in the newly created Los Alamos laboratory. So I came, as an administrative assistant to Robert Oppenheimer. Only later was I given the job of writing a wartime history. I was in fact the sole representative of my trade at Los Alamos, and the label "philosopher" usually caught curious attention. But Stan ignored it. He had come as a new member of the Theoretical Division, although no one (he slyly suggested) knew quite why. I later guessed that he had indeed been invited for no particular reason other than the urging of John von Neumann. Stan's version was characteristic: "Physicists don't know what to do with mathematicians."

It was the beginning of a long personal and family friendship. But here I shall restrict my recollections to associations of the thinner, more mathematical kind. We soon discovered one strong common interest, in the foundations and uses of probability theory. Some of Stan's work (Lomnicki and Ulam 1934) had preceded that of Kolmogorov on the measure-theoretic formulation of probability. Mine had been on the conceptual foundations, battled over since the time of Bernoulli and Leibnitz and closer to the philosopy of physics.

One day Stan threw a problem at me, as if to bring our academic discussions back to the concerns of a wartime laboratory. In the chain reaction that was to power the atomic bomb, some fraction of the neutrons liberated by a fission induce other fissions, which in turn liberate more neutrons that induce more fissions, and so on. Suppose the number of induced fissions per fission is a random variable that can take on the values $i = 0, 1, 2, \ldots$ with probability $p_i$. (That is, $p_0$ is the probability that the neutrons from a single fission induce no further fissions, $p_1$ is the probability that they induce one further fission, and so on.) What then is the probability distribution of the number of fissions occurring in the $n$th "generation" of such a process started by a single fission? Although we didn't know it at the time, the same problem—stated differently—had been solved long before. One earlier version had been posed in terms of the proliferation of a family name through male descendants. Assume that each male Jones produces $i$ male offspring with probability $p_i$, $i = 0, 1, 2, \ldots$ (and that this probability does not vary from one generation to another). What then is the probability that a given Jones has $k$ males in the $n$th generation of his descendants?

I spent several evenings on the problem. By persistence rather than insight I found the very simple solution (Hawkins and Ulam 1944). A lot of algebraic solvent evaporated and left behind an unexpected little crystal of a formula, the sort of outcome that makes you ask why it hadn't been obvious all along.

Let $f(x)$ be the Laplace generating function of the sequence of probabilities $\{p_0, p_1, p_2, \ldots\}$. (That is, let $f(x)$ be the function to which the infinite series $p_0 + p_1x + p_2x^2 + \cdots$ converges.) Then the probability that Jones has $k$ grandsons (or $k$ second-generation male descendants) is the coefficient of $x^k$ when $f_2(x) \equiv f\left(f(x)\right)$ is expanded in powers of $x$. And in general the probability that Jones has $k$ $n$th-generation male descendants is the coefficient of $x^k$ when $f_n(x) \equiv f\left(f_{n-1}(x)\right)$ is similarly expanded. Thus, to the biological process, that of reproduction, there corresponds an algebraic one, that of iteration, in which the argument of a function is replaced by the function itself. I'll mention other related results and further applications later, but this was the essence of our first venture into what was to develop into the theory of branching (we said "multiplicative") processes.*

Stan was delighted with my solution, and I, the rank amateur, was flattered. He already knew quite a lot about the deceptively simple operation of iteratively substituting a function for its own argument, and I got a lesson or two. In the course of these discussions, we got on to such topics as space-filling curves, turbulence, and what have recently come to be called catastrophes, in which deterministic laws lead rigorously to results we can only describe as chaotic. A good many years later when we were reminiscing about all of this, I complained that we had almost been pioneers in such matters. Why hadn't we pursued them? Stan's reply: "It's because there are so many of them guys and so few of us!"

*I should also mention a prior Los Alamos paper by S. Frankel, which may lie buried in the 1943 series of Los Alamos reports. Frankel had thought in terms of a continuous time parameter instead of discrete generations. That approach leads to a one- parameter family of generating functions embedding our $f_n(x)$. The problem actually has an even earlier origin. It was discussed by Darwin's cousin Francis Galton in 1889 and then by A. Lotka in 1939. Later, in 1945, Erwin Schrödinger addressed the problem, and I recall seeing the title of a relevant Russian paper (obviously declassified!) of about the same date. A section of Feller's classic text on probability theory (Feller 1968) is devoted to branching processes; a full development is that of T. E. Harris (Harris 1963).

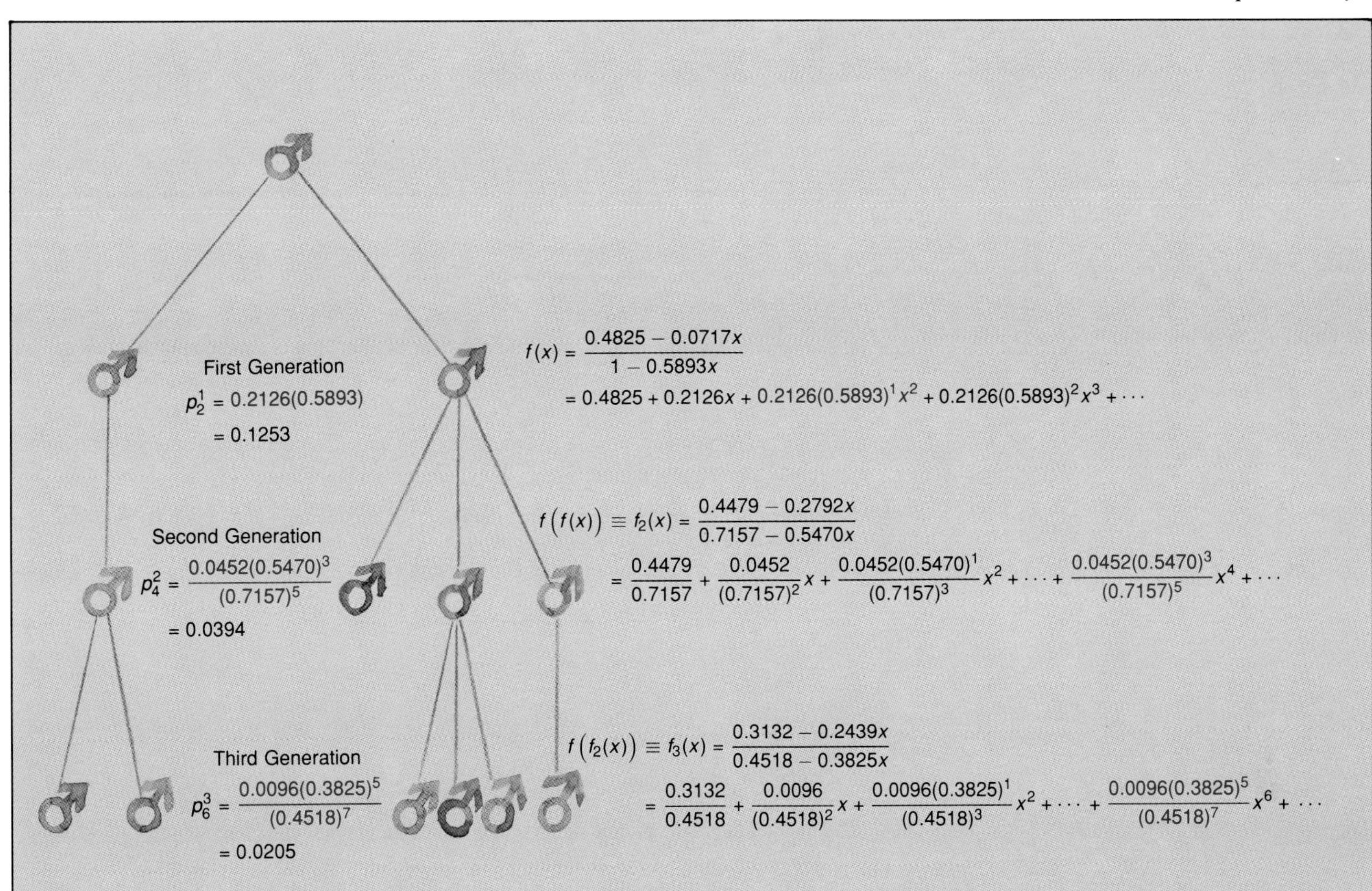


# A Biological Chain Reaction

A study in the thirties by A. Lotka showed that the probability $p_i^1$ of an American male having $i$ male children was described by the sequence of probabilities

$$\{p_0^1, p_1^1, p_2^1, \ldots\} = \{0.4825, 0.2126, 0.2126(0.5893), 0.2126(0.5893)^2, \ldots\}.$$

Assuming that this sequence is also applicable to American males of later generations, we can treat the production of male descendants (and the proliferation of a surname, assuming further that male children bear the surname of their father) as a simple branching process—a chain reaction. Let $f(x)$ be the generating function for Lotka's sequence of probabilities; that is,

$$f(x) \equiv 0.4825 + 0.2126x + 0.2126(0.5893) + 0.2126(0.5893)^2 + \cdots$$
$$= \frac{0.4825 - 0.0717x}{1 - 0.5893x}.$$

Then $p_k^n$, the probability that the $n$th-generation descendants of an American male includes $k$ males, is the coefficient of $x^k$ in the expansion of $f_n(x) \equiv f\big(f_{n-1}(x)\big)$. Illustrated here are the first three generations of male descendants of some ancestral male; the first generation includes two males, the second four, and the third six. Listed for each generation are $f_n(x)$ and its expansion in powers of $x$. The highlighted probabilities $p_k^n$ depend only on $n$ and $k$ and not on the particular family tree. ■

I know very little in detail of the wide range of Stan's work and his repertoire. In this memoir I shall confine myself to matters we corresponded about or worked on jointly. I do this partly because some of these may not be otherwise known and partly because they affected my own mathematical avocation in a way that throws some light on the character of Stan Ulam, teacher. I never sat in on any of his courses, to be sure, though I sometimes heard him lecture. The teaching I shall speak of is that I occasionally received, over many years, one-to-one. In talking about all this I shall refer to some work of mine that shows the nature of the Ulam influence; it is minor work but still a mirror of our associations. And I enjoy bringing these pieces together for the first time.

Stan was indeed a superb teacher, of a kind not very common. One part of his secret was a quite extraordinary talent for turning forbidding topics into attractive problems, attractive because they seemed promising, seemed to open up some larger area. Another part is a quality I am tempted to describe as meritorious laziness. Though Stan could, on occasion, himself engage in intense and concentrated work, as a teacher he would give you the challenge and then—let you do the work. I remember feeling a bit resentful. I did all the work on that first little paper, and he could have added more! But what he really added was to my confidence. For Stan no ego was invested.

Later, when I was at the University of Colorado, Stan and I both did some further work on branching processes. He, with C. J. Everett, had generalized the whole scheme by including "particles" of different types (Everett and Ulam 1948). This generalization, in its physical applications, allowed offspring and progenitors to differ from each other, for example, in their spatial or dynamical, and hence also in their reproductive, characteristics.

My own related work was inspired partly by a conversation we had about one of the great and vital mysteries of mathematics. The Greeks got on to it, long before Euclid, in the discovery that geometrical facts could be represented arithmetically, while those of arithmetic could be seen in the mirror of geometry. In our own day the pendulum has swung far toward the arithmetical, whether analytic or digital, side. Rather typically Stan took the "wrong" side, that of geometry. "Draw a curve," he said, "of a nice simple function. Now draw another curve parallel to it. The relation is very simple to *see* and understand, but algebraically it can be quite messy." How is it possible that relationships that are so complicated in one domain can be mapped into another where they appear so simple, or vice versa?

The generating-function transformation I had used in that first problem of ours is an elegant elementary example; it belongs to a wider family with many applications in applied mathematics, including probability theory. We had extended its use a bit, and it was Stan's challenge to extend it further, as he did in the work with Everett. To me the challenge was to explore the relevance of this transformation to other operations of a stochastic nature. Long known of course is the fact that addition of independent random variables corresponds to multiplication of their generating functions. What could one say about other arithmetical operations—division, say, or the logarithm—when random variables take the place of simple numbers?

Consider the following example. Physics students learn that the number of alpha particles emitted per unit time by a bit of uranium is a random variable (call it $F$) described by the Poisson distribution, whose Laplace generating function $f(x)$ is

$e^{-\lambda(1-x)}$, where $\lambda$ is a decay constant characteristic of uranium. The time between emission of successive alpha particles is also a random variable (call it $G$); it is described by the exponential distribution, whose generating function $g(x)$ is $\lambda \int_0^\infty e^{-\lambda u} x^u du = \lambda/(\lambda - \ln x)$. Now $F$ and $G$ are reciprocals of each other. What then is the relation between $f$ and $g$, the generating functions of their probability distributions? The answer is that they satisfy the remarkably simple relation

$$f^{-1}(x)g(x^{-1}) = f(x^{-1})g^{-1}(x) = 1,$$

where $f^{-1}$ and $g^{-1}$ denote the inverses of $f$ and $g$: $f^{-1}(x) = (\ln x + \lambda)/\lambda$ and $g^{-1}(x) = e^{\lambda(1-1/x)}$.

One can show that this functional relation holds quite generally. Whatever the probability distribution may be for $A$ per unit $B$, its generating function and that of the distribution for the reciprocal measure $B$ per unit $A$ will satisfy the above identity. (A mathematical nicety is that the inverses of such functions always exist.) Thus one can easily calculate means, variances, and higher moments of one distribution from those of the other.

A related topic is the "random logarithm": Find the probability distribution, for example, of the time required for a chain reaction to produce a given population size. I was in fact looking for some old notes on these matters, which Stan had asked about, when I learned of his death.

**SIEVE OF ERATOSTHENES**

After the war I was absented from weaponeering—first from choice and then by the F.B.I. I became politically opposed to the arms race that supported it, but not to wartime friends. Over the following years Stan and I corresponded or talked about a good many different topics, and again it was he who got me thinking about some of these. As I write now, I realize they all concerned iterative processes, deterministic or stochastic or mixed, that seemed to lie beyond the range of "standard methods." So although it might seem a bit of a jump to go from chain reactions to prime numbers, both fitted that general category. There is an iterative definition of the prime numbers, the sieve of Eratosthenes. The process is completely deterministic, but the way the primes are scattered among the other numbers has a very chancy look that has stimulated generations—centuries—of investigation.

First there came from Stan some rolls of print-out: very long lists of primes, of twin primes, of successive differences between them, and so forth. All these of course were computer-generated. Others may have computed even longer lists; Stan was one of the first to do so. But soon the pattern changed. The theory of primes is a high-order specialty for number theorists, and a happy hunting ground for amateurs like me. Stan was neither, or both. I think he may have been the first to think of the sieve of Eratosthenes as merely one among many sorts of iterative processes whose products lie beyond the range of standard methods. He thought of a good many other ways of generating number sequences that were more or less prime-like in their frequency and distribution. It was a flanking maneuver: If you can't solve the original problem, think of others that resemble it, and may be easier. Some of Stan's schemes seemed to me far-fetched, and I said so. His reply: "Yes, but I am the village idiot!"

Indeed, I think that Stan often did not care whether he got to the essence of a

## SOME RANDOM NUMBERS

**The random numbers below are a small fraction of the 10,000 generated in the early 1900s by L. H. C. Tippett, then a member of the Biometric Laboratory at University College, London. Tippett generated the numbers, which were used in statistical sampling procedures, by selecting 40,000 single digits from census reports and combining them by fours. The collection of numbers was originally handwritten; the excerpt here is reprinted, with permission, from a version published in 1959 by Cambridge University Press (*Random Sampling Numbers*, Tracts for Computers, edited by E. S. Pearson, Number 15).**

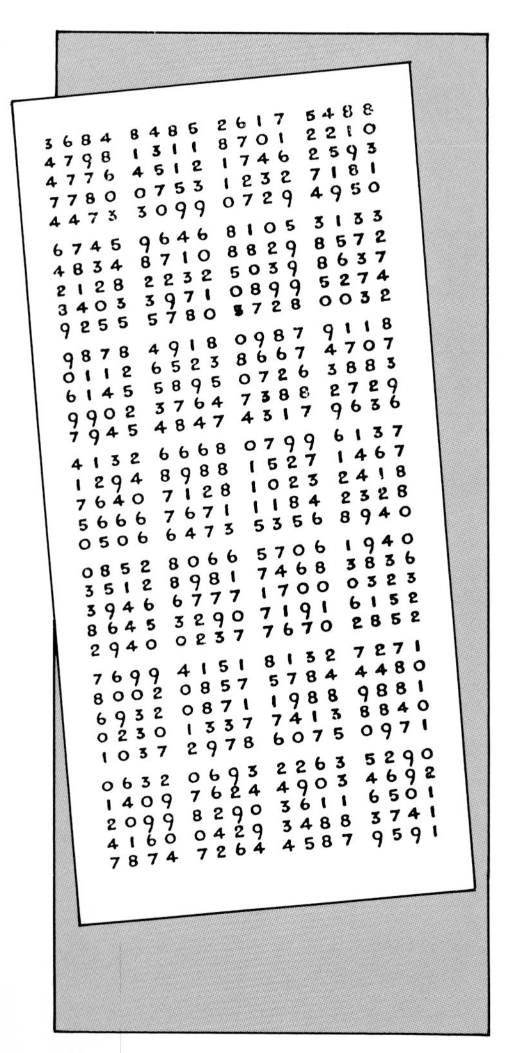

| | | | |
|---|---|---|---|
| 3684 | 8485 | 2617 | 5488 |
| 4798 | 1311 | 8701 | 2210 |
| 4776 | 4512 | 1746 | 2593 |
| 7780 | 0753 | 1232 | 7181 |
| 4473 | 3099 | 0729 | 4950 |
| 6745 | 9646 | 8105 | 3133 |
| 4834 | 8710 | 8829 | 8572 |
| 2128 | 2232 | 5039 | 8637 |
| 3403 | 3971 | 0899 | 5274 |
| 9255 | 5780 | 3728 | 0032 |
| 9878 | 4918 | 0987 | 9118 |
| 0112 | 6523 | 8667 | 4707 |
| 6145 | 5895 | 0726 | 3883 |
| 9902 | 3764 | 7388 | 2729 |
| 7945 | 4847 | 4317 | 9636 |
| 4132 | 6668 | 0799 | 6137 |
| 1294 | 8988 | 1527 | 1467 |
| 7640 | 7128 | 1023 | 2418 |
| 5666 | 7671 | 1184 | 2328 |
| 0506 | 6473 | 5356 | 8940 |
| 0852 | 8066 | 5706 | 1940 |
| 3512 | 8981 | 7468 | 3836 |
| 3946 | 6777 | 1700 | 0323 |
| 8645 | 3290 | 7191 | 6152 |
| 2940 | 0237 | 7670 | 2852 |
| 7699 | 4151 | 8132 | 7271 |
| 8002 | 0857 | 5784 | 4480 |
| 6932 | 0871 | 1988 | 9881 |
| 0230 | 1337 | 7413 | 8840 |
| 1037 | 2978 | 6075 | 0971 |
| 0632 | 0693 | 2263 | 5290 |
| 1409 | 7624 | 4903 | 4692 |
| 2099 | 8290 | 3611 | 6501 |
| 4160 | 0429 | 3488 | 3741 |
| 7874 | 7264 | 4587 | 9591 |

particular problem. There are many things you can do with problems besides solving them. First you must define them, pose them. But then of course you can also *refine* them, *depose* them, or *repose* them, even *dissolve* them! A given problem may send you looking for analogies, and some of these may lead you astray, suggesting new and different problems, related or not to the original. Ends and means can get reversed. You had a goal, but the means you found didn't lead to it, so you found a new goal they did lead to. It's called play. Cyril Smith has argued persuasively, and with good historical evidence, that play, not utility, has long been the mother of invention (Smith 1981). Utility has been only the nursemaid. Creative mathematicians play a lot; around any problem really interesting they develop a whole cluster of analogies, of playthings.

One of Stan's playthings, his "lucky numbers," got considerable attention (Gardiner, Lazarus, Metropolis, and Ulam 1956). These numbers, which are generated by a sieve quite like that for the primes, have no particular arithmetical properties; they are just lucky to survive the sieving. A number of us got involved in studying their long-run distribution, which turned out to very close to that of the primes (Hawkins and Briggs 1957).

It is here I should mention an important one of Stan's contributions in the general grouping of nonstandard iterative processes. There was nothing lazy about his pursuit of a really good idea. "Pursuit" is probably the wrong word; it implies he already had the idea on the run. In the beginning it maybe was more like a roundup, a nudging together of possible example after possible example. I recall his ruminations about the Monte Carlo method in 1944, when already he was talking about it. More than fifty years ago there began to appear small compilations of random numbers. I remember the incredulity of a good physicist friend when I showed him such a listing. He knew Jahnke-Emde, of course, an old book of tables of almost every then-standard function. But *random* numbers? Statisticians were the wave of the future in those days. They alone used random numbers—for honest sampling procedures—even though nobody quite knew what "random" meant. But Stan foresaw that when high-speed computers should come along, they might well be used to imitate various deterministic or stochastic processes. Stan's phrase was "playing the game." The question was how to provide a computer with random numbers. I favored the built-in alpha counter, but that violated the sensibilities of mathematicians. There is indeed a grey area between chance and determinism, occupied by pseudo-random sequences. Even that early Stan was exploring it.

I think the most original part of Stan's early thinking about such matters was the idea that you could transform the equations describing a completely deterministic process into a mathematical form that also describes a stochastic one, and then you could get approximate solutions by playing the game repeatedly on a high-speed computer. But Monte Carlo is not my topic, except again as it affects my picture of Stan Ulam, teacher.

One of the topics we got on to later was a Monte Carlo approach to the theory of prime numbers. Pick a number at random from the neighborhood of some number $N$. The probability of picking a prime there is about $\frac{1}{\ln N}$, their approximate frequency. Why not reverse the process and produce a *random* sequence of numbers that mimic the primes? Try out each number $N = 2, 3, 4, 5, \ldots$ against a game of chance for which the probability of "winning" is $\frac{1}{\ln N}$, and select it for the sequence only if it wins. The properties of such a sequence, or rather of many such sequences examined together, might throw light on some aspects of prime-number theory. The statistician Harald Cramer, it turned out, had already written about that. My own next bright idea was to turn the sieve of Eratosthenes itself into a Monte Carlo device. Drop out half the numbers beyond 2, namely those that "lose" a game of chance for which the probability of losing is $\frac{1}{2}$. Let $N$ be the first survivor. Now drop out $\frac{1}{N}$ of the numbers beyond $N$, those that lose another game of chance for which the probability of losing is $\frac{1}{N}$. Keep repeating the process indefinitely, each time basing the sieving on the first survivor. One can play this game on a computer, which I did. But the theory of these "random primes" turned out to be not too difficult (Hawkins 1974), and it showed that the prime-number sequence could be regarded as one of an infinite family of sequences very much like it in their average properties. It supports some familiar conjectures about the primes and suggest others. The best result, I think, is that the famous and unproved Riemann hypothesis turns out to be true of "almost all" sequences generated by the random sieve. This hypothesis is a more recondite example of the kind of transformation I have talked about. It concerns the zeros of a certain function in the complex domain and, if true, implies a whole batch of propositions in number theory. Many of these can be proved independently, and none have been disproved. But some seem to be beyond the range of simple methods. That the Riemann hypothesis can be shown to be true of the random primes, and thus of almost all prime-like sequences, surely makes even more unlikely the possibility that the primes themselves should prove an exception.

Here, finally, I should mention another component of Stan's work, one that I can also trace back to early Los Alamos days. It grew later to very substantial proportions. One beginning I recall was to discuss a stochastic branching process that requires the "mating" of two "particles" from one generation in order to produce "offspring" for the next: sexual reproduction. Here the branching goes in both time directions, backward genealogically and forward by descent. The theory of this branching is essentially nonlinear. "Sex," Stan said, "is quadratic!" I had indeed examined one kind of nonlinear stochastic process, a chain reaction in which depletion of fuel, or of nutrient in the case of bacterial reproduction, is a factor. This led to a stochastic version of the well-known logistic curve of growth, which at first rises exponentially and then tapers off to a zero or negative slope. My work had a certain mathematical interest because it showed that the statistical fluctuations in such a nonlinear process can also change its average character; they don't "average out."

Such work as this might have stayed in abeyance except for Stan's development of other and much broader interests, namely in mathematical models of growth and reproduction. I remember approaching him with my own new-found interest in Claude Shannon's work on information theory and in the discovery of Watson and Crick. I wanted to define a measure of biological complexity, or organization, in information-theory terms, and we were immediately at loggerheads. He wanted to insist that very simple instructions could produce very complex patterns and I that such simplicity would nevertheless limit the *variety* of such patterns. Each of us was defending a different meaning of "complex." I already knew of his work (or play) with computer-generated growth patterns (Ulam 1962) but hadn't realized fully the range of ideas he was bent on exploring. Once more it was that flanking move. The genetic instruction

### "SEX," Stan said, "IS QUADRATIC"

**In this quote Stan was expressing a broad mathematical view of sex as a branching process in which some interaction, or "mating," between "male" and "female" members of a species is required for reproduction. An example is the deadly mating of male and female black widow spiders.**

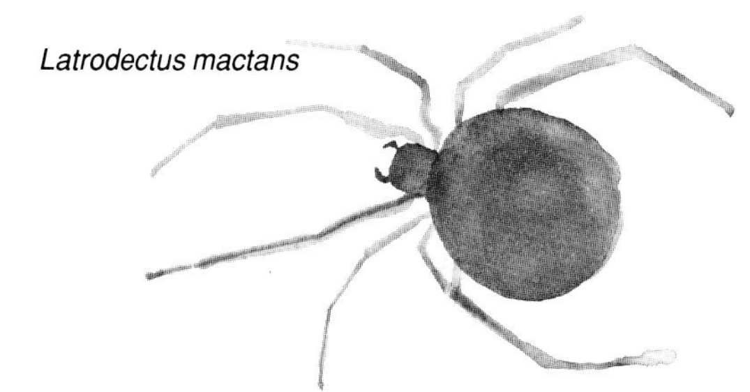
*Latrodectus mactans*

**Oddly enough, the animal kingdom includes some species, namely, a few of the tapeworms, that reproduce without any mating.**

Tapeworm

**The demography of a sexually reproducing species depends on (among other factors) a product of the male and female populations—hence the adjective"quadratic." For mathematical simplicity mating is often assumed to be random, as it is for the ornamental ginkgo, or maidenhair, tree.**

*Ginkgo biloba*

of biological growth and reproduction is a vast and still mostly uncharted domain for investigation. But once more the "village idiot" could invent all kinds of very simple processes bordering that domain. The idea of "growing" elaborate dendritic patterns, "organisms," by the endless repetition of a few simple "genetic" instructions, applied in each cycle to the results of previous cycles, was another in the category of iterative processes that lay beyond the range of standard methods. It later became the basis for the famous "game of life"—was Stan its first inventor? I don't know. I connect this work also with Stan's important work on the nature of and approach to equilibrium in even slightly nonlinear iterative processes. In the years following he became quite deeply involved in more realistic problems of genetics, but I mostly lost touch.

One of these problems, now well known and used in molecular genetics, came from Stan's deep familiarity with measure theory. Suppose a deck of cards can be shuffled only by several allowable operations. Knowing these and the end result of a shuffling, find the smallest number of allowable operations that accomplishes the given result, and call it the "distance" between the two orderings. Two decks of cards, or two nucleic acid strands, might appear very different in an item-by-item comparison yet be by shuffling history very close. Stan was a visiting professor at the University of Colorado's medical school when he worked on this, and I have a nice story from Theodore Puck. Stan got so interested in the mathematics (now *not* an iterative process) that he seemed to be ignoring the relevant biology. Reproached, he mended his ways. But he began his final talk on the subject with an imperative: "Ask not what mathematics can do for biology; ask rather what biology can do for mathematics!"

In the sixties and seventies I became more and more concerned with practical and theoretical work relating to elementary-school education in mathematics and science, to "school-doctoring." Toward this new career of mine Stan was—tolerant. We enjoyed good conversations but little time for shared work. It was only last year that I was suddenly recalled to our earliest association, catching up on some work he had done in population genetics and related matters. With characteristic initial disregard for humdrum scholarship, he had reinvented *and* extended some of the existing theory, developed first by R. A. Fisher and Sewall Wright.

I had known generally about this work but had missed one small paper, one in which he and Jan Mycielski formulated the basic theory of stochastic pairing, the branching process involved in sexual reproduction (Mycielski and Ulam 1969). Its main focus was not, however, on the fluctuational aspect of the process but on the average distribution and evolution of mutations within a species. The paper set forth three measures of the "distance" between two individuals. I shall mention only one of these, proposed by Mycielski. It is simply the sum, over the present generation and all past generations, of symmetric differences in genealogy; that is, the number of entries present in one family tree and absent from the other, plus the number present in the other and absent from the one. Since sexual reproduction is already a stochastic process, this measure is genetically crude (for example, it ignores sibling diversity). But it is surely a plausible first (or if you wish, zeroth) approximation—a measure of purely genealogical, not yet of genetic, distance.

Stan had done (as he often had for other problems) some Monte Carlo simulations assuming a constant population size of $2N$, random pairing between the $N$ males and females in each generation, and two offspring, one male and female, per pairing. The simulations had told him that the genealogical distance (as defined above) between two randomly selected individuals from the same generation was, on the average, about $4N$. It was shown subsequently, by Kahane and Marr, that the average distance is precisely $4(N-1)$. Intrigued by all this and some explanation by Jan Mycielski, I found myself recovering some of our ancient lore, or reinventing it, and realized that we had been within an inch of this more recent thought, and then lost it, forty years ago!

Stan was an accomplished Latinist, but he deferred to my own (rather slight) knowledge of Greek. "Tell me," he once asked, "why is it that people are always saying 'Eureka!' and never—what is the Greek?—'I *lost* it!' After all, it's much more likely." I got the Greek for him, something like "Ὀλωλα!." It is by chance pronounced very like the modern French "Oh là là!," which can have a related meaning, "I am undone!" At any rate it did come back, and we had indeed lost it.

Not long after Stan retired from Los Alamos and came to Boulder, we touched on the subject again. If we trace the branching of ancestors back far enough, we of course share almost all of them. All men overwhelmingly are indeed almost-brothers; they are $N$-fold $n$th cousins. We jokingly speculated about the distance back to some common ancestor—Stan a Polish Jew and I an Anglo-American. Very likely less than twenty generations. All this was play, rediscovering the obvious along a pathway paved with numbers. I also remember that Stan the set-theorist poked his head out here with a "little remark." In a population of fixed finite size and infinite duration, everyone could be assigned an integer as a proper name, but the set of all sets of ancestors, of the genealogical "names," would be uncountable. We also noted more mundanely that the backward count of ancestors would fit some logistic curve, going up exponentially at first and then flattening out.

When I read the above paper, I was reminded of all this and noticed that the random pairing process was after all quite tractable. ("Noticed" is one of Stan's words; for me it represented hours of fussing about.) If, among the $2N$-member population that existed $n$ generations ago, your mother had $r$ female ancestors and your father had $s$, then the probability $q_k$ that they share $k$ female ancestors in that generation is given by

$$q_k = \frac{\binom{r}{k}\binom{N-r}{s-k}}{\binom{N}{s}}. \qquad (1)$$

The sequence of probabilities defined by Eq. 1 is the hypergeometric distribution, a standard textbook entry. The applicability of this distribution to the problem at hand can perhaps be more easily seen by recasting it in textbook terms. Suppose we have an $N$-element set (the ancestral female population), $r$ members of which are white (your mother's female ancestors) and the remainder are non-white (say black). Suppose we choose, randomly from among that set, an $s$-element set (your father's female ancestors, an identification allowed by the assumption of random pairing). The problem then is to find the probability $q_k$ that exactly $k$ elements of the chosen set are white (shared ancestors). Now the $k$ white elements can be selected for the $s$-element set in $\binom{r}{k}$ ways and the $(s-k)$ black elements in $\binom{N-r}{s-k}$ ways. Since any choice of $k$ white elements can be combined with any choice of black elements, the $k$ white elements can be selected

**GROWTH CURVE FOR EXPECTED NUMBER OF FEMALE ANCESTORS *n* GENERATIONS AGO**

**As *n* increases, the expected number of female ancestors departs more and more from $2^n$ because of the increase in the number of shared female ancestors.**

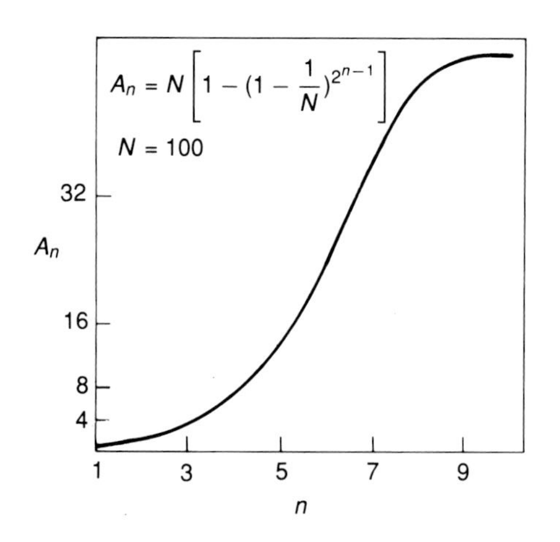


in a total of $\binom{r}{k}\binom{N-r}{s-k}$ ways. To obtain $q_k$, this number is then divided by $\binom{N}{s}$, the total number of ways of choosing the $s$-element set. (I should add that if the words "mother" and "father" are interchanged, along with $r$ and $s$, the answer is the same; though the resulting formula for $q_k$ will look different, it is not different in value.)

Using Eq. 1, we can now deduce the probability $p_{n+1,t}$ that, $(n + 1)$ generations back, you yourself have $t = r+s-k$ female ancestors. Let $p_{n,r}$ and $p_{n,s}$ be the respective probabilities that, among that generation, your mother has $r$ female ancestors and that your father has $s$. Since various values for $r$, $s$, and $k = r + s - t$ can yield a particular $t$ value, $p_{n+1,t}$ is a sum over those variables:

$$p_{n+1,t} = \sum_{s=2}^{t}\sum_{r=2}^{t} p_{n,r}p_{n,s}\frac{\binom{r}{r+s-t}\binom{N-r}{t-r}}{\binom{N}{s}}.$$

The Laplace generating function for this sequence of probabilities, call it $f_{n+1}(x)$, is therefore given by

$$f_{n+1}(x) = \sum_{t}\sum_{s}\sum_{r} p_{n,r}p_{n,s}\frac{\binom{r}{r+s-t}\binom{N-r}{t-r}}{\binom{N}{s}}x^t. \qquad (2)$$

Equation 2 does not lend itself to derivation of an elegant recurrence relation between $f_{n+1}(x)$ and $f_n(x)$, but it does provide such a relation between $A_{n+1}$ and $A_n$, where $A_n$ is the expected, or average, number of female ancestors $n$ generations back. This relation is

$$A_{n+1} = 2A_n - A_n^2/N, \qquad (3)$$

in which

$$A_n = N\left[1 - \left(1 - \frac{1}{N}\right)^{2^{n-1}}\right]. \qquad (4)$$

(Interestingly enough, the right side of Eq. 4 is also the answer to a much simpler problem: If $2^n$ objects are distributed randomly among $N$ boxes, what is the expected number of non-empty boxes?) If we identify the term $A_n^2/N$ as the average number of shared female ancestors $n$ generations back, then Eq. 4 defines just the logistic curve Stan and I had seen to describe the expected loss of ancestry; the difference between $2^n$ and $A_n$ (the average number of your female ancestors $n$ generations back) is just $A_{n-1}^2/N$ (the average number of female ancestors among that generation shared by your parents).

Now Mycielski's definition of the expected genealogical distance between two randomly chosen individuals of the same generation can be written $\sum_0^\infty 2(A_n - A_n^2/N)$. We can evaluate this distance by using Eqs. 3 and 4:

$$\sum_1^\infty 2(A_n - A_n^2/N) = 2\sum_1^\infty (A_{n+1} - A_n) = 2(N - 1),$$

which, when doubled to include male ancestors, is just the result obtained by Kahane and Marr.

After I had "noticed" these simple relations (with Jan Mycielski's forbearance), I went to some of the literature of mathematical demography and population genetics and learned, of course, that it dealt with much more recondite problems, which I was loth to become involved in. Not equipped to make judgements, I nevertheless wondered why it seemed to skip over these simple zeroth approximations. And then I realized why I wished to talk about all this in a personal memoir about Stan Ulam. He, the "village idiot," the one who had the necessary "don't-know-how," did *not* skip over them. It was his style to value the art of successive approximation, of evading the big complexities until he was ready for them, the art sometimes called common sense. Many of his computer simulations were rough sketches of this kind yet could lead into deep water, such as his work on iterated nonlinear transformations. [See "Iteration of Maps, Strange Attractors, and Number Theory—An Ulamian Potpourri."]

With a few further modifications this mathematical genealogy begins to resemble a real biological story, possibly our own, and with all kinds of further questions in tow. I bring a closure to this writing by mentioning two such modifications, neither of which is so complex as to obscure the essential simplicity. The first recapitulates our early work on branching processes (Hawkins and Ulam 1944). Such processes appear *within* the scheme of sexual reproduction as soon as we shift from pure genealogy to genetics and to an interest in evolution. I shall describe briefly the simplest example. The second modification is necessary to give context for the first. It generalizes the original scheme, moving it away from the unrealistic assumption of random pairing toward a pattern of "assortative" mating. This move is curiously parallel to the later work of Everett and Ulam on branching processes in several dimensions (Everett and Ulam 1948).

Genetically considered, sexual reproduction is not only quadratic but also bi-quadratic: Each partner contributes to an offspring half of a diploid genome. But once inherited, the genetic makeup of the offspring remains constant, apart from mutations. Consider then the fate, within our model, of any individual genetic token, taken to be the only one of its kind. It will or will not be transmitted to an offspring with probability $\frac{1}{2}$. So the probability of its transmission to 0, 1, or 2 offspring is the coefficient of the corresponding power of $x$ in the generating function $g(x) = (\frac{1}{2} + \frac{1}{2}x)^2$. Its appearance in subsequent generations is described by a simple chain reaction with $g_n(x) = g\left(g_{n-1}(x)\right)$, one just at the level of transition from a subcritical to a supercritical condition. In any later generation the expected number of descendants with the token is a constant, namely 1. The probability that the token eventually disappears is 1, but its expected lifetime is infinite. The model itself forbids any evolutionary consequences. All of the model's essential properties are preserved, however, by allowing a *variation* of family size, insisting only on a mean value of 2. (Indeed even a slow exponential rate of population growth leaves essentials unchanged.) Then inheritance of any given "bad" gene will be decreased, and that token will have a finite expected lifetime. For a "good" gene the chain goes supercritical; with probability greater than $\frac{1}{2}$ the number of descendants with the "good" gene will grow exponentially with time and eventually dominate the population.

In such a way we can mimic stochastic adaptation. That is a necessary condition for evolution, but not sufficient. *Diveregent* adaptation is also necessary. If different environmental conditions face two subpopulations, "good" genetic changes in one might be "bad" in the other. If the two are long separated, genealogical distances become very great, and the original gene pool may finally fission into those of separate species.

For such reasons we may consider a pattern of assortative mating that involves random pairing within subpopulations and rates of migration between them that decrease with some measure of distance. Successive generations in one subpopulation will gradually acquire more ancestors in the others. In the long run complete mixing will occur, but genealogical distances can now spread over a wide range. If the rate of mutation is assumed to be low but constant, genetic distances will increase with genealogical.

All this seemed at first quite difficult to mathematize, but surprisingly it is not. Shared ancestries and genealogical distances can be expressed in closed algebraic forms that depend only on the rates of diffusion between the subpopulations and their sizes. I leave the subject at this point. Stan's work in biomathematics went further in other areas, but this extension of early work I think would have pleased him.

I mentioned above that Stan was a bit standoffish about my involvement in work relating to the education of children. I was playing with them instead of him, my mathematical mentor! But I heartily forgive him. Some of what I had learned from him, that very spirit of play, I could take to the struggles for better science and mathematics teaching in the schools. Children don't have to be taught how to engage in serious play, usually, but teachers and other "educators" frequently do. They too often have *lost* the art, overwhelmed by mistaken notions of some puritan or utilitarian origin. Stan never lost it. ■

**David Hawkins** earned his academic degrees in philosophy: an A.B. and M.A. from Stanford University and a Ph.D. from the University of California, Berkeley. (The title of his doctoral dissertation, "A Causal Interpretation of Probability," reflects a combined interest in the humanities and science that continues to this day.) In 1943, after short teaching stints at Stanford and Berkeley, he joined the newly created Los Alamos laboratory, serving first as administrative aide to J. Robert Oppenheimer and later as historian. A year at George Washington University was followed in 1947 by a move, which proved permanent, to the University of Colorado, Boulder. He is now a Distinguished Professor Emeritus at that institution. Hawkins has devoted much of his professional life to projects concerning the teaching of mathematics and science. In 1970 he helped create the University of Colorado's Mountain View Center for Environmental Education, an advisory center for preschool and elementary teachers, and is still a participant in its activities. He has enjoyed leaves of absence at several colleges and universities in the United States and abroad and has been honored with a fellowship at the Institute for Advanced Study, a MacArthur Fellowship, membership in the Council of the Smithsonian Institution, and chairmanship of the Colorado Humanities Program. In addition to numerous journal articles, he has written four books: *Science and the Creative Spirit: Essays on Humanistic Aspects of Science* (Harcourt Brown, editor; 1958), *The Language of Nature: An Essay in the Philosophy of Science* (1964), *The Informed Vision: Essays on Learning and Human Nature* (1974), and *The Science and Ethics of Equality* (1977).

## Further Reading

David Hawkins. 1947. Manhattan District History: Project Y, The Los Alamos Project. Los Alamos Scientific Laboratory report LAMS–2532, vol. 1. Also in *Project Y: The Los Alamos Story*. The History of Modern Physics, 1800–1950, vol. 2. Los Angeles: Tomash Publishers, 1983.

Z. Lomnicki and S. Ulam. 1934. Sur la théorie de la mesure dans les espaces combinatoires et son application au calcul des probabilités. I. Variables indépendantes. *Fundamenta Mathematicae* 23: 237–278. Also in *Stanislaw Ulam: Sets, Numbers, and Universes*, edited by W. A. Beyer, J. Mycielski, and G.-C. Rota. Cambridge, Massachusetts: The MIT Press, 1974.

D. Hawkins and S. Ulam. 1944. Theory of multiplicative processes: Part 1. Los Alamos Scientific Laboratory report LA–171.

William Feller. 1968. *An Introduction to Probability Theory and Its Applications*. Third edition. New York: John Wiley & Sons, Inc.

Theodore E. Harris. 1963. *The Theory of Branching Processes*. Die Grundlehren der Mathematischen Wissenschaften in Einzeldarstellungen, vol. 119. Berlin: Springer-Verlag.

C. J. Everett and S. Ulam. 1948. Multiplicative systems, I. *Proceedings of the National Academy of Sciences* 34: 403–405. Also in *Stanislaw Ulam: Sets, Numbers, and Universes*, edited by W. A. Beyer, J. Mycielski, and G.-C. Rota. Cambridge, Massachusetts: The MIT Press, 1974.

Cyril S. Smith. 1981. *A Search for Structure: Selected Essays on Science, Art, and History*. Cambridge, Massachusetts: The MIT Press.

Verna Gardiner, R. Lazarus, N. Metropolis, and S. Ulam. 1956. On certain sequences of integers defined by sieves. *Mathematics Magazine* 29: 117–122. Also in *Stanislaw Ulam: Sets, Numbers, and Universes*, edited by W. A. Beyer, J. Mycielski, and G.-C. Rota. Cambridge, Massachusetts: The MIT Press, 1974.

D. Hawkins and W. E. Briggs. 1957. The lucky number theorem. *Mathematics Magazine* 31: 277–280.

D. Hawkins. 1974. Random sieves: Part II. *Journal of Number Theory* 6(3): 192–200.

S. Ulam. 1962. On some mathematical problems connected with patterns of growth of figures. In *Proceedings of the Symposium on Mathematical Problems in the Biological Sciences*, pp. 215–224. American Mathematical Society Symposia in Applied Mathematics, vol. 14. Providence, Rhode Island: American Mathematical Society.

Jan Mycielski and S. M. Ulam. 1969. On the pairing process and the notion of genealogical distance. *Journal of Combinatorial Theory* 6: 227–234. Also in *Stanislaw Ulam: Sets, Numbers, and Universes*, edited by W. A. Beyer, J. Mycielski, and G.-C. Rota. Cambridge, Massachusetts: The MIT Press, 1974.

Stanislaw M. Ulam. 1986. *Science, Computers, and People: From the Tree of Mathematics*. Edited by Mark C. Reynolds and Gian-Carlo Rota. Boston: Birkhäuser. This posthumous work of Stan's, which I received after writing this memoir, contains references to some of the matters I have discussed, notably in the essays on biomathematical topics.

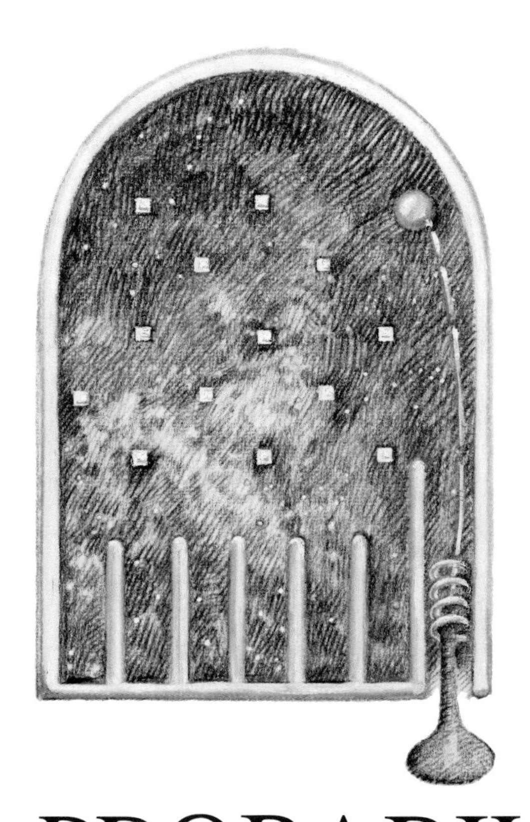

# PROBABILITY *and* NONLINEAR SYSTEMS

*by R. Daniel Mauldin*

Stan Ulam, at sixty-five, was vigorous, handsome, full of ideas. It was the spring of 1974, and he had come to lecture at the University of Florida, where I was a young assistant professor. I had known Stan by reputation for several years. In fact, the very first paper that I read as a part of my German language requirement in graduate school was his landmark 1930 paper on measurable cardinals, "Zur Masstheorie in der allgemeinen Mengenlehre." But listening to him in person was quite an inspiration. He did not lecture in the usual sense but presented snapshots of mathematical ideas, a style reminiscent of Steinhaus, one of Stan's teachers in Poland. Afterwards, several of us talked with him for a remarkably long time. I was immediately impressed with his ability to take up a mathematical topic and breathe new life into the subject.

## *Part I* *An Introduction*

The following year Stan took a position at Florida. His weekly seminar was similar to his book *A Collection of Mathematical Problems*. A topic would be brought up for discussion, and if it appeared to intrigue someone, we would return to it at a slightly deeper level. Stan soon became a stimulating source of encouragement to the younger mathematicians, and to me he became a mentor. As always, he was very generous in sharing his ideas. Throughout his life Stan nourished mathematics in that manner.

At first he would listen to us for a very short time—and then expound his own ideas. Eventually, however, our conversations became a witty (on his part) and very productive exchange. Like a master of reflecting boundaries, he would bounce ideas back to us from an endless variety of angles, especially humorous ones. The amplification of an idea could occur in a time span varying from a coffee conversation to a number of years. Although we would repeatedly go over the same topics, it wasn't exactly like working the beads on a rosary. Every so often an idea would undergo some adjustment or transformation, and something new, perhaps unexpected, would emerge. I don't know whether it was always his way to have short, quick discussions of some central idea, but that is certainly the impression one gets from perusing his comments and problems in *The Scottish Book*. (This famous notebook of problems was jotted down at the Scottish Café in Lwów during the 1930s and first published in this country in 1957. See "Excerpts from *The Scottish Book*.")

Ulam's incredible feel for mathematics was due to a rare combination of intuitions, a common feature of almost all great mathematicians. He had a very good sense of combinatorics and orders of magnitude, which included the ability to make quick, crude, but in-the-ballpark estimates. Those talents, combined with the more ordinary abilities to analyze a problem by means of logic, geometry, or probability theory, already made him very unusual. Besides, he had a good intuition for physical phenomena, which motivated many of his ideas.

Ulam's intuition, as exhibited in numerous problems formulated over a span of more than fifty years, covered an enormous range of subjects. The problems on computing, physical systems, evolution, and biology were stimulated by new developments in those fields. Many others seemed to spring from his head. He usually had some prime examples in mind that motivated his choice of mathematical model or method. In this regard one of his favorite quotes, from Shakespeare's *Henry VIII*, was

Things done without example
in their issue
Are to be feared.

In approaching a complicated problem Stan first searched for simplicity. He had no patience for complicated theories about simple objects, much less complex objects. That philosophical dictum happened to match his personality. He could not hold still for the time it would take to learn, let's say, modern abstract algebraic geometry, nor could he put up with the generalities of category theory. Also, he was familiar with, and early in his career obtained fundamental results in, measure and probability theories. That background led him to approach many problems by placing them in a probabilistic framework. Instead of considering just one possible outcome of a process, one can consider an infinite number of possible outcomes at once by randomizing the process. Then one can apply the powerful tools of probability, such as the laws of large numbers, to determine the likelihood of a given outcome. The famous Monte Carlo method is a perfect example of that approach. In fact, one of the favorite sayings of Erdös and Ulam, both of whom worked in combinatorics (in which the number of outcomes is finite) and probability, was

The infinite we do right away;
the finite takes a little longer.

Stan's interest in probability dates back to the early 1930s, when he and Łomnicki proved several theorems concerning its foundations. In particular, they showed how to construct consistent probability measures for systems involving infinite (as opposed to finite) sequences of independent random variables and, more generally, for Markov processes. (In Markov processes probabilities governing the future depend only on the present and are independent of the past.) At about the same time Kolmogorov, independently, proved his consistency theorem, which includes the Ulam and Łomnicki results as well as many more. Those results guarantee the existence of a probability measure on classes of objects generated by various random processes. The objects might be infinite sequences of numbers or more general geometrical or topological objects, such as the homeomorphisms (one-to-one, onto maps) discussed in detail later in this article. Stan's interest in probability continued after World War II, when he and Everett wrote fundamental papers on "multiplicative" processes (better known as branching processes). Those papers were stimulated by the need to calculate neutron multiplication in fission and fusion devices. (David Hawkins, in "The Spirit of Play," discusses some of the earliest work that Stan and he did on branching processes.)

Stan's background in probability made him a leader among the outstanding group of intellects who, during the late 1940s and early 1950s, recognized the potential value of the computer for doing experimental mathematics. They realized that the computer was an ideal tool for analyzing stochastic, or random, processes. While formal theorems gave rules on how to determine a probability measure on a space of objects, the computer opened up the possibility of generating those objects at random. Simply stated constructions that yield complicated objects could be implemented on the computer, and if one was lucky, *demonstrable* guesses could be made about their asymptotic, or long-term, behavior. That was the approach Stan took in studying deterministic as well as random recursions. In addition he invented cellular automata (lattices of cells and rules for evolution at each cell) and used them to simulate growth patterns on the computer.

The experimental approach to mathematics has since become very popular and has tremendously enhanced our vision of complex physical, chemical, and biological systems. Without the fortuitous conjunction of the computer and probability theory, it is very unlikely that we would have reached today's understanding of those nonlinear systems. Such systems present a challenge analogous to that Newton would have faced if the earth were part of a close binary or tertiary star system. (One can speculate whether Newton could have ever unraveled the law of gravitation from the complicated motions of such a system.) At present researchers are trying to formulate limiting laws governing the long-term dynamics of nonlinear systems that are analogous to the major limiting theorems in classical probability theory. The attempt to construct appropriate probability measures for such systems is one of the topics I will discuss in more depth.

Other interests that Ulam maintained throughout his life were logic and set theory. I remember a conference on large cardinal numbers in New York a few years ago. Stan was the honored participant. More than fifty years earlier he had shown that if a nontrivial probability measure can be defined on all subsets of the real numbers, then the cardinal number, or "size," of the set of all the subsets exceeded the wildest dreams of the time. (See "Learning from Ulam: Measurable Cardinals, Ergodicity, and Biomathematics.") But that large cardinal of his is minuscule compared with the cardinals of today. After listening to some of the conference talks, Stan said that he felt like Woody Allen in *Sleeper* when he woke up after a nap of many years and was confronted with an unbelievably large number on a McDonald's hamburger sign.

There is a serious aspect to that remark. Stan felt that a split between mathematics and physics had developed during this century. One factor was the trauma that shook the foundations of mathematics when Cantor's set theory was found to lead to paradoxes. That caused mathematics to enter a very introspective phase, which continues to this day. A tremendous effort was devoted to axiomatizing mathematics and raising the level of rigor. Physics, on the other hand, experienced an outward expansion and development. (The situation is somewhat reversed today, as internal issues concerning the foundations of physics receive attention.) As a result, university instruction of mathematicians has become so rigorous and demanding that the mathematical training of scientists has been taken over by other departments. Consequently, instruction in "applied" mathematics, or mathematical methods, is often at a fairly low level of rigor, and, even worse, some of the important mathematical techniques developed during this century have not made their way into the bag of tools of many physical scientists. Stan was very interested in remedying the situation and believed the Center for Nonlinear Studies at Los Alamos could play a significant role.

Stan was associated, either directly or through inspiration, with the three research problems described in Part III of this article. Each is an example of how a probabilistic approach and computer simulation can be combined to illuminate features of nonlinear systems. Since some background in modern probability theory is needed to follow the solutions to the problems, Part II provides a tutorial on that subject, which starts with a bit of history and concludes with several profound and useful theorems. Fortunately Mark Kac and Stan Ulam gave a very insightful summary of the development of probability theory in their book *Mathematics and Logic: Retrospect and Prospects*. I have adapted and extended their discussion to meet the needs of this presentation but have retained their broad perspective on the history of mathematics and, in some cases, their actual words.

## *Excerpts from the* SCOTTISH BOOK

These excerpts from *The Scottish Book: Mathematics from the Scottish Café* are reprinted with permission of Birkhäuser Boston. That 1981 edition of problems from "the book" kept at the Scottish Café was edited by R. Daniel Mauldin. The two earlier English-language editions of the problems were edited by Stan Ulam, the first being a 1957 mimeographed version of Ulam's own translation into English from the languages originally inscribed in "the book."

Problems 18 and 19 are still unsolved, and the work stimulated by Problem 43 has played a major role in understanding the consequences of the axiom of choice.

# 18

**ULAM**

Let a steady current flow through a curve in space which is closed and knotted. Does there exist a line of force which is also knotted (knotted = nonequivalent through any homeomorphism of the whole space $R_3$ with the circumference of a circle)?

# 153

**MAZUR**

PRIZE: *A live goose*

*November 6, 1936*

# 19

**ULAM**

Is a solid of uniform density which will float in water in every position a sphere?

**Commentary.** The two-dimensional version of the problem concerns a cylinder of uniform density which floats in every position, having the axis parallel to the water surface, and compatible with Archimedes' law. H. Auerbach ... showed that in the case of density 1/2, the cylinder need not be circular, or even convex, and gave a class of examples. We reproduce his illustration of two of them (Fig. 19.1).

Figure 19.1. Two possible solutions. The line segment rotates within the curve, and in each position cuts off half the area and half the perimeter.

# 43

**MAZUR**

PRIZE: *One bottle of wine, S. ULAM*

Definition of a certain game. Given is a set $E$ of real numbers. A game between two players $A$ and $B$ is defined as follows: $A$ selects an arbitrary interval $d_1$; $B$ then selects an arbitrary segment (interval) $d_2$ contained in $d_1$; then $A$ in his turn selects an arbitrary segment $d_3$ contained in $d_2$ and so on. $A$ wins if the intersection $d_1, d_2, \ldots, d_n \ldots$ contains a point of the set $E$; otherwise, he loses. If $E$ is a complement of a set of first category, there exists a method through which $A$ can win; if $E$ is a set of first category, there exists a method through which $B$ will win.

**Problem**. It is true that there exists a method of winning for the player $A$ only for those sets $E$ whose complement is, in a certain interval, of first category; similarly, does a method of win exist for $B$ if $E$ is a set of first category?

*Addendum.* Mazur's conjecture is true. S. Banach, August 4, 1935

Modifications of Mazur's Game

(1) There is given a set of real numbers $E$. Players $A$ and $B$ give in turn the digits 0 or 1. $E$ wins if the number formed by these digits in a given order (in the binary system) belongs to $E$. For which $E$ does there exist a method of win for player $A$ (player $B$)?

Ulam

(2) There is given a set of real numbers $E$. The two players $A$ and $B$ in turn give real numbers which are positive and such that a player always gives a number smaller than the last one given. Player $A$ wins if the sum of the given series of numbers is an element of the set $E$. The same question as for (1). Banach

**Commentary**. The first published paper on general finite games with perfect information is Zermelo's .... Here, in Problem 43, we have the first interesting definition of an infinite one. ...

# A TUTORIAL *on* PROBABILITY, MEASURE, *and the laws of* LARGE NUMBERS

*Part II*
PROBABILITY and NONLINEAR SYSTEMS

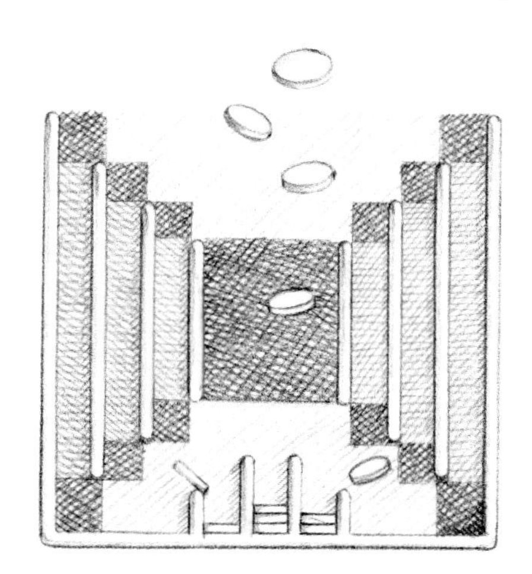

As mentioned in the introduction, Stan Ulam contributed to the measure-theoretic foundations that allow one to define a probability when the number of possible outcomes is infinite rather than finite. Here I will explain why this extension is so necessary and so powerful and then use it to introduce the laws of large numbers. Those laws are used routinely in probability and its applications (several times, for example, during solution of the problems discussed in Part III). Following the logic of Kac and Ulam I begin at the beginning.*

## Early Probability Theory

Probability theory has its logical and historical origins in simple problems of counting. Consider games of chance, such as tossing a coin, rolling a die, or drawing a card from a well-shuffled deck. No specific outcome is predictable with certainty, but all *possible* outcomes can usually be listed or described. In many instances the number of possible outcomes is finite (though perhaps exceedingly large). Suppose we are interested in some subset of the outcomes (say, drawing an ace from a deck of cards) and wish to assign a number to the likelihood that a given outcome belongs to that subset. Our intuitive notion of probability suggests that that number should equal the ratio of the number of outcomes yielding the event (4, in the case of drawing an ace) to the number of all possible events (52, for a full deck of cards).

This is exactly the notion that Laplace used to formalize the definition of probability in the early nineteenth century. Let $A$ be a subset of the set $\Omega$ of all possible outcomes, and let $P(A)$ be the probability that a given outcome is in $A$. For situations such that $\Omega$ is a *finite* set and all outcomes in $\Omega$ are *equally probable*, Laplace defined $P(A)$ as the ratio of the number $\nu(A)$ of elements in $A$ to the total number $\nu(\Omega)$ of elements of $\Omega$; that is,

$$P(A) = \frac{\nu(A)}{\nu(\Omega)}.$$

However, the second condition makes the definition circular, for the concept of *probability* then is dependent upon the concept of *equiprobability*. As will be described later, the more modern definition of probability does not have this difficulty.

For now let us illustrate how Laplace's definition reduces the calculation of probabilities to counting. Suppose we toss a fair coin (one for which heads and tails

*The material quoted in this tutorial from *Mathematics and Logic* has been reprinted with permission from Encyclopedia Britannica, Inc.

are equally probable) $n$ times and want to know the probability that we will obtain exactly $m$ heads, where $1 \leq m \leq n$. Each outcome of $n$ tosses can be represented as a sequence, of length $n$, of $H$'s and $T$'s ($HTHH \ldots THH$, for example), where $H$ stands for heads and $T$ for tails. The set $\Omega$ of all possible outcomes of $n$ tosses is then the set of all possible sequences of length $n$ containing only $H$'s and $T$'s. The total number of such sequences, $\nu(\Omega)$, is $2^n$. How many of these contain $H$ exactly $m$ times? This is a relatively simple problem in counting. The first $H$ can occur in $n$ positions, the second in $n-1$ positions, ..., and the $m$th in $(n-m+1)$ positions. So if the $H$'s were an ordered sample $(H_1, H_2, \ldots, H_m)$, the number of sequences with $m$ $H$'s would equal $n(n-1)(n-2)\ldots(n-m+1)$. But since all the $H$'s are the same, we have overcounted by a factor of $m!$ (the number of ways of ordering the $H$'s). So the number of sequences of length $n$ containing $m$ $H$'s is

$$\frac{n(n-1)\ldots(n-m+1)}{m!} = \frac{n!}{m!(n-m)!}.$$

(The number $n!/m!(n-m)!$, often written $\binom{n}{m}$, is the familiar binomial coefficient, that is, the coefficient of $x^m y^{n-m}$ in the expansion of $(x+y)^n$). Since the number of sequences with exactly $m$ $H$'s is $\binom{n}{m}$ and the total number of sequences is $2^n$, we have by Laplace's definition that the probability $P(m,n)$ of obtaining $m$ heads in $n$ tosses of a fair coin is

$$P(m,n) = \frac{n!}{m!(n-m)!}\,\frac{1}{2^n}.$$

Consider now a coin that is "loaded" so that the probability of a head in a single toss is 1/6 (and the probability of a tail in a single toss is 5/6). Suppose again we toss this coin $n$ times and ask for the probability of obtaining exactly $m$ heads. To describe the equiprobable outcomes in this case, one can resort to the artifice of thinking of the coin as a six-faced die with an $H$ on one face and $T$'s on all the others. Using this artifice to do the counting, one finds that the probability of $m$ heads in $n$ tosses of the loaded coin is

$$P(m,n) = \frac{n!}{m!(n-m)!}\left(\frac{1}{6}\right)^m\left(\frac{5}{6}\right)^{n-m}.$$

Suppose further that the coin is loaded to make the probability of $H$ irrational ($\sqrt{2}/2$, for example). In such a case one is forced into considering a many-faced die and passing to an appropriate limit as the number of faces becomes infinitely large. Despite this awkwardness the general result is quite simple: If the probability of a head in one toss is $p$, $0 \leq p \leq 1$, and the probability of a tail is $1-p \equiv q$, then the probability of $m$ heads in $n$ tosses is

$$P(m,n) = \frac{n!}{m!(n-m)!}\,p^m q^{n-m}.$$

Building on earlier work of de Moivre, Laplace went further to consider what happens as the number of tosses gets larger and larger. Their result, that the number of heads tossed obeys the so-called standard normal distribution of probabilities, was a major triumph of early probability theory. (The standard normal distribution function,

(a) **STANDARD NORMAL DISTRIBUTION FUNCTION**

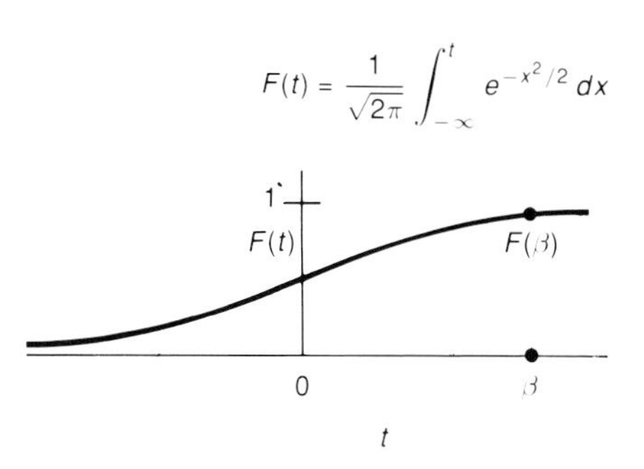


(b) **STANDARD NORMAL DENSITY FUNCTION**

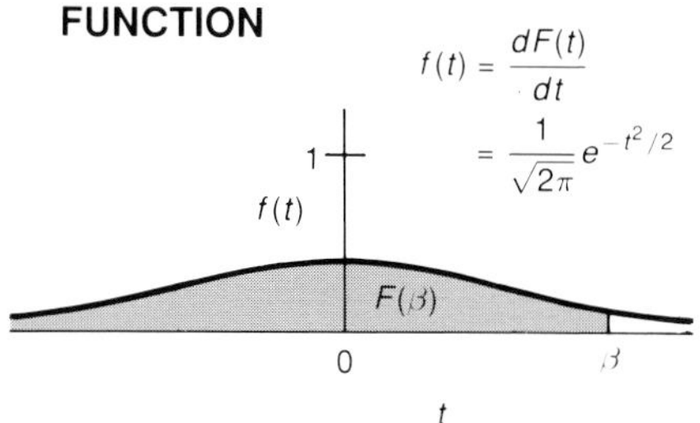


**Fig. 1. Almost two centuries ago Laplace showed that the number $N_H$ of heads obtained in a large number $n$ of tosses of a coin (fair or loaded) follows the standard normal distribution of probabilities. More precisely, he showed that the probability of $N_H$ being equal to or less than $np + t\sqrt{np(1-p)}$ (where $p$ is the probability of a head in a single toss and $t$ is some number) can be approximated, for large $n$, by the standard normal distribution function $F(t)$ shown in (a). The derivative of a distribution function (when it exists) is called a frequency, or density, function. Shown in (b) is the density function $f(t)$ for the standard normal distribution function. Note that the value of the distribution fuction at some particular value of $t$, say $\beta$, is equal to the area under the density function from $-\infty$ to $\beta$.**

**BERTRAND'S PARADOX**

*What is the probability P that a randomly chosen chord of a circle is longer than the side of the equilateral triangle inscribed within the circle?*

**This question cannot be answered by using Laplace's definition of probability, since the set of all possible chords is infinite, as is the set of desired chords (those longer than the side of the inscribed equilateral triangle). However, the question might be approached in the two ways depicted here and described in the text. Although both approaches seem reasonable, each leads to a different answer!**

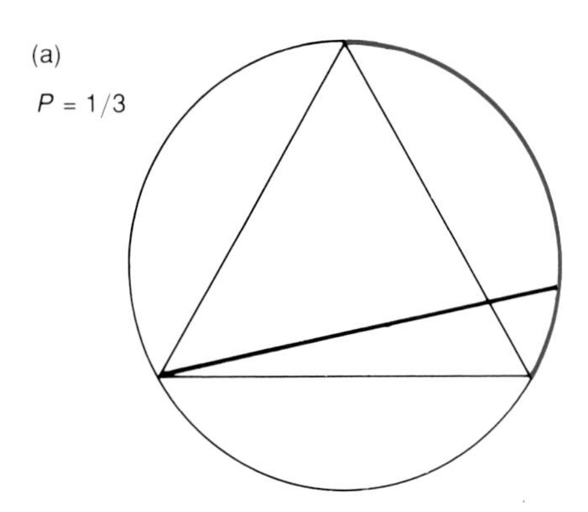


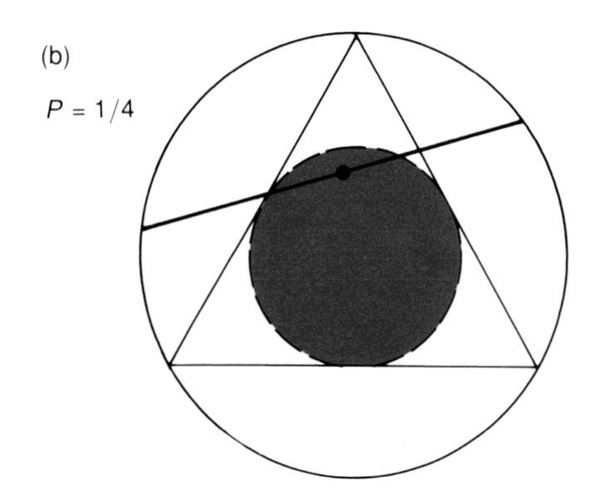


**Fig. 2.**

call it $F(t)$, is given by

$$F(t) = \frac{1}{\sqrt{2\pi}} \int_{-\infty}^{t} e^{-x^2/2} dx;$$

the function $dF/dt = \left(1/\sqrt{2\pi}\right) e^{-t^2/2}$ is called the standard normal density function.)

The de Moivre-Laplace result can be stated as follows. As $n$ gets larger and larger, the probability that $N_H$, the number of heads tossed, will be less than or equal to $np + t\sqrt{pqn}$ (where $t$ is some number) is approximated better and better by the standard normal distribution function. Symbolically,

$$\lim_{n \to \infty} P(N_H \leq np + t\sqrt{npq}) = \frac{1}{\sqrt{2\pi}} \int_{-\infty}^{t} e^{-x^2/2} dx.$$

In other words, $P(N_H \leq np + t\sqrt{npq})$ is approximated by the area under the standard normal density function from $-\infty$ to $t$, as shown in Fig. 1. (In modern terminology $N_H$ is called a random variable; this term and the terms distribution function and density function will be defined in general later.)

The de Moivre-Laplace theorem was originally thought to be just a special property of binomial coefficients. However, many chance phenomena were found empirically to follow the normal distribution function, and it thus assumed an aura of universality, at least in the realm of independent trials and events. The extent to which the normal distribution is universal was determined during the 1920s and 1930s by Lindeberg, Feller, and others after the measure-theoretic foundations of probability had been laid. Today the de Moivre-Laplace theorem (which applies to independent trials, each governed by the same probabilities) and its extension to Poisson schemes (in which each independent trial is governed by different probabilities) are regarded simply as special cases of the very general central limit theorem. Nevertheless they were the seeds from which most of modern probability theory grew.

## Bertrand's Paradox

The awkwardness and logical inadequacy of Laplace's definition of probability made mathematicians suspicious of the whole subject. To make matters worse, attempts to extend Laplace's definition to situations in which the number of possible outcomes is infinite resulted in seemingly even greater difficulties. That was dramatized by Bertrand, who considered the problem of finding the probability that a chord of a circle chosen "at random" be longer than the side of an equilateral triangle inscribed in the circle.

If we fix one end of the chord at a vertex of the equilateral triangle (Fig. 2a), we can think of the circumference of the circle as being the set $\Omega$ of all possible outcomes and the arc between the other two vertices as the set $A$ of "favorable outcomes" (that is, those resulting in chords longer than the side of the triangle). It thus seems proper to take 1/3, the ratio of the length of the arc to the length of the circumference, as the desired probability.

On the other hand we can think of the chord as determined by its midpoint and thus consider the interior of the circle as being the set $\Omega$ of all possible outcomes. The set $A$ of favorable outcomes is now the shaded circle in Fig. 2b, whose radius is one-half that of the original. It now seems equally proper to take 1/4 for our probability, the ratio of the area of the smaller circle to that of the original circle.

That two seemingly appropriate ways of solving the problem led to different answers was so striking that the example became known as "Bertrand's paradox." It is not, of course, a logical paradox but simply a warning against uncritical use of the expression "at random." One must specify exactly how something is to be done at random.

Coming as it did on top of other ambiguities and uncertainties, Bertrand's paradox greatly strengthened the negative attitude toward anything having to do with chance and probability. As a result, probability theory all but disappeared as a mathematical discipline until its spectacular successes in physics (in statistical mechanics, for example) revived interest in it early in the twentieth century. In retrospect, the logical difficulties of Laplace's theory proved to be minor, but clarification of the foundations of probability theory had a distinctly beneficial effect on the subject.

**AXIOM OF ADDITIVITY**

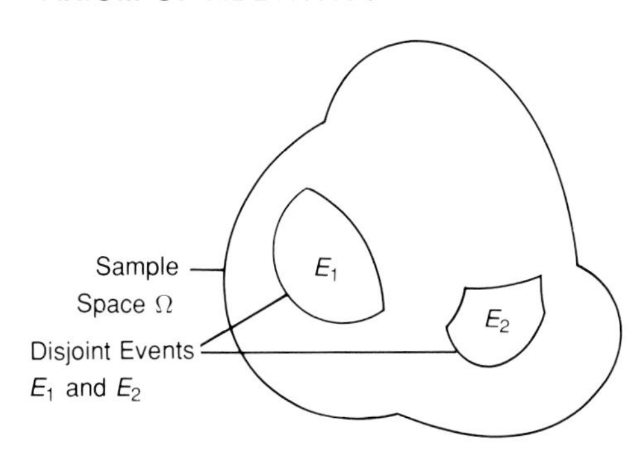


Probability of ($E_1$ or $E_2$) = Probability of $E_1$ + Probability of $E_2$

## Axioms of Modern Probability Theory

The contemporary approach to probability is quite simple. From the set $\Omega$ of all possible outcomes (called the sample space), a collection of subsets (called elementary events) is chosen whose probabilities *are assumed to be given once and for all.* One then tries to calculate the probabilities of more complicated events by the use of two axioms.

**AXIOM OF COMPLEMENTARITY**

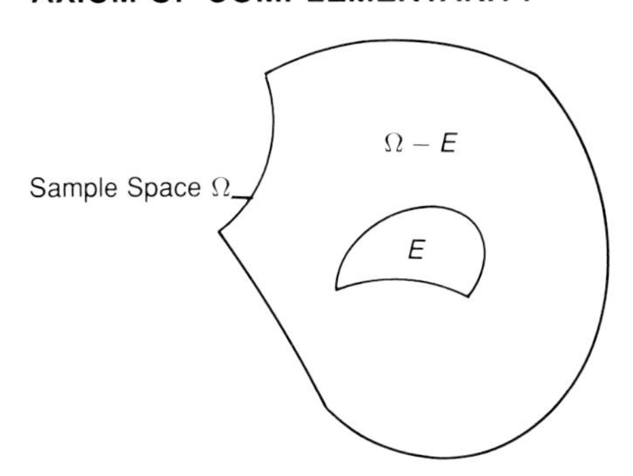


Probability of (not $E$) = Probability of ($\Omega - E$) = 1 − Probability of $E$

**Axiom of additivity:** If $E_1$ and $E_2$ are events, then "$E_1$ or $E_2$" is an event. Moreover, if $E_1$ and $E_2$ are disjoint events, (that is, the subsets corresponding to $E_1$ and $E_2$ have no elements in common), then the probability of the event "$E_1$ or $E_2$" is the sum of the probabilities of $E_1$ and $E_2$, provided, of course, that $E_1$ and $E_2$ can be assigned probabilities. Symbolically,

$$P(E_1 \cup E_2) = P(E_1) + P(E_2) \text{ provided } E_1 \cap E_2 = \emptyset.$$

**Axiom of complementarity:** If an event $E$ can be assigned a probability, then the event "not $E$" also can be assigned a probability. Moreover, since the whole sample space $\Omega$ is assigned a probability of 1,

$$P(\text{not } E) = P(\Omega - E) = 1 - P(E).$$

Why these axioms? What is usually required of axioms is that they should codify intuitive assumptions and that they be directly verifiable in a variety of simple situations. The axioms above clearly hold in all situations to which Laplace's definition is unambiguously applicable; they are also in accord with almost every intuition one has about probabilities, except possibly those involved in quantum mechanics (Feynman 1951).

As we will see in the section on measure theory, the axioms of additivity and complementarity have an impressive mathematical content. Nevertheless they are too general and all-embracing to stand alone as a foundation for a theory so rich and fruitful as probability theory. An additional axiom of "countable additivity" is required. That axiom is the basis for the limiting theorems presented below and their application through approximating forms. Finally, at the heart of the subject is the selection of elementary events and the decision on what probabilities to assign them. Here nonmathematical considerations come into play, and we must rely upon the empirical world to guide us toward promising areas of exploration. These considerations also lead to a central idea in modern probability theory—independence.

## The Definition of Independence

Let us return to the experiment of tossing a coin $n$ times. In attempting to construct any realistic and useful theory of coin tossing, we must first consider two entirely different questions: (1) What kind of coin is being tossed? (2) What is the tossing mechanism? The simplest assumptions are that the coin is fair and the tosses are "independent." Since the notion of independence is central to probability theory, we must discuss it in some detail.

Events $E$ and $F$ are independent in the ordinary sense of the word if the occurrence of one has no influence on the occurrence of the other. Technically, the two events (or, for that matter, any finite number of events) are said to be independent if the *rule of multiplication of probabilities* is applicable; that is, if the probability of the joint occurrence of $E$ and $F$ is equal to the product of their individual probabilities,

$$P(E \cap F) = P(E)\,P(F).$$

Kac and Ulam justified this definition of independence as follows:

"In other words, whenever $E$ and $F$ are independent, there should be a *rule* that would make it possible to calculate Prob. $\{E \text{ and } F\}$ provided only that one knows Prob. $\{E\}$ and Prob. $\{F\}$. Moreover, this rule should be *universal*; it should be applicable to every pair of independent events.

Such a rule takes on the form of a function $f(x,y)$ of two variables $x$, $y$, and we can summarize by saying that whenever $E$ and $F$ are independent we have

$$\text{Prob. } \{E \text{ and } F\} = f(\text{Prob. } \{E\}, \text{Prob. } \{F\})$$

Let us now consider the following experiment. Imagine a coin that can be 'loaded' in any way we wish (*i.e.*, we can make the probability $p$ of H any number between 0 and 1) and a four-faced die that can be 'loaded' to suit our purposes also. The faces of the die will be marked 1,2,3,4 and their respective probabilities will be denoted $p_1, p_2, p_3, p_4$; each $p_i$ is nonnegative and $p_1 + p_2 + p_3 + p_4 = 1$. We must now assume that whatever independence means, it should be possible to toss the coin and the die independently. If this is done and we consider (*e.g.*) the event 'H and (1 or 2)' then on the one hand

$$\text{Prob. } \{\text{H and } (1 \text{ or } 2)\} = f(p, p_1 + p_2)$$

while on the other hand, since the event 'H and (1 or 2)' is equivalent to the event '(H and 1) or (H and 2),' we also have

$$\text{Prob. } \{\text{H and}(1 \text{ or } 2)\} = \text{Prob. } \{\text{H and} 1\} + \text{Prob. } \{\text{H and } 2\} = f(p, p_1) + f(p, p_2)$$

Note that we have used the axiom of additivity repeatedly. Thus

$$f(p, p_1 + p_2) = f(p, p_1) + f(p, p_2)$$

for all $p, p_1, p_2$ restricted only by the inequalities

$$0 \leq p \leq 1, \quad 0 \leq p_1, \quad 0 \leq p_2, \quad p_1 + p_2 \leq 1$$

If one assumes, as seems proper, that $f$ depends continuously on its variables, it follows that $f(x, y) = xy$ and hence the probability of a joint occurrence of independent events should be the *product* of the individual probabilities.

This discussion (which we owe to H. Steinhaus) is an excellent illustration of the kind of informal (one might say 'behind the scenes') argument that precedes a formal definition. The argument is of the sort that says in effect: 'We do not really know what independence is, but whatever it is, if it is to make sense, it must have the following properties ...' Having drawn from these properties appropriate consequences (*e.g.*, that $f(x, y) = xy$ in the above discussion), a mathematician is ready to tighten things logically and to propose a *formal definition*."

Having now defined independence as the applicability of the rule of multiplication of probabilities, let us again derive the probability of obtaining $m$ heads in $n$ tosses of a coin loaded so that $p$ is the probability of a head in a single toss and $q = 1 - p$ is the probability of a tail. If the tosses are assumed to be independent, the probability of obtaining a *specified* sequence of $m$ heads (and $(n - m)$ tails) is $p^m q^{n-m}$ (by the rule of multiplication of probabilities). Since there are $\binom{n}{m}$ such sequences, the probability of the event that exactly $m$ out of $n$ *independent* tosses will be heads is

$$P(n, m) = \binom{n}{m} p^m q^{n-m}.$$

(Here we have applied the axiom of additivity). We have arrived at this formula, first developed almost two centuries ago, by using the modern concept of independence rather than Laplace's concept of equiprobability.

## Probability and Measure Theory

As soon as we consider problems involving an infinite (rather than a finite) number of outcomes, we can no longer rely on counting to determine probabilities. We need instead the concept of measure. Indeed, probabilities are measures; that is, they are numerical values assigned to sets in some collection of sets, namely to sets in the sample space of all possible outcomes. The realization, during the early part of this century, that probability theory could be cast in the mold of measure theory made probability theory respectable by supplying a rigorous framework. It also extended the scope of probability theory to new, more complex problems.

Before presenting the general properties of a measure, let us consider two problems involving an infinite number of outcomes. One is the problem that led to Bertrand's paradox, namely, find the probability that a chord of a circle chosen at random is longer than the side of an inscribed equilateral triangle. For that problem the event $A$, or subset $A$, of chords that are longer and the sample space $\Omega$ of all chords could be depicted geometrically. Thus the relative sizes (measures) of the two sets could be compared even though each was an uncountable set. (The measures of those sets were either lengths or areas.) Another situation in which an infinity of outcomes needs to be considered is the following. Suppose two persons A and B are alternately tossing a coin and that A gets the first toss. What is the probability that A will be the first

*What is the probability that A will be the first to toss a head?*

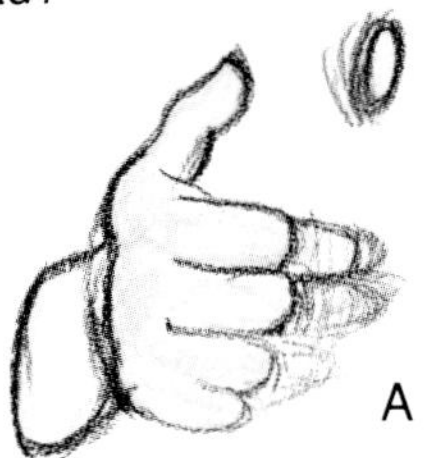


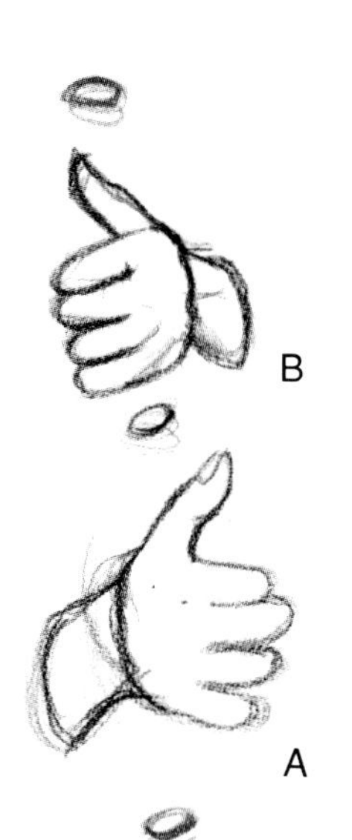


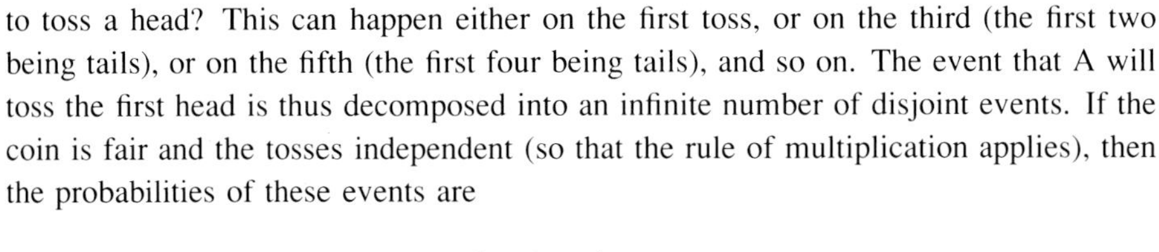

to toss a head? This can happen either on the first toss, or on the third (the first two being tails), or on the fifth (the first four being tails), and so on. The event that A will toss the first head is thus decomposed into an infinite number of disjoint events. If the coin is fair and the tosses independent (so that the rule of multiplication applies), then the probabilities of these events are

$$\frac{1}{2}, \frac{1}{2^3}, \frac{1}{2^5}, \ldots,$$

and the probability that A will toss the first head is simply the sum of a geometric series:

$$\frac{1}{2} + \frac{1}{2^3} + \frac{1}{2^5} + \cdots = \frac{2}{3}.$$

This result hinges on one very crucial proviso: that we can extend the axiom of additivity to an infinite number of disjoint events. This proviso is the third axiom of modern probability theory.

**Axiom of countable additivity:** If $E_1, E_2, E_3, \ldots$ is an infinite sequence of disjoint events, then $\bigcup_{i=1}^{\infty} E_i$ is an event and

$$P\left(\bigcup_{i=1}^{\infty} E_i\right) = \sum_{i=1}^{\infty} P(E_i).$$

Note that in solving the last problem we not only needed the axiom of countable additivity but also assumed that the probabilities used for finite sequences of trials are well defined on events in the space of infinite sequences of trials. Whether such probabilities could be defined that satisfy the axioms of additivity, complementarity, and countable additivity was one of the central problems of early twentieth-century mathematics. That problem is really the problem of defining a measure because, as we will see below, the axioms of probability are essentially identical with the required properties of a measure.

**Measure Theory.** The most familiar examples of measures are areas in a plane or volumes in three-dimensional Euclidean space. These measures were first developed by the Greeks and greatly extended by the calculus of Newton and Leibnitz. As mathematics continued to develop, a need arose to assign measures to sets less "tame" than smooth curves, areas, and volumes. Studies of convergence and divergence of Fourier series focused attention on the "sizes" of various sets. For example, given a trigonometric series $\sum a_n \cos nt + b_n \sin nt$, can one assign a measure to the set of $t$'s for which the series converges? (Cantor's set theory, which ultimately became the cornerstone of all of modern mathematics, originated in his interest in trigonometric series and their sets of convergence.) For another example, how does one assign a measure to an uncountable set, such as Cantor's middle-third set? (See "Cantor's Middle-Third Set".) Answers to such questions led to the development of measure theory.

The concept of measure can be formulated quite simply. One wants to be able to assign to a set $A$ a *nonnegative* number $\mu(A)$, which will be called the measure of $A$, with the following properties.

**Property 1:** If $A_1, A_2, \ldots$ are disjoint sets that are *measurable*, that is, if each $A_i$ can be assigned a measure $\mu(A_i)$, then their *union* $A_1 \cup A_2 \cup \ldots$ (that is, the set consisting of the elements of $A_1, A_2, \ldots$) is also measurable. Moreover,

$$\mu(A_1 \cup A_2 \cup \ldots) = \mu(A_1) + \mu(A_2) + \cdots.$$

**Property 2:** If $A$ and $B$ are measurable and $A$ is contained in $B$ ($A \subset B$), then $B - A$ (the set composed of elements that are in $B$ but *not* in $A$) is also measurable. By property 1 then, $\mu(B - A) = \mu(B) - \mu(A)$.

Two additional properties are assumed for measures on sets in a Euclidean space.

**Property 3:** A certain set $E$, the unit set, is assumed to have measure 1: $\mu(E) = 1$.

**Property 4:** If two measurable sets are congruent (that is, a rigid motion maps one onto the other), their measures are equal.

When dealing with sets of points on a line, in a plane, or in space, one chooses $E$ to be an interval, a square, and a cube, respectively. These choices are dictated by a desire to have the measures assigned to tame sets agree with those assigned to them previously in geometry or calculus.

Can one significantly enlarge the class of sets to which measures can be assigned in accordance with the above properties? The answer is a resounding yes, provided (and it is a crucial proviso) that in property 1 we allow *infinitely many* $A$'s. When we do, the class of measurable sets includes all (well, almost all—perhaps there may be some exceptions ...) the sets considered in both classical and modern mathematics.

Although the concept of countable additivity had been used previously by Poincaré, the explicit introduction and development of countably additive measures early in this century by Émile Borel and Henri Lebesgue originated a most vigorous and fruitful line of inquiry in mathematics. The Lebesgue measure is defined on sets that are closed under countably infinite unions, intersections, and complementations. (Such a collection of sets is called a $\sigma$-field.) Lebesgue's measure satisfies all four properties listed above. Lebesgue's measure on the real line is equivalent to our ordinary notion of length.

But how general is the Lebesgue measure? Can one assign it to every set on the line? Vitali first showed that even the Lebesgue measure has its limitations, that there are sets on the line for which it cannot be defined. The construction of such nonmeasurable sets involves the use of the celebrated axiom of choice. Given a collection of disjoint sets, one can choose a single element from each and combine the selected elements to form a new set. This innocent-sounding axiom has many consequences that may seem strange or paradoxical. Indeed, in the landmark paper on measurable cardinals mentioned at the beginning of this article, Ulam showed (with the aid of the axiom of choice) that if a nontrivial measure satisfying properties 1 through 3 can be defined on all subsets of the real line, then the cardinality of the real numbers is larger than anyone

# CANTOR'S MIDDLE-THIRD SET

During the last quarter of the nineteenth century, Georg Cantor introduced a series of concepts that now form the cornerstone of all modern mathematics—set theory. Those concepts arose from Cantor's attempt to depict the sets of convergence or divergence of, say, trigonometric series. Many such sets have pathological properties that are illustrated by his famous construction, the "middle-third" set. This set is described by the following recursion. Consider the closed unit interval $[0, 1]$. First remove the middle-third open interval, obtaining two intervals $[0, 1/3]$ and $[2/3, 1]$. Next remove from each of these intervals its middle-third interval. We now have four closed subintervals each of length 1/9. Continue the process. After $n$ steps we will have $2^n$ closed subintervals of $[0,1]$ each of length $1/3^n$. From each of these we will remove the middle-third interval of length $1/3^{n+1}$. Continue the process indefinitely. Cantor's middle-third set, K, consists of all numbers in $[0,1]$ that are never removed.

This set possesses a myriad of wonderful properties. For example, K is uncountable and yet has Lebesgue measure zero. To see that K has measure zero, consider the set $\{[0, 1] - \mathrm{K}\}$, which consists of the open intervals that were removed at some stage. At the $n$th stage $2^{n-1}$ open intervals of length $1/3^n$ were removed from the remainder. So, by the countable additivity of measure,

$$\mu([0, 1] - \mathrm{K}) = 1/3 + 2/3^2 + \cdots + 2^{n-1}/3^n + \ldots = (1/3)\left(1 + 2/3 + (2/3)^2 + \ldots\right) = 1.$$

Now, from the axiom of complementarity, $\mu(\mathrm{K}) = 0$, which is what we wanted to prove.

The construction of a nonzero measure on Cantor's middle-third set is discussed in the section of this article entitled Problem 2. Geometry, Invariant Measures, and Dynamical Systems. ■

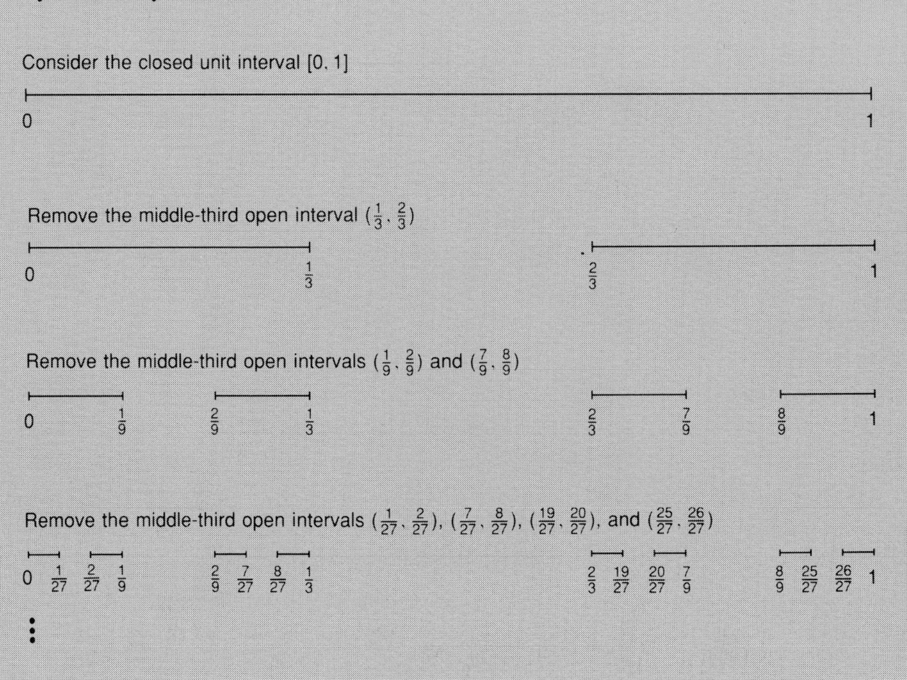

imagined. (See "Learning from Ulam: Measurable Cardinals, Ergodicity, and Biomathematics.") Another example is the Banach-Tarski paradox.

Banach and Tarski proved that each of two solid spheres $S_1$ and $S_2$ of *different* radii can be decomposed into the same finite number of sets, say $S_1 = A_1 \cup A_2 \cup \ldots \cup A_n$ and $S_2 = B_1 \cup B_2 \cup \ldots \cup B_n$, such that all the $A_i$'s and all the $B_i$'s are among themselves pairwise disjoint and yet $A_i$ is congruent to $B_i$ for all $i$. It is therefore impossible to define meaures for these sets, since their union in one fashion yields a certain sphere and their union in a different fashion yields a sphere of different size! That such a construction is possible rests on the complicated structure, earlier pointed out by Hausdorff, of the group of rigid motions of three-dimensional Euclidean space.

**BANACH-TARSKI PARADOX**

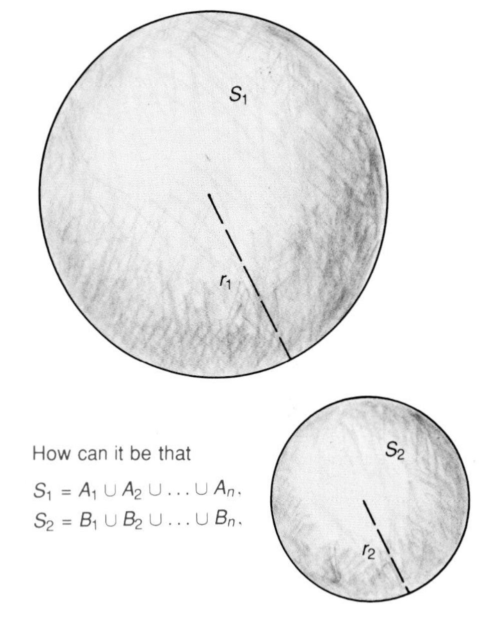


How can it be that
$S_1 = A_1 \cup A_2 \cup \ldots \cup A_n$,
$S_2 = B_1 \cup B_2 \cup \ldots \cup B_n$,

and $A_i$ is congruent to $B_i$ for $1 \leq i \leq n$?

We close this section on measure theory with a few comments from Kac and Ulam.

"Attempts to generalize the notion of measure were made from necessity. ... For example, one could formulate theorems that were valid for all real numbers *except* for those belonging to a specific set. One wanted to state in a rigorously defined way that the set of these exceptional points is in some sense small or negligible. One could 'neglect' merely countable sets as small in the noncountable continuum of all points but in most cases the exceptional sets turned out to be noncountable, though still of Lebesgue *measure* 0. In the theory of probability one has many statements that are valid 'with probability one' (or 'almost surely'). This simply means that they hold for 'almost all' points of an appropriate set; *i.e.*, for all points except for a set of measure 0. In statistical mechanics one has important theorems that assert properties of dynamic systems that are valid only for *almost all* initial conditions.

One final remark:

The notion or concept of measure is surely close to the most primitive intuition. The axiom of choice, that simply permits one to consider a new set $Z$ obtained by putting together an element from each set of a family of disjoint sets, sounds so obvious as to be nearly trivial. And yet it leads to the Banach-Tarski paradox!

One can see why a critical examination of the logical foundation of set theory was absolutely necessary and why the question of existence of mathematical constructs became a serious problem.

If to exist is to be merely free from contradiction as Poincaré decreed, we have no choice but to learn to live with unpleasant things like nonmeasurable sets or Banach-Tarksi decompositions."

## Consistency Theorems for Probability Measures

Now let us return to probability theory and consider the construction of countably additive probability measures. To see that a finitely additive measure cannot always be extended to a countably additive measure, consider the set $\Omega$ of integers and take as elementary events the subsets $A$ of $\Omega$ such that either the set $A$ is finite or the set $\Omega - A$ is finite. Set

$$\mu(A) = \begin{cases} 0 & \text{if } A \text{ is finite} \\ 1 & \text{if } \Omega - A \text{ is finite.} \end{cases}$$

So, $\mu(\Omega) = 1$ and $\mu$ satisfies the axioms of finite additivity and complementarity.

## MAPPING ELEMENTARY EVENTS ONTO THE UNIT INTERVAL

Let each elementary event be one of the sets of all infinite binary sequences with the first two digits fixed Then there are four elementary events.

0

$$E_1 = \left\{ \begin{array}{c} .000000\ldots, \\ \vdots \\ .001111\ldots \end{array} \right\} \text{ maps to } [0, \tfrac{1}{4})$$

$\frac{1}{4}$

$$E_2 = \left\{ \begin{array}{c} .010000\ldots, \\ \vdots \\ .011111\ldots \end{array} \right\} \text{ maps to } [\tfrac{1}{4}, \tfrac{1}{2})$$

$\frac{1}{2}$

$$E_3 = \left\{ \begin{array}{c} .100000\ldots, \\ \vdots \\ .101111\ldots \end{array} \right\} \text{ maps to } [\tfrac{1}{2}, \tfrac{3}{4})$$

$\frac{3}{4}$

$$E_4 = \left\{ \begin{array}{c} .110000\ldots, \\ \vdots \\ .111111\ldots \end{array} \right\} \text{ maps to } [\tfrac{3}{4}, 1]$$

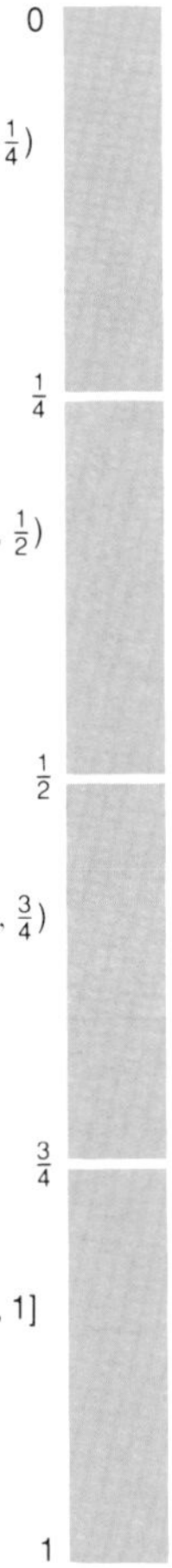

1

$$\sum_{i=1}^{4} P(E_i) = \text{Length of the unit interval}$$

However, if $\mu$ were countably additive, then we would have the contradiction

$$1 = \mu(\Omega) = \mu\left(\bigcup_{n=1}^{\infty}\{n\}\right) = \sum_{n=1}^{\infty}\mu(\{n\}) = 0.$$

Now consider the problem of defining a countably additive probability measure on the sample space $\Omega$ of all infinite two-letter sequences (each of which represents the outcome of an infinite number of independent tosses of a fair coin). Take as an elementary event a set $E$ consisting of all sequences whose first $m$ letters are specified ($m = 1, 2, \ldots$). Since there are $2^m$ such elementary events, we use the axiom of finite additivity to assign a probability $P$ of $1/2^m$ to each such event. Can this function $P$, which has been defined on the elementary events, be extended to a countably additive measure defined on the $\sigma$-field generated by the elementary events? Ulam and Łomnicki proved such an extension exists for any infinite sequence of independent trials. Kolmogorov obtained the ultimate consistency results by giving necessary and sufficient conditions under which an extension can be made from a finitely additive to a countably additive measure, including the case of non-independent trials. These extensions put the famous limiting laws of probability theory, such as the laws of large numbers, on solid ground.

In the case of coin tossing we have chosen our elementary events to be sets of infinite sequences whose first $m$ digits are fixed and have assigned them each a probability of $1/2^m$ in agreement with the finitely additive measure. Now we will show that the measure defined by these choices is equivalent to Lebesgue's measure on the unit interval [0,1] and is therefore a well-defined countably additive measure. First associate the digit 1 with a head and the digit 0 with a tail and encode each outcome of an infinite number of tosses as an infinite sequence of 1's and 0's (10110..., for example), which in turn can be looked upon as the binary representation of a real number $t$ ($0 \leq t \leq 1$). In this way we establish a correspondence between real numbers in [0,1] and infinite two-letter sequences; the correspondence can be made one-to-one by agreeing once and for all on which of the two infinite binary expansions to take when the choice presents itself. (For instance, we must decide between .01000... and .00111... as the binary representation of $1/4$.)

The use of the binary system is dictated not only by considerations of simplicity. As one can easily check, the crucial feature is that each elementary event maps into an interval whose *length* is equal to the corresponding *probability* of the event. In fact, fixing the first $m$ letters of a sequence corresponds to fixing the first $m$ binary digits of a number, and the set of real numbers whose first $m$ binary digits are fixed covers the interval between $\ell/2^m$ and $(\ell+1)/2^m$, where $\ell$ is $0, 1, 2, \ldots$, or $2^m - 1$, depending on how the first $m$ digits are fixed. Clearly the length of such an interval, $1/2^m$, is equal to the probability of the corresponding elementary event. Thus the probability measure in the sample space $\Omega$ of all infinite two-letter sequences maps into the ordinary Lebesgue measure on the interval [0,1] and is therefore equivalent to it.

The space of all infinite sequences of 0's and 1's is *infinite-dimensional* in the sense that it takes infinitely many "coordinates" to describe each "point" of the space. What we did was to construct a certain countably additive measure in the space that was "natural" from the point of view of independent tosses of a fair coin.

THE INSTITUTE FOR ADVANCED STUDY
SCHOOL OF MATHEMATICS
FINE HALL
PRINCETON, NEW JERSEY

Oct. 3., 1936.

Dear Ulam,

Many thanks for your kind letter—both Mariette and I are delighted to see you soon in Princeton again. Oct. 10—or any other date thereabouts which suits you—will be very convenient to us—and we expect that you *will* stay with us, and stay as long as possible.—

I agree wholeheartedly with your plans to write an up-to-date presentation of measure-theory. Carathéodorys exposition, which is perhaps the *relatively* best one existing, is hopelessly obsolete. A thoroughly modern one as much combinatorial and as little topological as possible, making extensive use of finite and infinite direct products, and—above all—interpreting measure much more as probability and much less as volume, would really be a very good thing. At least I often felt how badly such a thing is lacking in the present literature. What would be the style of your treatise, and its length? I will be very glad if you can let me see any part of your mscr. In the lectures I gave here on "linear operations" in 1933/34 and 34/35, I tried to deal with measure somewhat in the above spirit, but I was badly handicapped by the fact that measure was not my primary topic there.

I [am] looking forward, too, with great interest for your mscr. on the general product operation.

I am expecting to discuss several mathematical questions, when you come here, those you mentioned, and a few others. By then I will have unearthed my two last year's mscr's, too, which you mentioned—we are unpacking now, so the excavations do not proceed very quickly.

Expecting to see you soon again, and with the very best greetings from Mariette, too, I am yours as ever *John von Neumann*

*Note: With the help of J. D. Bernstein, Ulam started a book on measure theory while he was at Wisconsin. That collaboration was interrupted by Stan's war years at Los Alamos and was never resumed. The idea of presenting measure theory from the combinatorial or probabilistic perspective is now a common practice. A good example is P. Billingsley's* Probability and Measure.

This approach immediately suggests extensions to more general infinite-dimensional spaces in which each coordinate, instead of just being 0 or 1, can be an element of a more general set and need not even be a number. Such extensions, called product measures, were introduced by Łomnicki and Ulam in 1934. (Stan's idea of writing a book on measure theory emphasizing the probabilistic interpretation of measure is the subject of the accompanying letter from von Neumann to Ulam.) Measures for sets of curves have also been developed. The best known and most interesting of these was introduced by Norbert Wiener in the early 1920s and motivated by the theory of Brownian motion. Mathematicians have since found new and unexpected applications of the Wiener measure in seemingly unrelated parts of mathematics. For example, it turns out that the Wiener measure of the set of curves emanating from a point $p$ in space and hitting a three-dimensional region $R$ is equal to the electrostatic potential at $p$ generated by a charge distribution that makes the boundary of the "conductor" $R$ an equipotential surface on which the potential is equal to unity. Since the calculation of such a potential can be reduced by classical methods to solving a specific differential equation, we establish in this way a significant link between classical analysis and measure theory.

## Random Variables and Distribution Functions

Having introduced the measure-theoretic foundations of probability, we now turn to a convenient formalism for analyzing problems in probability. In many problems the possible outcomes can be described by numerical quantities called random variables. For example, let $X$ be the random variable describing the outcome of a single toss of a fair coin. Thus, set $X$ equal to 1 if the toss yields a head and to 0 if the toss yields a tail. This is an example of an elementary random variable; that is, $X$ is a function with a constant value on some elementary event and another constant value on the complementary event. In general a random variable is a real-valued function defined on the sample space $\Omega$ that can be constructed from elementary random variables by forming algebraic combinations and taking limits. For example, $N_H$, the number of heads obtained in $n$ tosses of a coin, is a random variable defined on the sample space consisting of all sequences of $T$'s and $H$'s of length $n$; its value is equal to $\sum_{i=1}^{n} X_i$, where $X_i = 1$ if the $i$th toss is a head and $X_i = 0$ otherwise.

In evaluating the outcomes of a set of measurements subject to random fluctuations, we are often interested in the mean, or expected, value of the random variable being measured. The expected value $E(X)$ (or $m$) is defined as

$$E(X) \equiv \int_{\Omega} X(\omega)\, dP(\omega),$$

where $X(\omega)$ is the value of $X$ at a point $\omega$ in the sample space and $P(\omega)$ is the probability measure defined on the sample space. In the case of a fair coin, $P(X = 1) = 1/2$ and $P(X = 0) = 1/2$, so the expected value of $X$ is a simple sum:

$$E(X) = \sum x_i P_i = (0 \times \frac{1}{2}) + (1 \times \frac{1}{2}) = \frac{1}{2}.$$

The expected value of a random variable $X$ is most easily determined by knowing its distribution function $F$. This function, which contains all the information we need to know about a random variable, is defined as follows:

$$F(t) \equiv P(X \leq t),$$

where the set $X \leq t$ is the set of all points $\omega$ in $\Omega$ such that $X(\omega) \leq t$. The form of this function is particularly convenient. It allows us to rewrite $E(X)$, which is a Lebesgue integral over an abstract space, as a familiar classical integral over the real line:

$$E(X) \equiv \int_{\Omega} X(\omega)\, dP(\omega) = \int_{-\infty}^{\infty} t\, dF(t).$$

Furthermore, if $X$ has a density function $f(t) \equiv dF(t)/dt$, then

$$E(X) = \int_{-\infty}^{\infty} t f(t)\, dt.$$

The expected value is one of the two commonly occurring averages in probability and statistics; the other is the variance of $X$, denoted by $\sigma^2(X)$ or $\text{var}(X)$. The variance is defined as the expected value of the square of the deviation of $X$ from its mean:

$$\sigma^2(X) = \text{var}(X) \equiv E\left((X - E(X))^2\right) = E(X^2) - \left(E(X)\right)^2.$$

The standard, or root-mean-square, deviation of $X$ is defined as $\sigma(X) = \sqrt{\text{var}(X)}$.

Figures 3 and 4 illustrate two distribution functions, the binomial distribution function for the number of heads obtained in five tosses of a fair coin and a normal distribution function with a positive mean.

**BINOMIAL DISTRIBUTION FUNCTION**

**Fig. 3. The distribution function $F(t)$ for the number of heads obtained in $n$ independent tosses of a fair coin is a binomial distribution, so called because the probability of obtaining $k$ heads in $n$ tosses of the coin is given by a formula involving binomial coefficients, namely $\binom{n}{k}\frac{1}{2^n}$. Shown here is the binomial distribution function for the number of heads obtained in five tosses of the coin. The value of $F(t)$ equals the probability that the number of heads is equal to or less than $t$.**

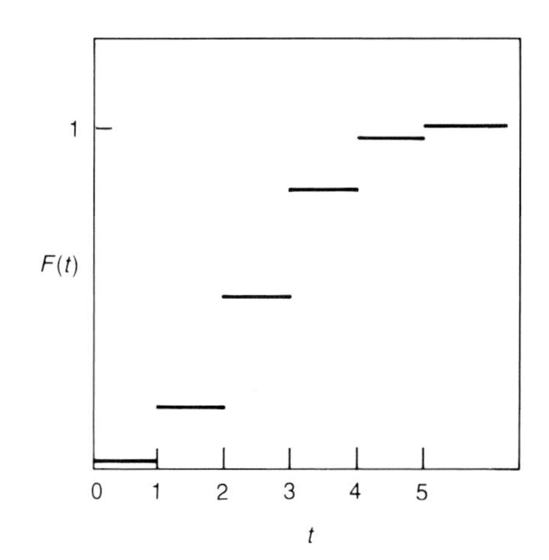


$$F(t) = \sum_{0}^{k} \binom{n}{k} \frac{1}{2^n} \text{ if } k \leq t \leq k+1,$$
$$k = 0, 1, 2, 3, 4$$
$$= 1 \text{ if } t \geq 5$$

## The Laws of Large Numbers

A historically important problem in probability theory and statistics asks for estimates on how a random variable deviates from its mean, or expected, value. A simple rough estimate is, of course, its root-mean-square deviation. An estimate of a different nature was obtained by the nineteenth-century mathematician Chebyshev. This estimate, known as Chebyshev's inequality, gives an upper limit on the probability that a random variable $Y$ deviates from its mean $E(Y)$ by an amount equal to or greater than $a$ $(a > 0)$:

**Chebyshev's inequality:** $P(|Y - E(Y)| \geq a) \leq \text{var}(Y)/a^2.$

This fundamental inequality will lead us to the famous laws of large numbers, which tell us about average values for infinite sequences of random variables. We begin by returning again to the coin loaded in such a way that $p$ is the probability of a head in a single toss. If this coin is tossed a large number of times $n$, shouldn't the frequency of heads, $N_H/n$, be approximately equal to $p$, at least in some sense?

This question can be answered on several levels. Let $X_i$ be the random variable describing the outcome of the $i$th toss. Set $X_i = 1$ if the $i$th toss is a head and $X_i = 0$ if

(a) **NORMAL DISTRIBUTION FUNCTION**

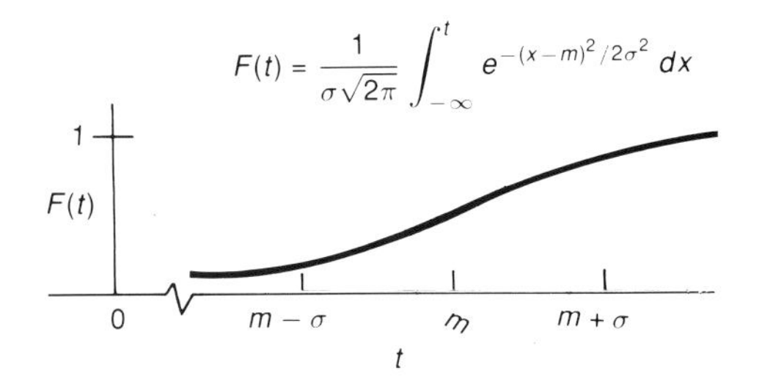


(b) **NORMAL DENSITY FUNCTION**

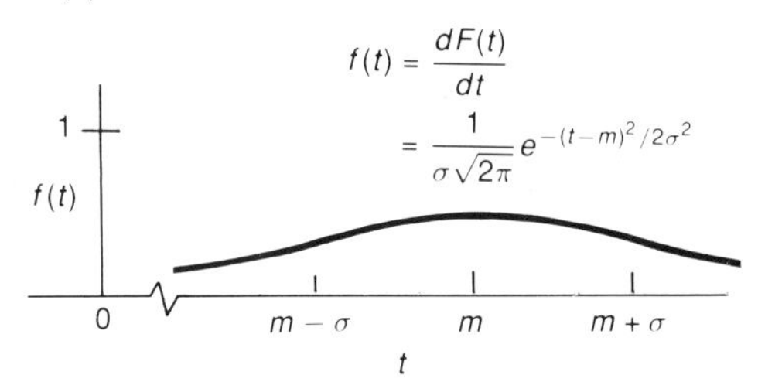


**Fig. 4. So many random variables can be described, at least approximately, by the distribution function shown in (a) that it is known as the normal distribution function. Examples of such random variables include the number of heads obtained in very many tosses of a coin and, as a general experimental fact, accidental errors of observation. The value of $F(t)$ equals the probability that the value of the random variable is equal to or less than $(t - m)/\sigma$, where $m$ is the mean, or expected, value of the random variable and $\sigma$ is its standard deviation. (The mean here is assumed to be positive.) Shown in (b) is the normal density function $f(t) \equiv dF(t)/dt$, which gives the probability that the value of the random variable is $(t - m)/\sigma$.**

the $i$th toss is a tail. Then $N_H = X_1 + \cdots + X_n$. Also, the distribution function for each $X_i$ is the same, namely,

$$F(t) = \begin{cases} 0 & t < 0 \\ 1 - p & 0 \leq t < 1 \\ 1 & 1 \leq t. \end{cases}$$

(Random variables that have the same distribution function are said to be identically distributed.) Now the expected value of $N_H/n$ is easy to compute:

$$E\left(N_H/n\right) = E\left((X_1 + \ldots + X_n)/n\right) = (1/n)\sum_{i=1}^{n} E(X_i) = (1/n)(np) = p.$$

Thus, on the simplest level our guess is right: The frequency of heads, $N_H/n$, is approximately equal to $p$ in the sense that the expected value of $N_H/n$ is $p$. But surely, even in a very long series of tosses, it would be foolish to expect $N_H/n$ to exactly equal $p$ (and $N_T/n$ to exactly equal $1 - p$). What one is looking for is a statement that holds only *in the limit as the number of tosses becomes infinite.*

Bernoulli proved such a theorem: As $n$ gets larger and larger, the probability that $N_H/n$ differs from its expected value $p$ by more than a positive amount $\epsilon$ tends to 0:

$$\lim_{n\to\infty} P_n(|N_H/n - p| \geq \epsilon) = 0,$$

where $P_n$ is the probability measure on $\Omega_n$, the space of all sequences of $H$'s and $T$'s of length $n$. No matter what positive $\epsilon$ is chosen, the probability that the difference between the frequency of heads and $p$, the probability of a head in a single trial, exceeds $\epsilon$ can be made arbitrarily small by tossing the coin a sufficiently large number of times.

Let us see how Bernoulli's theorem follows from Chebyshev's inequality. First, notice that $\text{var}(X_i) = p(1 - p)$ for all $i$. Second, the random variables $X_1, \ldots, X_n$ are independent (the outcome of the $i$th toss has no influence on the outcome of $j$th). Now, from the fact that $E(XY) = E(X)E(Y)$ for independent random variables, we get

$$\text{var}(X_1 + \cdots + X_n) = \sum_{i=1}^{n} \text{var}(X_1) = np(1 - p).$$

So, by Chebyshev's inequality

$$P_n(|N_H/n - p| \geq \epsilon) \leq np(1 - p)/n^2\epsilon^2 = p(1 - p)/\epsilon^2 n.$$

Thus, for each $\epsilon > 0$

$$\lim_{n\to\infty} P_n(|N_H/n - p| \geq \epsilon) = 0.$$

Notice that the measure-theoretic background of Bernoulli's theorem is trivial (at least as far as coin-tossing is concerned), since the events of interest correspond to finite sets. That is, for each $n$ we need only estimate how many trials of length $n$ there are such that the number of heads differs from $np$ by more than $\epsilon n$. Nevertheless, the simple argument just given can be generalized to prove the famous weak law of large numbers.

**Weak law of large numbers:** Let $X_1, X_2, X_3, \ldots$ be independent, identically distributed random variables such that $\text{var}(X_1) < \infty$. Then for each $\epsilon > 0$

$$\lim_{n \to \infty} P\left(|(X_1 + \ldots + X_n)/n - E(X_1)| \geq \epsilon\right) = 0.$$

In other words, for any positive $\epsilon$ the probability that the deviation between the frequency in $n$ trials and the expected value in a single trial exceeds $\epsilon$ can be made arbitrarily small by considering a sufficiently large number of trials.

For our coin-tossing example $N_H/n$ approximately equals $p$ in another sense also. Suppose one asks for the probability that the frequency of heads (in the limit as the number of tosses becomes infinite) is actually equal to $p$. The answer was obtained by Borel in 1909:

$$P\left(\lim_{n \to \infty} N_H/n = p\right) = 1.$$

Notice the complexity of the question. In order to deal with it, the sample space $\Omega$ is now by necessity the set of all *infinite* two-letter sequences $\omega$ and the subset of interest is the set $A$ of those sequences for which

$$\lim_{n \to \infty} \frac{N_n(\omega)}{n} = p,$$

where $N_n(\omega)$ is the number of $H$'s among the first $n$ letters of the infinite sequence $\omega$. It takes some work just to show that $A$ is an event in the sample space $\Omega$. Unlike the question that led to the weak law of large numbers, this question required the full apparatus of modern probability theory. An extension of Borel's result by Kolmogorov is known as the strong law of large numbers.

**Strong law of large numbers:** Let $X_1, X_2, X_3, \ldots$ be independent, identically distributed random variables such that $E(|X_1|) < \infty$. Then

$$P\left(\lim_{n \to \infty} (X_1 + \cdots + X_n)/n = E(X_1)\right) = 1.$$

**An Application of the Strong Law of Large Numbers.** Let us illustrate the power of the strong law of large numbers by using it to answer the following question: What is the probability that, in an infinite sequence of tosses of a fair coin, two heads occur in succession?

We will first answer this question using only the rules governing the probabilities of independent events. In particular, we will use the axioms of countable additivity and complementarity and the rule of multiplication of probabilities. Let $A_k$ be the event that a head occurs on the $(2k-1)$th toss and on the $(2k)$th toss. Each $A_k$ is an elementary event, and $P(A_k) = 1/4$. Now, by the axiom of countable additivity, $\bigcup_{k=1}^{\infty} A_k$ is an event; in particular, it is the event that, for *some* $k$, heads occur on the $(2k-1)$th and $2k$th tosses. By the axiom of complementarity,

$$P\left(\Omega - \bigcup_{k=1}^{\infty} A_k\right) = P\left(\bigcap_{k=1}^{\infty}(\Omega - A_k)\right) = \lim_{n \to \infty} P\left(\bigcap_{k=1}^{n}(\Omega - A_k)\right).$$

Since the events $A_1, A_2, A_3, \ldots$ are independent, the events $\Omega - A_1, \Omega - A_2, \Omega - A_3, \ldots$ are also independent, and we can apply the rule of multiplication of probabilities:

$$P\left(\Omega - \bigcup_{k=1}^{\infty} A_k\right) = \lim_{n\to\infty} \prod_{k=1}^{n} P(\Omega - A_k) = \lim_{n\to\infty} (3/4)^n = 0.$$

Finally, by the axiom of complementarity, $P\left(\bigcup A_k\right) = 1$; that is, there exists, with probability 1, some $k$ such that the $(2k - 1)$th and $(2k)$th tosses are heads.

Now we will answer the same question by using the strong law of large numbers. Let $X_i$ be the random variable such that

$$X_i(\omega) = \begin{cases} 1 & \text{if } \omega \in A_i \\ 0 & \text{if } \omega \notin A_i. \end{cases}$$

Then $X_1, X_2, X_3, \ldots$ is a sequence of independent random variables. Also, they all have the same distribution: $(P(X_i = 1) = 1/4, P(X_i = 0) = 3/4$, and $E(X_i) = 1/4$. Therefore, according to the strong law of large numbers,

$$\lim_{n\to\infty} (X_1 + \ldots + X_n)/n = 1/4 \text{ with probability } 1.$$

This result is stronger than that obtained above. It guarantees, with probability 1, the existence of infinitely many $k$'s such that heads occur on the $(2k - 1)$th and $(2k)$th tosses; further, the set of all such $k$'s has an arithmetic density of 1/4.

Borel's theorem marked the beginning of the development of modern probability theory, and Kolmogorov's extension to the strong law of large numbers greatly expanded its applicability. To quote Kac and Ulam:

> "Like all great discoveries in mathematics the strong law of large numbers has been greatly generalized and extended; in the process it gave rise to new problems, and it stimulated the search for new methods. It was the first serious venture outside the circle of problems inherited from Laplace, a venture made possible only by developments in measure theory. These in turn were made possible only because of polarization of mathematical thinking along the lines of set theory."

The polarization Kac and Ulam were referring to concerns the great debate at the turn of the century about whether the infinite in mathematics should be based upon Cantor's set theory and its concomitant logical difficulties. The logical problems have been met, and today we use Cantor's theory with ease.

**The Monte Carlo Method.** One of Stan Ulam's great ideas, which was first developed and implemented by von Neumann and Metropolis, was the famous Monte Carlo method. It can be illustrated with Chebyshev's inequality. Suppose that we need to quickly get a rough estimate of $\int_1^\infty (\sin x)/x^3\, dx$. Setting $t = 1/x$, the problem then is to estimate $\int_0^1 t \sin(1/t)\, dt$. Let $y_1, \ldots, y_n$ be independent random variables each uniformly distributed on [0,1]. That is, for all $i$, $P(a < y_i < b) = b - a$, where $(a, b)$ is a subinterval of [0,1]. Now set $f(t) = t \sin(1/t)$ and for each $i$ let $X_i = f(y_i)$. Then

$X_1, \ldots, X_n$ is a sequence of independent identically distributed random variables. Also,

$$|E(X_i)| = \left| \int_0^1 t \sin(1/t)\, dt \right| < \int_0^1 |t \sin(1/t)|\, dt < 1,$$

and

$$\operatorname{var}(X_1) = \int_0^1 t^2 \sin^2(1/t)\, dt - \left(E(X_i)\right)^2 < \int_0^1 t^2 \sin^2(1/t)\, dt < \int_0^1 t^2 dt = 1/3.$$

By Chebyshev's inequality we have

$$P\left( \left| (1/n) \sum_{i=1}^{n} X_i - \int_0^1 t \sin(1/t)\, dt \right| \geq a \right) \leq \operatorname{var}\left( (1/n) \sum_{i=1}^{n} X_i \right) / a^2 \leq \operatorname{var}(X_1)/na^2.$$

Thus if $n$ is large, $(1/n)\sum_{i=1}^{n} X_i$ is, with high probability, a good estimate of the value of the integral. For example, if $a = 0.005$ and $n = 134,000$, then

$$P\left( \left| (1/n) \sum_{i=1}^{n} X_i - \int_0^1 t \sin(1/t)\, dt \right| \geq 0.005 \right) < 0.1.$$

In other words, if we chose 134,000 numbers $y_1, \ldots, y_{134,000}$ independently and at random from [0,1], then we are 90 percent certain that $(1/134,000)\sum_{i=1}^{134,000} y_i \sin(1/y_i)$ differs from the integral by no more than 0.005. So, if we can statistically sample the unit interval with numbers $y_1, \ldots, y_n$, then

$$(1/n) \sum_{i=1}^{n} y_i \sin(1/y_i) \approx \int_1^{\infty} \frac{\sin\, t}{t^3}\, dt.$$

(The reader may well wonder why such a large number of sample points is required to be only 90 percent certain of the value of the integral to within only two decimal places. The answer lies in the use of Chebyshev's inequality. By using instead the stronger central limit theorem, which will be introduced below, many fewer sample points are needed to yield a similar estimate.)

The Monte Carlo method is a wonderful idea and, of course, tailor-made for computers. Although it might be regarded simply as an aspect of the more ancient statistical sampling technique, it had many exciting new aspects. Three of these are (1) a scope of application that includes large-scale processes, such as neutron chain reactions; (2) the capability of being completely implemented on a digital computer; and (3) the idea of generating random numbers and random variables. How do we mechanically produce numbers $y_1, \ldots, y_n$ in [0,1] such that the $y_i$'s are independent and identically distributed? The answer is we don't. Instead, so-called pseudo-random numbers are generated. Many fascinating problems surfaced with the advent of Monte Carlo. Dealing with them is one of the major accomplishments of the group of intellects gathered at Los Alamos in the forties and the fifties. (See "The Beginning of the Monte Carlo Method.")

## Central Limit Theorem

We close this tutorial by returning to the de Moivre-Laplace theorem and interpreting it in the modern context. Let $X_i$ be a random variable describing the outcome of the $i$th toss of a coin; set $X_i = 1$ if the $i$th toss is a head and $X_i = 0$ otherwise. Let $S_n$ be the number of heads obtained in $n$ tosses; that is $S_n = X_1 + \cdots + X_n$. Then the de Moivre-Laplace theorem can be stated as follows:

$$\lim_{n \to \infty} P\left(\frac{S_n - np}{\sqrt{np(1-p)}} \leq t\right) = \frac{1}{\sqrt{2\pi}} \int_{-\infty}^{t} e^{-x^2/2}\, dx.$$

Now $np = nE(X_1) = E(S_n)$ and $\sqrt{np(1-p)} = \sqrt{n}\sigma(X_1) = \sigma(S_n)$. So if we "renormalize" $S_n$ by setting $Y_n = \big(S_n - E(S_n)\big)/\sigma(S_n)$, each $Y_n$ has a mean of 0 and a standard deviation of 1. Then the equation above tells us that the distribution function of $Y_n$ tends to the standard normal distribution. The central limit theorem is a generalization of this result to any sequence of identically distributed random variables. We state the central limit theorem formally.

**Central limit theorem:** Let $X_1, X_2, X_3, \cdots$ be a sequence of independent, identically distributed random variables with $E(X_1) = m$ and $\text{var}(X_i) = \sigma^2 < \infty$. Set $S_n = X_1 + \cdots + X_n$. Then

$$\lim_{n \to \infty} P\left(\frac{S_n - nm}{\sqrt{n}\sigma} \leq t\right) = \frac{1}{\sqrt{2\pi}} \int_{-\infty}^{t} e^{-x^2/2}\, dx.$$

Thus the normal distribution is the universal behavior in the domain of independent trials under renormalization. Its appearance in so many areas of science has led to many debates as to whether it is a "law of nature" or a mathematical theorem.

Thanks to the developments in modern probability theory, we begin our investigations with many powerful tools at our disposal. Those tools were forged during a period of tremendous upheavals and turmoil, a time when very careful analysis carried the day. At the heart of that analysis lay the concept of countable additivity. Stan Ulam played a seminal role in developing these tools and presenting them to us.

# PROBABILISTIC APPROACHES *to* NONLINEAR PROBLEMS

## **Problem 1.** *Energy Redistribution: An Exact Solution to a Nonlinear, Many-Particle System*

Ulam's talent for seeing new approaches to familiar problems is evident in one he posed concerning the distribution of energy in physical systems. Will the energy distribution of an isolated system of $N$ interacting particles always evolve to some limiting energy distribution? And, if so, what is the distribution? (Note that this question differs from the one asked in statistical mechanics. There one assumes that at equilibrium the system will have the most probable distribution. One then derives that the most probable distribution is the Boltzmann distribution, the density of which is $e^{-E/kT}$.)

Obviously, following the evolution of a system of $N$ interacting particles in space and time is a very complex task. It was Stan's idea to simplify the situation by neglecting the spatial setting and redistributing the energy in an abstract random manner. What insights can one gain from such a simplification? One can hope for new perspectives on the original problem as well as on the standard results of statistical mechanics. Also, even if the simplification is unrealistic, one can hope to develop some techniques of analysis that can be applied to more realistic models. In this case David Blackwell and I were able to give an exact analysis of an abstract, highly nonlinear system by using a combination of the machinery of probability theory and higher order recursions (Blackwell and Mauldin 1985). We hope that the technique will be useful in other contexts.

Let us state the problem more clearly and define what we mean by redistributing energy in an "abstract random manner." Assume we have a vast number of indistinguishable particles with some initial distribution of energy, and that the average energy per particle is normalized to unity. Further, let us assume the particles interact only in pairs as follows: At each step in the evolution of the system, pair all the particles at random and let the total energy of each pair be redistributed between the members of the pair according to some fixed "law of redistribution" that is independent of the pairs. Iterate this procedure. Does the system have a limiting energy distribution and, if so, how does it depend on the redistribution law?

**The Simplest Redistribution Law.** To begin we will consider the simplest redistribution law: each particle in a random pair gets one-half the total energy of the pair. If the number of particles in the system is finite, it is intuitively clear that under iteration the total energy of the system will tend to become evenly distributed—all the particles

**SIMPLEST LAW FOR ENERGY REDISTRIBUTION**

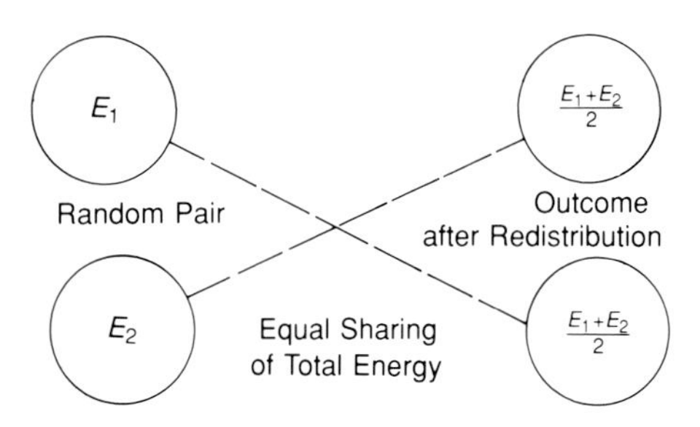


**LIMITING ENERGY DISTRIBUTION**

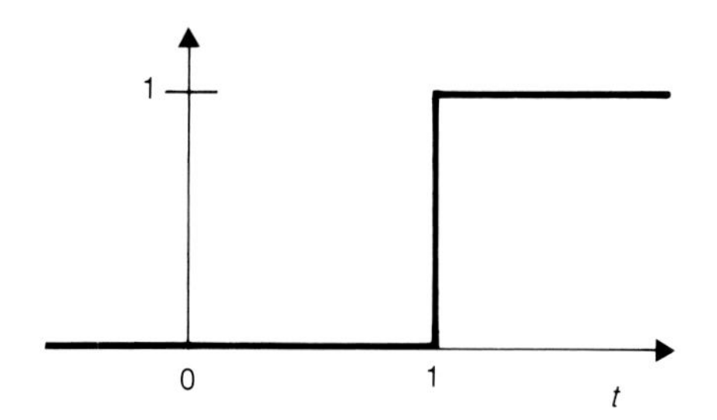


**Fig. 5. Consider a system of $N$ particles with some arbitrary initial distribution of energy. Assume that the initial mean energy is 1 and that the particles interact in pairs. Assume further that the total energy of an interacting pair is redistributed so that each member of the pair acquires one-half the total energy of the pair. Then with probability 1 the system reaches a limiting energy distribution described by a step function with a step height of 1 at $t$ = 1. That is, the probability that the energy per particle is less than $t$ equals 0 for $t < 1$ and equals 1 (the initial mean energy) for $t \geq 1$.**

will tend to have the same energy. So, a system with only finitely many particles has a limiting distribution of energy, namely, a step function with a jump of size 1 at $t = 1$, and moreover, no matter what the initial distribution of energy is, the system tends to this distribution under iteration.

Even for a system with a continuum of particles, our observations for the finite case still hold. In order to see this, we formalize the problem in terms of probability theory.

Let $X$ be a random variable corresponding to the initial energy of the particles. Thus, the distribution function $F_1$ associated with $X$ is the initial distribution of energy: $F_1(t) = P(X < t)$ is the proportion of particles with energy less than $t$. Our arguments and analysis will be based only on the knowledge of the energy distribution function and how it is transformed under iteration by the redistribution law. In terms of distribution functions, our normalization condition, that the average energy per particle is unity, means that the expected value of $X$, $\int_0^\infty t\, dF_1(t)$, equals 1.

We seek a random variable $T(X)$ corresponding to the energy per particle after applying the redistribution law once. To say that the indistinguishable particles are paired *at random* in the redistribution process means that, given one particle in the pair, we know nothing about the energy of the second except that its distribution function should be the initial distribution function $F_1$. In other words, we can describe the energy of the randomly paired particles by two *independent* random variables $X_1$ and $X_2$, each having the same distribution as $X$. Thus the simplest redistribution law, according to which paired particles share the total energy of the pair equally, can be expressed in terms of $T(X)$, $X_1$, and $X_2$ as

$$T(X) = \frac{X_1 + X_2}{2}.$$

The new distribution of energy, call it $F_2$, that describes the random variable $T(X)$ will be a convolution of the distributions of $X_1/2$ and $X_2/2$. Since $X_1$ and $X_2$ both have the distribution $F_1$, the distribution $F_2$ of $T(X)$ is given by

$$F_2(t) = P\left(\frac{X_1 + X_2}{2} \leq t\right) = P(X_1 + X_2 \leq 2t) = \int_{-\infty}^{t} F_1(2t - x)\, dF_1(x).$$

To carry out the second iteration, we repeat the process. The energy $T^2(X) = T\left(T(X)\right)$ will have the same distribution as $(Y_1+Y_2)/2$, where $Y_1$ and $Y_2$ are *independent* and each is distributed as $T(X)$. In other words, if we let $X_1$, $X_2$, $X_3$, and $X_4$ be independent and distributed as $X_1$, then $Y_1$ is distributed as $(X_1 + X_2)/2$, and $Y_2$ is distributed as $(X_3 + X_4)/2$. The energy is distributed as $T^2(X) = (X_1 + X_2 + X_3 + X_4)/4$.

After $n$ iterations the energy per particle will have the same distribution as $T^n(X) = (X_1 + \cdots + X_{2^n})/2^n$, where the $X_i$'s are independent and distributed as $X$. This expression for $T^n(X)$ is exactly the expression that appears in the strong law of large numbers (see page 71). Therefore the strong law tells us that the limiting energy of each particle $\omega$ as $n \to \infty$ is

$$\lim_{n\to\infty} T^n\left(X(\omega)\right) = \lim_{n\to\infty} \frac{X_1(\omega) + \cdots + X_{2^n}(\omega)}{2^n} = E(X_1) = 1, \text{ almost surely,}$$

where $E(X_1)$ is the expected value of the initial distribution Thus, after $n$ iterations of this random process, the energies of almost all particles converge to unity. In terms of distribution functions, we say that in the space of all "potential" actualizations of this iterative random process, almost surely, or with probability 1, the limiting distribution of energy will be a step function with a jump of size 1 at $t = 1$ (Fig. 5).

Notice that for this simplest redistribution law (1) the redistribution operator $T$ is a simple linear operator and (2) even so, the strong law of large numbers is needed to determine the limiting behavior.

**More Complicated Redistribution Laws.** Stan proposed more interesting laws of redistribution. The redistribution operator $T$ for each of these laws is nonlinear, and different techniques are needed to analyze the system. For example, after pairing the particles, choose a number $\alpha$ between 0 and 1 at random. Then instead of giving each particle one-half the total energy of the pair, let us give one particle $\alpha$ times the total energy of the pair and give the other particle $(1 - \alpha)$ times the total energy. The energy $T(X)$ will then have the same distribution as $U(X_1 + X_2)$, where $U$ is uniformly distributed on [0,1] (that is, all values between 0 and 1 are equally probable) and $U, X_1$, and $X_2$ are independent. What happens to this system under iteration is a much more complicated matter. For one thing, unlike the redistribution operator in the simplest case, the operator $T$ is now highly *nonlinear* and the law of large numbers is not available as a tool. A new approach is required. To get an idea of what to expect, Stan first used the computer as an experimental tool. From these studies he correctly guessed the limiting behavior (Ulam 1980): no matter what the initial distribution of energy is, we have convergence to the *exponential* distribution (Fig. 6).

Let me indicate how Blackwell and I proved this conjecture. We used a classical *method of moments* together with an analysis of a *quadratic recursion*. For now let us assume that a stable limiting distribution exists and let $X$ have this distribution. Then $T(X) = U(X_1 + X_2)$ has the same distribution. So, calculating $m_n$, the $n$th moment of $X$ (that is, the expected value of $X^n$), we have

$$m_n = E(X^n) = E\left(T(X)^n\right) = E\left((U(X_1 + X_2))^n\right) = E\left(U^n(X_1 + X_2)^n\right).$$

By independence and the binomial theorem

$$m_n = E(U^n)E\left((X_1 + X_2)^n\right) = \frac{1}{n+1}E\left(\sum_{p=0}^{n}\binom{n}{p}X_1^p X_2^{n-p}\right) = \frac{1}{n+1}\sum_{p=0}^{n}\binom{n}{p}E(X_1^p X_2^{n-p}).$$

Since $X_1$ and $X_2$ are independent, the expected value of each product is equal to the product of the expected values, $E(X_1^p X_2^{n-p}) = E(X_1^p)E(X_2^{n-p})$. Substituting this into the equation above and using the definition of moments, we have

$$m_n = \frac{1}{n+1}\sum_{p=0}^{n}\binom{n}{p}m_p m_{n-p} = \frac{2}{n+1}m_0 m_n + \sum_{p=1}^{n-1}\binom{n}{p}m_p m_{n-p}.$$

Using the fact that $m_0 = 1$, we solve for $m_n$:

$$m_n = \frac{1}{n-1}\sum_{p=1}^{n-1}\binom{n}{p}m_p m_{n-p}.$$

**RANDOM LAW FOR ENERGY REDISTRIBUTION**

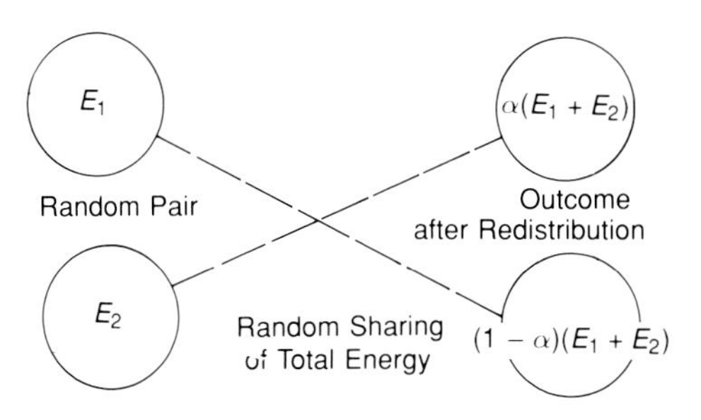


**LIMITING ENERGY DISTRIBUTION**

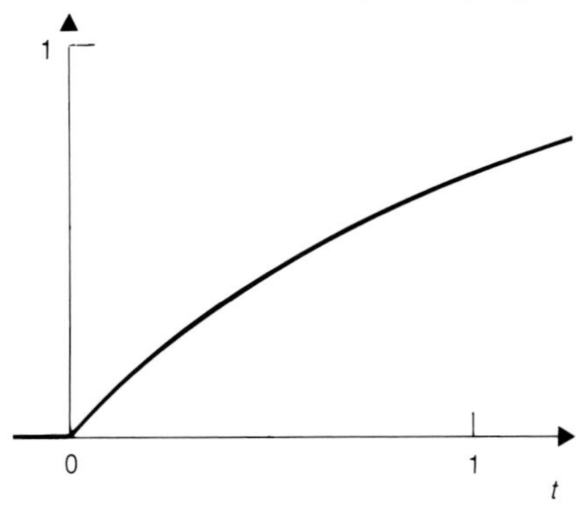


**Fig. 6. Consider a system identical to the one described in Fig. 5 except that the total energy of an interacting pair is redistributed randomly between the members of the pair. In particular, assume that one particle receives a randomly chosen fraction $\alpha$ of the total energy and the other particle receives the remainder. The system still reaches a limiting energy distribution, one equal to 0 for $t < 0$ and equal to $1 - e^{-t}$ for $t \geq 0$.**

This is a quadratic recursion formula. Substituting the initial condition $m_1 = 1$, we find that $m_2 = 2$ and $m_3 = 6$. An induction argument shows that $m_n = n!$ for all $n$. But $n!$ is the $n$th moment of the exponential distribution! Of course, our assumption is that a stable distribution and all its moments exist. It takes some work to prove that this assumption is indeed true and that no matter what initial distribution one starts with, the distribution of the iterates converges to the exponential.

It should not be too surprising that our result agrees in its general form with the Boltzmann distribution of statistical mechanics. After all, both are derived from similar assumptions. The Boltzmann distribution is derived from the assumptions that (1) energy and the number of particles are conserved, (2) all energy states are equally probable, and (3) the distribution of energy is the most probable distribution. In our problem we also assumed conservation of energy and number of particles. Moreover, taking $U$ in our redistribution law to be the uniform distribution makes all energy states equally probable. The difference is that the iteration process selects the most probable distribution with no a priori assumption that the most probable distribution will be reached.

We can go further and replace $U$ by any random variable with a symmetric distribution on [0,1]. The symmetric condition insures that the particles are indistinguishable. We call the distribution of $U$ the redistribution law. Again, one obtains a quadratic recursion formula. Blackwell and I analyzed this formula and showed that for every such $U$ the system tends toward a stable limiting distribution. In other words, there is an attractive fixed point in the space of all distributions. Moreover, there is a one-to-one correspondence between the stable limiting distribution and the redistribution law that yields it.

**Momentum Redistribution.** There is a corresponding momentum problem. Assume we have a vast number of indistinguishable particles (all of unit mass) with some initial distribution of momentum. Let us assume that the particles interact in pairs as follows. At each step in the evolution of the system, pair all the particles at random and let the total momentum of each pair be redistributed between the members of the pair according to some law of redistribution that is independent of the pairs. Of course, we wish to conserve energy and momentum. These conservation laws place severe constraints on the possibilities. If $\mathbf{v}_1$ and $\mathbf{v}_2$ are the initial velocity vectors of two particles in a pair and $\mathbf{v}_1'$ and $\mathbf{v}_2'$ are the velocity vectors after collision, then by momentum conservation

$$\mathbf{v}_1 + \mathbf{v}_2 = \mathbf{v}_1' + \mathbf{v}_2'$$

and by energy conservation

$$\|\mathbf{v}_1\|^2 + \|\mathbf{v}_2\|^2 = \|\mathbf{v}_1'\|^2 + \|\mathbf{v}_2'\|^2.$$

Consider this process in the center-of-mass frame of reference. Let $\lambda_i$ be the fraction of the total kinetic energy that particle $i$ has after collision and let $\mathbf{u}_i$ be the unit vector in the direction of the velocity of particle $i$. Then

$$\lambda_1 + \lambda_2 = 1$$

and

$$\sqrt{\lambda_1}\mathbf{v}_1 + \sqrt{\lambda_2}\mathbf{v}_2 = 0.$$

From these equations it follows that $\lambda_1 = \lambda_2 = 1/2$ and $\mathbf{v}_2 = -\mathbf{v}_1$. What this means is that all we can do is choose in the center-of-mass frame a new direction vector for one of the two colliding particles. Everything else is then determined. The other particle goes in the opposite direction, and the total kinetic energy in the center-of-mass frame is divided evenly between the two particles. Thus, the only element of randomness is in how the new direction vector is chosen. If all directions are assumed to be equiprobable, then it can be shown that no matter what the initial distribution of velocity is, the system tends under iteration to a limiting distribution that is the standard normal distribution in three-dimensional Euclidean space $\Re^3$. We have thus rederived the Maxwell-Boltzmann distribution of velocities. Here again we can go further and consider more complicated redistribution laws.

Suppose one allows ternary collisions instead of binary collisions. Then there are more degrees of freedom, and the problem again becomes interesting mathematically. The results of our analysis show that the situation is much like the redistribution of energy in that the limiting distribution of velocity depends on the law of redistribution of velocity.

## **Problem 2.** *Geometry, Invariant Measures, and Dynamical Systems*

The intimate relationship among geometry, measures, and dynamical systems that was elucidated in the last century continues to deepen and hold our attention today. Poincaré made several monumental contributions to this development in his treatise *Les Méthodes Nouvelles de la Mécanique Céleste*. One major issue he considered concerned the stability of motion in a gravitational field such as that of our solar system. Would small perturbations from any given set of initial orbits lead to a collision of the planets? A tremendous amount of work had been done on this dynamical system, but the governing system of differential equations remained unsolved. Faced with this situation, Poincaré made a wonderful flanking maneuver by introducing "qualitative" methods that involved measures.

For the setting consider the motion of $N$ bodies and the corresponding phase space $S$, whose $6N$ coordinates code the position and momentum of each of the $N$ bodies. The phase space is a subset of Euclidean $6N$-space and each point of $S$ corresponds to a state of the system. Consider $T$, the time-one map of $S$. That is, if $s$ is the initial state of the system, then $T(s)$ is the state of the system one time unit later. Now, various notions of stability can be given in terms of the properties of $T$. One of these is recurrence, or, as Poincaré said, "stabilité à la Poisson." A state $s$ is said to be recurrent provided that if the system is ever in $s$, then it will return arbitrarily close to $s$ infinitely often. Formally, $s$ is recurrent provided that for every open region $U$ about $s$ there are infinitely many positive integers $n$ such that $T^n(s)$ is in $U$. Poisson had earlier attempted to show this kind of stability for the restricted three-body problem. Poincaré used the fundamental tenet of measure theory, countable additivity, to prove that the set of all points $s$ in the phase space for which recurrence *does not* occur is of measure zero.

**Recurrence Theorem:** Let $B = (s \in S \,|\, s$ is not recurrent). Then $B$ has measure zero.

Poincaré's proof of this theorem (see "The Essence of Poincaré's Proof of the Re-

# POINCARÉ'S PROOF *of the* RECURRENCE THEOREM

Let us indicate the essential ingredients of the argument that $B$, the set of points in phase space that are not recurrent, has measure zero. Assume $S$ is a surface of constant energy and the volume (measure) of $S$ is finite, $\nu(S) < \infty$. Let $U_1, U_2, U_3, \ldots$ be an infinite sequence of open balls in $S$ such that each point of $S$ lies in one of the balls (no matter how small the radii of the balls may be). For each $n$ let $B_n$ be the set of points in $U_n$ that are *not* recurrent; that is, $B_n$ consists of all points $s \in U_n$ such that $T^p(s) \in U_n$ for only finitely many positive integers $p$. Now consider the set $B = \bigcup_{n=1}^{\infty} B_n$, that is, the set of *all* points that are not recurrent. Since the measure $\nu$ is *assumed to be countably additive,* we have $\nu(B) \leq \sum_{n=1}^{\infty} \nu(B_n)$. Poincaré also assumed that the notion of volume could be extended to sets $B_n$ that are more complicated than open regions. Given these assumptions we can prove that $B$ has measure zero if we show that $\nu(B_n) = 0$ for each $n$.

The argument goes like this. Fix $n$ and let $U = U_n$. Let

$$C = U - \bigcup_{p=1}^{\infty} T^{-p}(U).$$

It is easy to show that, for each $k, s \in T^{-k}(C)$ if and only if $T^k(s) \in U$ and $T^i(s) \notin U$ for all $i > k$. Consequently,

$$T^{-i}(C) \cap T^{-j}(C) = \emptyset \text{ if } 0 \leq i < j,$$

and

$$B_n = \bigcup_{k=0}^{\infty} T^{-k}(C).$$

Therefore, since the sets $T^{-k}(C)$ are pairwise disjoint and the measure is countably additive,

$$\nu(B_n) = \sum_{k=0}^{\infty} \nu\left(T^{-k}(C)\right).$$

Since $T$ is volume-preserving, the sets $T^{-k}(C)$ all have the same measure, $a$. If $a > 0$, then $\nu(S)$ would be infinite, which is a contradiction. Thus each $B_n$ has measure zero, and therefore $B$ also has measure zero. ■

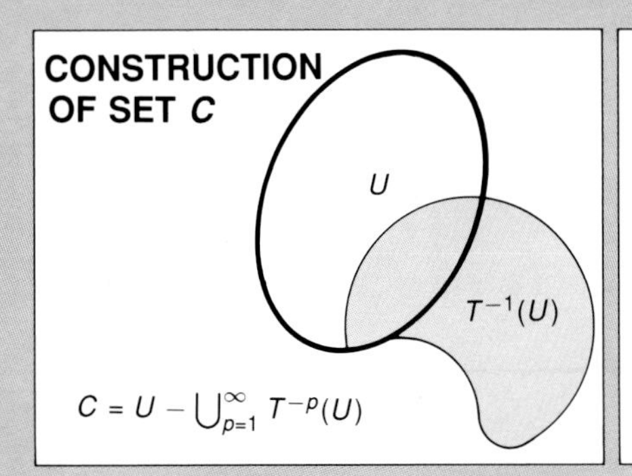


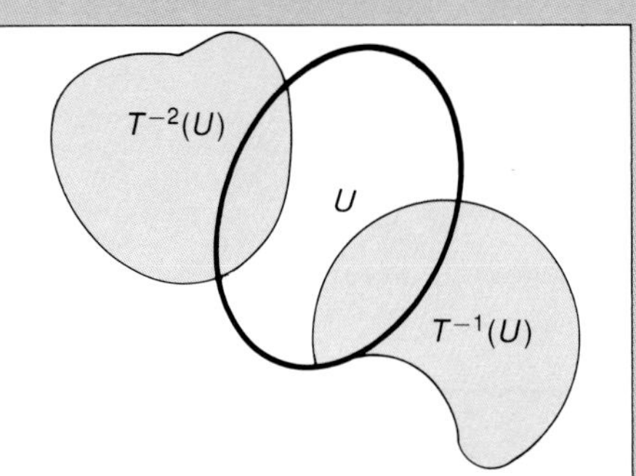


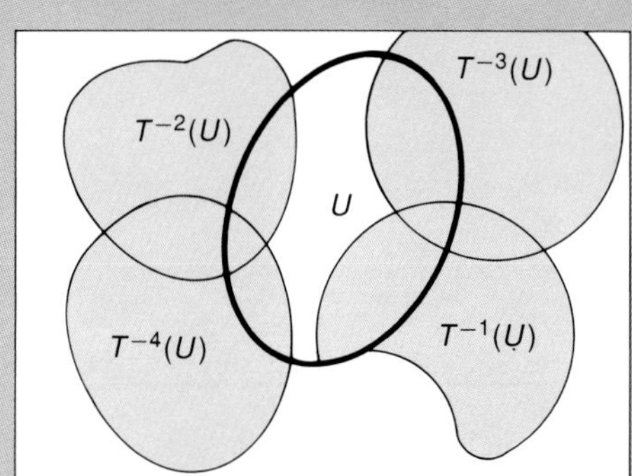


currence Theorem") is a shining jewel that made clear to the mathematical world the importance of countable additivity in the development of measure.

But what measure did Poincaré have in mind here? After all, there is an entire grab bag of measures on the subsets of $S$. In the case of the $N$-body problem, since the system is a Hamiltonian system, the *geometry* of the phase space clearly indicates the correct measure. Let us see why. Liouville had proved the seminal result that if the map $T$ that describes the time evolution of the system is a Hamiltonian, then $T$ is volume-preserving in the phase space. That is, if $U$ is an open set or region, then $\nu(U) = \nu\left(T(U)\right)$, where $\nu(E)$ is the volume of $E$. Poincaré carried out his analysis on a "surface of constant energy." Since the $N$-body problem is a conservative system, the function $T$ leaves the total energy invariant and therefore maps each such surface into itself. Moreover, since $T$ is a Hamiltonian, it is volume-preserving on this surface. Consequently, the geometric structure of the surface determines the appropriate measure to use. Since the surface is a manifold, by definition there is a positive integer $m$ such that each point of $S$ lies in a region that is geometrically the same as a piece of Euclidean $m$-dimensional space. So, the measure to use on the manifold $S$ is the one we naturally associate with Euclidean $m$-dimensional space, namely, $m$-dimensional volume.

## Geometry and Dynamical Systems

To summarize, the $N$-body problem is a classical dynamical system in which the time-one map $T$ is a continuous one-to-one map of the phase space $X$ onto itself. The inverse map, $T^{-1}$, is also continuous. Thus, $T$ is a *homeomorphism*. There is a natural measure on the phase space $X$ that is invariant under $T$. From one point of view, this measure is the volume element corresponding to the dimension of the phase space. From another viewpoint the natural invariant measure expresses the fact that the system is a Hamiltonian system. In the phase space $X$ a surface $S$ of constant energy forms an invariant set, and again there is an invariant measure on $S$ corresponding to our ordinary notion of volume. The set $B$ of all points that are not recurrent is also an invariant set with respect to $T$. However, it is not at all clear that we can define some natural invariant measure on $B$ that is both nonzero and invariant under $T$. Many dynamical systems being studied today "live" on invariant sets that, like $B$, are not manifolds. Instead they are "pathological" sets, sets that at one time were thought to be the private domain of the purest and most abstract mathematicians. The examples range from Cantor sets to nowhere-differentiable curves to indecomposable continua. Many of these pathological invariant sets are "strange attractors" of dynamical systems; the system is "attracted" in the sense that it will eventually end up on the set from any starting point. (The discovery of one of the first strange attractors is described in the section Cubic Maps and Chaos of the article "Iteration of Maps, Strange Attractors, and Number Theory—An Ulamian Potpourri.")

**Properties of Invariant Sets.** Let us now indicate some of the problems and techniques used in studying such sets in the context of *dynamical systems*. We will consider discrete dynamical systems, that is, systems in which the time evolution is described by discrete steps. We consider a function $T$ that maps a space $X$ into itself and the iterates of $T$, that is, $T^1, T^2, T^3, \ldots$, where $T^{n+1}(x) = T\big(T^n(x)\big)$. We are interested in an *invariant* set—a subset $M$ of $X$ such that $T(M) \subset M$. The simplest invariant set consists of a fixed point $x$ such that $T(x) = x$; a more complicated invariant set is a periodic orbit, a set consisting of the points $x, T(x), \ldots, T^{n-1}(x)$, and $T^n(x) = x$. Invariant sets are further classified according to how points near the invariant set behave under $T$. An invariant set $M$ is called an *attractor* if there is a region $U$ surrounding $M$ such that if $x \in U$, then $T^n(x)$ gets closer and closer to $M$ as $n$ increases. On the other hand, $M$ is called a *repeller* if there is a region $U$ surrounding $M$ such that if $x \in (U - M)$, then $T^n(x)$ is not in $M$ for $n$ sufficiently large. For example, if $X$ is the real number line, then 0 is an attracting fixed point for $T(x) = x/2$ and a repelling fixed point for $T^{-1}(x) = 2x$. The intrinsic properties of an invariant set are also of interest. For example, one might want to know whether there is a point $x$ of $M$ such that the *orbit* of $x$, that is, $x, T(x), T^2(x), \ldots$, is dense in $M$. If $T$ is an irrational rotation of the plane, then the unit circle is invariant and the orbit of every point on the circle is dense in the circle. Another possibility is that $T$ is *topologically mixing* on $M$; that is, for every region $U$ of $M$ there is some $n$ such that $M \subset T^n(U)$.

One central problem we will look at in some depth is the construction of "natural" or useful invariant measures for the sets $M$. In particular we want a measure $\mu$ such that $\mu(X - M) = 0$ and $\mu\left(T^{-1}(B)\right) = \mu(B)$ for each measurable subset $B$ of $M$. That is, the measure is zero for points outside the invariant set $M$ and is invariant with respect to the inverse of $T$.

THE "TRIANGLE FUNCTION"

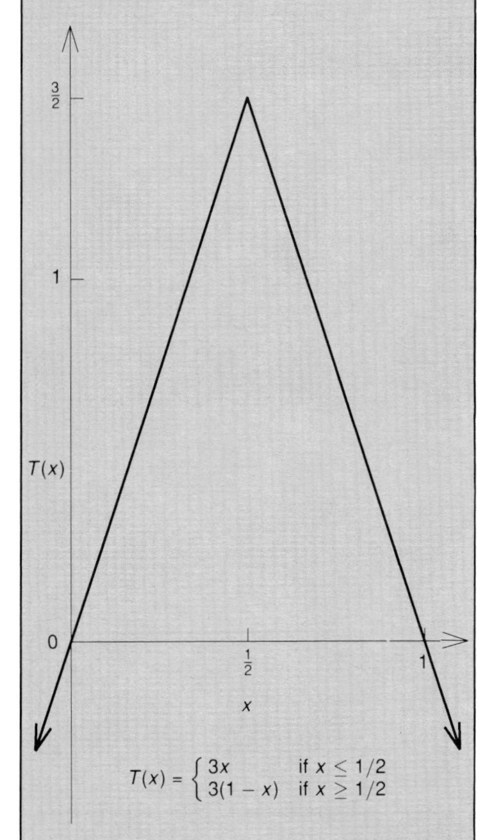


**Fig. 7. The transformation $T(x)$ maps the set of real numbers into itself. That is, it establishes a correspondence (a two-to- one correspondence) between the real numbers and a subset of the real numbers, those less than or equal to 3/2.**

**Cantor's Set as an Invariant Set.** Let us consider a simple example of a map whose invariant set is Cantor's middle-third set. Let $X$ be the real number line and let $T(x) = (3/2)(1 - |2x - 1|)$. Then $T$ is a two-to-one map of $X$ into itself, the "triangle function" whose graph is shown in Fig. 7. This transformation can also be written in the following form:

$$T(x) = \begin{cases} 3x & \text{if } x \leq 1/2 \\ 3(1-x) & \text{if } x \geq 1/2. \end{cases}$$

Now consider what happens to $x$ under the iterates of $T$. If $x < 0$, then $T^n(x) = 3^n x$ and $T^n(x) \to -\infty$. If $1 < x$, then $T(x) < 0$ and higher iterates are given by $3^n(1-x)$. Again, $T^n(x) \to -\infty$. Thus, the iterates of all points outside the interval [0,1] are repelled. On the other hand, $x = 0$ is a fixed point, and, since $T(1/4) = 3/4$ and $T(3/4) = 1/4$, the set $\{1/4, 3/4\}$ forms a periodic orbit of order 2. It turns out that there is a natural invariant set under the iterates of $T$ that lies in the interval [0,1]. To find it we consider successive iterations of $T$ and keep track of the parts of the interval [0,1] that are mapped outside the interval by each interation. The first few iterations of $T$ are illustrated in Fig. 8 and are described below. If $x$ is in the open interval $(1/3, 2/3)$, $T(x) > 1$, and thus $T$ maps this open interval out of the interval [0,1]. The two intervals $J_1 = [0, 1/3]$ and $J_2 = [2/3, 1]$ are each mapped onto [0,1]. Thus, $J_1 \cup J_2$ consists of all points remaining in the interval [0,1] after one iteration. What points of $J_1$ remain in [0,1] after the second iteration? The middle third of $J_1$, namely $(1/9, 2/9)$ is mapped out of the interval [0,1] by the second iteration of $T$, and the two subintervals $J_{11} = [0, 1/9]$ and $J_{12} = [2/9, 1/3]$ make up the points of $J_1$ that remain in [0, 1] after two iterations of $T$. Similarly, the middle third of $J_2, (7/9, 8/9)$, is mapped out of [0,1] by $T^2$, and the two subintervals of $J_2$, $J_{21} = [2/3, 7/9]$ and $J_{22} = [8/9, 1]$, make up the points of $J_2$ that remain in [0,1] under $T^2$. Continuing this analysis, we find that the points of [0,1] that remain in [0,1] after $n$ iterations of $T$ consist of $2^n$ intervals. Moreover, they are precisely the same $2^n$ intervals that appear in the construction of Cantor's famous middle-third set. Thus, Cantor's middle-third set, call it $M$, is invariant under $T$, and if $x \notin M$, then for some $k$, $T^k(x)$ is not in [0,1]. Thus, if $x \notin M, T^n(x) \to -\infty$. The Cantor set is a repellent invariant set of $T$, and this map is also topologically mixing on $M$.

**Hausdorff Measure and Dimension.** If we think of $T$ as an analog of a dynamical system whose motion in phase space is restricted to a Cantor set we might like to find a natural measure on this set. Our problem is: Which one of the many possible invariant measures is useful? One clue for determining the appropriate measure for the $N$-body problem was the fact that the phase space is a manifold and we therefore know the *dimension* of the space. We could then use the corresponding volume in the Euclidean space of that dimension to guide us to the correct measure. But what do we do with the Cantor set of our example? What is its dimension? In the early part of this century Felix Hausdorff developed an approach for determining the dimension of a general metric space (a space with a notion of a metric, or distance, between points) in terms of measures associated with the metric. It is perhaps surprising at first that the dimension of a space may not be an integer. Such spaces have been christened *fractals* by Mandelbrot, and he has provided many examples of their occurrence in physical phenomena. The idea behind Hausdorff's generalization of dimension is very simple

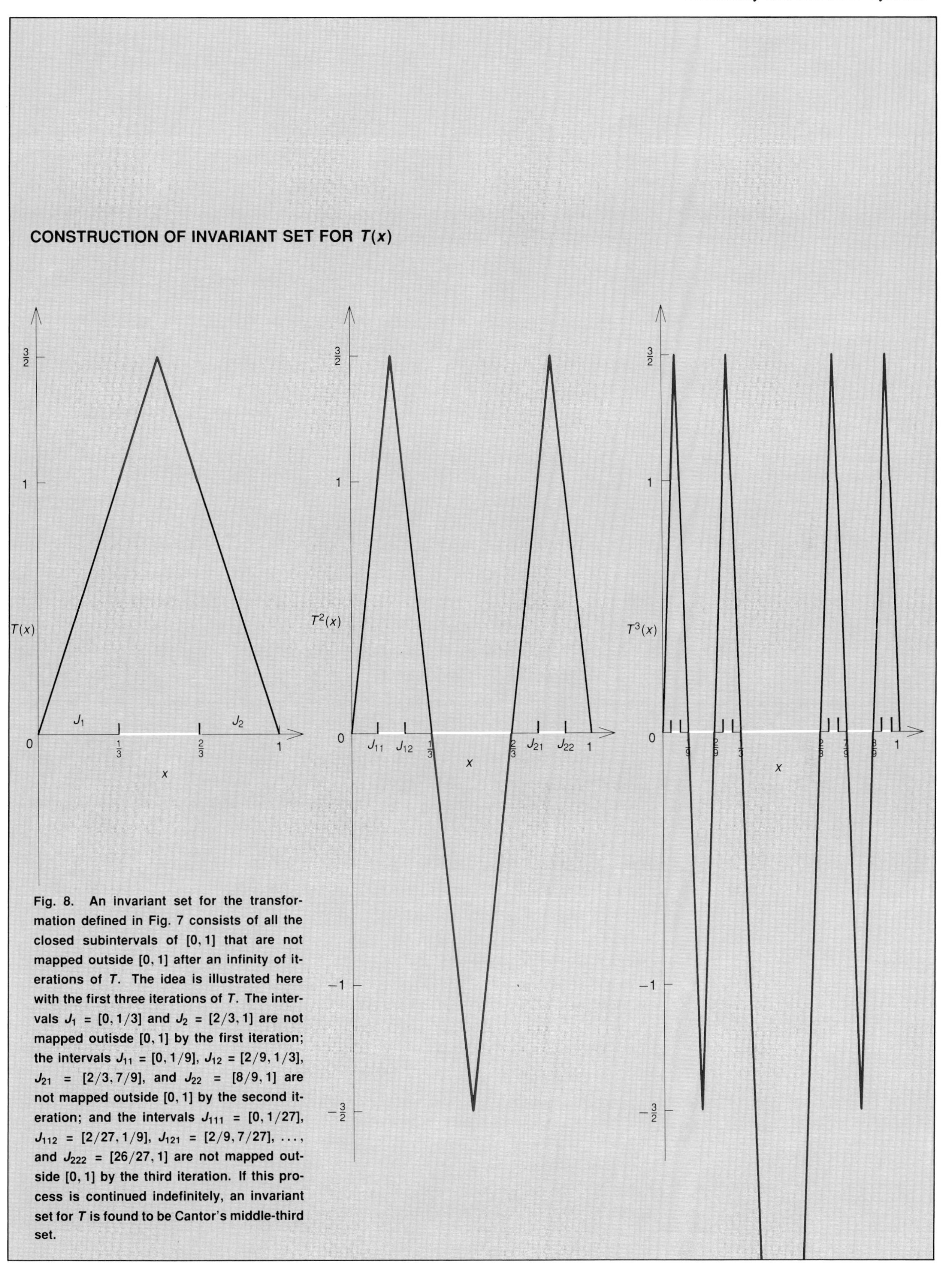


Fig. 8. An invariant set for the transformation defined in Fig. 7 consists of all the closed subintervals of $[0, 1]$ that are not mapped outside $[0, 1]$ after an infinity of iterations of $T$. The idea is illustrated here with the first three iterations of $T$. The intervals $J_1 = [0, 1/3]$ and $J_2 = [2/3, 1]$ are not mapped outisde $[0, 1]$ by the first iteration; the intervals $J_{11} = [0, 1/9]$, $J_{12} = [2/9, 1/3]$, $J_{21} = [2/3, 7/9]$, and $J_{22} = [8/9, 1]$ are not mapped outside $[0, 1]$ by the second iteration; and the intervals $J_{111} = [0, 1/27]$, $J_{112} = [2/27, 1/9]$, $J_{121} = [2/9, 7/27]$, ..., and $J_{222} = [26/27, 1]$ are not mapped outside $[0, 1]$ by the third iteration. If this process is continued indefinitely, an invariant set for $T$ is found to be Cantor's middle-third set.

and is based on the idea of *self-similarity* or *scaling*.

Let's take the simplest example, the unit square. We could say that the dimension of the unit square is 2 for the following reason. Consider any scaling transformation $f(x) = \lambda x$, where $x$ is a point in the plane. The transformation $f$ is called a similarity map of the plane and the image of the unit square under $f$ will be a square whose area is $\lambda^2$. The *power* to which we raise the scaling exponent to obtain the measure of the image set is the dimension of the original set. Exactly the same reasoning shows that the unit cube in Euclidean $n$ space has dimension $n$.

The generalization to more complicated metric spaces is straightforward. Consider a general metric space $X$. A map $f$ is a *similarity* map of a subset $E$ of $X$ if the distance between points in $E$ scale by a factor $r$ under the action of the map. In other words there is a number $r$ such that for all $x$ and $y$ in $E$, $\operatorname{dist}\big(f(x), f(y)\big) = r \operatorname{dist}(x, y)$. Hausdorff defined for each number $\beta \geq 0$ a measure $H^\beta$ on $X$ that obeys the scaling law of Hausdorff measures.

**Scaling law of Hausdorff measures:** If $E \subset X$ and $f$ is a similarity map of $E$ onto $f(E)$ with similarity ratio $r$, then $H^\beta\big(f(E)\big) = r^\beta H^\beta(E)$.

While the measures $H^\beta$ are defined on the metric space for all values of $\beta > 0$, Hausdorff showed that there is one and only one measure $H^\alpha$ for which a "jump" occurs. He called $\alpha$ the dimension of the metric space.

**Hausdorff dimension theorem:** For each metric space $X$, there is a number $\alpha$ such that if $\beta < \alpha$, then $H^\beta(X) = \infty$ and if $\alpha < \beta$, then $H^\beta(X) = 0$. The number $\alpha$ is called the Hausdorff dimension of $X$.

How do Hausdorff's definitions of measure and dimension compare with our ordinary notions in Euclidean space? It turns out that the Hausdorff dimension of $n$-dimensional Euclidean space is $n$ (which it should be, of course) and the associated Hausdorff measure $H^n$ is the same as our usual definition of volume element. Thus, $H^\alpha$ is a natural generalization to a space of dimension $\alpha$ of our ordinary notions of measure, or volume element, in Euclidean space. Once the Hausdorff dimension $\alpha$ of a space is known, we have a natural measure on the space, namely $H^\alpha$. So the first problem is to determine the dimension of the space under consideration.

**Hausdorff Dimension of Cantor's Middle-Third Set.** As an example, we will show that the self-similarity properties of the middle-third Cantor set $C$ define its Hausdorff dimension as $\log 2/\log 3$. (In fact, Hausdorff proved this in his original paper.)

Consider the two similarity maps $f_1(x) = x/3$ and $f_2(x) = x/3 + 2/3$. Then $f_1(C) = C \cap [0, 1/3]$ and $f_2(C) = C \cap [2/3, 1]$. So $C = f_1(C) \cup f_2(C)$. Since $f_1(C)$ and $f_2(C)$ are disjoint and $H^\alpha$ is a measure,

$$H^\alpha(C) = H^\alpha\big(f_1(C)\big) + H^\alpha\big(f_2(C)\big).$$

By the scaling law, $H^\alpha\big(f_1(C)\big) = (1/3)^\alpha H^\alpha(C)$ and $H^\alpha\big(f_2(C)\big) = (1/3)^\alpha H^\alpha(C)$. Therefore

$$H^\alpha(C) = (1/3)^\alpha H^\alpha(C) + (1/3)^\alpha H^\alpha(C) = (2/3^\alpha) H^\alpha(C).$$

Cancelling $H^\alpha(C)$, we have

$$1 = 2/3^\alpha, \text{ or } \alpha = \log 2/\log 3.$$

We conclude that the Hausdorff dimension of $C$ is $\log 2/\log 3$. Of course, this is only a heuristic argument (because we cannot cancel $H^\alpha(C)$ unless $H^\alpha(C)$ is positive and finite), but it can be justified.

Returning to our example $T(x) = (3/2)(1 - |2x - 1|)$, we have shown that the invariant set $M$ is Cantor's middle-third set and that the Hausdorff dimension of $M$ is $\alpha = \log 2/\log 3$. In fact $\mu = H^\alpha$, Hausdorff's volume element in dimension $\alpha$, is an invariant measure on $M$.

Our analysis of this example is typical of the analyses of many discrete dynamical systems. We found an invariant set $M$ that is constructed by an algorithm that analyzes the behavior of points near $M$. The first application of the algorithm yields nonoverlapping closed regions $J_1, \ldots, J_n$ the second yields nonoverlapping subregions $J_{i1}, \ldots, J_{in}$ in each $J_i$, and so forth. Finally, the invariant set $M$ is realized as

$$M = \bigcap_{k=1}^{\infty} \left( \bigcup_{i_j \leq n} J_{i_1 \ldots i_k} \right).$$

In this example the construction is *self-similiar*; that is, there are scaling ratios $t_1, \ldots, t_n$ such that a region at iteration $k$, $J_{i_1 \ldots i_k}$ and a subregion at level $k+1$, $J_{i_1 \ldots i_{k+1}}$, are geometrically similar with ratio $t_{i_{k+1}}$. (In our example $t_1 = t_2 = 1/3$.) When such similarity ratios exist, one can use a fundamental formula due to P. A. P. Moran for calculating the Hausdorff dimension of the invariant set.

**Theorem:** If $M = \bigcap_{k=1}^{\infty} \left( \bigcup_{i_j \leq n} J_{i_1 \ldots i_k} \right)$, then $\dim(M) = \alpha$, where $\alpha$ is the solution of $t_1^\alpha + \cdots + t_n^\alpha = 1$. Moreover, $0 < H^\alpha(M) < +\infty$.

That is, $\alpha$ is the Hausdorff dimension of $M$, and $H^\alpha$ is a well-defined finite measure on $M$.

**Random Cantor Sets.** One of my current interests centers on analyzing the invariant sets obtained when the dynamical system experiences some sort of random perturbation. The perturbation introduces a perturbation in the algorithm used to construct the invariant set. Thus we randomize the algorithm, and the scaling ratios $t_1, t_2, \ldots, t_n$, instead of having fixed or deterministic values as before, are now random variables that have a certain probability distribution. One theorem of Williams and mine (Mauldin and Williams 1986) is that the Hausdorff dimension of the final "perturbed" set $M$ is, with probability 1, the solution of

$$E\left(t_1^\alpha + \cdots + t_n^\alpha\right) = 1,$$

where $E(t_1^\alpha + t_2^\alpha + \cdots)$ is the expected value of the sum of the $\alpha$th powers of the scaling ratios. Note that this formula reduces to Moran's formula in the deterministic case.

As an example suppose our randomly perturbed system produces Cantor subsets of [0,1] as follows. First, choose $x$ at random according to the uniform distribution on [0,1]. Then between $x$ and 1 choose $y$ at random according to the uniform distribution on $[x, 1]$. We obtain two intervals $J_1 = [0, x]$ and $J_2 = [y, 1]$. Now in each of these intervals repeat the same procedure (independently in each interval). We obtain two

**THE GOLDEN MEAN**

**Fig. 9. (a) Consider a rectangle with sides of length $A$ and $B$, $A < B$. Let $r$ denote the ratio of $A$ to $B$. Divide this rectangle into a square of side $A$ and a new rectangle. If the ratio of the lengths of the sides of the new rectangle, $(B - A)/A$, also equals $r$, then both the original rectangle and the new rectangle are golden rectangles and $r$ is equal to the golden mean $m$. (The numerical value of $m$, $(\sqrt{5} - 1)/2$, is obtained by solving the two simultaneous equations $r = A/B$ and $r = (B - A)/A$.) (b) The process of dividing a golden rectangle into a square and a new golden rectangle can, of course, be continued indefinitely. It can be shown that the logarithmic spiral given in polar coordinates by $\log \rho = m\theta$ passes through two opposite vertices of each successively smaller square. This fact may help explain why the Hausdorff dimension of the random Cantor sets described in the text is equal to the golden mean.**

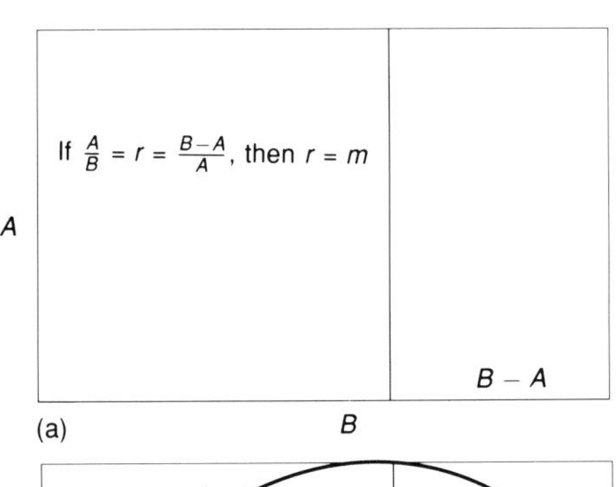


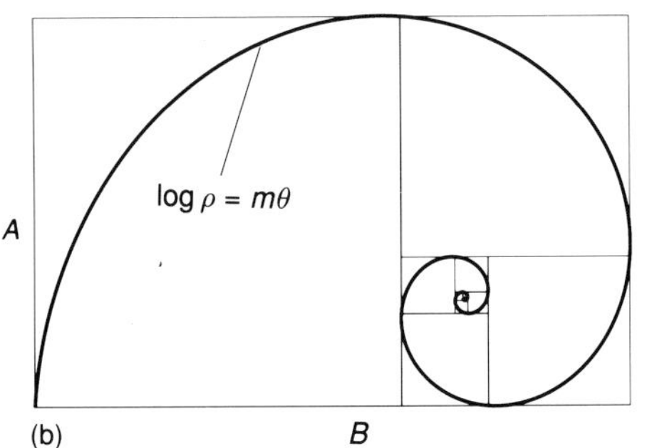


subintervals of $J_1$, $J_{11}$ and $J_{12}$, and two subintervals of $J_2$, $J_{21}$ and $J_{22}$. Continue this process. We will obtain a random Cantor set, and its Hausdorff dimension $\alpha$ is, with probability 1, the solution of $E\left(t_1^\alpha + t_2^\alpha\right) = 1$, or

$$\int_0^1 \left( x^\alpha + \frac{1}{1-x} \int_x^1 (1-y)^\alpha dy \right) dx = 1.$$

A little calculus shows that

$$\alpha = \frac{\sqrt{5}-1}{2}, \text{ the golden mean!}$$

A problem left for the reader: Why should the golden mean (Fig. 9) arise as the dimension of these randomly constructed Cantor sets?

## Problem 3. *Computer Experiments and Random Homeomorphisms*

One topic Stan and I discussed several times was whether one could "randomize" dynamical systems in some way. Is it possible to define a probability measure on a wide class of dynamical systems such that meaningful statements could be made, for instance, about the probability that a system would become turbulent or about the expected time to the "onset of chaos"? To get started on this very ambitious problem, we discussed how we would go about generating homeomorphisms at random. For simplicity, let us generate homeomorphisms of the unit interval [0,1] onto itself. Thus, we wish to build continuous, strictly increasing maps $h$ with $h(0) = 0$ and $h(1) = 1$. One algorithm for doing this randomly follows.

Set $h(0) = 0$ and $h(1) = 1$. Choose $h(1/2)$ according to the uniform distribution on [0,1]. Continue by choosing $h(1/4)$ and $h(3/4)$ according to the uniform distribution on $[0, 1/2]$ and $[1/2, 1]$, respectively. In general, once the values of $h(i/2^n)$ have been determined for $i = 0, 1, \ldots, 2^n$, choose $h\left((2i+1)/2^{n+1}\right)$ according to the uniform distribution on $\left[h(i/2^n), h(i+1)/2^n\right]$. This simple algorithm is easily implemented on a computer. (It needs no more than fifty lines of FORTRAN.) If the computer's random-number generator is fairly good, general properties of these functions can be guessed. However, to show that this algorithm defines an associated probability measure $P$ on $\Omega$, the set of all homeomorphisms of [0,1] onto [0,1], is no small task. First we need to define a class of elementary events and the probabilities associated with them. An elementary event in the sample space $\Omega$ comes naturally from the random algorithm. For a positive integer $n$, consider the dyadic grid on [0,1] given by the points $1/2^n$, $2/2^n, \ldots, (2^n - 1)/2^n$. Over each grid point $i/2^n$ construct a "gate", an interval $(a_i, b_i)$ such that $a_i < b_i \leq a_{i+1}$. An elementary event consists of all elements $h$ of $\Omega$ that pass through all the gates: $a_i < h(i/2^n) < b_i$, for $i = 1, 2, \ldots, 2^n - 1$ (Fig. 10).

The probability assigned to an elementary event is defined by induction on $n$. For example, if $n = 1$, an elementary event consists of all $h$ that pass through a single gate: $a < h(1/2) < b$. Since the random algorithm chooses $h(1/2)$ uniformly, the probability assigned to this event is the length of the interval, $b - a$. If $n > 1$, the probability of an elementary event is determined from the conditional probabilities given by the algorithm. For example, the distribution function of the random variable $h(3/4)$ is $P\left(h(3/4) \leq t\right)$. To calculate this distribution function, we first find the conditional probability that $h(3/4) \leq t$, given that $h(1/2) = s$. It follows directly from the construction algorithm that

$$P\left(h(3/4) \leq t | h(1/2) = s\right) = \begin{cases} 1 & \text{if } 1 \leq t \\ (t-s)/(1-s) & \text{if } s < t < 1 \\ 0 & \text{if } t \leq s. \end{cases}$$

So,

$$\begin{aligned} P\left(h(3/4) \leq t\right) &= \int_0^1 P\left(h(3/4) \leq t | h(1/2) = s\right) ds \\ &= \int_0^t P\left(h(3/4) \leq t | h(1/2) = s\right) ds \\ &= \int_0^t \left((t-s)/(1-s)\right) ds \\ &= t + (1-t)\ln(1-t). \end{aligned}$$

The distribution of $h(3/4)$ is shown in Fig. 11.

The exact formulas for the probabilities assigned to various elementary events are quite complicated. What is required is to determine that probabilities of the form

$$P\left(h(1/2^n) \leq t_1, h(2/2^n) \leq t_2, \ldots, h\left((2^n - 1)/2^n\right) \leq t_{2^n - 1}\right)$$

satisfy Kolmogorov's consistency theorem. We have shown that these conditions are indeed satisfied and therefore a probability measure $P$ is defined on the homeomorphisms of [0,1]. To see what these homeomorphisms look like, we used the computer. Figure 12 shows a few samples from our computer studies in which the values of $h(i/2^n)$ are computed for $n = 10$.

S. Graf, S. C. Williams, and I studied this method in detail (Graf, Mauldin, and Williams 1986). For example, we examined a large number of the computer studies and guessed that with probability 1 the derivative of a random homeomorphism at the origin is 0. This conjecture turned out to be correct. The argument is essentially the following. First, since $h$ is increasing and $h(0) = 0$, it is enough to show that

$$\lim_{n \to \infty} \frac{h(1/2^n) - h(0)}{1/2^n} = \lim_{n \to \infty} 2^n h(1/2^n) = 0.$$

Second, set

$$\Psi_n(h) = \frac{h(1/2^n)}{h(1/2^{n+1})},$$

where $n = 1, 2, 3, \ldots$. It is intuitively clear and can be proved that $\Psi_1, \Psi_2, \Psi_3, \ldots$ are independent random variables, all uniformly distributed on [0,1]. Set $X_n = \ln \Psi_n$. The $X_n$'s are independent and identically distributed, and $E(X_n) = \int_0^1 \ln t \, dt = -1$. Therefore, by the strong law of large numbers,

$$\lim_{n \to \infty} (1/n) \sum_{p=1}^{n} X_p = -1.$$

## CONSTRUCTION OF ELEMENTARY EVENTS

**Fig. 10. In the study of random homeomorphisms described in the text, an elementary event is defined as the set of all homeomorphisms $h$ that pass through $2^n - 1$ "gates" consisting of open intervals $(a_i, b_i)$ over the grid points $i/2^n$ ($i = 1, 2, \ldots, 2^n - 1$). The $a_i$'s and $b_i$'s are restricted by the conditions $a_i < b_i < a_{i+1}$. Shown here is one possible set of gates for $n = 2$ and a member of the corresponding elementary event.**

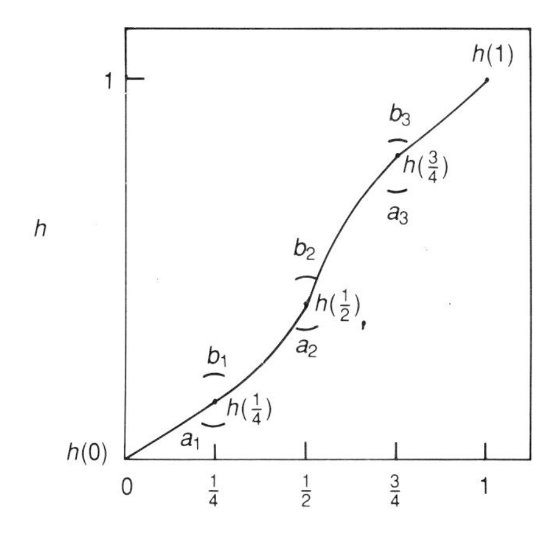


## DISTRIBUTION FUNCTION FOR $h(3/4)$

**Fig. 11. As demonstrated in the text, $F(t) \equiv P\left(h(3/4) \leq t\right)$ equals 0 for $t < 0$ and equals $t + (1 - t)\ln(1 - t)$ for $t \geq 0$. Shown here is the graph of that distribution function.**

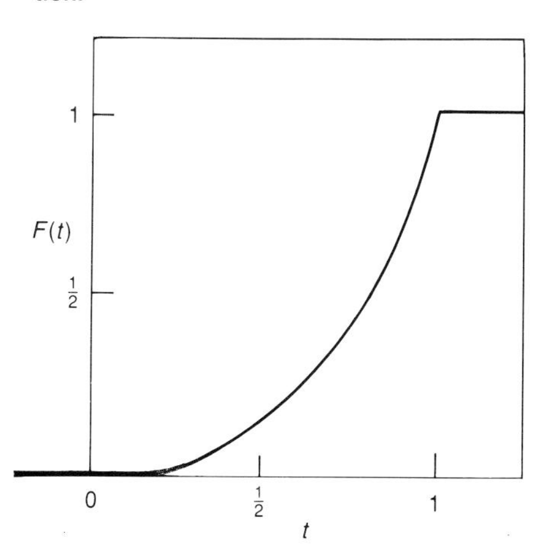

## COMPUTER-GENERATED RANDOM HOMEOMORPHISMS

Fig. 12. Each of the graphs here is a random homeomorphism passing through a set of points $h(1/2^{10})$, $h(2/2^{10})$, ..., $h(1023/2^{10})$. The sets of points were generated by a computer according to the algorithm described in the text. Such graphs provide experimental data about the properties of the homeomorphisms as a class.

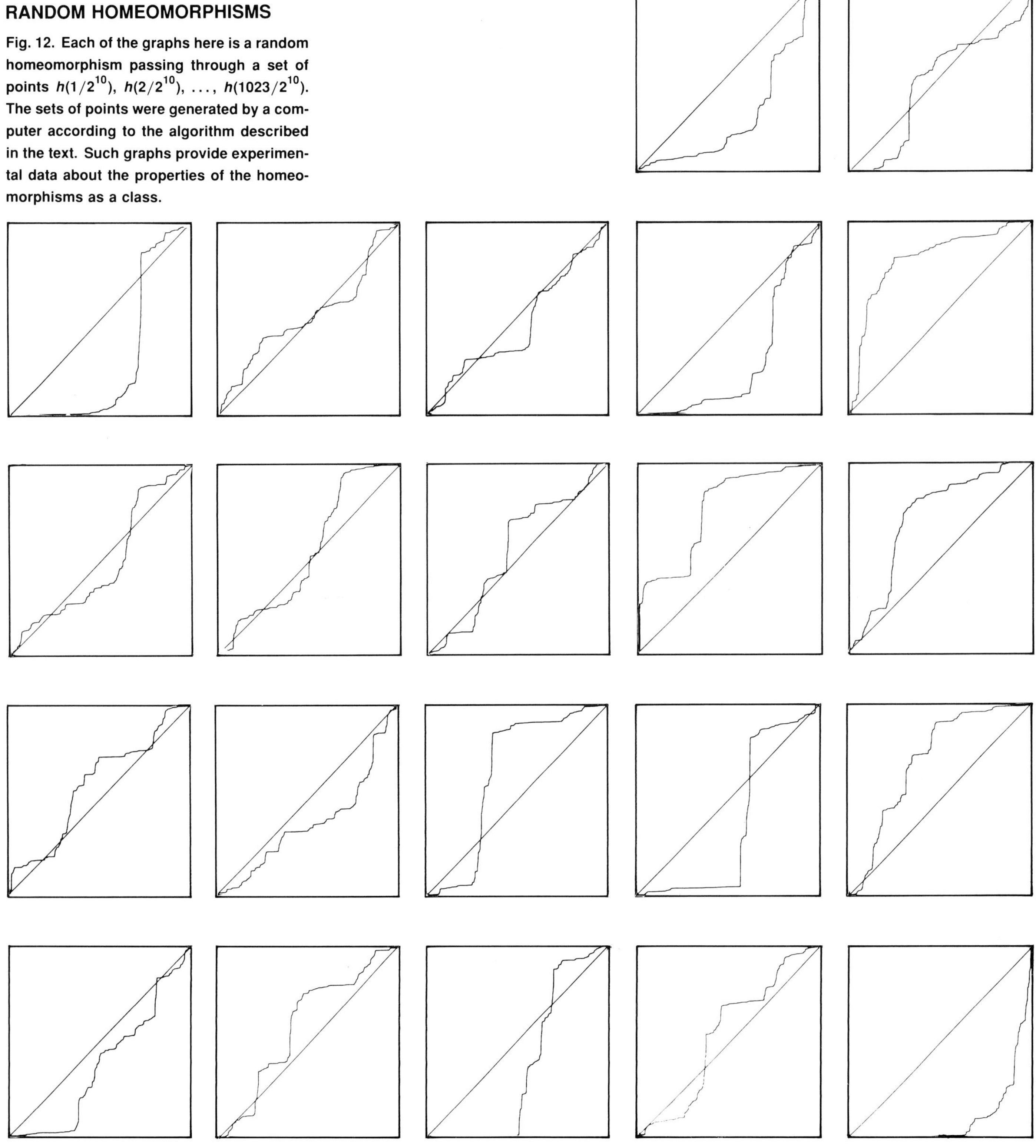

Multiplying both sides by $n$ we have, with probability 1,

$$-\infty = \lim_{n\to\infty} \sum_{p=1}^{n} X_p = \lim_{n\to\infty} \sum_{p=1}^{n} \ln \Psi_p = \lim_{n\to\infty} \ln \prod_{p=1}^{n} \Psi_p .$$

Exponentiating we get

$$0 = \lim_{n\to\infty} \prod_{p=1}^{n} \Psi_p = \lim_{n\to\infty} 2^n h(1/2^n),$$

which is what we wanted to show.

We have also shown that, with probability 1, a random homeomorphism has a derivative of 0 almost everywhere, that is, everywhere except for a subset of [0,1] with Lebesgue measure 0. Consequently, with probability 1, a random homeomorphism is not smooth. Therefore this approach will not yield answers to questions concerning the transition from smooth to turbulent, or chaotic, behavior. As often happened with Stan's problems, the original question, which was motivated by physics, would eventually become a purely mathematical problem.

By the way, our original studies on an Apple computer illustrate the pitfalls of working with numerical results. From looking at the graphs we guessed that the set of fixed points for these homeomorphisms is a Cantor set. When we were unable to prove this conjecture, Tony Warnock conducted more highly resolved computer studies on a Cray. The results suggested not that the fixed points are a Cantor set but rather that a high proportion of the random homeomorphisms have an odd number of fixed points (see the accompanying table). This time we guessed that, with probability 1, a random homeomorphism has a finite odd number of fixed points. Indeed we were able to prove this; however, the proof is too complicated to outline here.

A few closing comments on this problem. First, the procedure for generating a random homeomorphism can also be viewed as a procedure for generating a distribution function at random. Thus, we have a probability measure on the space of probability measures! This viewpoint was thought of and developed earlier by Dubins and Freedman. Second, Stan and I did consider the generation of random homeomorphisms on other spaces. For example, the algorithm for generating homeomorphisms of the circle reads almost exactly like that for generating homeomorphisms of the interval. (However, in that case we don't know whether there is a positive probability of generating homeomorphisms with no periodic points. This is an interesting possibility.) Third, it is possible to bootstrap oneself up from generating homeomorphisms of the interval to generating homeomorphisms of the square, the cube, and so on. These possibilities are described in Graf, Williams, and Mauldin 1986. Finally Stan had some wild ideas about "crossing" random homeomorphisms with something like Brownian motion to produce flows at random.

That wildness was the joy of being with Stan Ulam. His boundless imagination opened up one's mind to the endless possibilities of creating. It was my good fortune to have known Stan for some ten years as a deep personal friend, a most stimulating collaborator, and an endless source of inspiration. ■

## FIXED POINTS OF RANDOM HOMEOMORPHISMS

**Listed here are computer-generated sets of data on the number of fixed points possessed by each of (a) 5000 and (b) 10,000 of the random homeomorphisms (*h*'s) defined in the text. Note the predominance of homeomorphisms with odd numbers of fixed points. That observation led us to conjecture, and to prove, that, with probability 1, any such random homeomorphism has a finite odd number of fixed points.**

| Number $k$ of Fixed Points | Number of $h$'s with $k$ Fixed Points | |
|---|---|---|
| | (a) | (b) |
| 0 | 185 | 510 |
| 1 | 1332 | 2868 |
| 2 | 196 | 544 |
| 3 | 876 | 1835 |
| 4 | 179 | 418 |
| 5 | 605 | 1138 |
| 6 | 143 | 283 |
| 7 | 410 | 751 |
| 8 | 114 | 174 |
| 9 | 259 | 464 |
| 10 | 75 | 136 |
| 11 | 187 | 276 |
| 12 | 52 | 95 |
| 13 | 114 | 190 |
| 14 | 32 | 50 |
| 15 | 61 | 80 |
| 16 | 20 | 25 |
| 17 | 48 | 59 |
| 18 | 23 | 13 |
| 19 | 38 | 25 |
| 20 | 9 | 9 |
| 21 | 6 | 21 |
| 22 | 7 | 5 |
| 23 | 19 | 12 |
| 24 | 3 | 3 |
| 25 | 1 | 2 |
| 26 | 2 | 1 |
| 27 | 2 | 1 |
| 28 | 0 | 1 |
| 29 | 0 | 2 |

## Further Reading

The first five works are general; the remainder are those cited in reference to specific topics.

A. I. Khinchin. 1949. *Mathematical Foundations of Statistical Mechanics.* New York: Dover Publications, Inc.

Mark Kac. 1959. *Probability and Related Topics in Physical Sciences.* New York: Interscience Publishers, Inc.

Mark Kac and Stanislaw M. Ulam. 1968. *Mathematics and Logic: Retrospect and Prospects.* New York: Frederick A. Praeger, Inc. Also in Volume 1 of *Britannica Perspectives.* Chicago: Encyclopedia Britannica, Inc.

R. Daniel Mauldin, editor. 1981. *The Scottish Book: Mathematics from the Scottish Café.* Boston: Birkhäuser Boston.

R. D. Mauldin and S. M. Ulam. 1987. Problems and games in mathematics. *Advances in Applied Mathematics* 8: 281–344.

Richard P. Feynman. 1951. The concept of probability in quantum mechanics. In *Proceedings of the Second Berkeley Symposium on Mathematical Statistics and Probability,* edited by Jerzy Neyman. Berkeley and Los Angeles: University of California Press.

David Blackwell and R. Daniel Mauldin. 1985. Ulam's redistribution of energy problem. *Letters in Mathematical Physics* 60: 149. (This entire issue is devoted to Stan Ulam.)

S. Ulam. 1980. On the operations of pair production, transmutations, and generalized random walk. *Advances in Applied Mathematics* 1: 7–21.

R. D. Mauldin and S. C. Williams. 1986. Random recursive constructions. *Transactions of the American Mathematical Society* 295: 325–346.

S. Graf, R. Daniel Mauldin, and S. C. Williams. 1986. Random homeomorphisms. *Advances in Mathematics* 60: 239.

**R. Daniel Mauldin** received his Ph.D. in mathematics from the University of Texas in 1969. In 1977, after eight years at the University of Florida, he joined the faculty at North Texas State University, where he is currently the Decker Science Fellow. He is a frequent visitor to the Laboratory. Some of his current research interests involve deterministic and random recursions and the asymptotic geometrical and measure-theoretic properties of objects defined by these processes. He is a member of the American Mathematical Society and an editor of its Proceedings.

# Iteration of *MAPS*,

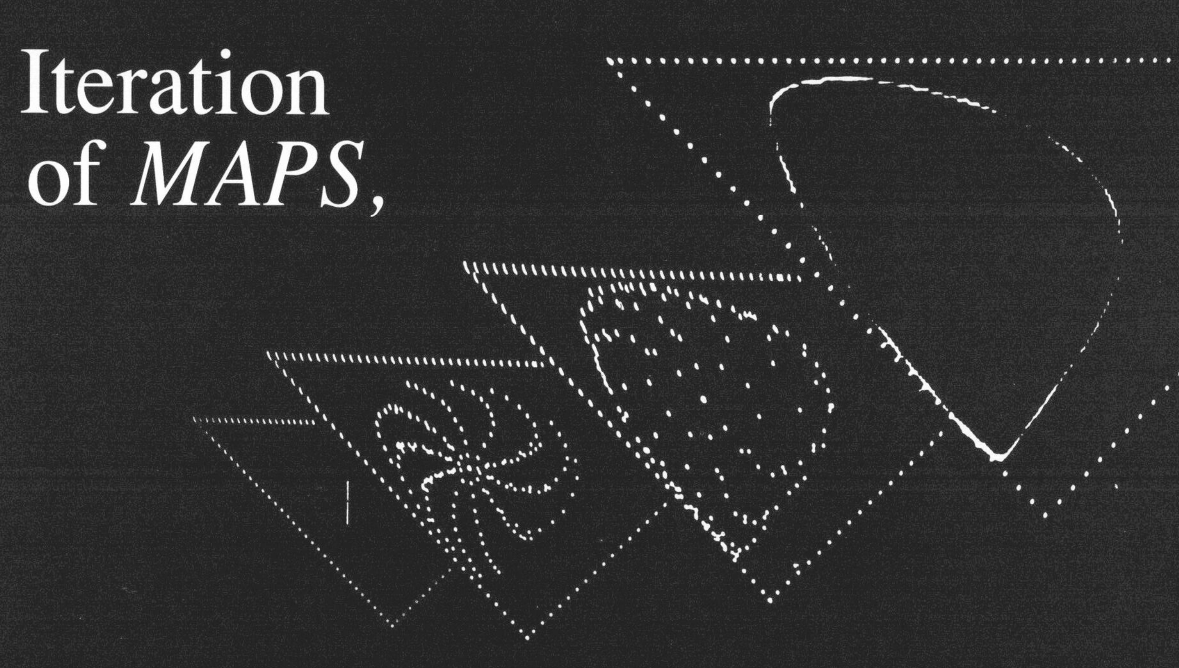

# *STRANGE ATTRACTORS*, and *NUMBER THEORY*

## — *AN ULAMIAN POTPOURRI*

***by Paul R. Stein***

I first met Stan Ulam during the war, when I was at Los Alamos as a GI, working in Hans Bethe's Theoretical Division. Our friendship was social rather than professional, for at that time I had little to contribute. I returned to Los Alamos in 1950 and was immediately caught up in the weapons program, spending much of my time in the East helping to run problems on computers in Washington, Philadelphia, and Aberdeen. What time remained was spent in Santa Monica consulting with the Rand Corporation—and courting my future wife. Fortunately, in 1953 I managed to get married, and that, of course, settled me down. The next six years witnessed my gradual conversion, under Stan's tutelage, from physicist to mathematician.

Our collaboration started in a low key. At first it was limited to discussions—rather one-sided, as I recall. I listened as Stan aired his prejudices concerning mathematical biology as it then was (circa 1955): "It is all foolishness, don't you think?" I was in no position to counter these remarks, and soon he had me more or less believing them. One argument he advanced more than once (and which I no longer believe) was about the human eye. Stan could not imagine that something so complex could have evolved by random processes in the time available, even granting the effect of natural selection. Neither of us, however, could think of a practicable calculation to settle the question, so we turned to simpler matters.

The first mathematical problem we undertook together, with the aid of an IBM 704 computer, concerned the evolution of large populations under the assumption of random mating, to which we added the effect of mutation. (This description of the problem may tempt the reader to interpret what follows in terms of Mendelian genetics. That topic, however, had already been treated mathematically in great detail, and our interest lay rather in investigating mathematical and computational approaches to other examples of evolutionary processes.) Stan made it very clear that he wanted nothing to do with the customary approach via differential equations (à la Sewall Wright); instead, everything was to be based on point-wise iteration. I heartily agreed.

We characterized the "type" of an individual in the population by a pair of integer indices $(i,j)$, with $i,j = 1,2,\ldots,N$. The number of males of type $(i,j)$ was assumed to equal the number of females of that type; in fact, males and females were not distinguished, so, despite the use of the word "mating," the problem involved no sex (and none of the mathematical complications that go with it). The fraction of individuals of type $(i,j)$ in the $n$th generation of the population was denoted by $x_{ij}^{(n)} = x_{ji}^{(n)}$. Random mating then changes the population fractions from generation to generation according to the equation

$$x_{ij}^{(n+1)} = \sum_{p,q,r,s} \gamma_i^{pr} \gamma_j^{qs} x_{pq}^{(n)} x_{rs}^{(n)} . \tag{1}$$

The summation in Eq. 1 was carried out under the restrictions of a "mating rule," namely, that progeny of type $(i,j)$ result from mating between individuals of type $(p,q)$ and $(r,s)$ only if

$$\min(p,r) \leq i \leq \max(p,r)$$

and

$$\min(q,s) \leq j \leq \max(q,s). \tag{2}$$

(Here $\min(u,v)$ and $\max(u,v)$ mean, respectively, the smaller and the larger of the two integers $u$ and $v$.) In other words, the indices of an offspring fall within the ranges defined by those of its parents.

For technical reasons that I will not pursue here, we imposed simplifying conditions on the coefficients $\gamma_k^{uv}$ as follows:

$$\begin{aligned} \gamma_k^{uv} = \gamma_k^{vu} &> 0 \quad \text{if } \min(u,v) \leq k \leq \max(u,v), \\ &= 0 \quad \text{otherwise (in conformance with the mating rule);} \end{aligned} \tag{3}$$

$$\sum_{k=v}^{u} \gamma_k^{uv} = 1; \tag{4}$$

and

$$\sum_{k=v}^{u} k\, \gamma_k^{uv} = \frac{u+v}{2}. \tag{5}$$

Finally, we normalized the initial population fractions $x_{ij}^{(0)}$ by requiring that

$$\sum_{i,j=1}^{N} x_{ij}^{(0)} = 1. \tag{6}$$

It is easy to show that the normalization is preserved through all generations, or in other words that

$$\sum_{i,j=1}^{N} x_{ij}^{(n)} = 1 \quad \text{for all } n.$$

To include mutation we modified Eq. 1 by adding linear terms multiplied by a small positive number $\epsilon$:

$$x_{ij}^{(n+1)} = -\epsilon x_{ij}^{(n)} + \frac{\epsilon}{2}\left(x_{i-1,j}^{(n)} + x_{i,j-1}^{(n)}\right) + \sum_{p,q,r,s} \gamma_i^{pr} \gamma_j^{qs} x_{pq}^{(n)} x_{rs}^{(n)}. \tag{7}$$

(The added terms reflect the assumption that mutation causes type $(u, v)$ to give rise to types $(u + 1, v)$ and $(u, v + 1)$ with probability $\frac{1}{2}\epsilon$.)

We performed very many numerical experiments on the systems represented by Eq. 7, varying $\epsilon$ and using special sets of coefficients satisfying Eqs. 3, 4, and 5. Two particularly convenient coefficient sets were

$$\gamma_i^{jk} = \frac{1}{2^{|j-k|}} \binom{|j-k|}{i - \min(j,k)}$$

(where the term in parentheses is the usual binomial coefficient) and

$$\gamma_i^{jk} = \frac{1}{|j - k| + 1}.$$

Unfortunately, the detailed results of these experiments have disappeared over the thirty or so years since the computations were done. I seem to recall, however, that all the systems we looked at "converged"; in fact, after a sufficiently large number of generations, only a single type remained (survival of the fittest?). I also remember that the convergence was not usually monotone.

Although nothing of a detailed theoretical nature was discovered about the systems including mutation, the simpler systems without mutation (Eq. 1) could be analyzed exactly by elementary methods, even when individuals were distinguished by many indices rather than only two. In brief, each system, as defined by a set of initial population fractions, converged to a state determined entirely by that set. (Details of the analysis are given in Menzel, Stein, and Ulam 1959 and in Stein and Ulam 1964.)

Our next joint project was undertaken with more mathematical aims in view, although Stan never lost his strong interest in biology. (A good summary of Stan's contributions to that field can be found in a 1985 article by Beyer, Sellers, and Waterman. The reader should take note of the 1967 paper by Schrandt and Ulam. The study of growth patterns contained therein bears a close resemblance to some recent work on cellular automata.) After extensive discussion, we decided to study the behavior under iteration of a restricted class of quadratic transformations, or maps, of the plane. The idea was mainly Stan's, but I managed to contribute some practical suggestions.

At this point is seems appropriate to explain what it meant to collaborate with Stan. At some stage in his mathematical career, he apparently lost his taste for detailed mathematical work. Of course, his mind was always brimmming with ideas, most

### DOMAIN OF TWO-DIMENSIONAL MAPS

**Fig. 1. The restrictions $0 \leq x_i \leq 1$ ($i$ = 1, 2, 3) and $\sum_1^3 x_i = 1$ limit the domain of the $x_i$'s, and of the iterates of the two-dimensional quadratic and cubic maps discussed in the text, to the equilateral triangle shown in (a). For more convenient graphic display of the iterates, we introduced the variables $S = \frac{1}{2}(1 + x_1 - x_3)$ and $\alpha = \frac{1}{2}x_2 = \frac{1}{2}(1 - x_1 - x_3)$. These new variables, and the iterates of the maps, are limited (by the restrictions on the $x_i$'s) to the isosceles triangle shown in (b).**

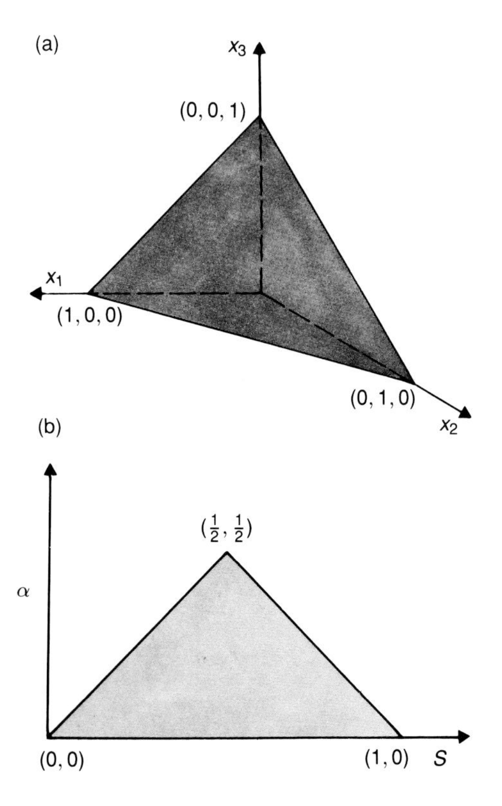


of them good; it was the collaborator's job to fill in the details. Stan was often of great help here with suggestions on how to evade difficulties, but he himself would not work out anything that required more than a few lines of calculation. In the late 1940s C. J. Everett and Stan wrote three brilliant papers on branching processes in $n$ dimensions—a technical tour de force. I recently asked Everett how he and Stan had worked together on those papers. Everett's reply was succinct: "Ulam told me what to do, and I did it." In my case collaboration with Stan usually involved a third person. I had given up programming after having had my fill of it during the first three years at Los Alamos. (In the last four years I have had to take it up again.) Among those who did my coding from time to time were Bob Bivins, Cerda Evans, Verna Gardiner, Mary Menzel, Dorothy Williamson, and in particular Myron Stein, who collaborated with me for many years until the pressure of his own work made it impossible.

The study Stan and I made of quadratic transformations used the programming skills of Mary Menzel; the results appeared in 1959 as "Quadratic Transformations, Part I"—there never was a Part II—under all three names. The computations were done on the Laboratory's own computer, MANIAC II (now defunct). In the following section I will describe that study in some detail; it will then be unnecessary to say much about the mechanical aspects of our later (and more exciting) generalization to cubic maps, since the underlying assumptions were the same.

## Quadratic Transformations à la Stein-Ulam

Consider three variables $x_1$, $x_2$, and $x_3$ restricted as follows:

$$0 \leq x_i \leq 1, \quad i = 1, 2, 3$$

and

$$x_1 + x_2 + x_3 = 1. \tag{8}$$

These restrictions limit the variables to the two-dimensional domain shown in Fig. 1a.

If we multiply out $(x_1 + x_2 + x_3)^2$, we get the six terms $x_1^2$, $x_2^2$, $x_3^2$, $2x_1x_2$, $2x_1x_3$, and $2x_2x_3$. We distribute these six terms among three nonidentical boxes, no box remaining empty. (The boxes correspond to the transformed variables $x_1'$, $x_2'$, and $x_3'$.) This distribution can be done in many ways, in fact, in 540 ways. (The distribution (4,1,1), that is, the distribution such that the first box contains four terms and the second and third boxes each contain one term, can be done in thirty ways, as can the distributions (1,4,1) and (1,1,4); the distributions (3,2,1), (3,1,2), (1,3,2), (1,2,3), (2,1,3), and (2,3,1) each in sixty ways; and the distribution (2,2,2) in ninety.) Let us choose the distribution (3,2,1) to construct an example of a quadratic map. We take three terms, say $x_1^2$, $2x_1x_2$, and $2x_2x_3$, and form their sum; then we sum two other terms, say $x_2^2$ and $x_3^2$, leaving the term $2x_1x_3$ to stand alone. The corresponding map is given by the equations

$$\begin{aligned} x_1' &= x_1^2 + 2x_1x_2 + 2x_2x_3 \\ x_2' &= x_2^2 + x_3^2 \\ x_3' &= 2x_1x_3. \end{aligned} \tag{9}$$

Iteration is carried out by setting $x_i$ equal to $x_i'$ (the first iterate) and substituting the new $x_i$'s back into the right side of Eq. 9 ad infinitum.

Biology has not quite disappeared from the problem. If the $x_i$'s are interpreted as population fractions, Eq. 9 represents the evolution of a population containing three types of individuals randomly mating according to the following rule:

> mating between types 1 and 1, 1 and 2, and 2 and 3 produces type 1;
> mating between types 2 and 2 and 3 and 3 produces type 2;
> and mating between types 1 and 3 produces type 3.

One could also write this rule as a table or a matrix, forms that are more revealing of the algebraic and group properties of the transformation.

Note that if we add up the three rows of Eq. 9, we get $x_1' + x_2' + x_3' = (x_1 + x_2 + x_3)^2$, which equals unity because of Eq. 8. Thus the normalization is preserved algebraically. Nevertheless, in carrying out the iterations on MANIAC II we found that $D$, the sum of the computed $x_i'$'s, could be slightly different from unity because of roundoff. Therefore it was necessary to renormalize after each iteration as follows: $x_i'/D \rightarrow x_i'$ for all $i$.

The "fixed points" of a transformation (more precisely the "first-order fixed points") are points that remain unchanged under iteration; they are solutions to the equations obtained by removing the primes on the equations defining the transformation. The fixed points for the map given by Eq. 9 are easily determined. First note that $x_3 = 2x_1x_3$ (obtained from the third row of Eq. 9) implies that $x_3 = 0$ or $x_1 = \frac{1}{2}$. These possibilities, together with $x_1 = x_1^2 + 2x_1x_2 + 2x_2x_3$ (obtained from the first row of Eq. 9) and the restriction $x_1 + x_2 + x_3 = 1$, lead to two "nodal" fixed points, (1,0,0) and (0,1,0), and one "internal" fixed point, $\left(\frac{1}{2}, \frac{1}{4}(2 - \sqrt{2}), \frac{1}{4}\sqrt{2}\right)$.

How does the map given by Eq. 9 behave under iteration? Experimentally, if we choose an initial point $(x_1, x_2, x_3)$ at random, it is highly probable that the successive iterates will converge to the map's internal fixed point. For some initial points, including those such that $x_3 = 0$ and $x_1 \neq 0$, the iterates converge to the nodal fixed point (1,0,0). (The other nodal fixed point is nonattractive: iterates diverge from (0,1,0) no matter how close to that point an initial point may be.) So this map has two attractive limit sets, or attractors, each characterized by its "basin of attraction" (the set of initial points that iterate to the attractor).

As mentioned above, there are many more maps of the present kind, which we called binary reaction systems. Fortunately, we needed to examine only those that are inequivalent, that is, those that cannot be transformed into each other by some permutation of the indices on the $x_i$'s and the $x_i'$'s (the order of the rows clearly does not matter). It turns out that precisely 97 of the possible 540 maps are inequivalent according to this criterion. The fixed points of all the inequivalent maps were worked out by hand (Stan himself verified some of those calculations), and their limiting behavior under iteration from several randomly chosen initial points was examined numerically. The latter was a very slow process in 1958: MANIAC II could perform only about fifty such iterations per second. Of course, MANIAC II was a stand-alone "dedicated" machine, and that helped make up for its lack of speed.

For more convenient graphic display of the results, we arbitrarily introduced two new variables

$$S = \frac{1}{2}(1 + x_1 - x_3) \text{ and } \alpha = \frac{1}{2}x_2 = \frac{1}{2}(1 - x_1 - x_3). \tag{10}$$

### DEPENDENCE OF LIMIT SET ON INITIAL POINT OF ITERATION

**Fig. 2. A few of our two-dimensional quadratic maps exhibited one of two limiting behaviors under iteration, depending on the location of the initial point. For example, the map defined by the equations below (in both $x_i$ and $S, \alpha$ coordinates) iterates to an internal fixed point ($S_0$ = 0.62448516, $\alpha_0$ = 0.09239627) from any initial point within the dark gray region of the reference triangle and to a nodal period of order 3 $\left((0,0) \rightarrow (\frac{1}{2}, \frac{1}{2}) \rightarrow (1,0)\right)$ from any initial point within any of the three light gray regions. The "separatrix" demarcating the basins of attraction of the two limit sets was determined experimentally.**

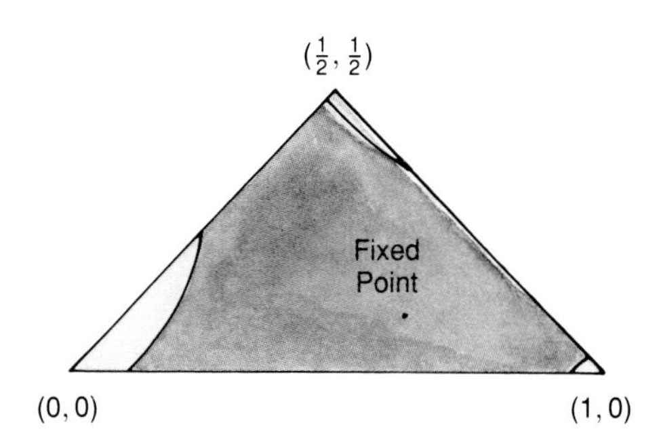


$$T_{x_i}: x_1' = x_2^2 + 2x_1x_2 + 2x_1x_3$$
$$x_2' = x_3^2 + 2x_2x_3$$
$$x_3' = x_1^2$$

$$T_{S,\alpha}: S' = \frac{1}{2}(1 + \alpha^2 - 3S^2) - \alpha + S + 3\alpha S$$
$$\alpha' = \frac{1}{2}(1 + S^2 - 3\alpha^2) + \alpha - S - \alpha S$$

The domain of these new variables is an isosceles triangle in the $S, \alpha$ plane, with unit base and half-unit height (Fig. 1b). Note that the vertices of this "reference triangle" correspond to the nodal points of the original domain.

What we found was less than overwhelming. One transformation had an internal periodic limit set of order 3 (that is, its limit set consisted of three internal points traversed in a certain order), four had internal periods of order 2, one showed no limiting behavior at all, and one converged to an internal fixed point as $\frac{1}{r}$, where $r$ is the distance of the iterate from the fixed point. In addition, a few maps had a "separatrix" (Fig. 2); that is, they showed one of two limiting behaviors (usually convergence to a fixed point or to a periodic limit set of order 2) depending on the location of the initial point. Everything else converged to a fixed point (not necessarily internal) or had nodal periods of order 2 or 3. The interested reader will find a description of the many generalizations we tried in Menzel, Stein, and Ulam 1959.

## Cubic Maps and Chaos

Although some interesting facts emerged from the study described above, Stan and I were disappointed at the lack of variety in the limiting behavior we observed. We even tried to enliven the situation by generalizing the generic map to the form

$$x_i' = d_{i1}x_1^2 + d_{i2}x_2^2 + d_{i3}x_3^2 + 2d_{i4}x_1x_2 + 2d_{i5}x_1x_3 + 2d_{i6}x_2x_3, \quad i = 1, 2, 3,$$

with the coefficients randomly chosen but restricted by $0 \le d_{ij} \le 1$ for all $i, j$ and $\sum_{i=1}^{3} d_{ij} = 1$. Of several hundred such systems investigated, almost all iterated to a fixed point; in other words, the special quadratic maps we had originally looked at were more interesting than the general case.

What to do? Stan and I had, simultaneously, the idea of looking at three-variable cubic maps of the same structure as our quadratics. That is, we would distribute the ten terms arising from expansion of $(x_1 + x_2 + x_3)^3$ among three boxes and construct the maps in the same way as before. A short calculation (see pp. 7–8 of Stein and Ulam 1964) showed that there were more than 9330 inequivalent maps of this type. (The

### THREE-DIMENSIONAL QUADRATIC MAPS WITH INFINITE LIMIT SETS

**Fig. 3. The limit sets of a small fraction of our quadratic transformations in four variables contain what appear to be infinite numbers of points. Shown below are three-dimensional projections of four such limit sets, which were obtained by photographing plots of successive iterates on an oscilloscope screen. The set of axes in the center of each display indicates the orientation of the limit set relative to the viewer, who is conceived of as stationed at a certain distance from the origin along the $x_2$ axis. The limit set for $T_a$ consists of two "curves," one in the $x_1, x_3$ plane and the other in a plane inclined at 45° to the $x_1, x_3$ plane. $T_a$ evidently transforms these planes into each other, since successive iterates lie alternately on the two curves. The limit sets for $T_b$, $T_c$, and $T_d$ are even more complicated, constituting implausibly tortuous curves in space.**

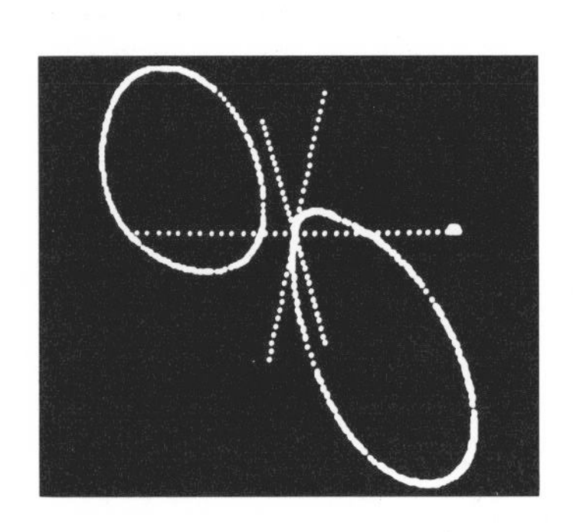

$$T_a: \begin{aligned} x_1' &= x_1^2 + x_2^2 + 2x_2x_4 \\ x_2' &= x_4^2 + 2x_1x_4 + 2x_3x_4 \\ x_3' &= x_3^2 + 2x_1x_2 + 2x_1x_3 \\ x_4' &= 2x_2x_3 \end{aligned}$$

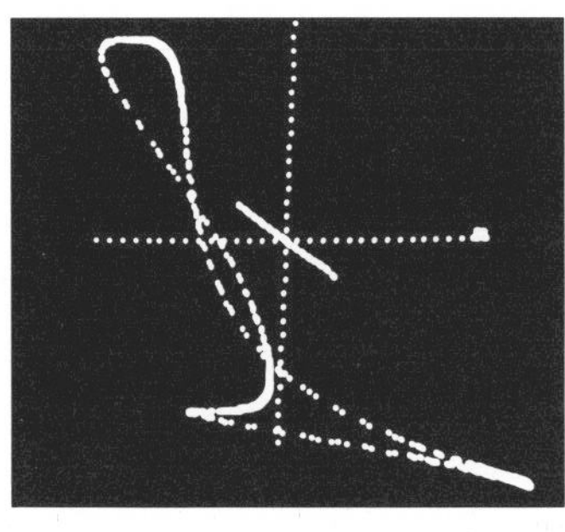

$$T_b: \begin{aligned} x_1' &= x_1^2 + x_3^2 + 2x_3x_4 \\ x_2' &= x_4^2 + 2x_1x_4 + 2x_2x_4 \\ x_3' &= x_2^2 + 2x_1x_2 \\ x_4' &= 2x_1x_3 + 2x_2x_3 \end{aligned}$$

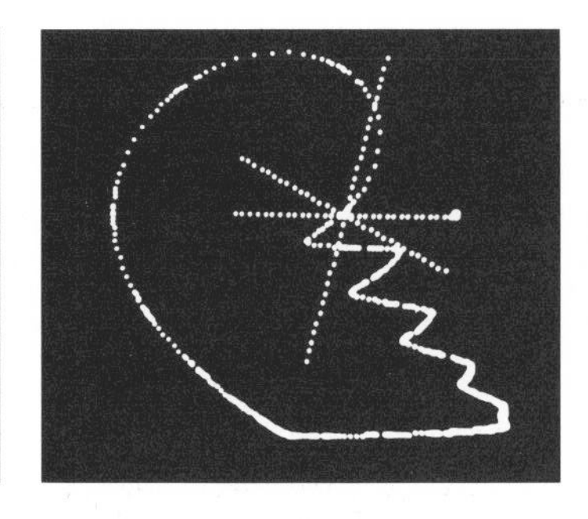

$$T_c: \begin{aligned} x_1' &= x_1^2 + x_3^2 + 2x_1x_2 \\ x_2' &= 2x_1x_3 + 2x_2x_4 + 2x_3x_4 \\ x_3' &= x_2^2 + 2x_2x_3 \\ x_4' &= x_4^2 + 2x_1x_4 \end{aligned}$$

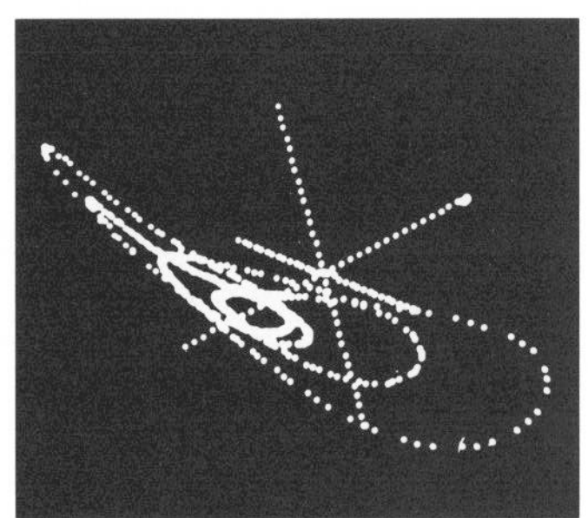

$$T_d: \begin{aligned} x_1' &= x_1^2 + 2x_1x_4 + 2x_2x_4 \\ x_2' &= x_3^2 + x_4^2 + 2x_2x_3 \\ x_3' &= x_2^2 + 2x_1x_2 + 2x_3x_4 \\ x_4' &= 2x_1x_3 \end{aligned}$$

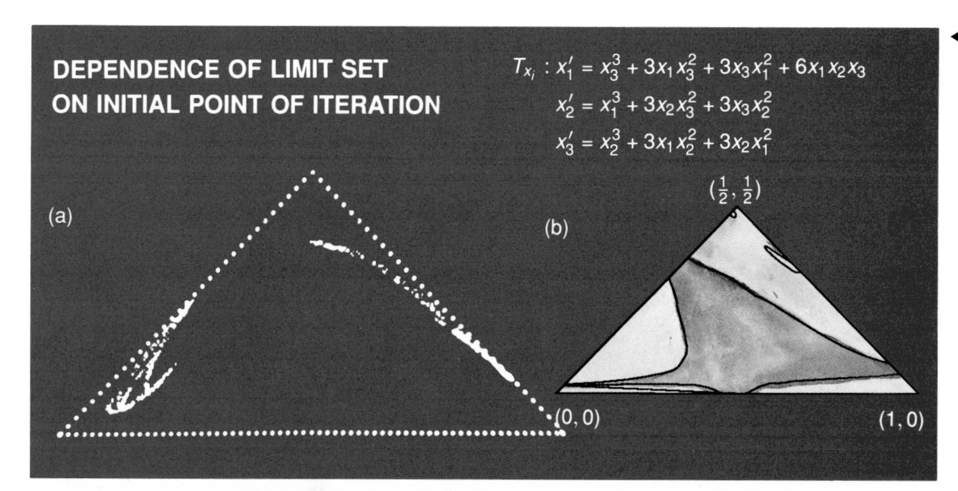


**Fig. 4.** Shown in (a) is one of two possible limiting behaviors for the map defined by the given equations, namely, convergence to a "mess," an apparently infinite number of points with a complex distribution and no discernible structure. The map iterates to this messy limit set from any initial point within any of the light gray regions in (b). If, however, the initial point lies within any of the dark gray regions, the map iterates to the fixed point $S_0 = 0.6259977$, $\alpha_0 = 0.1107896$. The complicated separatrix was determined experimentally.

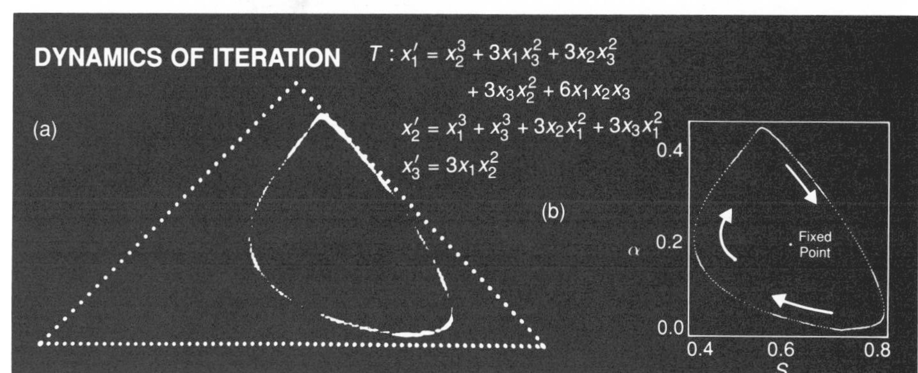


**Fig. 5.** The two-dimensional cubic map defined by the given equations iterates to an infinite limit set composing the closed curve shown in (a). (Whether the words "infinite" and "curve" can be applied here in the strict mathematical sense is not known.) When this map is iterated from some point $p$ in the limit set, successive iterates do not trace out the curve in an orderly fashion. However, the 71st, 142nd, 213th, ..., $(71n)$th, ... iterates of $p$, which are plotted in (b), do lie close to each other and trace out the curve in a clockwise direction. Various stages in the iteration of this map are featured in the art work on the opening page of the article. The first image (counted from background to foreground) shows the set of points at which the iterations were begun, namely twenty-one points uniformly distributed along a line segment whose midpoint is coincident with the nonattractive fixed point of the transformation. (The horizontal and vertical coordinates of this fixed point are approximately 0.6149 and 0.1944, respectively.) The second and third images, which are superpositions of the 8th through 15th and the 15th through 22nd sets of iterates, respectively, capture the dynamics of these early iterations. The final image, a superposition of the 1800th through 2700th sets of iterates (and the same as that in (a) here), shows the stable pattern to which the sets of iterates converge.

exact number turned out to be 9370, arrived at by a more complicated combinatorial calculation.) Perhaps among this plethora of possibilities we would find some systems that showed truly unexpected limiting behavior. I am happy to say that the results far exceeded our expectations.

We also considered transformation in three dimensions, specifically quadratics in four variables with $x_1 + x_2 + x_3 + x_4 = 1$. But 34,337 of these are inequivalent (not an easy fact to come by), so we were never able to give them the attention they deserved. (Figure 3 gives a glimpse of some interesting cases.) Unless someone writes a fast program to evaluate automatically the amusement value of limit sets, that is as far as such studies will ever go: the case that comes next (when ranked by the number of inequivalent maps) is that of quartics in three variables, and more than 3,275,101 of these are inequivalent (the exact number is unknown).

Returning to our study of cubic maps, we plotted the sets of points obtained by iteration on an oscilloscope screen in the reference triangle of Fig. 1b. "Hard copy" was obtained directly from the screen with a Polaroid camera mounted on the oscilloscope. This method, in addition to being cheaper, was more convenient than the current method, which involves a \$20,000 Tektronix terminal with a hard-copy device.

There is not enough space to give all the details of what we found; an extensive summary is given in Stein and Ulam 1964, and Figs. 4–7 show some interesting

## TWO AMUSING INFINITE LIMIT SETS

**Fig. 6. Examples of two-dimensional cubic maps with infinite limit sets constituting (a) a more irregular closed curve than that illustrated in Fig. 5 and (b) three separate closed curves.**

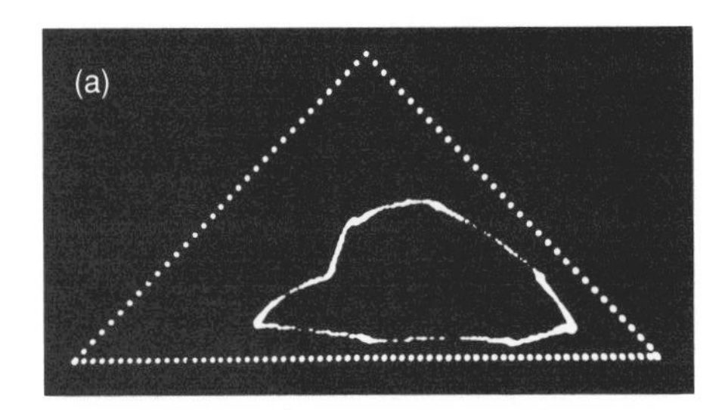


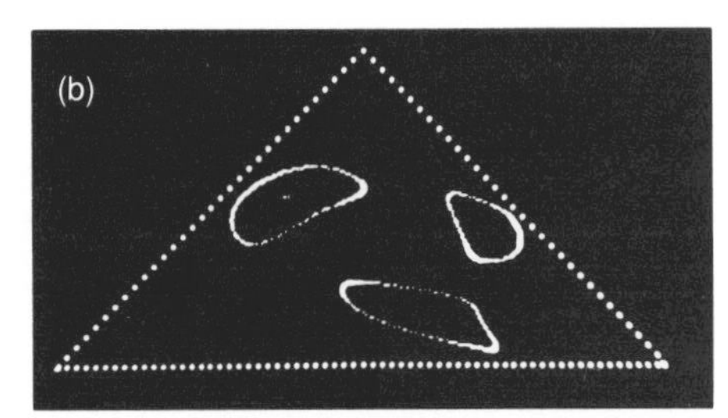


$$T_a : x_1' = x_1^3 + x_2^3 + x_3^3 + 3x_1x_2^2 + 3x_2x_1^2$$
$$x_2' = 3x_2x_3^2 + 3x_3x_2^2 + 6x_1x_2x_3$$
$$x_3' = 3x_1x_3^2 + 3x_3x_1^2$$

$$T_b : x_1' = 3x_1x_3^2 + 3x_2x_3^2 + 3x_3x_1^2 + 3x_3x_2^2$$
$$x_2' = x_1^3 + 3x_2x_1^2 + 6x_1x_2x_3$$
$$x_2' = x_2^3 + x_3^3 + 3x_1x_2^2$$

## A PARTICULARLY FASCINATING INFINITE LIMIT SET

**Fig. 7. The infinite limit set of the two-dimensional cubic map defined here consists of seven separate subsets. Each subset is invariant under the seventh power of the transformation; that is, if *p* is a point in any one of the subsets, the 7th, 14th, 21st, . . . , (7*n*)th, . . . iterates of *p* are also in that subset. Shown magnified in the inset are the 7th, 14th, 21st, . . . , 2695th iterates of a point in the outlined subset of the limit set.**

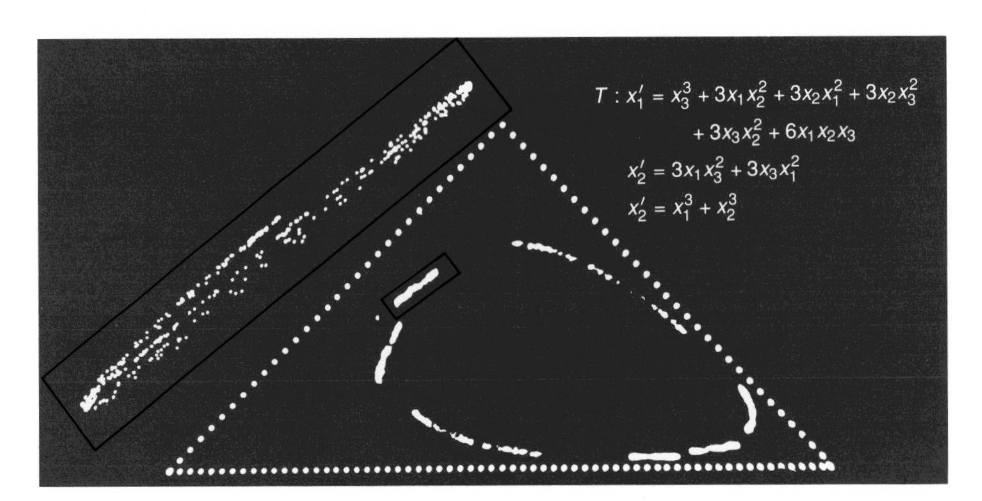


examples of limiting behavior. Again, a large majority of the transformations converged to fixed points or to periodic limit sets (some of quite high order). Of most interest to us, however, were 334 transformations that exhibited no periodic limiting behavior, suggesting that their limit sets contained infinite numbers of points. Some of these appeared to be closed curves or sets of closed curves, although to this day not one has been shown to satisfy the mathematical criteria for a curve. Others bore a striking resemblance to the night sky; at the time these strange limit sets were commonly referred to as messes.

The transformation that iterated to the mess shown in Fig. 8 was studied in great detail and received the special name $T_A$. For the record I give its definition here, both in $x_i$ and $S, \alpha$ coordinates.

$$T_A : x_1' = x_3^3 + 3x_1x_3^2 + 3x_2x_3^2 + 3x_3x_2^2 + 6x_1x_2x_3$$
$$x_2' = x_1^3 + x_2^3 + 3x_3x_1^2 \qquad (11a)$$
$$x_3' = 3x_1x_2^2 + 3x_2x_1^2$$

and

$$T_A : S' = S^3 - 6S^2\alpha - 3S\alpha^2 + 4\alpha^3 - \frac{3}{2}S^2 + 3S\alpha - \frac{3}{2}\alpha^2 + 1$$
$$\alpha' = -S^3 + 3S\alpha^2 + 2\alpha^3 + \frac{3}{2}S^2 - 3S\alpha + \frac{3}{2}\alpha^2. \qquad (11b)$$

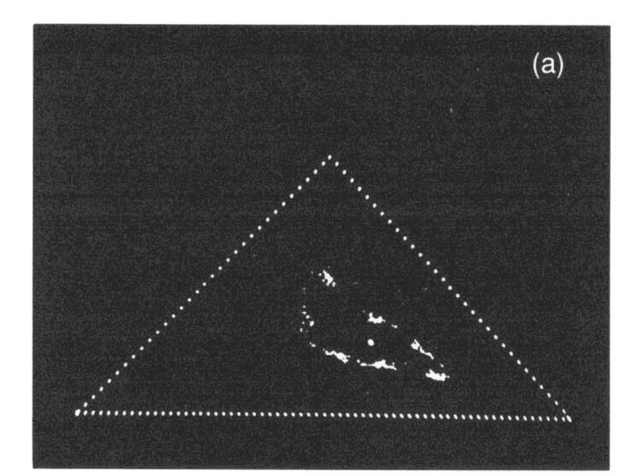


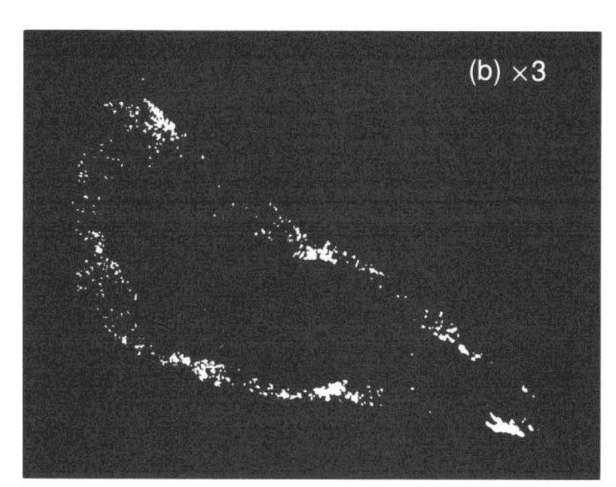


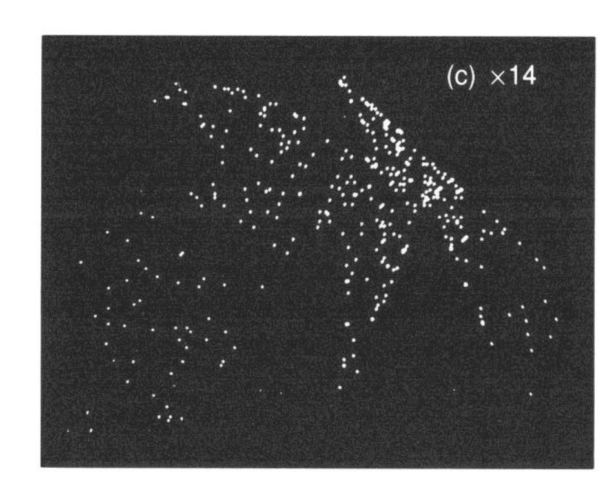


**THE INFINITE LIMIT SET OF $T_A$—A "MESS"**

**Fig. 8. The transformation $T_A$ (see text for defining equations) is one of our two-dimensional cubic maps that iterates to a mess. Shown in (a) is its messy limit set; its nonattractive fixed point has been superimposed on the photograph. The magnifications in (b) and (c) reveal ever greater complexities.**

$T_A$ has an unstable (nonattractive) internal fixed point; its approximate coordinates are $S_0 = 0.5885696$ and $\alpha_0 = 0.1388662$. Some twenty years after the appearance of our paper, $T_A$ was examined on a Cray computer by Erica Jen. The results strongly suggested that its limit set is what is today called a strange attractor, with a fractal (noninteger) dimension of about 1.7. The term "strange attractor" was coined by Ruelle and Takens in 1971 in the course of a study of turbulence. Strange attractors are now known to arise often during iteration of the nonlinear differential or difference equations used to describe phenomena in, for example, meteorology and fluid dynamics.

Several other messes have been classified as strange attractors by present-day criteria, the main one being sensitive dependence on initial conditions. That is, a limit set is a strange attractor if any two points within the set, no matter how close, move farther and farther apart under the action of the mapping. If the limit set is bounded away from infinity (as it is here), the points cannot keep moving apart, and the criterion then is that the relative positions of the limit points become uncorrelated—a feature of chaos. Unfortunately, no numerical experiment can *prove* that some limit set is a strange attractor. For example, what appears to be a strange attractor may actually be a periodic limit set of very high order. To my knowledge, rigorous measures of the likelihood that a computer-generated limit set is a strange attractor have not yet been developed.

Having said that, I shall pretend that some of our cubic maps do illustrate strange attractors. How can those maps be studied further? One way is to introduce another variable $\delta$ ($0 < \delta \leq 1$). Letting $S' = F(S, \alpha)$ and $\alpha' = G(S, \alpha)$ denote the defining equations of the map (cf. Eq. 11b), we write a new set of equations as follows:

$$\begin{aligned} S' &= (1 - \delta)S + \delta F(S, \alpha) \\ \alpha' &= (1 - \delta)\alpha + \delta G(S, \alpha). \end{aligned} \tag{12}$$

Note that $\delta = 1$ corresponds to the original map. (If $\delta = 0$, Eq. 12 reduces to the identity transformation.) So long as $\delta$ lies in the given range, the first-order fixed points are independent of this parameter. The original system may have a nonattractive fixed point; it cannot, of course, be found by iteration. If, however, the fixed point can be made attractive by decreasing $\delta$ (from unity), then iteration can be used, thus avoiding some messy algebra. In fact, a sufficient decrease in $\delta$ will—in almost all cases—decrease the absolute value of both eigenvalues of the Jacobian matrix of Eq. 12 to less than unity at the fixed point, which is precisely the criterion for the attractive

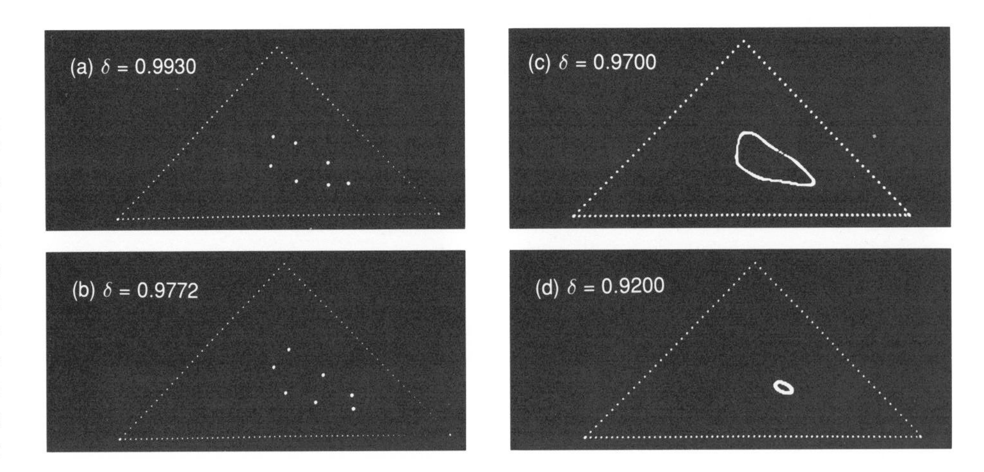


**EFFECT OF $\delta$ ON THE LIMIT SET OF $T_A$**

**Fig. 9. Our two-dimensional cubic maps can be generalized by introducing the parameter $\delta$ as described in the text. Shown here is the effect of varying this parameter on the messy limit set of $T_A$ (see Fig. 8). As $\delta$ is decreased from unity, the limit points at first coalesce into seven distinct bunches, forming what we call a pseudo-period. (a) Then at $\delta \approx 0.9930$, the infinite limit set becomes periodic (and hence finite), with an order of 7. (b) This configuration persists over a range of $\delta$ values, although the coordinates of the limit points vary. (c) Then at $\delta \approx 0.9770$, the periodic limit set changes into a closed curve. (d) As $\delta$ is decreased further, the curve becomes smaller and smaller. Finally, at $\delta \approx 0.9180$, the curve collapses to a single point, the nonattractive fixed point of the original transformation ($\delta$ = 1).**

character of such a point. This is the fact that motivated the introduction of $\delta$, but the effect of its variation turned out to be much more interesting than we expected. Decreasing $\delta$ may cause a remarkable change in the appearance of a messy limit set (Fig. 9). Points may start to cohere, forming a pattern of disjoint arcs. Further decrease of $\delta$ may lead to a periodic limit set of finite order, which persists over a range of $\delta$ values. As $\delta$ approaches the value at which the limit set collapses to the fixed point, the set may metamorphose into a closed curve (at least something that looks like a curve) that shrinks continuously with $\delta$. This behavior is typical; even more complex changes have been observed in some cases (Fig. 10).

Another way to study cubic maps with messes as their limit sets is to vary the coefficients. This is done just as it was for the quadratic maps, but the results are far more dramatic. Figure 11 shows a few examples of the fascinating behavior that has been observed. Here the coefficients constitute a twenty-parameter set, so exploration of all possibilities is not feasible; the usual practice is to vary the coefficients of one or two terms at a time. Much numerical work of that type was done at the Laboratory in 1984 and 1985 on a Cray computer, and many new strange attractors turned up. The aim of this work is to find some "structural" (geometric or algebraic) principle underlying the relatively bizarre phenomena our computer screens reveal.

## One-Dimensional Maps and Universality

The first part of this section is a historical note on the origins of a 1973 paper by Metropolis, Stein, and Stein. The paper dealt with a certain universal structure and hierarchy of the periodic limit sets that can arise in the iteration of one-dimensional maps; it has been cited by Mitchell Feigenbaum as a source of inspiration for his later work on the universal nature of the approach to chaos by "period doubling."

The origins of our paper lie in the work discussed above by Stan and me on cubic maps. We had found fifteen or sixteen that had the property of transforming a pair of sides of the $S, \alpha$ reference triangle into each other. It is clear that the "square" of such a map (the second iterate) transforms one side of the triangle into itself, and the map is therefore one-dimensional. We rewrote some of these as maps defined on the unit interval and iterated them on MANIAC II. In every case we obtained a periodic limit

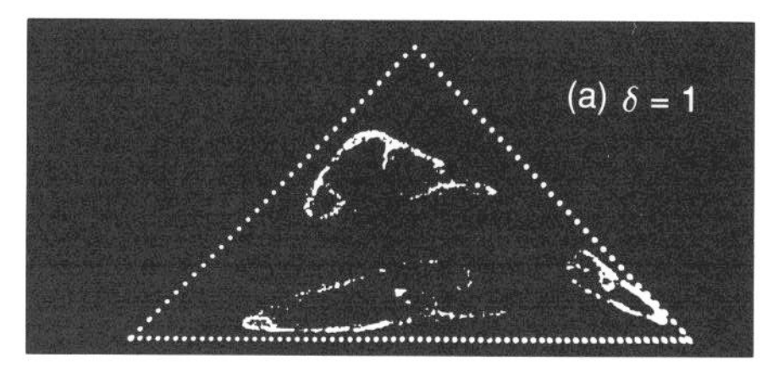


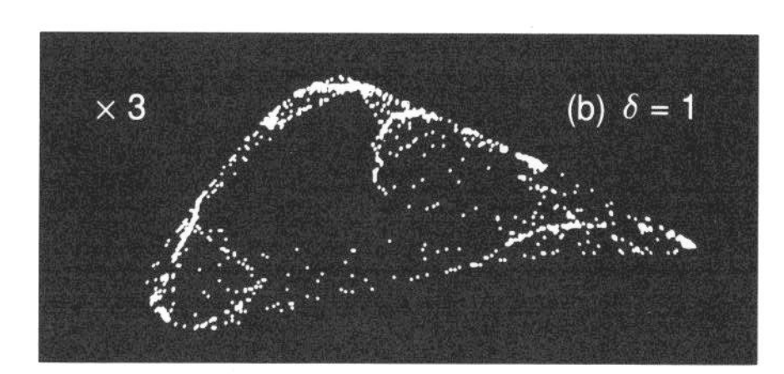


$T: x_1' = x_2^3 + 3x_1x_2^2 + 3x_2x_1^2 + 3x_2x_3^2 + 3x_3x_2^2,\ x_2' = 3x_1x_3^2 + 3x_3x_1^2 + 6x_1x_2x_3,\ x_3' = x_1^3 + x_3^3$

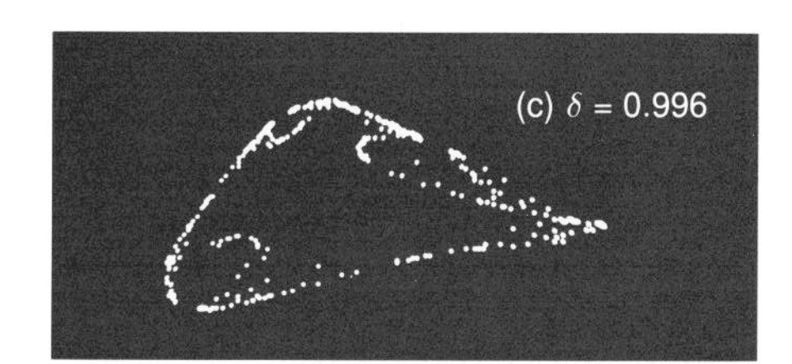


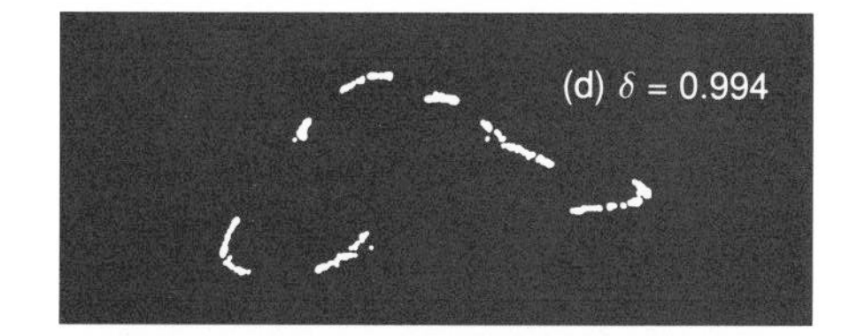


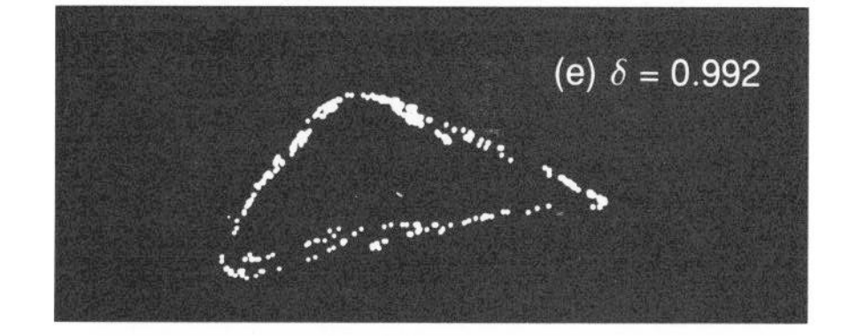


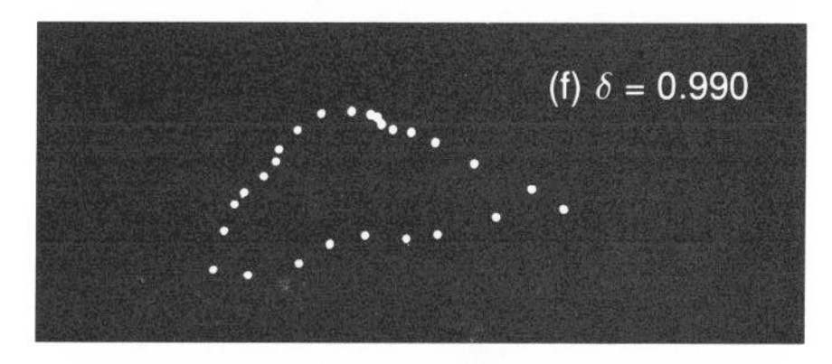


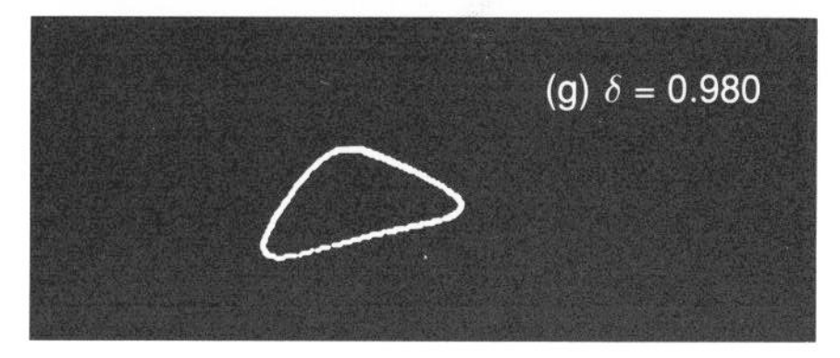


**ANOTHER EXAMPLE OF THE EFFECT OF $\delta$ ON A MESSY LIMIT SET**

**Fig. 10. An even more striking example of the effect of varying $\delta$ on a messy limit set. (a) The limit set for the original two-dimensional cubic map ($\delta$ = 1) consists of three separate pieces. Photographs (c) through (g) focus on the changes that occur in the piece shown in greater detail in (b); similar changes occur in the other two pieces. As $\delta$ is decreased monotonically from unity, the limit points (c) consolidate, (d) form a set of disjoint arcs, (e) disperse, (f) collapse to a periodic limit set of order 26, and (g) form a closed curve that eventually collapses to a single point.**

set of high order (1500 or thereabout). We had reasons for thinking that these results were spurious, caused by the limited precision of the machine, and that what we were seeing were artifacts. Indeed, when we iterated the two simpler maps

$$x' = 4x(1-x), \quad 0 \leq x \leq 1 \tag{13a}$$

and

$$x' = \sin \pi x, \quad 0 \leq x \leq 1, \tag{13b}$$

we also found high-order periods. For these maps, however, it was easy to prove that no such limit sets could exist, so our suspicions were confirmed. A year or two later the IBM 7030 ("Stretch") became available. With its larger word size, it failed to reproduce our impossible periods.

In 1970 Nick Metropolis and Myron Stein joined me in an attempt to find out what was really going on in all these one-dimensional examples. Of course, we could not resist generalizing the problem slightly by introducing a parameter $\lambda$, essentially the height of the map in a plot of $x'$ versus $x$. For instance, instead of Eqs. 13a and 13b we wrote

$$x' = \lambda x(1-x), \quad 0 \leq x \leq 1 \text{ and } 3 < \lambda < 4 \tag{14a}$$

and

$$x' = \lambda \sin \pi x, \quad 0 \leq x \leq 1 \text{ and } \approx 0.71 < \lambda < 1. \tag{14b}$$

The restrictions on $\lambda$ insure that the iterates of the maps lie within the specified $x$ interval and that the nonzero first-order fixed points of the maps are nonattractive. (Equation 14a, the "parameterized parabola," is well known in ecology as the logistic equation. It is a transform of a quadratic map studied in the early sixties by the Finnish mathematician P. Myrberg. Had we been aware of his study, considerable time would have been saved.)

Equations 14a and 14b are examples of maps of the general form

$$T_\lambda(x) : x' = \lambda f(x),$$

where $f(x)$ is defined on the interval [0,1] and has a single maximum (at which $dx'/dx = 0$). For simplicity we placed the maximum at $x = \frac{1}{2}$ and at first restricted ourselves to functions symmetric about that point. This restriction does not affect the results presented in the "MSS" paper (a name due to Derrida, Gervois, and Pomeau). We also required $f(x)$ to be strictly concave; relaxing this requirement can have drastic effects, as we learned later.

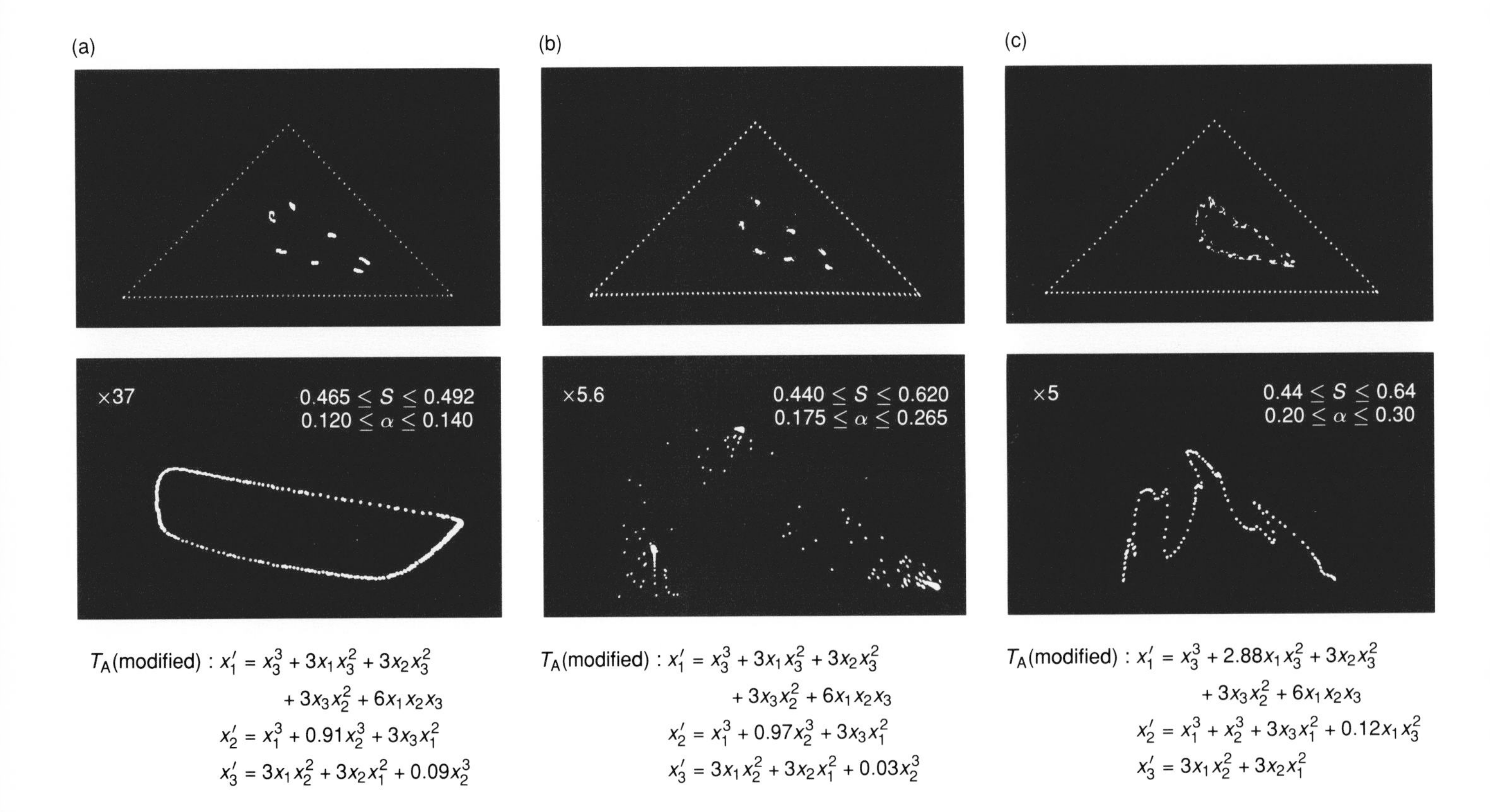


$T_A$(modified) : $x_1' = x_3^3 + 3x_1x_3^2 + 3x_2x_3^2 + 3x_3x_2^2 + 6x_1x_2x_3$
$x_2' = x_1^3 + 0.91x_2^3 + 3x_3x_1^2$
$x_3' = 3x_1x_2^2 + 3x_2x_1^2 + 0.09x_2^3$

$T_A$(modified) : $x_1' = x_3^3 + 3x_1x_3^2 + 3x_2x_3^2 + 3x_3x_2^2 + 6x_1x_2x_3$
$x_2' = x_1^3 + 0.97x_2^3 + 3x_3x_1^2$
$x_3' = 3x_1x_2^2 + 3x_2x_1^2 + 0.03x_2^3$

$T_A$(modified) : $x_1' = x_3^3 + 2.88x_1x_3^2 + 3x_2x_3^2 + 3x_3x_2^2 + 6x_1x_2x_3$
$x_2' = x_1^3 + x_2^3 + 3x_3x_1^2 + 0.12x_1x_3^2$
$x_3' = 3x_1x_2^2 + 3x_2x_1^2$

## EFFECT OF COEFFICIENTS ON THE LIMIT SET OF $T_A$

**Fig. 11. Our two-dimensional cubic maps can also be generalized by varying the coefficients. In the examples above the transformation $T_A$ was modified by varying only two coefficients. The photographs show the dramatic effect of such modifications on the messy limit set of $T_A$ (Fig. 8). The modification given in (a) changes the mess into a seven-member set of closed curves, one of which is shown in detail. The very similar modification given in (b) changes the mess into a pseudo- period of order 7, that is, into seven distinct bunches of points, three of which are shown in detail. The modification given in (c) results in a remarkably different but still messy limit set.**

In addition to the parabola and the sine, we also studied two other functions satisfying the conditions given above. One, a sixth-degree polynomial, was the transform to the unit interval of one of the one-dimensional cubic maps mentioned previously; the other was a trapezoid (in the $x',x$ plane).

For all four maps we calculated the periodic limit sets of order $k$ that begin and end with $x = \frac{1}{2}$. These correspond to $\lambda$ values that are solutions of

$$T_\lambda^{(k)}\left(\frac{1}{2}\right) = \frac{1}{2}$$

and are necessarily attractive because of the condition that $dx'/dx|_{x=\frac{1}{2}} = 0$. (This condition guarantees what is referred to as superstability.) To characterize the limit sets in a function-independent way, we used the minimum distinguishing information, namely, the positions of the successive iterates relative to $x = \frac{1}{2}$. For this purpose we employed the letters $R$ and $L$ ("right" and "left"). For example, when $k = 5$, all our maps have three distinct periodic limit sets of order 5, each associated with a different value of $\lambda$. Naturally, for different functions the $\lambda$ values are different, as are the actual values of the iterates, but the $R, L$ (or MSS) patterns are identical. The three patterns for $k = 5$, in order of increasing $\lambda$, are $\frac{1}{2} \to R \to L \to R \to R \to \frac{1}{2}$, $\frac{1}{2} \to R \to L \to L \to R \to \frac{1}{2}$, and $\frac{1}{2} \to R \to L \to L \to L \to \frac{1}{2}$. Omitting the initial and final $\frac{1}{2}$'s, we may write these patterns in simplified form as $RLR^2$, $RL^2R$, and $RL^3$.

The identity of the MSS patterns and their ordering on $\lambda$ was found to hold among all of our four functions for all values of $k$ such that $2 \leq k \leq 15$. We immediately noted the phenomenon that Feigenbaum later called period doubling. As an example, consider the period of order 2, the pattern of which is $R$. The patterns of its first two doublings are $RLR$ ($k = 4$) and $RLR^3LR$ ($k = 8$). A simple rule relates the pattern $P$ of a given period and that of its doubling: if $P$ contains an odd (even) number of $R$'s, the pattern of its doubling is $PLP$ ($PRP$). Note that $P$ must be an MSS pattern; that is, it must begin and end at the $x$ value for which $x'$ is maximum. (Obviously, not every $R, L$ succession is such a pattern.)

Period doublings are, of course, ordered on increasing $\lambda$. The $\lambda$ values corresponding to two successive doublings, $\lambda_1$ and $\lambda_2$, are "contiguous" in the sense that no $\lambda$ between $\lambda_1$ and $\lambda_2$ corresponds to a periodic limit set beginning at $\frac{1}{2}$.

Our initial work indicated that a large class of maps generates the same sequence of patterns ordered on increasing $\lambda$. Later experiments on some fifty additional maps confirmed this conclusion. It is still not known exactly, however, how this "large class" (almost certainly infinite) should be defined.

One of the most interesting results presented in the MSS paper is an algorithm for generating the MSS sequence. No iterations are needed, and no functions are explicitly specified. The algorithm is purely logical; given a limiting value $k_{\max}$ for the period order, it produces all MSS patterns with $k \leq k_{\max}$ in the canonical ordering (that is, on increasing $\lambda$). An independent proof of this algorithm is given for trapezoidal maps in Louck and Metropolis 1986. Others have found new algorithms for generating the MSS sequence, but, in my opinion, none of these are substantially simpler than ours.

Since the publication of these results, many mathematicians and physicists have studied one-dimensional maps, but much more work has been done on Feigenbaum's "quantitative" universality than on the "structural" universality represented by the MSS sequence. A few years ago Bill Beyer, Dan Mauldin (of North Texas State University), and I initiated new attacks on some of the problems suggested by MSS. We also considered a few new questions. One of these has to do with maps that exhibit a multiple appearance of some MSS patterns. If a map is strictly concave, it is our conjecture that each pattern occurs for just one value of $\lambda$. We found that something else can happen otherwise. Consider the "indented trapezoid" map shown in Fig. 12, which is not strictly concave. For certain ranges of the parameters $b$ and $c$, the same MSS pattern corresponds to three different $\lambda$ values. (This phenomenon implies that Feigenbaum's quantitative universality, which hinges on the occurrence of period doublings at unique $\lambda$ values, is not applicable to certain maps and hence is less than truly universal.)

Our multiplicity, as we called it, is more than an interesting mathematical fact. It has helped in understanding the latest results of an extensive study of the Belousov-Zhabotinskii reaction by H. L. Swinney and his collaborators. (The B-Z reaction, the oxidation of malonic acid by an acidic bromate solution in the presence of a cerous ion catalyst, is an oscillating chemical system, that is, a system in which the concentrations of the chemical species do not vary monotonically with time but instead oscillate, sometimes chaotically, sometimes periodically.) In 1982 Simoyi, Wolf, and Swinney had identified certain members of the MSS sequence in the periodic concentration variations of the bromide ion, one of some thirty chemical entities involved in the reaction. In addition they found that the MSS patterns observed were ordered on a parameter $\tau$ (the residence time of the reactants in the reaction vessel, which is inversely proportional to their rate of flow through the vessel) in exactly the same manner as the patterns in the MSS sequence are ordered on $\lambda$. Several years later

## THE "INDENTED TRAPEZOID" MAP

**Fig. 12. Because it is not strictly concave, the "indented trapezoid" map exhibits "multiplicity"; that is, it does not exhibit a one-to-one correspondence between MSS patterns (see text) and $\lambda$ values. In particular, for certain ranges of the parameters $b$ and $c$, some of the patterns correspond to three values of $\lambda$.**

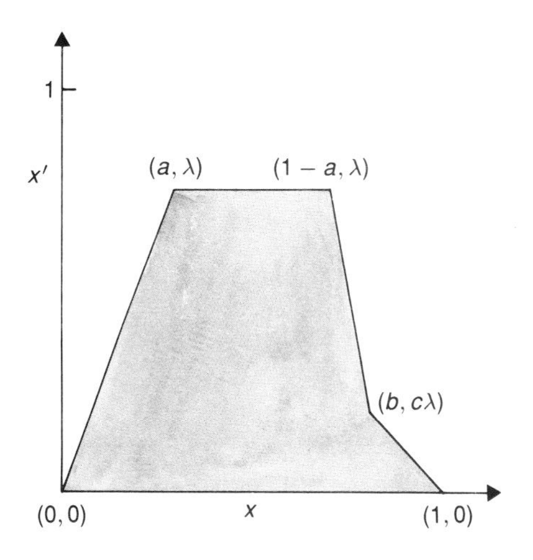


$$\frac{1}{2} < b < 1$$
$$a + b \neq 1$$
$$c > \frac{1-b}{a}, \quad \frac{1}{2} < a + b < 1$$
$$0 < c < \frac{1-b}{a}, \quad 1 < a + b < \frac{3}{2}$$

$$\begin{aligned} T: x' &= \frac{\lambda}{a}x, \quad 0 \leq x \leq a \\ &= \lambda, \quad a \leq x \leq (1-a) \\ &= \lambda\frac{(c-1)x + c(a-1)}{a+b-1}, \\ &\quad (1-a) \leq x \leq b \\ &= c\lambda\frac{1-x}{1-b}, \quad a \leq x \leq 1 \end{aligned}$$

Coffman, McCormick, and Swinney made further measurements on the system, this time controlling the flow rate much more precisely. Again they found members of the MSS sequence, but some of the patterns occurred for three values of $\tau$. At that time they knew nothing of our recent work on the indented trapezoid and suspected that their strange results were due to some systematic error. How they came to learn of multiplicity was a matter of pure chance. Swinney was visiting North Texas State (where Mauldin teaches mathematics) to give a talk. His hosts, looking for some way to amuse him between lunch and the colloquium, brought him to Mauldin's office. To pass the time, Dan started to discuss our discovery. Swinney immediately realized that what he and his colleagues had seen was not, after all, an artifact. They went on to identify the analogue of the indentation parameter $c$ as trace impurities in one of the reactants. It is certainly gratifying when some purely mathematical construct helps to explain physical reality.

## Number Theory

Stan Ulam's name seems to have disappeared from these pages; it is time to bring it back, if only briefly. Stan was not a number theorist, but he knew many number-theoretical facts, some of them quite recondite. As all who knew him will remember, it was Stan's particular pleasure to pose difficult, though simply stated, questions in many branches of mathematics. Number theory is a field particularly vulnerable to the "Ulam treatment," and Stan proposed more than his share of hard questions; not being a professional in the field, he was under no obligation to answer them.

Stan was very much interested in "sieve" methods—the sieve of Eratosthenes to generate the primes is the most famous—but from an experimental rather than an analytic viewpoint. He was always trying to invent new sieves that would generate sequences of numbers that were in some sense prime-like. His greatest success was the "lucky number" sieve (the name is derived from a story in Josephus's *History of the Jewish War*). In Eratosthenes's sieve one crosses out 1 from a list of the integers and then, keeping 2 (the first prime), crosses out all of its other multiples. The first survivor after 2 is 3, so next one crosses out all of its higher multiples, and so on. In the lucky number sieve one first crosses out every second number, that is, *all* the evens; in fact, one throws them out of the list, which is consequently collapsed. The first survivor after 1 is 3, so, again starting from the beginning, one throws out every third number, collapsing the list further. The next survivor is 7, so one then throws out every seventh number, and so on. The first ten lucky numbers are 1, 3, 7, 9, 13, 15, 21, 25, 31, and 33. All lucky numbers less than 3,750,000 were known by the early sixties. (Compared to the sieve for the primes, the lucky number sieve is rather slow.) Perhaps progress has been made, but I doubt that the range has been increased by a factor of 100 to match our current knowledge of the primes.

Although the lucky numbers are clearly not a multiplicative basis for the integers, they do have some prime-like properties. For example, their asymptotic distribution is, to first order, the same as that of the primes (Hawkins and Briggs). The luckies are, however, somewhat sparser than the primes, as, if I am not mistaken, Stan predicted. (Expressions for $p_n$, the $n$th prime, and $\ell_n$, the $n$th lucky number are

$$p_n = n \ln n + n \ln(\ln n) + \text{higher order terms}$$

and $$\ell_n = n \ln n + \frac{1}{2} n \left(\ln(\ln n)\right)^2 + \text{higher order terms.}$$

Of course these expressions make sense only for large $n$. Nevertheless, if we (recklessly) disregard the higher order terms, they imply that $\ell_n > p_n$ for $n > 1619$. In fact, however, we find that $\ell_n > p_n$ for $11 \leq n \leq 3,750,000$.) The distribution of the lucky numbers is similar to that of the primes in another respect: there seem to be an infinite number of lucky "twins," that is, luckies whose difference is 2. The evidence for this is far from overwhelming because the lucky sieve is hard to implement on a computer.

What I learned from Stan's ventures into number theory was that amateurs can make useful contributions to the field. That moved me to launch an attack on my favorite classical problem, the Goldbach conjecture. This is the statement, made by Christian Goldbach in a letter to his friend Euler, that every even integer equal to or greater than 6 is the sum of two odd prime numbers in at least one way. It remains unproven to this day, although very few mathematicians have doubts about its truth. Curiously, the analogous problem for odd integers, namely, that from some point on, each is the sum of three odd primes, was proved by Vinogradov in 1937. His original proof is long and difficult; it may have been at least a decade before its correctness was generally admitted. As for the Goldbach conjecture, the best result to date is that all sufficiently large even integers can be expressed as the sum of a prime and an integer that has at most two prime factors. This result, due to J.-R. Chen, is considered to be the greatest triumph ever achieved by sieve methods. That the Goldbach "property" is true for lucky numbers was conjectured by Stan, and work by Myron Stein and me gives some support. Stan's conjecture should not be too surprising in view of a 1970 study by Everett and me, which shows that almost all sequences with overall prime-like distributions have both the twin property and the Goldbach property. (Here "almost all" is to be understood in a measure-theoretic sense.)

In the mid sixties Myron Stein and I decided to look at the Goldbach problem numerically. We started by examining the so-called Goldbach curve, that is, the plot versus the even numbers of the total number of ways of expressing each as the sum of two primes. The curve is rather bumpy, usually peaking locally at multiples of 6 (as explained in the introduction to Stein and Stein 1964). Clearly many more primes exist than are necessary to constitute an additive 2-basis for the even numbers. This motivated us to look for sparse subsets of the primes possessing that property ("S" bases). We found a good algorithm (the S algorithm) for producing such subsets; each is completely determined by the choice of a smallest prime $p_0$. Our S bases cover almost all the evens from $2p_0$ to 10,000,000, leaving uncovered only a few low evens at the start. The sparseness achieved is striking; with one exception the S bases consist of less than 1.6 percent of the primes less than 10,000,000. The exception is the basis beginning with 7, which contains roughly twice as many primes as any other (a fact still unexplained). Our conclusion from this work is that the Goldbach property does not critically involve the famous prime property of being a multiplicative basis for the integers.

In concluding I must mention that the above investigation of the Goldbach problem moved the number theorist Daniel Shanks to convey on those involved the title "Los Alamos School of Experimental Number Theory." As to this new institution, there is no doubt that Stan Ulam was the founder. ■

**Paul R. Stein** has been a staff member at the Laboratory since 1950, working on problems that range from mathematical biology to nonlinear transformations and experimental number theory.

## Further Reading

M. T. Menzel, P. R. Stein, and S. M Ulam. 1959. Quadratic transformations: Part I. Los Alamos Scientific Laboratory report LA–2305.

P. R. Stein and S. M. Ulam. 1964. Non-linear transformation studies on electronic computers. *Rozprawy Matematyczne* 39. Also in *Stanislaw Ulam: Sets, Numbers, and Universes*, edited by W. A. Beyer, J. Mycielski, and G.-C. Rota. Cambridge, Massachusetts: The MIT Press, 1974.

M. L. Stein and P. R. Stein. 1964. Tables of the number of binary decompositions of all even numbers $0 < 2n < 200,000$ into prime numbers and lucky numbers. Los Alamos Scientific Laboratory report LA–3106, volumes 1 and 2.

M. L. Stein and P. R. Stein. 1965. Experimental results on additive 2-bases. *Mathematics of Computation* 19: 427–434.

C. J. Everett and P. R. Stein. 1969. On random sequences of integers. Los Alamos Scientific Laboratory report LA–4268.

C. J. Everett and P. R. Stein. 1970. On random sequences of integers. *Bulletin of the American Mathematical Society* 76: 349–350.

R. G. Schrandt and S. M. Ulam. 1967. On recursively defined geometrical objects and patterns of growth. Los Alamos Scientific Laboratory report LA–3762. Also in *Essays on Cellular Automata*, edited by Arthur W. Burks. Urbana: University of Illinois Press, 1970.

N. Metropolis, M. L. Stein, and P. R. Stein. 1973. On finite limit sets for transformations on the unit interval. *Journal of Combinatorial Theory* 15: 25–44.

Mitchell J. Feigenbaum. 1980. Universal behavior in nonlinear systems. *Los Alamos Science* 1(1): 4–27.

Reuben H. Simoyi, Alan Wolf, and Harry L. Swinney. 1982. One-dimensional dynamics in a multicomponent chemical reaction. *Physical Review Letters* 49: 245-248.

William A. Beyer, Peter H. Sellers, and Michael S. Waterman. 1985. Stanislaw M. Ulam's contributions to theoretical biology. *Letters in Mathematical Physics* 10: 231–242.

W. A. Beyer, R. D. Mauldin, and P. R. Stein. 1986. Shift-maximal sequences in function iteration: Existence, uniqueness, and multiplicity. *Journal of Mathematical Analysis and Applications* 113: 305–362.

K. Coffman, W. D. McCormick, and Harry L. Swinney. 1986. Multiplicity in a chemical reaction with one-dimensional dynamics. *Physical Review Letters* 56: 999–1002.

J. D. Louck and N. Metropolis. 1986. *Symbolic Dynamics of Trapezoidal Maps*. Dordrecht, Holland: D. Reidel Publishing Co.

# *Learning from* Ulam

## •Measurable Cardinals •Ergodicity •Biomathematics

***by Jan Mycielski***

Stanislaw M. Ulam was one of a group of distinguished Polish mathematicians who immigrated to the United States around 1939. Other members of the group who come to my mind were Natan Aronszajn, Stefan Bergman, Samuel Eilenberg, Mark Kac, Otto Nikodym, Alfred Tarski, and Antoni Zygmund. Ulam was born in 1909 in Lwów, which was then within the boundaries of Poland. The society in which he grew up and was educated was almost completely obliterated during World War II; the surviving population was dispersed within the present boundaries of Poland. Lwów is now a Ukrainian city, and only its buildings remain to remind one that for several centuries it was an outpost of Polish culture.

Lwów was the birthplace of functional analysis, and Ulam, although he cannot be called a functional analyst, had strong connections to the group of mathematicians (Banach, Mazur, Steinhaus, and Schauder) who created it. Stan had a superb memory, and he beautifully described those times in his book *Adventures of a Mathematician*, in an article on the Scottish Café in the 1981 edition of *The Scottish Book*, and in an obituary for Stefan Banach. He received an excellent classical education at the Sixth Gymnasium in Lwów and throughout his life could quote Greek and Roman poetry. Encouraged by his parents to study "something useful," he went to the Lwów Polytechnic School to become an engineer, but he quickly became interested by a lively group of mathematicians and began to spend more and more time with them.

One of the people who attracted him to mathematics was the famous topologist K. Kuratowski. Ulam told me about the following "little invention" he made while attending Kuratowski's course in calculus. Over a non-negative, decreasing function on the positive portion of the real line, build stairs of equal depth (Fig. 1). The problem is to prove that the shaded area (the sum of the areas between the function and the steps) is finite. Ulam said: "Shift the shaded pieces horizontally to the left until they all find themselves within the first column." Since the area of the first column is finite, the sum of the shaded pieces is also finite. Kuratowski was very happy to hear this original thought from his student. Perhaps that was the pebble that turned Ulam into a mathematician.

At the age of twenty, Ulam published his first paper: "A Remark on the Gener-

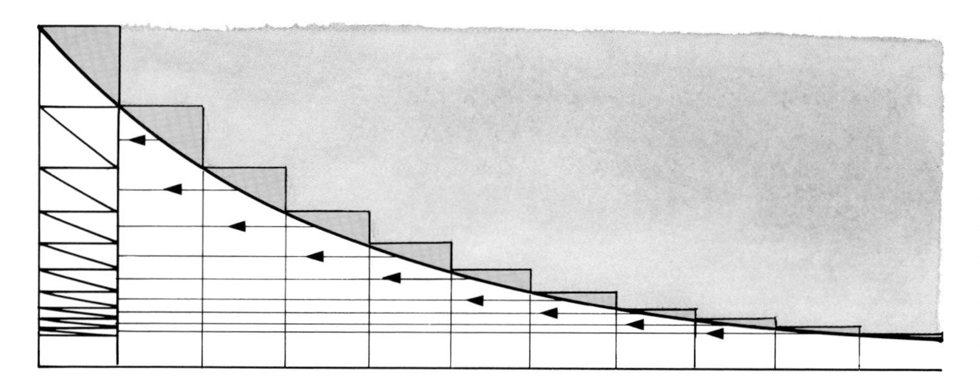

**A "LITTLE INVENTION"**

**Fig. 1. K. Kuratowski posed the following problem to his calculus class: Over a non-negative, decreasing function on the positive part of the real line, construct a step function with steps of equal depth. Prove that the area of the shaded regions between the two functions is finite. Ulam's solution was to move each shaded region into the first column, the area of which is finite.**

alized Bernstein's Theorem." That short paper solves a problem posed by Kuratowski. It belongs to the theory of jigsaw puzzles (also called the theory of equivalence by finite decomposition) and is one of the earliest applications of graphs in set theory. It appears in the 1974 volume *Stanislaw Ulam: Sets, Numbers, and Universes*, which contains more than half of Ulam's hundred or so then-published papers. We can learn a lot from that volume. I will try to describe some of what I have learned, but first let me record some memories from our numerous conversations over the years.

Ulam liked to consider amusing objects and processes. It didn't matter to him whether or not they were real or imaginary, but they had to be intrinsically interesting, not just tools. Consequently most of his work has a directness similar to the directness of an observation of nature. That distinguishes his work from the majority of mathematical papers, which elaborate existing theories. In fact, in his later life he became quite critical of such mathematical investigations, which he regarded as too abstruse or unimaginative. He would even remark that the study of specific subjects, such as advanced chapters of algebra, algebraic topology, or analysis, was motivated by the history of mathematics rather than by the interest or notoriety of their problems. I would reply that mathematics is also an art, motivated by its internal beauty, and that only persistent study may reveal that beauty. He would agree only that his opinion was not easy to interpret correctly. In the end I am sure that there is wisdom in what he said, if only because he discovered several facts that are fundamental to modern mathematical culture, and I can hardly imagine discoveries of that nature in the areas he was criticizing.

## Measurable Cardinals

I will now try to tell you about one of Ulam's important discoveries. It pertains to the foundations of mathematics and to the theory of large cardinal numbers. To give it the proper perspective, let me recall that Euclid was the first to organize the mathematics of his time into an axiomatic theory. That means he started from certain basic principles called axioms that he accepted without proof, and from them he obtained by pure deduction all the mathematical knowledge of his time. The system of Euclid became the accepted definition of mathematics until the time of Newton and Leibnitz. After the discovery of calculus, it became apparent that the development of mathematics within the system of Euclid is very unwieldy, and the system had to be abandoned. For a few centuries mathematics was in a sense unruly. Axiomatic organization returned to it around the turn of this century with the discoveries of Frege, Cantor, and Zermelo. Frege developed logic, Cantor invented and developed set theory, and Zermelo gave axioms for Cantor's set theory. Soon it became clear that all modern mathematics can be smoothly developed within set theory. Gradually it also became apparent that there is a whole hierarchy of larger and larger set theories, and one of the best ways to classify them is to see how large are the infinite cardinal numbers that can be shown to exist in those theories. (By a famous definition of Cantor, two sets $A$ and $B$, finite or infinite, have the same cardinal number if and only if there exists a one-to-one function mapping $A$ onto $B$). One might think that very large cardinal numbers are rather exotic and abstract objects whose existence is not of great mathematical interest. But by a famous theorem Gödel proved in 1931 (the so-called second incompleteness theorem), it follows that the larger the cardinal number whose existence can be proven in a given set theory, the more theorems can be proved in that theory, even theorems pertaining to such elementary operations as the addition and multiplication of integers. This fact was not yet known at the time Ulam made his discovery. His motivation was different—he was attracted by the mystery of the very large cardinal numbers for its own sake. Let me try to explain his theorem.

The smallest infinite cardinal number is called $\aleph_0$ (aleph zero). It is the cardinality of the set of integers. Clearly $\aleph_0$

has the following property: If we multiply less than $\aleph_0$ cardinal numbers each of which is less than $\aleph_0$, then the product is also less than $\aleph_0$. Well, of course, this tells only that the product of finitely many integers is finite. Thus we can say that $\aleph_0$ is inaccessible by products. This property is also called *strong inaccessibility*: $\aleph_0$ is strongly inaccessible. Are there any cardinal numbers larger than $\aleph_0$ (such cardinals are called uncountable) that are also strongly inaccessible? It turns out that this problem cannot be solved. The axioms of set theory do not imply the existence of such cardinals, and one can only postulate their existence as an axiom, which is what Felix Hausdorff did. Indeed, a set theory in which we accept this axiom is stronger (in the sense that it gives rise to more theorems of arithmetic) than the original set theory of Cantor and Zermelo.

To explain the work of Ulam we need the concept of a measure. For a set of points on the plane, area is a measure, and for a set of points in three-dimensional space, volume is a measure. In general given any set $X$, a measure is a function $\mu$ that attaches to subsets of $X$ some non-negative numbers in such a way that the following condition is satisfied:

$C_0$: If $A$ and $B$ are disjoint subsets of $X$, then $\mu(A \cup B) = \mu(A) + \mu(B)$.

There are many variants of the concept of measure. The version that is the most important for mathematical analysis says that $\mu(A)$ must be defined for all subsets $A$ of $X$ that are in a collection $\Omega$ of subsets such that $\Omega$ is closed under countable unions and complementations. That is, if $A_i \in \Omega$ for $i = 1, 2, \ldots$, then the union of the $A_i$'s is in $\Omega$, $\bigcup_{i=1}^{\infty} A_i \in \Omega$, and if $A_i \in \Omega$, then its complement, or $X - A_i$, is in $\Omega$. Moreover the measure $\mu$ must be countably additive; that is, if $A_i \in \Omega$, then $\mu\left(\bigcup_{i=1}^{\infty} A_i\right) = \sum_{i=1}^{\infty} \mu(A_i)$, provided the $A_i$'s are disjoint.

Ulam considered special measures that satisfy some additional conditions:

$C_1$: $\mu(A) = 0$ whenever $A$ consists of just one element of $X$;

$C_2$: $\mu(X) = 1$ (that is, the measure of the whole space is 1); and

$C_3$: $\Omega$ is the set of *all* subsets of $X$.

Measures that satisfy $C_0$, $C_1$, $C_2$, and $C_3$ are called *universal* measures. Readers familiar with Lebesgue's measure may recall that it is not a universal measure since the collection $\Omega$ on which it is defined is not the set of *all* subsets of $[0,1]$. On the other hand, Lebesgue's measure is invariant under translations, whereas the set $X$ is just an abstract set without any transformations upon which $\mu$ could be assumed to be invariant. Even in this abstract setting it is very difficult to construct a universal measure. For example, if $X$ is countable, no such measure exists since condition $C_1$ plus countable additivity forces $\mu(X)$ to equal 0, contrary to $C_2$.

Ulam proved two fundamental results about universal measures. The first tells that no universal measure exists for many uncountable sets. In particular, for many consecutive cardinals larger than $\aleph_0$ (for example, $\aleph_1, \aleph_2, \ldots, \aleph_\omega, \aleph_{\omega+1}, \ldots$), sets of those cardinalities do not have universal measures.

To explain Ulam's second result, we restrict the concept of universal measure still further by adding the following condition:

$C_4$: $\mu(A) = 0$ or $\mu(A) = 1$ for all subsets $A$ of $X$.

The cardinal number of a set that has a countably additive measure satisfying $C_1$, $C_2$, $C_3$, and $C_4$ is called an Ulam cardinal. Again, we can ask whether any Ulam cardinals exist. Ulam's famous theorem is that if such a cardinal does exist, then it is strongly inaccessible. This result implies that if we consider two set theories, one in which we assume the existence of uncountable strongly inaccessible cardinals and the second in which we assume the existence of Ulam cardinals, then the second theory is at least as strong as the first. Today we know many interesting theorems that follow from postulating the existence of Ulam cardinals. In particular, thirty years after Ulam's paper on measurable cardinals, William Hanf and Alfred Tarski proved that the least uncountable strongly inaccessible cardinal is smaller than the least Ulam cardinal. Thus a set theory in which Ulam cardinals exist is strictly stronger than one in which only uncountable strongly inaccessible cardinals exist. Many more results of this sort have been discovered since. The theory of large cardinals has become very rich, but Ulam's paper remains one of its keystones.

## Ergodic Theory and Topology

Between 1929 and 1938 Ulam published about twenty papers. What distinguishes those from the papers of other members of the Polish school before 1939 was his interest in topological groups, especially the groups of homeomorphisms of spheres.

A homeomorphism of a space $X$ is a transformation of $X$ onto itself that is one-to-one and continuous and whose inverse is also continuous. Of course such transformations constitute a group under composition. It is not obvious how to introduce a natural topology or even metrization into such a group. The following formula was often proposed (for example, it appears in Banach's classic book *Théorie des Operations Linéaires*): The distance between two homeomorphisms $f$

and $g$ of a compact space $X$, $\mathrm{Dist}(f,g)$, is given by

$$\mathrm{Dist}(f,g) = \max_{x\in X} \mathrm{dist}\big(f(x), g(x)\big) + \max_{x\in X} \mathrm{dist}\big(f^{-1}(x), g^{-1}(x)\big),$$

where dist denotes the distance in $X$.

The surprising property of this formula is that it converts the space of homeomorphisms of $X$ into a complete metric space. In other words, if a sequence of homeomorphisms satisfies the condition of Cauchy, then it has a limit that is a homeomorphism. The fact that the space of homeomorphisms can be treated as a complete metric space is very important because for such spaces there exist very natural definitions of largeness or smallness of subsets. The small ones are called meager (or of the first category) and the large ones comeager (or complements of meager). These topological concepts were invented by Baire. Several brilliant applications of these notions were made by Banach and Mazur. A very famous one was made by John C. Oxtoby and Ulam around 1941. Let me try to describe it here.

Take a glass of water, gently stir its contents, and let the water stop moving. Each particle of water has an initial and final position. The operation has thus defined a transformation of the interior of the glass into itself. Since water is viscous, this transformation is continuous and its inverse is also continuous. So we have here a homeomorphism. Moreover, since water is incompressible, the homeomorphism is volume-preserving. Homeomorphisms with that property constitute a complete subspace of the space of all homeomorphisms. If our transformation had been a simple rotation, then the altitudes of the particles of water and their distances from the central axis of the glass would not have changed. Many parts of the water would have remained invariant; that is, such parts would have been mapped into themselves. Even if we had applied the rotation many times, the water would never have been mixed. Are there any volume-preserving homeomorphisms that do mix? Such transformations, which are called ergodic, or metrically transitive, must exist if the ergodic hypothesis of statistical mechanics is correct. However, the existence of such transformations had remained an open question since the work of Poincaré and G. D. Birkhoff. Oxtoby and Ulam, in their paper entitled "Measure-Preserving Homeomorphisms and Metrical Transitivity," showed not only that such homeomorphisms exist but also that the set of ergodic homeomorphisms is comeager, that is, large in the sense of category. More precisely, any homeomorphism of that comeager set has the property that its application to any proper part of our glass of water deflects its boundary (Fig. 2). Thus the homeomorphism mixes the water in the sense that no part returns to its initial position. The Oxtoby-Ulam theorem remains one of the high points of the mathematical theory concerning ergodic properties of dynamical systems. The introduction to their paper, excerpted on the following page, explains the connection to the ergodic hypothesis. (These excerpts may be better understood after reading "The Ergodic Hypothesis: A Complicated Problem of Mathematics and Physics," as well as the section entitled Problem 2. Geometry, Invariant Measures, and Dynamical Systems in the article "Probability and Nonlinear Systems," all in this issue.)

We must caution, however, that application of mathematical theorems to the real world is sometimes a delicate problem. As you know, a sequence of heads and tails obtained by consecutive tosses of a fair coin has the property that the frequency of heads converges to 1/2 as the number of tosses becomes large. One can say (and prove in precise mathematical terms) that if we choose a sequence at random from the space of all such sequences, then, with probability 1, the limiting frequency of heads in this sequence is 1/2. Unfortunately, in another sense, namely that of category, almost all sequences (namely a comeager set) do *not* have any limiting value for the frequency of heads! So the very sense in which almost all volume-preserving homeomor-

**EFFECT OF AN ERGODIC TRANSFORMATION**

Fig. 2. If $h$ is an ergodic transformation, every surface $S$ separating the water is deflected by $h$ from its original position.

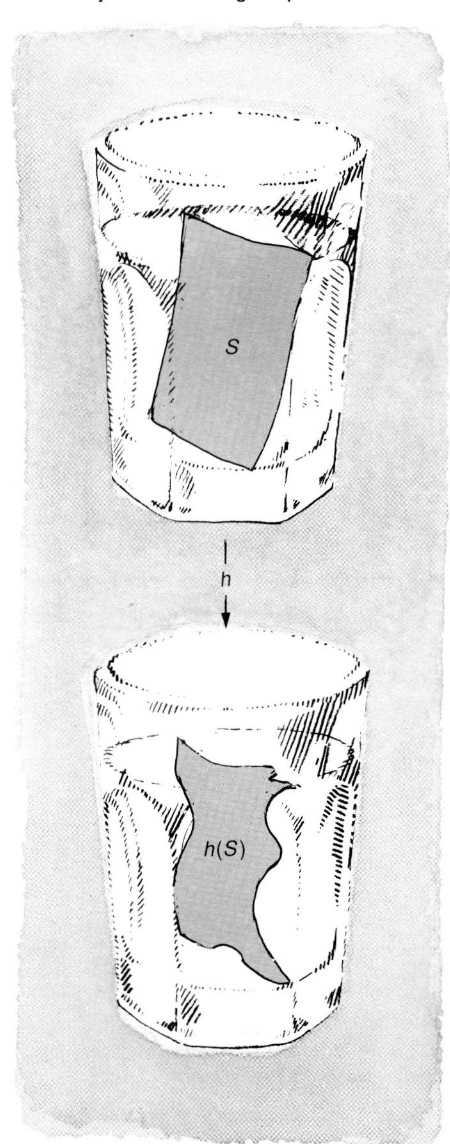

phisms of a cube are ergodic suggests the physically false result that almost all sequences of heads and tails lack a well-defined frequency of heads. Can we then trust the theorem of Ulam and Oxtoby as an expression of the truth of the ergodic hypothesis in physics? Stan and I often discussed this question. We thought that the answer is yes, but what is really needed is a new theorem in which almost all, in the sense of category, is replaced by some other more reliable sense. (I have outlined an idea of such a new theorem or conjecture in two papers in *Journal of Symbolic Logic*, one in volume 46 (1981) and the other in volume 51 (1986), but I do not know how to prove it.

Ulam and Jósef Schreier obtained another interesting result about the group of homeomorphisms of a spherical surface. They proved that there exist two special homeomorphisms such that every homeomorphism in the group can be approximated with arbitrary accuracy (relative to the distance defined above) by appropriate iterative compositions of those two homeomorphisms and their inverses.

Kuratowski and Ulam proved an extension of the theorem of Fubini to the context of Baire's category that is often very useful.

An interesting feature of Ulam's work followed from his great ability to collaborate with others. Almost all of his papers are co-authored with other mathematicians or physicists. He had many ideas, and he was very successful in stirring the imagination and enthusiasm of others. His most important collaborators were Jósef Schreier, John C. Oxtoby, and C. J. Everett. He invented a large number of original problems, some of which were solved by other mathematicians and even became famous theorems. One such conjecture was proved by K. Borsuk and is known today as the theorem on antipodes (the two points at the opposite ends of a diameter of a sphere are called antipodes of each other). It is sometimes called the ham-and-cheese sandwich theorem. It tells the following: For every continuous mapping of the spherical surface into the plane, there exist antipodes that are mapped into the same point on the plane. This theorem is equivalent to the following statement. Given three bodies (say ham, cheese, and bread), one can find a single plane that divides each body into two parts of equal volume. (Each body may consist of disjoint pieces, as does the bread in a sandwich, and the bodies may overlap, as shown in Fig. 3a.) Another equivalent statement is that at any time antipodal points can be found on the earth where the temperature and the barometric pressure are the same (Fig. 3b).

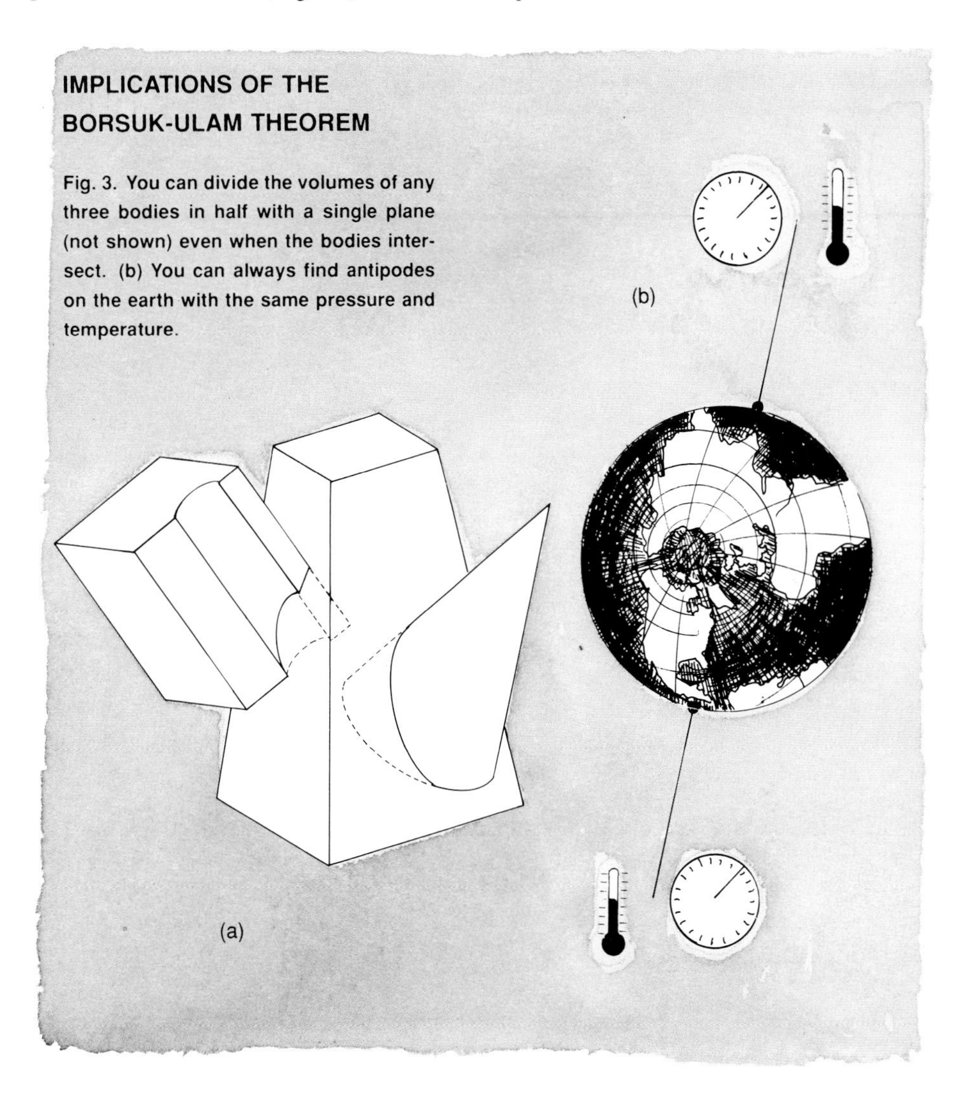


**IMPLICATIONS OF THE BORSUK-ULAM THEOREM**

Fig. 3. You can divide the volumes of any three bodies in half with a single plane (not shown) even when the bodies intersect. (b) You can always find antipodes on the earth with the same pressure and temperature.

## Topics in Biology and Some Applications of Computers

I began to collaborate with Stan Ulam in 1969 when he invited me to the University of Colorado in Boulder. We spoke frequently about the problems of the organization and function of the human brain and the structure of memory. He presented his ideas on this subject in the talk "Reflections on the Brain's Attempts to Understand Itself," which is posthumously published in this issue.

We also talked often about the problem of accumulation of mutations in a given species. As a result of our discussions I

# The Existence and Significance of Ergodic Transformations

## *Excerpts from the Introduction to Oxtoby and Ulam's "Measure-Preserving Homeomorphisms and Metrical Transitivity"*

In the study of dynamical systems one is led naturally to the consideration of measure-preserving transformations. A Hamiltonian system of $2n$ differential equations induces in the phase space of the system a measure-preserving flow, that is, a one-parameter group of transformations that leave invariant the $2n$-dimensional measure. ... If the differential equations are sufficiently regular the flow will have corresponding properties of continuity and differentiability. Thus the study of one-parameter continuous groups of measure-preserving automorphisms of finite dimensional spaces has an immediate bearing on dynamics and the theory of differential equations.

In statistical mechanics one is especially interested in time-averaging properties of a system. In the classical theory the assumption was made that the average time spent in any region of phase space is proportional to the volume of the region in terms of the invariant measure, more generally, that time-averages may be replaced by space-averages. To justify this interchange, a number of hypotheses were proposed, variously known as ergodic or quasi-ergodic hypotheses, but a rigorous discussion of the precise conditions under which the interchange is permissible was only made possible in 1931 by the ergodic theorem of Birkhoff. This established the *existence* of the time-averages in question, for almost all initial conditions, and showed that if we neglect sets of measure zero, the interchange of time- and space-averages is permissible if and only if the flow in the phase space is *metrically transitive*. A transformation or a flow is said to be metrically transitive if there do not exist two disjoint invariant sets both having positive measure. Thus the effect of the ergodic theorem was to replace the ergodic hypothesis by the hypothesis of metrical transitivity.

Nevertheless, in spite of the simplification introduced by the ergodic theorem, the problem of deciding whether particular systems are metrically transitive or not has proved to be very difficult. ...

... The known examples of metrically transitive continuous flows are all in manifolds, indeed in manifolds of restricted topological type, either toruses or manifolds of direction elements over surfaces of negative curvature. An outstanding problem in ergodic theory has been the existence question—can a metrically transitive continuous flow exist in an arbitrary manifold, or in any space that is not a manifold? In the present paper we shall obtain a complete answer to this question, at least on the topological level, for polyhedra of dimension three or more. It will appear that the only condition that needs to be imposed is a trivially necessary kind of connectedness. In particular, there exists a metrically transitive continuous flow in the cube, in the solid torus, and in any pseudo-manifold of dimension at least three. Since the phase spaces of dynamical systems have the required kind of connectedness, it follows that the hypothesis of metrical transitivity in dynamics involves no *topological* contradiction. More precisely, in any phase space there can exist a continuous flow metrically transitive with respect to the invariant measure associated with the system.

It must be emphasized, however, that our investigation is on the topological level. The flows we construct are continuous groups of measure-preserving automorphisms, but not necessarily differentiable or derivable from differential equations. Thus they correspond to dynamical systems only in a generalized sense.

It may be recalled that the original ergodic hypothesis of Boltzmann—that a single streamline passes through all points of phase space—had to be abandoned because it involved a topological impossibility. It was replaced by a quasi-ergodic hypothesis—that some streamline passes arbitrarily close to all points of phase space. But it is not obvious that even this weak hypothesis is topologically reasonable in general phase spaces, and in any case it is not sufficient to justify the interchange of time- and space-averages. It is therefore of some interest to know that the ergodic hypothesis in its modern form of metrical transitivity is at least free from any objection on topological grounds.... ■

*Editor's note: Despite the existence of ergodic transformations, the ergodicity of many familiar dynamical systems remains an open and thought-provoking question. For details see Patrascioiu's article, "The Ergodic Hypothesis: A Complicated Problem of Mathematics and Physics."*

Footnotes have been omitted from these excerpts, which are reprinted with permission from *Annals of Mathematics*.

proposed to study the "genealogical distance" $d(a, b)$ between two individuals $a$ and $b$, which is defined as follows. Count the number of ancestors of $a$ that are not ancestors of $b$, and add to it the number of ancestors of $b$ that are not ancestors of $a$. Assume that the size of the population is constant in time, that mating is random, and that $a$ and $b$ belong to the same generation. Ulam soon discovered by experimenting on a computer that under those conditions the expected value of $d(a, b)$ is twice the size of the population. Later Joseph Kahane and Robert Marr proved this conjecture (*Journal of Combinatorial Theory*, Series A, volume 13 (1972)). The smallness of this expected distance suggests that all profitable mutations are soon present in all individuals of subsequent generations.

Ulam liked to invent problems that could be studied by means of electronic computers. He was the first to realize that computers are ideal tools for watching the evolution of patterns governed by simple laws. He proposed many experiments of this type, the most famous of which is reported in the paper of Fermi, Pasta, and Ulam on dynamical evolution governed by nonlinear laws. Later he invented various simple rules to produce crystal-like growths in space. He also observed simple cases of "wars" between growing populations of crystals or cells. Nowadays many such processes are being investigated; Conway's "game of life" is a popular example. It is hoped that this approach will help us to understand certain qualitative features of natural evolution. For example, one can replace the complicated rules of chemistry governing real life by simpler rules and, through numerical simulation, watch the ways in which the patterns (objects) yielded by these rules grow and compete in complicated and surprising ways. (In my own work I am trying to explain human thought and learning, which we so often discussed together, by applying local rules of interaction that may define interesting processes. It is already known that the computations going on in the cerebral cortex are local in some sense.)

I have tried to give you glimpses of certain works of Stan Ulam. Of course, in this short article I have discussed only those that seem to me the most important or with which I was the most familiar.

Every creative mathematician must allow his imagination to flow in a free way. I think that Ulam did this more than others. He was drawn to work upon problems that suggested essentially new ideas and avoided the attractive pull of well-developed mathematics. Few mathematicians have the intelligence or the courage that Ulam had to think about important problems irrespective of whether their solutions are in sight. But this is the only course that can lead to outstanding achievements. ■

## Further Reading

W. A. Beyer, J. Mycielski, and G.-C. Rota, editors. 1974. *Stanislaw Ulam: Sets, Numbers, and Universes*. Cambridge, Massachusetts: The MIT Press.

A collection of declassified Los Alamos technical reports by Ulam and his collaborators is to be published soon by University of California Press.

Frank R. Drake. 1974. *Set Theory: An Introduction to Large Cardinals*. Amsterdam: North-Holland Publishing Company.

Jan Mycielski. 1985. Can mathematics explain natural intelligence? Los Alamos National Laboratory report LA-10492-MS. Also in *Physica D* 22(1986): 366–375.

Jan Mycielski and S. Swierczkowski. A model of the neocortex. Los Alamos National Laboratory report. To be published.

**Jan Mycielski** was born in Poland in 1932. After receiving his Ph.D. from the University of Wroclaw, he worked at the Institute of Mathematics of the Polish Academy of Science. In 1969 he immigrated to the United States and obtained a professorship at the University of Colorado, a position he has retained to the present. He has held visiting positions at the Centre National de la Recherche Scientifique, Case Western Reserve University, and the University of California, Berkeley. His research has resulted in over one hundred papers pertaining to mathematical logic, set theory, game theory, geometry, algebra, and learning systems. He is a member of the American Mathematical Society, the Mathematical Association of America, the Association for Symbolic Logic, and the Polish Mathematical Society.

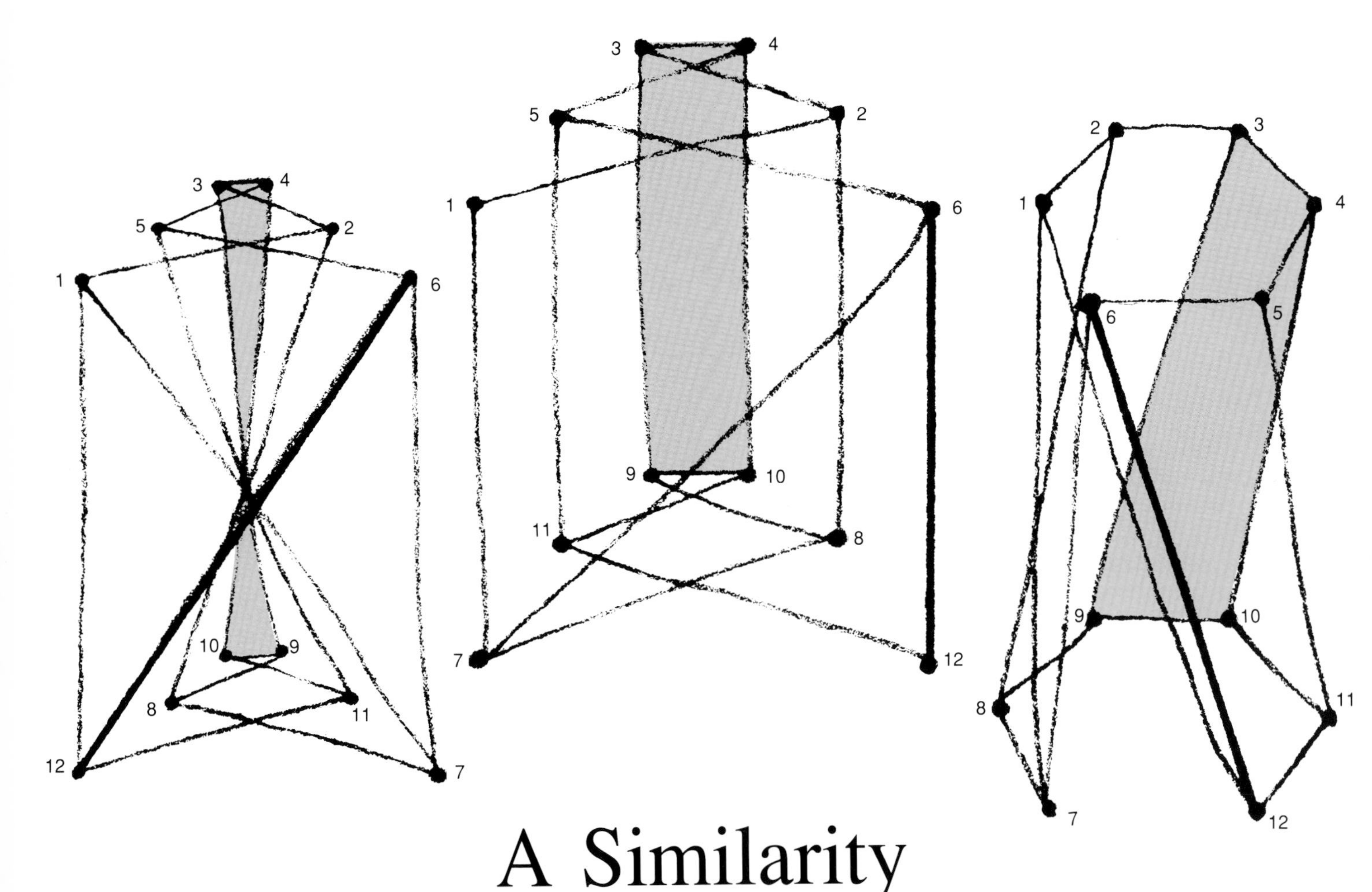


# A Similarity Measure for Graphs-

*reflections on a theme of Ulam*

**by Ronald L. Graham**

Among a variety of fundamental themes running through Stan Ulam's mathematical research, one that particularly intrigued him was that of *similarity*. He was constantly fascinated by the problem of quantifying exactly how *alike* (or *different*) two mathematical objects or structures were, and during his career he discovered many ingenious ways of doing so. A good example is the well-known Ulam distance between finite sequences, which has recently been applied so effectively in analysis of DNA sequences and recognition of speech (Sankoff and Kruskal 1983). (Also see "Sequence Analysis: Contributions by Ulam to Molecular Biology" in this issue.)

Here I will describe another measure of similarity suggested by Stan, one applicable to a wide assortment of combinatorial structures. Like many seeds planted by his fertile imagination, this similarity measure has taken root and flowered in the modern mathematical jungle.

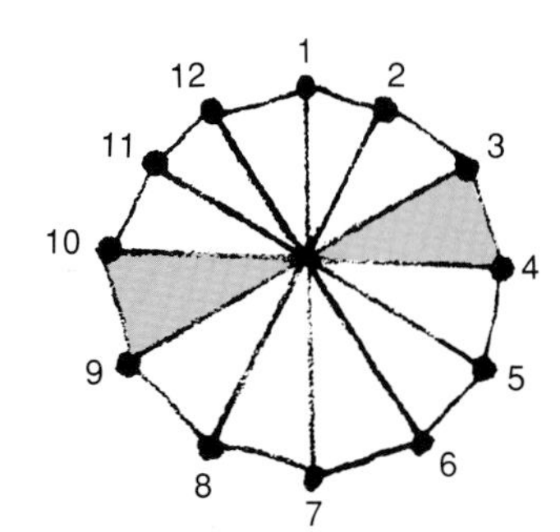


The story begins one morning in late July of 1977, during one of my aperiodic visits to Stan and Françoise's marvelous house on the outskirts of Santa Fe. Stan and I had just finished playing tennis, which not only generated a plentiful supply of perspiration (and consequent thirst) but also inevitably led to a lively discussion of the differences in the game at an altitude of over 7000 feet, where the balls are effectively more highly pressurized, the air resistance is diminished, less oxygen is available for demanding lungs, and so on.

Perhaps stimulated by trying to get a better grasp on understanding just how various aspects of the game (such as the

## EXAMPLES OF GRAPHS

**Fig. 1. Shown here are the pictorial and mathematical representations of two simple graphs. $G_1$ has the maximum number of edges for a given number $n$ of vertices (namely, $\binom{n}{2}$, or $\frac{1}{2}n(n-1)$). $G_2$ is an example of a graph consisting of disjoint subgraphs. (For simplicity in this and all other figures, the edges of graphs are depicted as straight lines.)**

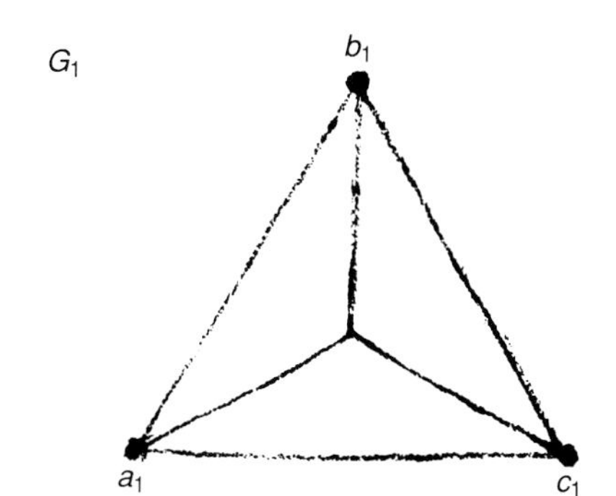


$V_1 = \{a_1, b_1, c_1, d_1\}$

$E_1 = \{(a_1, b_1), (a_1, c_1), (a_1, d_1), (b_1, c_1), (b_1, d_1), (c_1, d_1)\}$

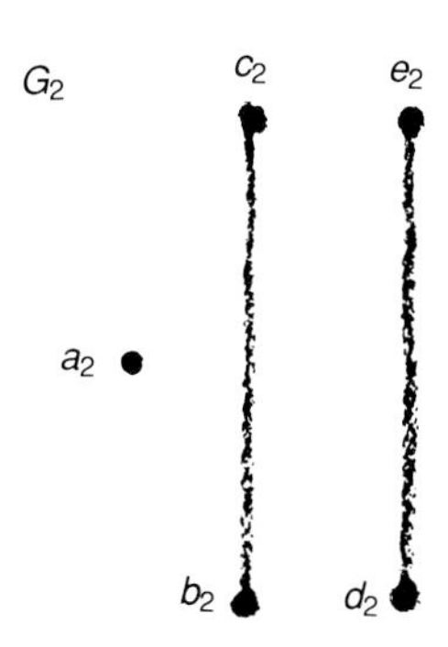


$V_2 = \{a_2, b_2, c_2, d_2, e_2\}$

$E_2 = \{(b_2, c_2), (d_2, e_2)\}$

## ISOMORPHISM OF GRAPHS

**Fig. 2. Let $T$ be the following one-to-one transformation of $V_1$ onto $V_2$: $a_1 \rightarrow a_2$, $b_1 \rightarrow d_2$, $c_1 \rightarrow b_2$, $d_1 \rightarrow e_2$, $e_1 \rightarrow c_2$, and $f_1 \rightarrow f_2$. Since $T(E_1) = E_2$, $G_1$ is isomorphic to $G_2$, or, symbolically, $G_1 \cong G_2$. Note that a necessary condition for isomorphism of two graphs is that they have the same number of vertices and edges. The relation of isomorphism, like that of equality in arithmetic, is reflexive, symmetric, and transitive. That is, $G \cong G$; if $G_1 \cong G_2$, then $G_2 \cong G_1$; and if $G_1 \cong G_2$ and $G_1 \cong G_3$, then $G_2 \cong G_3$.**

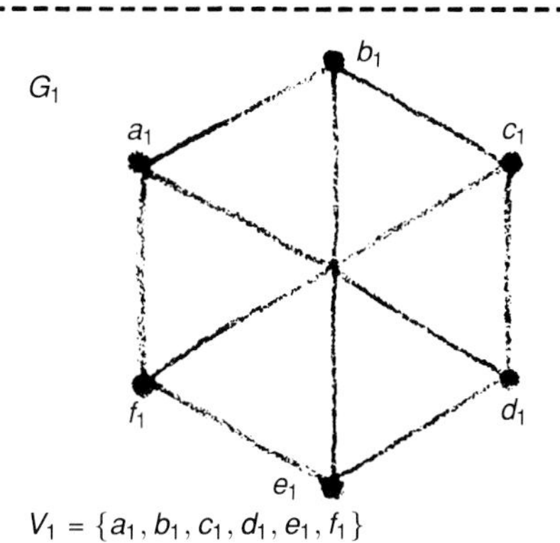


$V_1 = \{a_1, b_1, c_1, d_1, e_1, f_1\}$

$E_1 = \{(a_1, b_1), (a_1, d_1), (a_1, f_1), (b_1, c_1), (b_1, e_1), (c_1, d_1), (c_1, f_1), (d_1, e_1), (e_1, f_1)\}$

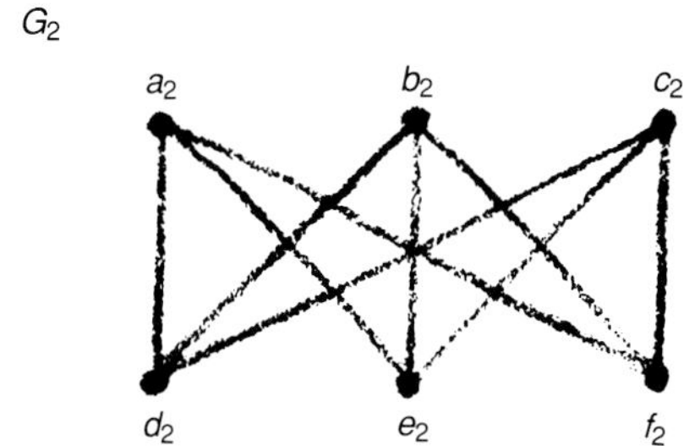


$V_2 = \{a_2, b_2, c_2, d_2, e_2, f_2\}$

$E_2 = \{(a_2, d_2), (a_2, e_2), (a_2, f_2), (b_2, d_2), (b_2, e_2), (b_2, f_2), (c_2, d_2), (c_2, e_2), (c_2, f_2)\}$

serve, the stroke, and the strategy) might change under varying conditions, Stan suddenly suggested, "Why not measure the difference between objects by trying to break them up into as few as possible pairwise equal pieces?" At first I didn't see quite what Stan was driving at (which happened fairly often), but after we talked it over, it became clear that here was an entirely new way of defining a measure of similarity between two (or more) combinatorial structures. In fact, it is very much akin to comparing two complex molecules by breaking them up into a number of pairwise identical fragments—the smaller the number of pieces needed, the more similar are the molecules.

Our first application of the approach was to a class of mathematical objects known as *graphs*. Simply speaking, a graph $G$ consists of a set $V$ of elements called the *vertices* of $G$ and a set $E$ of certain pairs of elements of $V$ called the *edges* of $G$. Graphs are often pictured by representing the vertices in $V$ as points and the edges as lines between the pairs of points in $E$ (Fig. 1).

Before proceeding to the main topic of the article, we need two more basic definitions—those for *isomorphism* of graphs and for a *partition* of the edge set of a graph.

Two graphs $G_1$ and $G_2$ are said to be isomorphic ($G_1 \cong G_2$) if, as shown in Fig. 2, a one-to-one transformation of $V_1$ onto $V_2$ effects a one-to-one transformation of $E_1$ onto $E_2$.

By a partition of the edge set $E$ of a graph $G$ is meant a set of pairwise disjoint subsets $E_i$ of $E$ such that $\bigcup_i E_i = E$ (Fig. 3). (The number of ways to partition an edge set depends, in a complicated way, on the number of edges of the graph, $e(G)$.)

We now come to the key definitions. Let $G$ and $G'$ be two graphs having the same number of edges. An *Ulam decomposition* of $G$ and $G'$ is a pair of par-

## PARTITIONS OF THE EDGE SET OF A GRAPH

**Fig. 3. The edge set $E$ of the graph $G$ can be partitioned (divided into subsets $E_i$ such that $\bigcup_i E_i = E$) in numerous ways, two of which are illustrated here.**

$E = \{(v_1, v_2), (v_1, v_3), (v_1, v_4), (v_2, v_3), (v_2, v_4), (v_3, v_4)\}$

Partition $P$

$E_1 = \{(v_1, v_2), (v_1, v_3), (v_2, v_3)\}$

$E_2 = \{(v_1, v_4), (v_2, v_4), (v_3, v_4)\}$

$E_1 \cup E_2 = E$

Partition $P'$

$E'_1 = \{(v_1, v_2)\}$ $E'_2 = \{(v_1, v_3)\}$ $E'_3 = \{(v_1, v_4)\}$

$E'_4 = \{(v_2, v_3)\}$ $E'_5 = \{(v_2, v_4)\}$ $E'_6 = \{(v_3, v_4)\}$

$E'_1 \cup E'_2 \cup E'_3 \cup E'_4 \cup E'_5 \cup E'_6 = E$

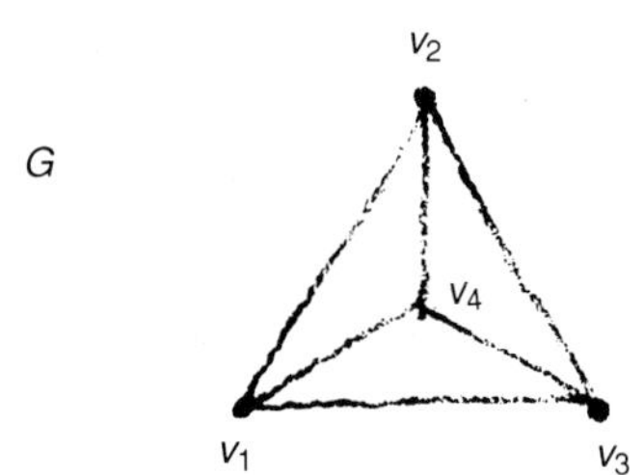


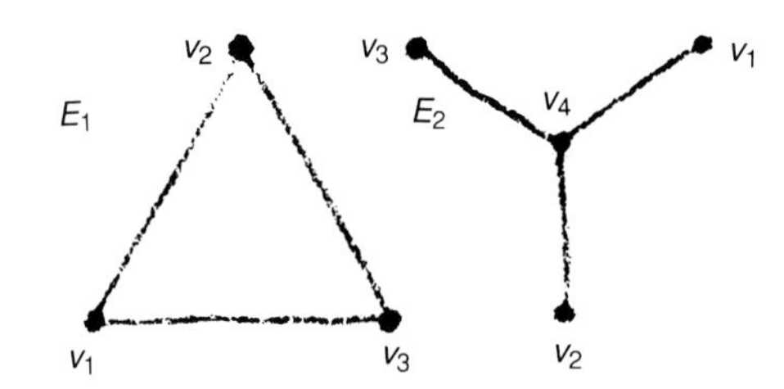


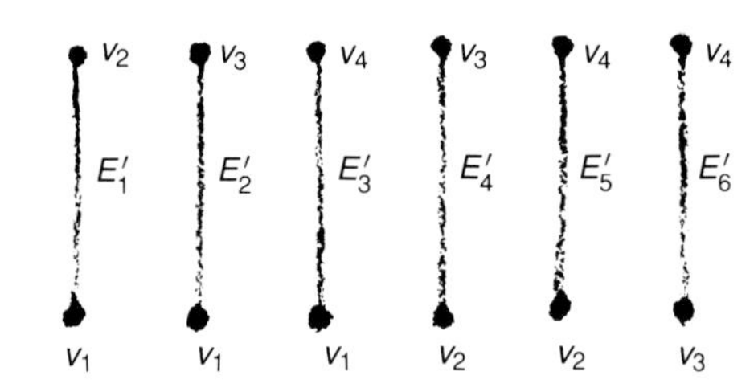


## EXAMPLES OF MINIMUM ULAM DECOMPOSITIONS

**Fig. 4. The minimum Ulam decompositions shown here illustrate that $U(G, G')$ is a measure of similarity for a pair of graphs that agrees with our intuitive notion of their resemblance to each other in the sense of connections among vertices: the two graphs in (a) bear less resemblance in that sense than do the two graphs in (b).**

(a)

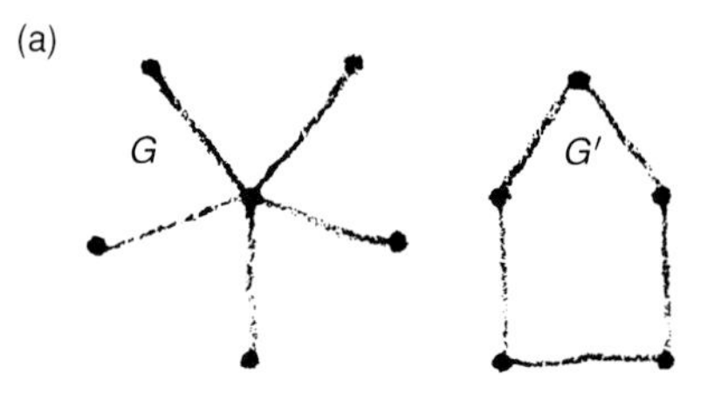


Minimum Ulam Decomposition of $G$ and $G'$

$U(G, G') = 3$

(b)

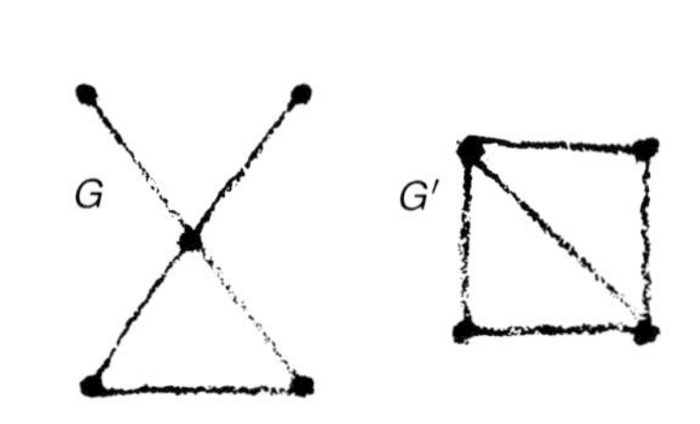


Minimum Ulam Decomposition of $G$ and $G'$

$U(G, G') = 2$

## EXAMPLES OF $\Gamma_{n,e}$

**Fig. 5. Two examples of $\Gamma_{n,e}$, the set of graphs each of which has $e$ edges and at most $n$ vertices. (Graphs with isolated vertices are not shown.) In both examples $n = 2e$, the maximum number of vertices for a given number of edges. $\Gamma_{n,e}$ for $n < 2e$ is a subset of $\Gamma_{2e,e}$.**

$\Gamma_{6,3}$

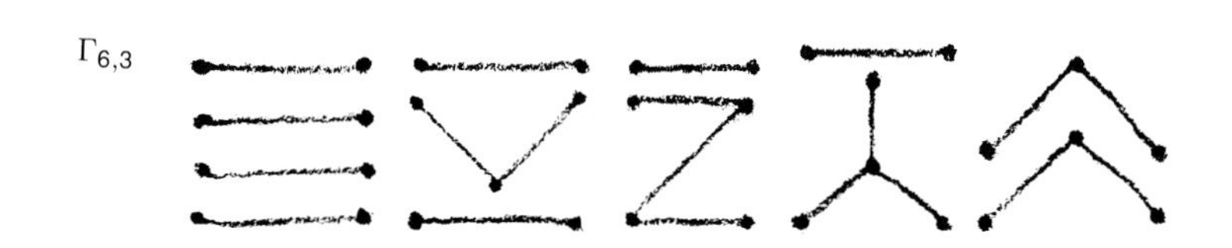

$\Gamma_{8,4}$

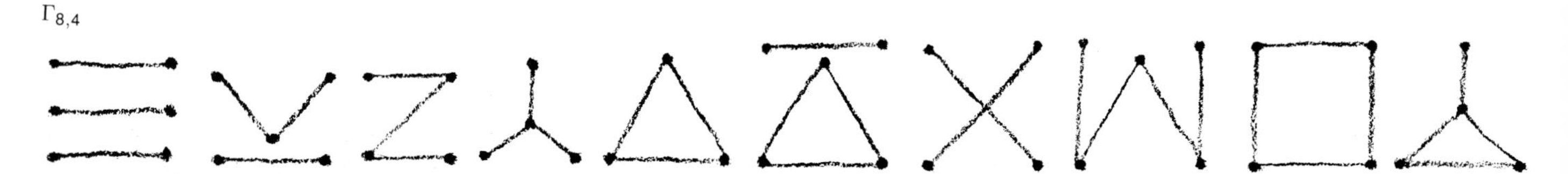

titions $\{E_1, \ldots, E_r\}$ and $\{E'_1, \ldots, E'_r\}$ of the respective edge sets of $G$ and $G'$ such that, as graphs, $E_i \cong E'_i$ for $1 \leq i \leq r$. Such a decomposition always exists since we can always choose each $E_i$ (and $E'_i$) to be a single edge (and by hypothesis $e(G) = e(G')$). Further, define $U(G, G')$ to be the *least* value of $r$ for which an Ulam decomposition of $G$ and $G'$ into $r$ parts exists. (Figure 4 shows such minimum Ulam decompositions for two pairs of graphs.) Thus $U(G, G')$ is a similarity measure for pairs of graphs: the smaller $U(G, G')$ is, the more alike $G$ and $G'$ are. In particular, $U(G, G') = 1$ exactly when $G$ and $G'$ are isomorphic (that is, when they differ only in the way their vertices are labeled).

We now extend our view from a single pair of graphs with the same number of edges to sets of graphs $\Gamma_{n,e}$ with $e$ edges and at most $n$ vertices (Fig. 5) and define the function $U(n)$:

$$U(n) \equiv \max_{G, G' \in \Gamma_{n,e}} U(G, G').$$

So in $U(n)$ we have a measure of the maximum dissimilarity among all pairs of graphs in $\Gamma_{n,e}$.

A fundamental question about $U(n)$ that occurs at the outset is this: How large can $U(n)$ ever be? To whet your appetite for the answer to this question, let us determine $U(G, G')$ for two examples. ($U(n)$ is of course equal to or greater than $U(G, G')$.) In the first example $G$ is a $k$-rayed "star" (that is, it consists of one vertex joined to each of $k$ others), and $G'$ consists of $k$ disjoint edges (Fig. 6a). Here $e = k$ and $n = 2k$. For such a pair of graphs, the only Ulam decomposition comes from taking each $E_i$ and $E'_i$ to be a single edge. Therefore

$$U(n) \geq U(G, G') = k = \frac{1}{2}n.$$

In the second, slightly more sophisticated example $G$ consists of a $3k$-rayed star, and $G'$ consists of $k$ disjoint triangles (Fig. 6b). Here $e = 3k$ and $n = 3k + 1$. What is $U(G, G')$ for such pairs of graphs? It is not difficult to see that the best we can do is to decompose each graph into $k$ disjoint 2-rayed stars and $k$ disjoint edges. Thus

$$U(n) \geq U(G, G') = 2k = \frac{2}{3}(n-1) \approx \frac{2}{3}n.$$

At this point one may well wonder whether further search will produce even more complicated examples from which even larger lower bounds on $U(n)$ can be deduced. That this *cannot* happen is the content of Theorem 1, which was the main result in our first paper on the subject (Chung, Erdös, Graham, Ulam,

**CALCULATION OF LOWER BOUNDS ON $U(n)$**

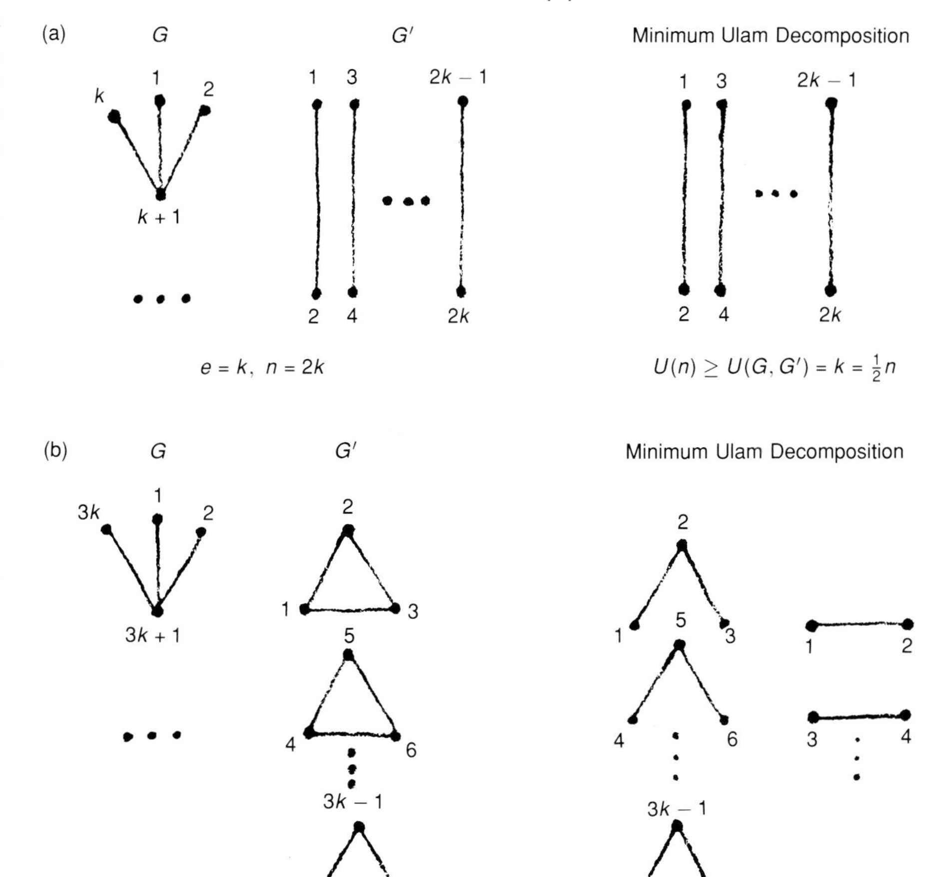


**Fig. 6. By considering pairs of graphs of the general form shown in (a), $U(n)$ is found to be equal to or greater than $\frac{1}{2}n$, whereas by considering pairs of graphs of the general form shown in (b), $U(n)$ is found to be greater than $\frac{2}{3}(n-1)$, which is greater than $\frac{1}{2}n$ (for $n > 4$). This situation might lead one to conclude that $U(n)$ has no upper bound other than $\binom{n}{2}$, the greatest possible number of edges possessed by any graph with $n$ vertices. However, as demonstrated in the text, $U(n)$ has an upper bound that is linear in $n$.**

and Yao 1979).

**Theorem 1**. For a suitable fixed constant $c$, $U(n) \leq \frac{2}{3}n + c$ for all $n$.

Our proof of Theorem 1 uses several ideas that are now standard items in the toolbox of every combinatorialist. One is the idea of a *greedy algorithm*. It seems only natural to try to remove the *largest* subgraph common to each of the two graphs for which one is seeking a minimum Ulam decomposition (although in many situations that myopic approach is far from optimal). Indeed, such a technique is quite effective for the problem at hand. However, it leads to the next question: Just how large can we expect (or guarantee) such a common subgraph to be? Here the second technique we want to mention comes in, namely, the so-called *probabilistic method*, which was pioneered so effectively by Paul Erdös. Suppose $G$ and $G'$ each have $n$ vertices and $e$ edges. What we will show is that they must share a common subgraph $H$ having at least $2e^2/n(n-1)$ edges. However, we won't be able to specify what $H$ is or how to get it—just that it exists! How do we do this? Every mathematical paper should have at least one proof, so here comes ours.

Label the vertices of $G$ and $G'$ by, say, $V = \{x_1, \ldots, x_n\}$ and $V' = \{x'_1, \ldots, x'_n\}$. Let $\Lambda$ denote the set of one-to-one mappings of $V$ onto $V'$. Thus, $\Lambda$ has $n!$ elements. If $y = \{x_i, x_j\}$ and $y' = \{x'_k, x'_l\}$ are given elements in $V$ and $V'$, respectively, there are exactly $2(n-2)!$ elements $\lambda \in \Lambda$ that map $y$ onto $y'$. (The factor of 2 counts the two possibilities $\lambda(x_i) = x'_k, \lambda(x_j) = x'_l$ and $\lambda(x_i) = x'_l, \lambda(x_j) = x'_k$.) Define the *indicator function* $i_\lambda(y, y')$:

$$i_\lambda(y, y') = \begin{cases} 1 & \text{if } \lambda \text{ maps } y \text{ onto } y'; \\ 0 & \text{otherwise.} \end{cases}$$

Now sum $i_\lambda(y, y')$ over all $\lambda \in \Lambda$ and all $y \in E, y' \in E'$:

$$S \equiv \sum_{\lambda \in \Lambda} \sum_{y,y'} i_\lambda(y, y')$$

$$= \sum_{y,y'} \sum_{\lambda \in \Lambda} i_\lambda(y, y')$$

$$= \sum_{y,y'} 2(n-2)!$$

$$= 2e^2(n-2)!.$$

In the first step we have interchanged the order of summation, and in the second we have used the previously noted fact about the number of $\lambda \in \Lambda$ that map any given $y \in E$ onto any given $y' \in E'$.

Now we note that since $S$ is a sum of $n!$ terms of the form $\sum_{y,y'} i_\lambda(y, y')$ (one for each $\lambda \in \Lambda$), at least one of those terms must equal or exceed their *average*, which of course is just $2e^2(n-2)!/n!$, or $2e^2/n(n-1)$. In other words, for *some* $\lambda \in \Lambda$, say $\lambda_0$, we have

$$\sum_{y,y'} i_{\lambda_0}(y, y') \geq \frac{2e^2}{n(n-1)}.$$

Having proved that the two graphs $G$ and $G'$ have a common subgraph $H$ with at least $2e^2/n(n-1)$ edges, suppose now that we remove $H$ from $G$ and $G'$, producing the graphs $G_1$ and $G'_1$, which have at most $e_1 = e - 2e^2/n(n-1)$ edges. Our theorem says that those graphs also have in common a subgraph $H_1$ with at least $2e_1^2/n(n-1)$ edges. Remove $H_1$ to produce $G_2$ and $G'_2$, and so on. It is not hard to show that after repeating the process $k$ times, we have graphs $G_k$ and $G'_k$ with at most $n(n-1)/2k$ edges. That in turn can be used to show that $U(n) < \sqrt{2}n$. To squeeze the last bit of juice out of the argument and show that $U(n) \leq \frac{2}{3}n + c$ requires more complicated considerations that we will not go into here.

It was only natural that we began to wonder next about what happens if instead of starting with two graphs, we start with *three* (or more). Indeed, defining $U(G_1, G_2, G_3)$ as the minimum value of $r$ for which an Ulam decomposition of $G_1$, $G_2$, and $G_3$ exists and $U_3(n)$ as $\max_{G_1,G_2,G_3 \in \Gamma_{n,e}} U(G_1, G_2, G_3)$, we soon saw that $U_3(n)$ was going to be larger than $U_2(n) \equiv U(n)$ by considering the example shown in Fig. 7. For those triplets of graphs, just as for the pairs of graphs shown in Fig. 6, the minimum Ulam decomposition consists of a certain number (approximately $\frac{1}{4}n$) of 2-rayed stars and a certain number (approximately $\frac{1}{2}n$) of disjoint edges. Thus

$$U_3(n) \geq U(G_1, G_2, G_3) \geq \frac{3}{4}n.$$

It has been shown (this time by much more complicated arguments) that $U_3(n)$, like $U_2(n)$, possesses an upper bound (Chung, Erdös, and Graham 1981):

**Theorem 2**. For a suitable fixed constant $c$, $U_3(n) \leq \frac{3}{4}n + c$ for all $n$.

As mathematicians are prone to do, we naturally began to look beyond $U_3(n)$ to the general case, namely

$$U_k(n) \equiv \max_{G_1,\ldots,G_k \in \Gamma_{n,e}} U(G_1, \ldots, G_k).$$

Here, however, something completely unexpected happened. We had been guessing what the coefficient of $n$ was going to be in the bound for $U_4(n)$ (why not $\frac{4}{5}$?) and more generally for $U_k(n)$ (could it be $\frac{k}{k+1}$?). We were quite unprepared for the following result, which was finally proved with the full arsenal of techniques we were rapidly accumulating (Chung, Erdös, and Graham 1981):

**Theorem 3**. For each $k \geq 3$, there is a fixed constant $c_k$ such that $U_k(n) \leq \frac{3}{4}n + c_k$ for all $n$.

In other words, the constant factor of $\frac{3}{4}$ that appears in the bound for $U_3(n)$ *does not increase* for values of $k$ greater than 3. It is as though the space of $n$-vertex graphs is in some sense "three-dimensional," and once you have three graphs that are maximally separated, then adding further graphs can cause no real additional trouble. In fact, the most striking result we were finally able to establish dealt with trying to decompose *all* graphs on $n$ vertices simultaneously into mutually isomorphic subgraphs. If $U^*(n)$ denotes the smallest number of subgraphs needed for such an Ulam decomposition, then we have the ultimate generalization of Theorem 1 (Chung, Erdös, and Graham 1983):

**Theorem 4.** For a suitable fixed constant $c^*$, $U^*(n) \leq \frac{3}{4}n + c^*$ for all $n$.

A key concept arising in these investigations is that of an *unavoidable* subgraph of a graph. To be precise, we say that $H$ is $(n, e)$-unavoidable if any graph with $n$ vertices and $e$ edges contains $H$ as a subgraph. For example, *any* $n$-vertex graph is $(n, \binom{n}{2})$-unavoidable (since there is only one graph with $n$ vertices and $\binom{n}{2}$ edges and that graph includes all possible edges). Also, a $d$-rayed star is $(n, \frac{1}{2}n(d-1)+1)$-unavoidable, where $n \geq d+1$ and $n$ must be even if $d$ is even (Fig. 8). Many beautiful results on unavoidable graphs have been proved in recent years; indeed, that subject is developing, primarily under the leadership of F. R. K. Chung, into a central area of graph theory.

We mention finally the concept of a universal graph, a concept related to that of an unavoidable graph and one motivated in part by the problem of finding Ulam decompositions. If $\mathbf{F}$ is a family of graphs, we say that a graph $G$ is $\mathbf{F}$-universal if every $F \in \mathbf{F}$ occurs as a subgraph of $G$. The connection between these two ideas is the following. Let $\bar{G}$ denote the complementary graph of the graph $G$; that is, $\bar{G}$ is a graph with the same vertices as $G$ and exactly (and only) the edges that $G$ does not have. Thus, for a graph with $n$ vertices,

$$e(G) + e(\bar{G}) = \binom{n}{2}.$$

Now if $\mathbf{F}(i, j)$ denotes the family of all graphs with $i$ vertices and $j$ edges, then the statement

$$H \text{ is } (n, e)\text{-unavoidable}$$

is equivalent to the statement

$$\bar{H} \text{ is } \mathbf{F}(n, \tbinom{n}{2} - e)\text{-universal.}$$

Figure 9 illustrates this equivalence.

Much is now known about $\mathbf{F}$-universal graphs for special classes of $\mathbf{F}$. For example, if $\mathbf{F} = \mathbf{T}_n$, the family of all *trees* (connected graphs containing no closed loops) with $n$ vertices, then the minimum possible number $s(\mathbf{T}_n)$ of edges in a $\mathbf{T}_n$-universal graph satisfies

$$\frac{1}{2}n \ln n < s(\mathbf{T}_n) < \frac{7}{\ln 4} n \ln n.$$

**Fig. 7. By considering the graphs $G_1$ (a $9k^2$-rayed star), $G_2$ ($3k^2$ disjoint triangles), and $G_3$ ($\frac{1}{2}3k(3k+1)$ disjoint edges and a $3k$-sided polygon with each vertex connected to every other), one can deduce that $U_3(n)$ is equal to or greater than about $\frac{3}{4}n$. (The graphs shown here illustrate the case $k = 2$.)**

## CALCULATION OF A LOWER BOUND ON $U_3(n)$

$G_1$

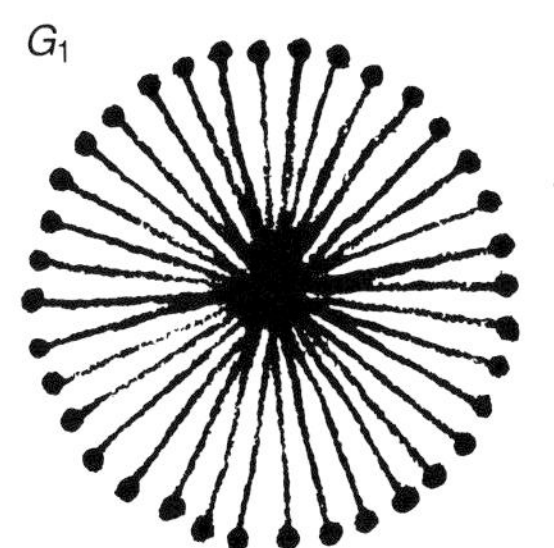

$e(G_1) = 9k^2$
$n(G_1) = 9k^2 + 1$

$G_2$

$e(G_2) = 9k^2$
$n(G_2) = 9k^2$

$e = 9k^2, \quad n = 3k(3k+2)$

$G_3$

$$e(G_3) = \frac{1}{2}3k(3k-1) + \frac{1}{2}3k(3k-1) = 9k^2$$

$$n(G_3) = 3k + 3k(3k+1) = 3k(3k+2)$$

Minimum Ulam Decomposition of $G_1$, $G_2$, and $G_3$

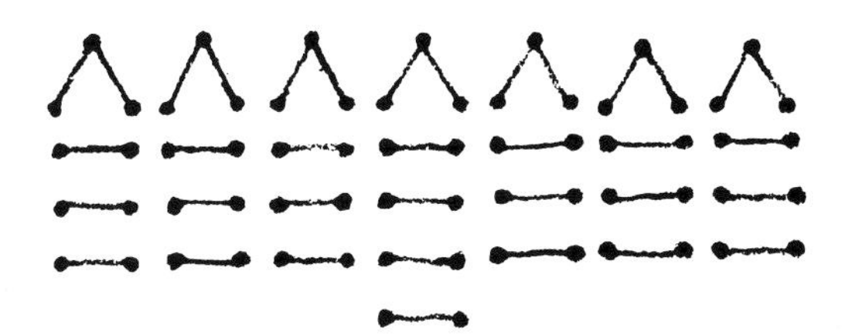

$$U(G_1, G_2, G_3) = \begin{cases} \frac{3}{4}k(3k-1) + \frac{3}{2}k(3k+1) = \frac{3}{4}(9k^2+k) & \text{if } k = 4j \text{ or } 4j-1 \\ \frac{3}{4}k(3k-1) - \frac{1}{2} + \frac{3}{2}k(3k+1) + 1 = \frac{3}{4}(9k^2+k) + \frac{1}{2} & \text{otherwise} \end{cases}$$

$$\approx \frac{3}{4}n$$

**A $d$-RAYED STAR IS $(n, \frac{1}{2}n(d-1))$-UNAVOIDABLE**

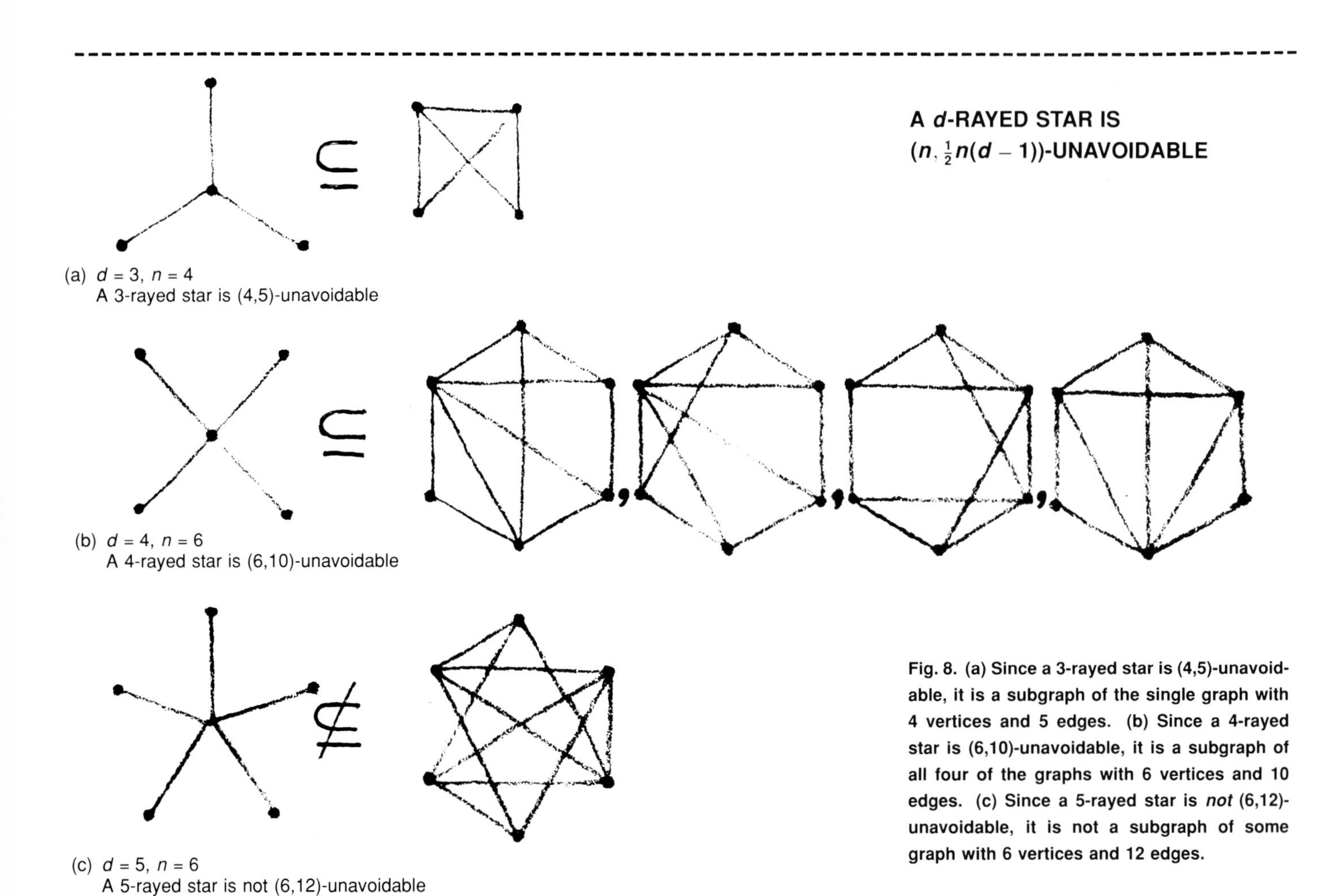


**Fig. 8. (a) Since a 3-rayed star is (4,5)-unavoidable, it is a subgraph of the single graph with 4 vertices and 5 edges. (b) Since a 4-rayed star is (6,10)-unavoidable, it is a subgraph of all four of the graphs with 6 vertices and 10 edges. (c) Since a 5-rayed star is *not* (6,12)-unavoidable, it is not a subgraph of some graph with 6 vertices and 12 edges.**

## UNAVOIDABLE AND UNIVERSAL GRAPHS

**Fig. 9. The graph $H$ and its complement $\bar{H}$ illustrate that $H$ is $(n, e)$-unavoidable if and only if $\bar{H}$ is $F(n, \binom{n}{2} - e)$-universal. $H$ is (6, 10)-unavoidable; that is, any graph with 6 vertices and 10 edges contains $H$ as a subgraph (see Fig. 8). Therefore $\bar{H}$ is F(6, 5)-universal; that is, any graph with 6 vertices and 5 edges (such as a 5-rayed star) is contained in $\bar{H}$.**

$H$

$H$ is 6, 10)-unavoidable

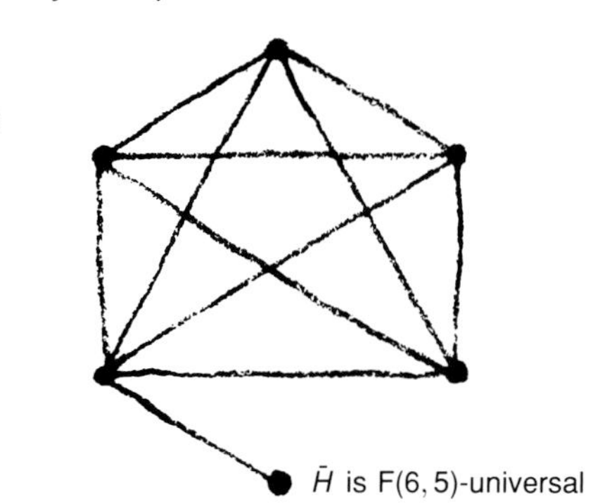


That result, and many other similar results (which have interesting applications to the design of VLSI chips, for example) can be found in Chung and Graham 1978, 1979, 1983; Chung, Graham, and Pippenger 1978; Bhatt and Leighton 1984; and Bhatt and Ipsen 1985. The basic idea here is that a silicon chip (or wafer) can have a universal graph $G$ for some class of graphs, say for all trees with twenty vertices. When a particular tree $T$ is needed for connecting various components on the chip, the appropriate edges of $G$ can be "activated" to realize $T$.

In the spirit of the current algorithmic trend in mathematics, we might ask how hard it is in practice to find the minimuml Ulam decomposition for two graphs $G$ and $G'$. In general that is almost surely a difficult computational problem. More precisely, the question "Is $U(G, G') = 2$?" has been shown (Yao 1979) to belong to the notorious class of *NP-complete* problems, an intensively studied collection of thousands of computational problems, including the well-known traveling salesman and graph coloring problems (see Garey and Johnson 1979). Computer scientists believe, although no one has yet been able to prove, that an NP-complete problem is inherently intractable as the general instance size of the problem increases. The resolution of that question remains as perhaps the outstanding problem in theoretical computer science.

It's interesting to note that the related question "Is $U(G, G') = 1$?" (or "Is $G \cong G'$?") is *not* known to belong to the class of NP-complete problems, and indeed, many people believe that an efficient algorithm does exist for its solution. A fuller treatment of such matters can be found in Garey and Johnson.

In conclusion, all of the preceding questions can also be asked about numerous other combinatorial and algebraic structures, such as directed graphs, hypergraphs, partially ordered sets, finite metric spaces, and so on. Some work on those topics can be found in Chung, Graham, and Erdös 1981; Chung, Graham, and Shearer 1981; Babai, Chung, Erdös, Graham, and Spencer 1982; Chung, Erdös, and Graham 1982; Chung 1983; and Chung and Erdös 1983. Clearly, however, that area of research remains mostly unexplored—and is one more example of the prolific mathematical legacy left to us by Stan Ulam. ■

**Ronald L. Graham** received his Ph.D. in mathematics from the University of California, Berkeley. He is currently Director of the Mathematical Sciences Research Center at AT&T Bell Laboratories and University Professor of Mathematical Sciences at Rutgers University. He is a member of the National Academy of Sciences and a Fellow of the American Academy of Arts and Sciences. He has held visiting faculty positions at the University of California, Stanford University, the California Institute of Technology, and Princeton University. His professional interests include combinatorics, graph theory, number theory, geometry, and various areas of theoretical computer science. He is the author of *Rudiments of Ramsey Theory* (American Mathematical Society, 1981) and the co-author (with Bruce L. Rothschild and Joel H. Spencer) of *Ramsey Theory* (John Wiley & Sons, 1980) and (with P. Erdös) of *Old and New Problems and Results in Combinatorial Number Theory* (L'Enseignement Mathématique, Université de Genève, 1980). He also likes to point out that he is a Past President of the International Jugglers Association.

## Further Reading


F. R. K. Chung and R. L. Graham. 1978. On graphs which contain all small trees. *Journal of Combinatorial Theory, Series B* 24: 14–23.

F. R. K. Chung, R. L. Graham, and N. Pippenger. 1978. On graphs which contain all small trees, II. In *Combinatorics*, edited by A. Hajnal and Vera T. Sós, 213–223. Amsterdam: North-Holland Publishing Company.

F. R. K. Chung, P. Erdös, R. L. Graham, S. M. Ulam, and F. F. Yao. 1979. Minimal decompositions of two graphs into pairwise isomorphic subgraphs. In *Proceedings of the Tenth Southeastern Conference on Combinatorics, Graph Theory, and Computing (Florida Atlantic University, Boca Raton, Florida, April 2–6, 1979)*, 3–18. Congressus Numerantium, volume 23. Winnipeg, Manitoba: Utilitas Mathematica Publishing Incorporated.

F. R. K. Chung and R. L. Graham. 1979. On universal graphs. *Annals of the New York Academy of Sciences* 319: 136–140.

Michael R. Garey and David S. Johnson. 1979. *Computers and Intractability: A Guide to the Theory of NP-Completeness.* New York: W. H. Freeman and Company.

F. Frances Yao. 1979. Graph 2-isomorphism is NP-complete. *Information Processing Letters* 9: 68–72.

F. R. K. Chung, P. Erdös, and R. L. Graham. 1981. Minimal decompositions of graphs into mutually isomorphic subgraphs. *Combinatorica* 1: 13–24.

F. R. K. Chung, R. L. Graham, and J. Shearer. 1981. Universal caterpillars. *Journal of Combinatorial Theory, Series B* 31: 348–355.

L. Babai, F. R. K. Chung, P. Erdös, R. L. Graham, and J. Spencer. 1982. On graphs which contain all sparse graphs. *Annals of Discrete Mathematics* 12: 21–26.

F. R. K. Chung, P. Erdös, and R. L. Graham. 1982. Minimal decompositions of hypergraphs into mutually isomorphic subhypergraphs. *Journal of Combinatorial Theory, Series A* 32: 241–251.

F. R. K. Chung. 1983. Unavoidable stars in 3-graphs. *Journal of Combinatorial Theory, Series A* 35: 252–262.

F. R. K. Chung and P. Erdös. 1983. On unavoidable graphs. *Combinatorica* 3: 167–176.

F. R. K. Chung and R. L. Graham. 1983. On universal graphs for spanning trees. *Journal of the London Mathematical Society, Second Series* 27: 203–211.

David Sankoff and Joseph B. Kruskal, editors. 1983. *Time Warps, String Edits, and Macromolecules: The Theory and Practice of Sequence Comparison.* Reading, Massachusetts: Addison-Wesley Publishing Company.

S. N. Bhatt and F. T. Leighton. 1984. A framework for solving VLSI graph layout problems. *Journal of Computer and System Sciences* 28: 300–343.

F. R. K. Chung, P. Erdös, and R. L. Graham. 1984. Minimal decomposition of all graphs with equinumerous vertices and edges into mutually isomorphic subgraphs. In *Finite and Infinite Sets*, edited by A. Hajnal, L. Lovász, and V. T. Sós, 171–179. Amsterdam: North-Holland Publishing Company.

S. N. Bhatt and I. Ipsen. 1985. Embedding trees in the hypercube. Yale University Research Report RR–443.

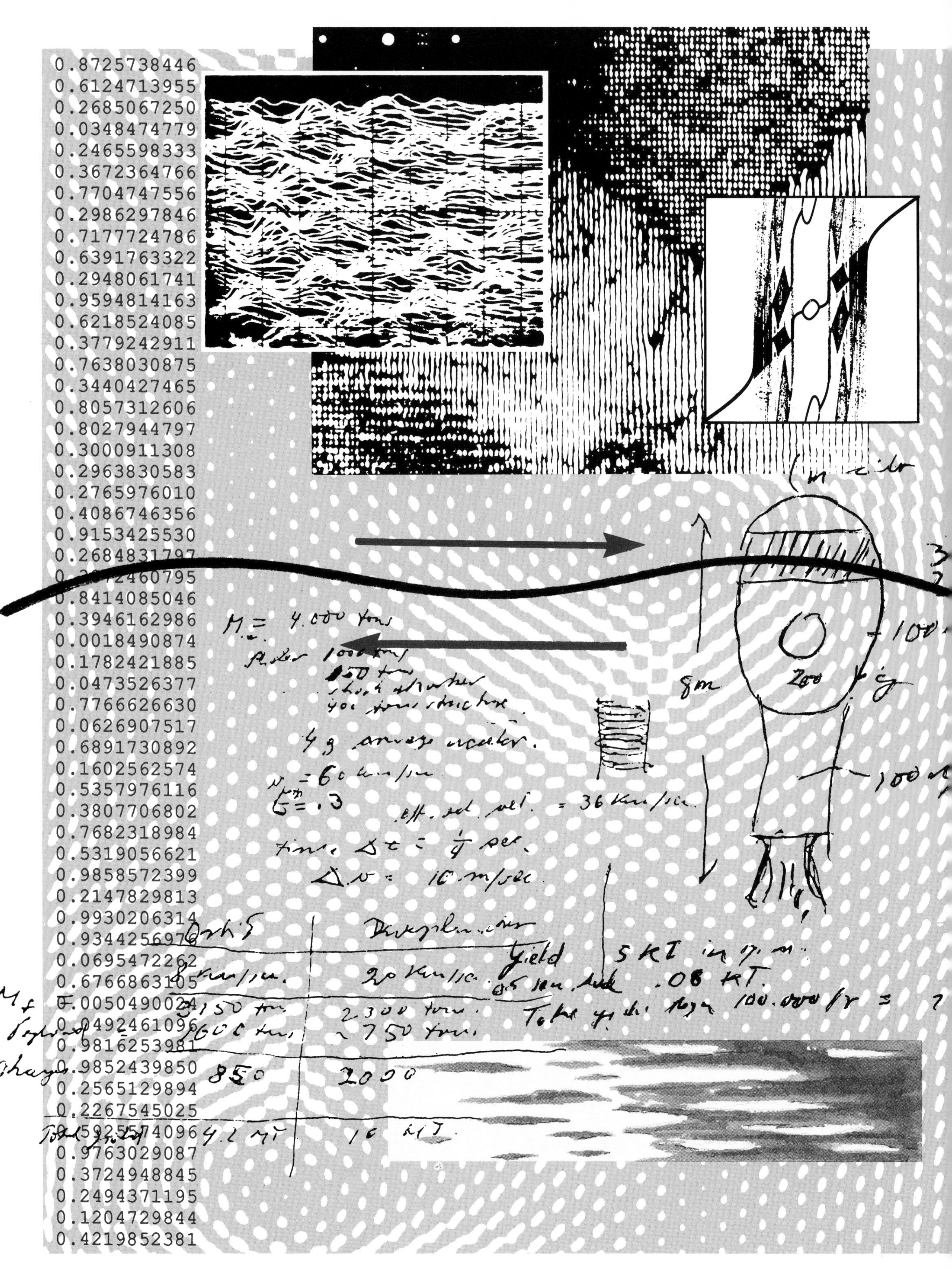

0.8725738446
0.6124713955
0.2685067250
0.0348474779
0.2465598333
0.3672364766
0.7704747556
0.2986297846
0.7177724786
0.6391763322
0.2948061741
0.9594814163
0.6218524085
0.3779242911
0.7638030875
0.3440427465
0.8057312606
0.8027944797
0.3000911308
0.2963830583
0.2765976010
0.4086746356
0.9153425530
0.2684831797
0.8414085046
0.3946162986
0.0018490874
0.1782421885
0.0473526377
0.7766626630
0.0626907517
0.6891730892
0.1602562574
0.5357976116
0.3807706802
0.7682318984
0.5319056621
0.9858572399
0.2147829813
0.9930206314
0.0695472262
0.6766863105
0.9816253981
0.9852439850
0.2565129894
0.2267545025
0.9763029087
0.3724948845
0.2494371195
0.1204729844
0.4219852381
M = 4.000 tons
8m
Yield 5 KT
.08 KT
100.000/yr
850
2000
4.2 MT
16 MT

*"Because of the novel problems which confronted its scientists during the wartime establishment of Los Alamos, the need arose for research and ideas in domains contiguous to its central purpose. This trend continues unabated to the present.*

*Problems of a complexity surpassing anything that had ever existed in technology rendered imperative the development of electronic computing machines and the invention of new theoretical computing methods. There, consultants like von Neumann played an important role in helping enlarge the horizon of the innovations, which required the most abstract ideas derived from the foundations of mathematics as well as from theoretical physics. [These ideas] were and still are invested in new, fruitful ways. ...*

*In at least two different and separate ways the availability of computing machines has enlarged the scope of mathematical research. It has enabled us ... to gather through heuristic experiments impressions of the morphological nature of various mathematical concepts such as the behaviour ... of certain non-linear transformations, the properties of some combinatorial systems and some topological curiosities of seemingly general behaviours. It has also enabled us to throw light on the behaviour of ... complicated systems ... [through] Monte Carlo type experiments and extensive but 'intelligently chosen' brute force approaches, in hydrodynamics for example."*

S. Ulam 1984

The remarks at left, written to preface Ulam's collected Los Alamos reports, suggest the context, the content, and the import of his influence at Los Alamos. The Los Alamos report on the hydrogen bomb, written with Edward Teller, is certainly of great interest, but classification precludes any discussion beyond that found in "Vita" and "From Above the Fray" in Part I. Weapons development, however, was only one among many Los Alamos projects in which Stan had a hand. In fact the list of his publications at the back of this volume reads in many parts like a history of ideas at Los Alamos. The reports authored by Stan and his many illustrious collaborators on the Monte Carlo method, hydrodynamics, nonlinear transformations, computer studies of nonlinear systems, and other diverse topics, as well as his informal talks and conversations, have left a legacy of ideas and possibilities that are still only partially explored. Stan was present at the opening of the computing era, and together with von Neumann, he understood perhaps better than others its revolutionary potential for exploring complex systems of all kinds.

Nick Metropolis, in "The Beginning of the Monte Carlo Method," describes from firsthand experience the early years of the computer revolution and the invention and first applications of the Monte Carlo method. As the name implies, this method turns an iterative process (such as a neutron chain reaction or repeated application of a deterministic transformation) into a game of chance. The computer plays the game over and over again to obtain good estimates of the average (or asymptotic) outcome of the process. In his 1950 paper "Random Processes and Transformations" Ulam outlined a variety of ways in which deterministic problems in mathematics and physics could be converted into equivalent random processes, or games of chance. His vision was prophetic and this method has taken hold in many areas; even particle physicists are applying Monte Carlo techniques to find solutions to complex problems in quantum field theory.

While Metropolis's article and the others on Monte Carlo are basically review, the remaining articles in this section describe current research on nonlinear systems whose inspiration or approach relate back to Stan. In each case computer experiments are used to gain insight into complex behavior.

Turbulence, the chaotic flow we observe in streams and waterfalls, in the oceans, and in the atmosphere, is one of the most difficult problems in nonlinear science. It has defied a fundamental description by mathematicians and physicists for over a hundred years, but its effects must be modeled if we are to achieve success in many technological programs. In this issue two articles deal with attempts to model this phenomenon by computer simulation.

"Instabilities and Turbulence" by Frank Harlow and his collaborators introduces us to the nature of turbulence, describes its disruptive effects in technological applications, and presents a new theory of turbulence transport that strips away the fine-scale details of this complicated phenomenon and comes to grips with its main effects on real physical systems. As Frank mentions in "Early Work in Numerical Hydrodynamics," the approach taken in turbulence transport theory is in line with Ulam's early insight about modeling turbulence—that one needs to model not the detailed shape of the fluid flow but rather the rate of energy flow from large to smaller and smaller length scales.

"Discrete Fluids" by Brosl Hasslacher introduces an alternate and completely novel approach to modeling complex fluid flows. The model is deceptively simple—a set of particles that live on a lattice of discrete points. They hop from point to point at constant speed and when they run into one another they change direction in a way that conserves momentum. That is all there is—but in a miraculous manner still not completely understood this model reproduces the whole spectrum of collective motion in fluids, from smooth to turbulent flow. This lattice gas automaton is a variation of the cellular automata, invented by Ulam and von Neumann. Its behavior embodies one of Ulam's favorite insights—that simple rules are capable of describing arbitrarily complex behavior (maybe even the behavior of the human brain). Although the model is simple, Brosl's three-part article is rich in ideas, suggesting a new paradigm for parallel computing and for modeling many other physical systems usually described by partial differential equations.

At this point the reader may be overwhelmed by the sprawling array of topics and approaches that appear to be part of nonlinear science. Why call it *a* science instead of many different ones? In the remarkably informative overview "Nonlinear Science—From Paradigms to Practicalities" David Campbell pulls together the common features and methodology that link disparate phenomena into a single new discipline. He identifies three major paradigms of this field—solitons and coherent structures, chaos and fractals, and complex configuration and patterns—and then follows through with many examples of physical and mathematical systems in which these paradigms play a significant role. The latter two paradigms have roots in Ulam's work with Stein, discussed in the mathematics section of "The Ulam Legacy," and all three have roots in the FPU problem, Fermi, Pasta, and Ulam's seminal paper on nonlinear systems. (Excerpts of this paper appear toward the end of the physics section).

The FPU results, apart from leading to the discovery of the soliton by Kruskal and Zabusky, were the first indication that the ergodic hypothesis of statistical mechanics may be wrong. In the last paper of this section Adrian Patrascioui, the 1986 Ulam Scholar at the Laboratory's Center for Nonlinear Studies, describes further results in this direction and speculates that a reconsideration of the ergodic hypothesis may lead to profound changes in our understanding of the meaning of quantum mechanics.

The field of nonlinear science is changing our understanding of the world around us. As experimental mathematics uncovers certain universal features of complex deterministic systems, it also brings us face to face with the limits of their predictability.

We hope this section illustrates a remark from Ulam's 1984 preface, that:

*"... a mathematical turn of mind, a mathematical habit of thinking, a way of looking at problems in different subfields [of science] ... can suggest general insights and not just offer the mere use of techniques."*

# THE BEGINNING *of the* MONTE CARLO METHOD

***by N. Metropolis***

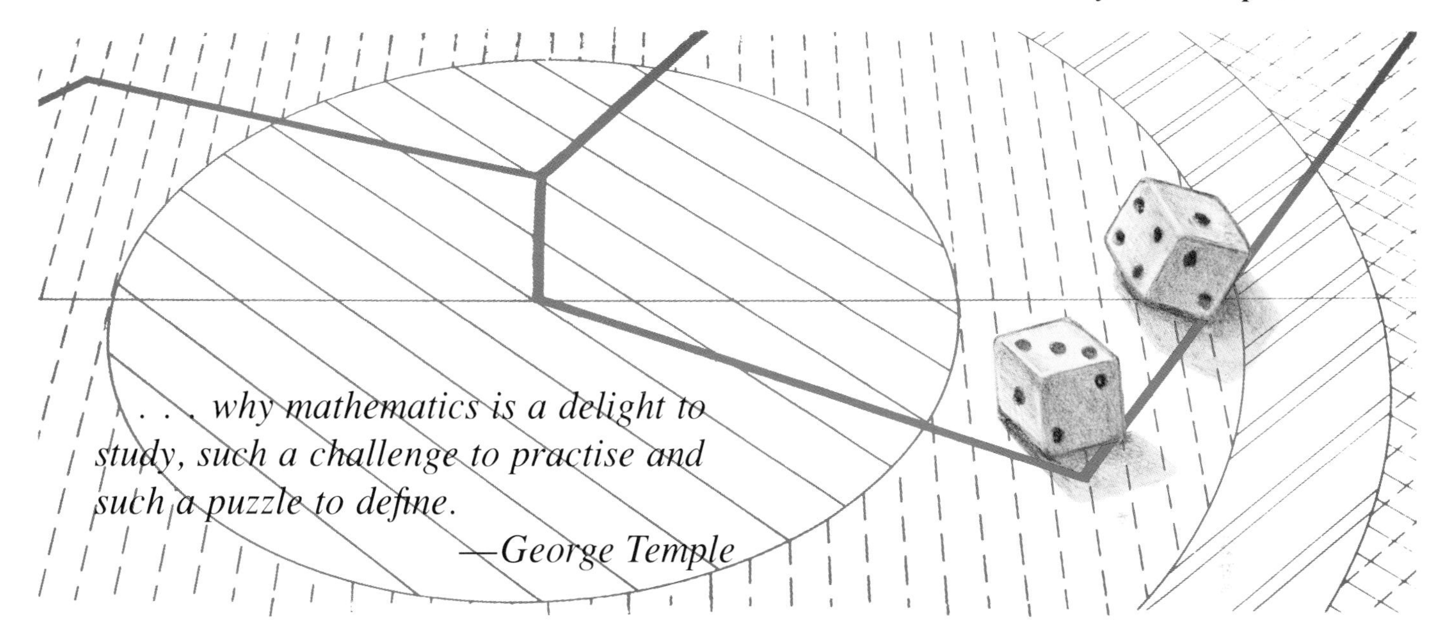


The year was 1945. Two earth-shaking events took place: the successful test at Alamogordo and the building of the first electronic computer. Their combined impact was to modify qualitatively the nature of global interactions between Russia and the West. No less perturbative were the changes wrought in all of academic research and in applied science. On a less grand scale these events brought about a renascence of a mathematical technique known to the old guard as statistical sampling; in its new surroundings and owing to its nature, there was no denying its new name of the Monte Carlo method.

This essay attempts to describe the details that led to this renascence and the roles played by the various actors. It is appropriate that it appears in an issue dedicated to Stan Ulam.

## Some Background

Most of us have grown so blasé about computer developments and capabilities —even some that are spectacular—that it is difficult to believe or imagine there was a time when we suffered the noisy, painstakingly slow, electromechanical devices that chomped away on punched cards. Their saving grace was that they continued working around the clock, except for maintenance and occasional repair (such as removing a dust particle from a relay gap). But these machines helped enormously with the routine, relatively simple calculations that led to Hiroshima.

**The ENIAC.** During this wartime period, a team of scientists, engineers, and technicians was working furiously on the first electronic computer—the ENIAC—at the University of Pennsylvania in Philadelphia. Their mentors were Physicist First Class John Mauchly and Brilliant Engineer Presper Eckert. Mauchly, familiar with Geiger counters in physics laboratories, had realized that if electronic circuits could count, then they could do arithmetic and hence solve, *inter alia*, difference equations—at almost incredible speeds! When he'd seen a seemingly limitless array of women cranking out firing tables with desk calculators, he'd been inspired to propose to the Ballistics Research Laboratory at Aberdeen that an electronic computer be built to deal with these calculations.

John von Neumann, Professor of Mathematics at the Institute for Advanced Study, was a consultant to Aberdeen and to Los Alamos. For a whole host of reasons, he had become seriously interested in the thermonuclear problem being spawned at that time in Los Alamos by a friendly fellow-Hungarian scientist, Edward Teller, and his group. Johnny (as he was affectionately called) let it be known that construction of the ENIAC was nearing completion, and he wondered whether Stan Frankel and I would be interested in preparing a preliminary computational model of a thermonuclear reaction for the ENIAC. He felt he could convince the authorities at Aberdeen that our problem could provide a more exhaustive test of the computer than mere firing-table computations. (The designers of the ENIAC had wisely provided for the capability of much more ambitious versions of firing tables than were being arduously computed by hand, not to mention other quite different applications.) Our response to von Neumann's suggestion was enthusiastic, and his heuristic arguments were accepted by the authorities at Aberdeen.

In March, 1945, Johnny, Frankel, and I visited the Moore School of Electrical Engineering at the University of Pennsylvania for an advance glimpse of the ENIAC. We were impressed. Its physical size was overwhelming—some 18,000 double triode vacuum tubes in a system with 500,000 solder joints. No one ever had such a wonderful toy!

The staff was dedicated and enthusiastic; the friendly cooperation is still remembered. The prevailing spirit was akin to that in Los Alamos. What a pity that a war seems necessary to launch such revolutionary scientific endeavors. The components used in the ENIAC were joint-army-navy (JAN) rejects. This fact not only emphasizes the genius of Eckert and Mauchly and their staff, but also suggests that the ENIAC was technically realizable even before we entered the war in December, 1941.

After becoming saturated with indoctrination about the general and detailed structure of the ENIAC, Frankel and I returned to Los Alamos to work on a model that was realistically calculable. (There was a small interlude at Alamogordo!) The war ended before we completed our set of problems, but it was agreed that we continue working. Anthony Turkevich joined the team and contributed substantially to all aspects of the work. Moreover, the uncertainty of the first phase of the postwar Los Alamos period prompted Edward Teller to urge us not only to complete the thermonuclear computations but to document and provide a critical review of the results.

**The Spark.** The review of the ENIAC results was held in the spring of 1946 at Los Alamos. In addition to Edward Teller, the principals included Enrico Fermi, John von Neumann, and the Director, Norris Bradbury. Stanley Frankel, Anthony Turkevich, and I described the ENIAC, the calculations, and the conclusions. Although the model was relatively simple, the simplifications were taken into account and the extrapolated results were cause for guarded optimism about the feasibility of a thermonuclear weapon.

Among the attendees was Stan Ulam, who had rejoined the Laboratory after a brief time on the mathematics faculty at the University of Southern California. Ulam's personality would stand out in any community, even where "characters" abounded. His was an informal nature; he would drop in casually, without the usual amenities. He preferred to chat, more or less at leisure, rather than to dissertate. Topics would range over mathematics, physics, world events, local news, games of chance, quotes from the classics—all treated somewhat episodically but always with a meaningful point. His was a mind ready to provide a critical link.

During his wartime stint at the Laboratory, Stan had become aware of the electromechanical computers used for implosion studies, so he was duly impressed, along with many other scientists, by the speed and versatility of the ENIAC. In addition, however, Stan's extensive mathematical background made him aware that statistical sampling techniques had fallen into desuetude because of the length and tediousness of the calculations. But with this miraculous development of the ENIAC—along with the applications Stan must have been pondering—it occurred to him that statistical techniques should be resuscitated, and he discussed this idea with von Neumann. Thus was triggered the spark that led to the Monte Carlo method.

Stanislaw Ulam

## The Method

The spirit of this method was consistent with Stan's interest in random processes—from the simple to the sublime. He relaxed playing solitaire; he was stimulated by playing poker; he would cite the times he drove into a filled parking lot at the same moment someone was accommodatingly leaving. More seriously, he created the concept of "lucky numbers," whose distribution was much like that of prime numbers; he was intrigued by the theory of branching processes and contributed much to its development, including its application during the war to neutron multiplication in fission devices. For a long time his collection of research interests included pattern development in two-dimensional games played according to very simple rules. Such work has lately emerged as a cottage industry known as cellular automata.

John von Neumann saw the relevance of Ulam's suggestion and, on March 11, 1947, sent a handwritten letter to Robert Richtmyer, the Theoretical Division leader (see "Stan Ulam, John von Neumann, and the Monte Carlo Method"). His letter included a detailed outline of a possible statistical approach to solving the problem of neutron diffusion in fissionable material.

Johnny's interest in the method was contagious and inspiring. His seemingly relaxed attitude belied an intense interest and a well-disguised impatient drive. His talents were so obvious and his cooperative spirit so stimulating that he garnered the interest of many of us. It was at that time that I suggested an obvious name for the statistical method—a suggestion not unrelated to the fact that Stan had an uncle who would borrow money from relatives because he "just had to go to Monte Carlo." The name seems to have endured.

The spirit of Monte Carlo is best conveyed by the example discussed in von Neumann's letter to Richtmyer. Consider a spherical core of fissionable material surrounded by a shell of tamper material. Assume some initial distribution of neutrons in space and in velocity but ignore radiative and hydrodynamic effects. The idea is to now follow the development of a large number of individual neutron chains as a consequence of scattering, absorption, fission, and escape.

At each stage a sequence of decisions has to be made based on statistical probabilities appropriate to the physical and geometric factors. The first two decisions occur at time $t = 0$, when a neutron is selected to have a certain velocity and a certain spatial position. The next decisions are the position of the first collision and the nature of that collision. If it is determined that a fission occurs, the number of emerging neutrons must be decided upon, and each of these neutrons is eventually followed in the same fashion as the first. If the collision is decreed to be a scattering, appropriate statistics are invoked to determine the new momentum of the neutron. When the neutron crosses a material boundary, the parameters and characteristics of the new medium are taken into account. Thus, a genealogical history of an individual neutron is developed. The process is repeated for other neutrons until a statistically valid picture is generated.

**John von Neumann**

**Random Numbers.** How are the various decisions made? To start with, the computer must have a source of uniformly distributed *psuedo*-random numbers. A much used algorithm for generating such numbers is the so-called von Neumann "middle-square digits." Here, an arbitrary $n$-digit integer is squared, creating a $2n$-digit product. A new integer is formed by extracting the middle $n$-digits from the product. This process is iterated over and over, forming a chain of integers whose properties have been extensively studied. Clearly, this chain of numbers repeats after some point. H. Lehmer has suggested a scheme based on the Kronecker-Weyl theorem that generates all possible numbers of $n$ digits before it repeats. (See "Random-Number Generators" for a discussion of various approaches to the generation of random numbers.)

Once one has an algorithm for generating a uniformly distributed set of random numbers, these numbers must be transformed into the nonuniform distribution $g$ desired for the property of interest. It can be shown that the function $f$ needed to achieve this transformation is just the inverse of the nonuniform distribution function, that is, $f = g^{-1}$. For example, neutron physics shows us that the distribution of free paths—that is, how far neutrons of a given energy in a given material go before colliding with a nucleus—decreases exponentially in the interval $(0, \infty)$. If $x$ is uniformly distributed in the open interval $(0, 1)$, then $f = -\ln x$ will give us a nonuniform distribution $g$ with just those properties.

The reader will appreciate many of the advantages of the Monte Carlo method compared to the methods of differential equations. For example, a neutron-velocity spectrum with various peaks and valleys is difficult to handle mathematically. For Monte Carlo one needs only to mirror the velocity spectrum in the probability distribution. Also, the Monte Carlo method is sufficiently flexible to account for hydrodynamic effects in a self-consistent way. In an even more elaborate code, radiation effects can be dealt with by following the photons and their interactions (see "Monte Carlo at Work").

Clearly, applications of the Monte Carlo method are much broader than so far outlined. (Although I emphasize the use of Monte Carlo in the study of physical systems, random sampling is also an efficient way to evaluate complicated and many-dimensional integrals. For an example, see the section entitled "The Monte Carlo Method" in "A Primer on Probability, Measure, and the Laws of Large Numbers.") Since its inception, many international conferences have been held on the various applications of the method. Recently, these range from the conference, "Monte Carlo Methods and Applications in Neutronics, Photonics, and Statistical Physics," at Cadarache Castle, France, in the spring of 1985 to the latest at Los Alamos, "Frontiers of Quantum Monte Carlo," in September, 1985.

## Putting the Method into Practice

Let me return to the historical account. In late 1947 the ENIAC was to be moved to its permanent home at the Ballistics Research Laboratory in Maryland. What a gargantuan task! Few observers were of the opinion that it would ever do another multiplication or even an addition. It is a tribute to the patience and skill of Josh Gray and Richard Merwin, two fearless uninitiates, that the move was a success. One salutary effect of the interruption for Monte Carlo was that another distinguished physicist took this occasion to resume his interest in statistical studies.

Enrico Fermi helped create modern physics. Here, we focus on his interest in neutron diffusion during those exciting times in Rome in the early thirties. According to Emilio Segrè, Fermi's student and collaborator, "Fermi had invented, but of course not named, the present Monte Carlo method when he was studying the moderation of neutrons in Rome. He did not publish anything on the subject, but he used the method to solve many problems with whatever calculating facilities he had, chiefly a small mechanical adding machine."*

In a recent conversation with Segrè, I learned that Fermi took great delight in astonishing his Roman colleagues with his remarkably accurate, "too-good-to-believe" predictions of experimental results. After indulging himself, he revealed that his "guesses" were really derived from the statistical sampling techniques that he used to calculate with whenever insomnia struck in the wee morning hours! And so it was that nearly fifteen years earlier, Fermi had independently developed the Monte Carlo method.

**Enrico Fermi**

It was then natural for Fermi, during the hiatus in the ENIAC operation, to dream up a simple but ingenious *analog* device to implement studies in neutron transport. He persuaded his friend and collaborator Percy King, while on a hike one Sunday morning in the mountains surrounding Los Alamos, to build such an instrument—later affectionately called the FERMIAC (see the accompanying photo).

The FERMIAC developed neutron genealogies in two dimensions, that is, in a plane, by generating the site of the "next collision." Each generation was based on a choice of parameters that characterized the particular material being traversed. When a material boundary was crossed, another choice was made appropriate to the new material. The device could accommodate two neutron energies, referred to as "slow" and "fast." Once again, the Master had just the right feel for what was meaningful and relevant to do in the pursuit of science.

**The First Ambitious Test.** Much to the amazement of many "experts," the ENIAC survived the vicissitudes of its 200-mile journey. In the meantime Richard Clippinger, a staff member at Aberdeen, had suggested that the ENIAC had sufficient flexibility to permit its controls to be reorganized into a more convenient (albeit static) stored-program mode of operation. This mode would have a capacity of 1800 instructions from a vocabulary of about 60 arithmetical and logical operations. The previous method of programming might be likened to a giant plugboard, that is to say, to a can of worms. Although implementing the new approach is an interesting story, suffice it to say that Johnny's wife, Klari, and I designed the new controls in about two months and completed the implementation in a fortnight. We then had the opportunity of using the ENIAC for the first ambitious test of the Monte Carlo method—a variety of problems in neutron transport done in collaboration with Johnny.

Nine problems were computed corresponding to various configurations of materials, initial distributions of neutrons, and running times. These problems, as yet, did not include hydrodynamic or radiative effects, but complex geometries and realistic neutron-velocity spectra were handled easily. The neutron histories were subjected to a variety of statistical analyses and comparisions with other approaches. Conclusions about the efficacy of the method were quite favorable. It seemed as though Monte Carlo was here to stay.

Not long afterward, other Laboratory

*quoted with permission of W. H. Freeman and Company from *From X-Rays to Quarks* by Emilio Segrè.

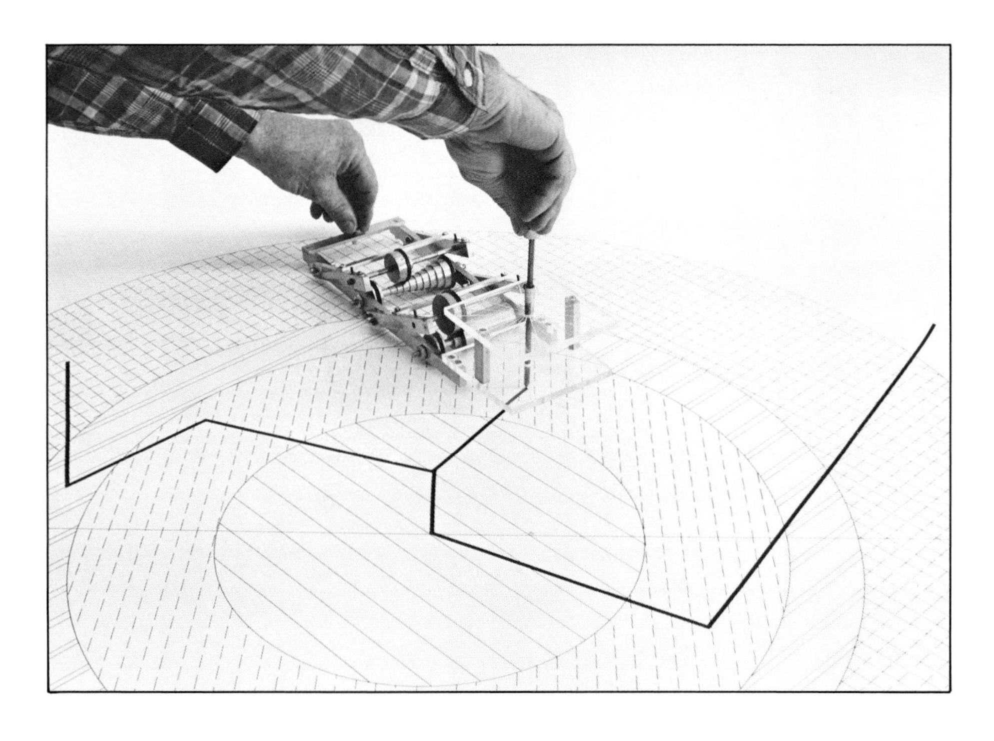

### THE FERMIAC

The Monte Carlo trolley, or FERMIAC, was invented by Enrico Fermi and constructed by Percy King. The drums on the trolley were set according to the material being traversed and a random choice between fast and slow neutrons. Another random digit was used to determine the direction of motion, and a third was selected to give the distance to the next collision. The trolley was then operated by moving it across a two-dimensional scale drawing of the nuclear device or reactor assembly being studied. The trolley drew a path as it rolled, stopping for changes in drum settings whenever a material boundary was crossed. This infant computer was used for about two years to determine, among other things, the change in neutron population with time in numerous types of nuclear systems.

staff members made their pilgrimages to ENIAC to run Monte Carlo problems. These included J. Calkin, C. Evans, and F. Evans, who studied a thermonuclear problem using a cylindrical model as well as the simpler spherical one. B. Suydam and R. Stark tested the concept of artificial viscosity on time-dependent shocks; they also, for the first time, tested and found satisfactory an approach to hydrodynamics using a realistic equation of state in spherical geometry. Also, the distinguished (and mysterious) mathematician C. J. Everett was taking an interest in Monte Carlo that would culminate in a series of outstanding publications in collaboration with E. Cashwell. Meanwhile, Richtmyer was very actively running Monte Carlo problems on the so-called SSEC during its brief existence at IBM in New York.

In many ways, as one looks back, it was among the best of times.

**Rapid Growth.** Applications discussed in the literature were many and varied and spread quickly. By midyear 1949 a symposium on the Monte Carlo method, sponsored by the Rand Corporation, the National Bureau of Standards, and the Oak Ridge Laboratory, was held in Los Angeles. Later, a second symposium was organized by members of the Statistical Laboratory at the University of Florida in Gainesville.

In early 1952 a new computer, the MANIAC, became operational at Los Alamos. Soon after Anthony Turkevich led a study of the nuclear cascades that result when an accelerated particle collides with a nucleus. The incoming particle strikes a nucleon, experiencing either an elastic or an ineleastic scattering, with the latter event producing a pion. In this study particles and their subsequent collisions were followed until all particles either escaped from the nucleus or their energy dropped below some threshold value. The "experiment" was repeated until sufficient statistics were accumulated. A whole series of target nuclei and incoming particle energies was examined.

Another computational problem run on the MANIAC was a study of equations of state based on the two-dimensional motion of hard spheres. The work was a collaborative effort with the Tellers, Edward and Mici, and the Rosenbluths, Marshall and Arianna (see "Monte Carlo at Work"). During this study a strategy was developed that led to greater computing efficiency for equilibrium systems obeying the Boltzmann distribution function. According to this strategy, if a statistical "move" of a particle in the system resulted in a decrease in the energy of the system, the new configuration was accepted. On the other hand, if there was an increase in energy, the new configuration was accepted only if it survived a game of chance biased by a Boltzmann factor. Otherwise, the old configuration became a new statistic.

It is interesting to look back over two-score years and note the emergence, rather early on, of experimental mathematics, a natural consequence of the electronic computer. The role of the Monte Carlo method in reinforcing such mathematics seems self-evident. When display units were introduced, the temptation to experiment became almost irresistible, at least for the fortunate few who enjoyed the luxury of a hands-on policy. When shared-time operations became realistic, experimental mathematics came of age. At long last, mathematics achieved a certain parity—the twofold aspect of experiment and theory—that all other sciences enjoy.

It is, in fact, the coupling of the subtleties of the human brain with rapid and reliable calculations, both arithmetical and logical, by the modern computer that has stimulated the development of experimental mathematics. This development will enable us to achieve Olympian heights.

## The Future

So far I have summarized the rebirth of statistical sampling under the rubric of Monte Carlo. What of the future—perhaps even a not too distant future?

The miracle of the chip, like most miracles, is almost unbelievable. Yet the fantastic performances achieved to date have not quieted all users. At the same time we are reaching upper limits on the computing power of a single processor.

One bright facet of the miracle is the lack of macroscopic moving parts, which makes the chip a very reliable bit of hardware. Such reliability suggests parallel processing. The thought here is not a simple extension to two, or even four or eight, processing systems. Such extensions are adiabatic transitions that, to be sure, should be part of the immediate, short-term game plan. Rather, the thought is massively parallel operations with thousands of interacting processors—even millions!

Already commercially available is one computer, the Connection Machine, with 65,536 simple processors working in parallel. The processors are linked in such a way that no processor in the array is more than twelve wires away from another and the processors are pairwise connected by a number of equally efficient routes, making communication both flexible and efficient. The computer has been used on such problems as turbulent fluid flow, imaging processing (with features analogous to the human visual system), document retrieval, and "common-sense" reasoning in artificial intelligence.

One natural application of massive parallelism would be to the more ambitious Monte Carlo problems already upon us. To achieve good statistics in Monte Carlo calculations, a large number of "histories" need to be followed. Although each history has its own unique path, the underlying calculations for all paths are highly parallel in nature.

Still, the magnitude of the endeavor to compute on massively parallel devices must not be underestimated. Some of the tools and techniques needed are:

- A high-level language and new architecture able to deal with the demands of such a sophisticated language (to the relief of the user);
- Highly efficient operating systems and compilers;
- Use of modern combinatorial theory, perhaps even new principles of logic, in the development of elegant, comprehensive architectures;
- A fresh look at numerical analysis and the preparation of new algorithms (we have been mesmerized by serial computation and purblind to the sophistication and artistry of parallelism).

Where will all this lead? If one were to wax enthusiastic, perhaps—just perhaps—a simplified model of the brain might be studied. These studies, in turn, might provide feedback to computer architects designing the new parallel structures.

Such matters fascinated Stan Ulam. He often mused about the nature of memory and how it was implemented in the brain. Most important, though, his own brain possessed the fertile imagination needed to make substantive contributions to the very important pursuit of understanding intelligence. ■

## Further Reading

S. Ulam, R. D. Richtmyer, and J. von Neumann. 1947. Statistical methods in neutron diffusion. Los Alamos Scientific Laboratory report LAMS–551. This reference contains the von Neumann letter discussed in the present article.

N. Metropolis and S. Ulam. 1949. The Monte Carlo method. *Journal of the American Statistical Association* 44:335–341.

S. Ulam. 1950. Random processes and transformations. *Proceedings of the International Congress of Mathematicians* 2:264–275.

Los Alamos Scientific Laboratory. 1966. Fermi invention rediscovered at LASL. *The Atom*, October, pp. 7–11.

C. C. Hurd. 1985. A note on early Monte Carlo computations and scientific meetings. *Annals of the History of Computing* 7:141–155.

W. Daniel Hillis. 1987. The connection machine. *Scientific American*, June, pp. 108–115.

**N. Metropolis** received his B.S. (1937) and his Ph.D. (1941) in physics at the University of Chicago. He arrived in Los Alamos, April 1943, as a member of the original staff of fifty scientists. After the war he returned to the faculty of the University of Chicago as Assistant Professor. He came back to Los Alamos in 1948 to form the group that designed and built MANIAC I and II. (He chose the name MANIAC in the hope of stopping the rash of such acronyms for machine names, but may have, instead, only further stimulated such use.) From 1957 to 1965 he was Professor of Physics at the University of Chicago and was the founding Director of its Institute for Computer Research. In 1965 he returned to Los Alamos where he was made a Laboratory Senior Fellow in 1980. Although he retired recently, he remains active as a Laboratory Senior Fellow Emeritus.

# STAN ULAM, JOHN VON NEUMANN, *and the* MONTE CARLO METHOD

**by Roger Eckhardt**

The Monte Carlo method is a statistical sampling technique that over the years has been applied successfully to a vast number of scientific problems. Although the computer codes that implement Monte Carlo have grown ever more sophisticated, the essence of the method is captured in some unpublished remarks Stan made in 1983 about solitaire.

"The first thoughts and attempts I made to practice [the Monte Carlo method] were suggested by a question which occurred to me in 1946 as I was convalescing from an illness and playing solitaires. The question was what are the chances that a Canfield solitaire laid out with 52 cards will come out successfully? After spending a lot of time trying to estimate them by pure combinatorial calculations, I wondered whether a more practical method than "abstract thinking" might not be to lay it out say one hundred times and simply observe and count the number of successful plays. This was already possible to envisage with the beginning of the new era of fast computers, and I immediately thought of problems of neutron diffusion and other questions of mathematical physics, and more generally how to change processes described by certain differential equations into an equivalent form interpretable as a succession of random operations. Later... [in 1946, I] described the idea to John von Neumann and we began to plan actual calculations."

Von Neumann was intrigued. Statistical sampling was already well known in mathematics, but he was taken by the idea of doing such sampling using the newly developed electronic computing techniques. The approach seemed especially suitable for exploring the behavior of neutron chain reactions in fission devices. In particular, neutron multiplication rates could be estimated and used to predict the explosive behavior of the various fission weapons then being designed.

In March of 1947, he wrote to Robert Richtmyer, at that time the Theoretical Division Leader at Los Alamos (Fig. 1). He had concluded that "the statistical approach is very well suited to a digital treatment," and he outlined in some detail how this method could be used to solve neutron diffusion and multiplication problems in fission devices for the case "of 'inert' criticality" (that is, approximated as momentarily static config-

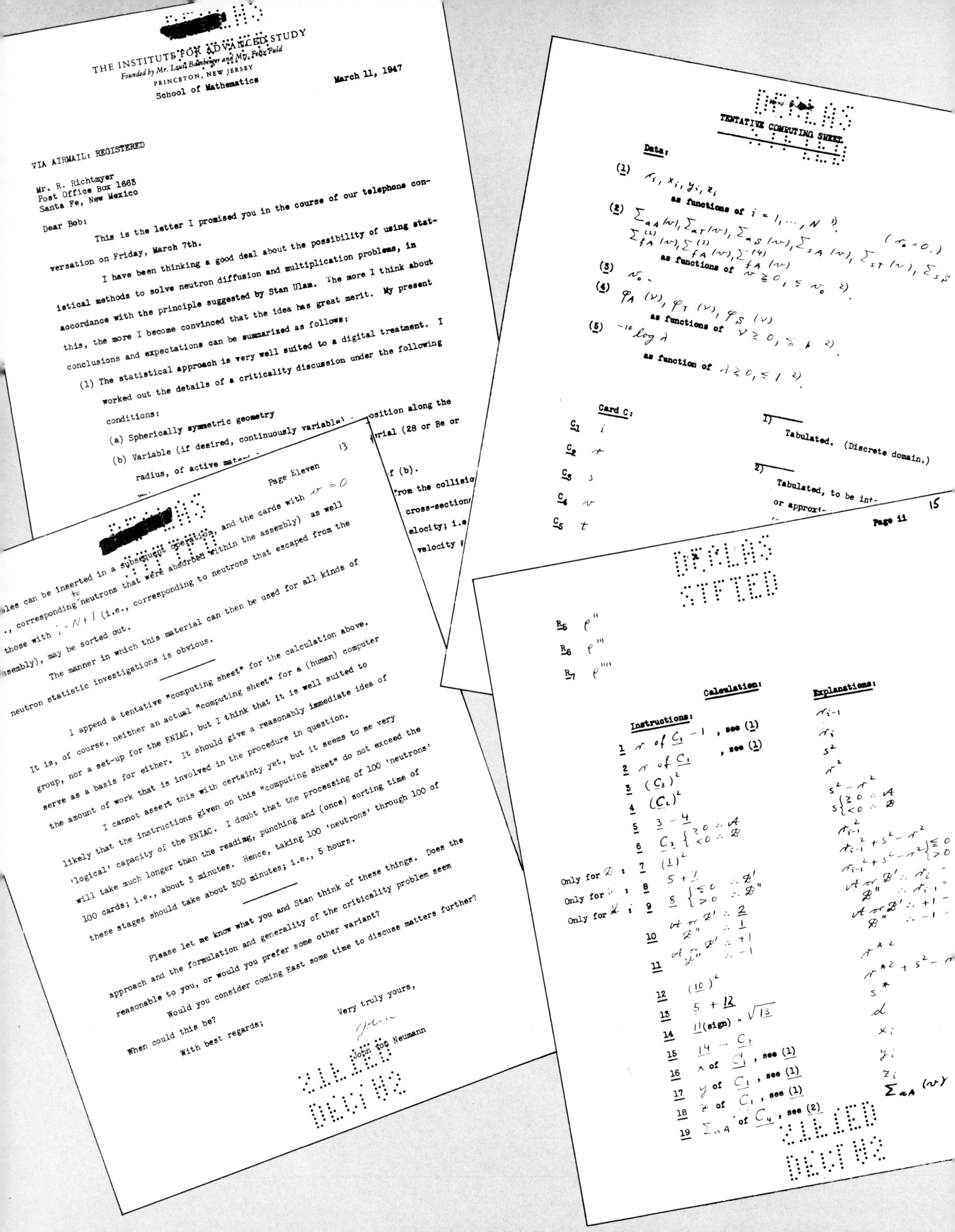

THE INSTITUTE FOR ADVANCED STUDY
Founded by Mr. Louis Bamberger and Mrs. Felix Fuld
PRINCETON, NEW JERSEY
School of Mathematics

March 11, 1947

VIA AIRMAIL: REGISTERED

Mr. R. Richtmyer
Post Office Box 1663
Santa Fe, New Mexico

Dear Bob:

This is the letter I promised you in the course of our telephone conversation on Friday, March 7th.

I have been thinking a good deal about the possibility of using statistical methods to solve neutron diffusion and multiplication problems, in accordance with the principle suggested by Stan Ulam. The more I think about this, the more I become convinced that the idea has great merit. My present conclusions and expectations can be summarized as follows:

(1) The statistical approach is very well suited to a digital treatment. I worked out the details of a criticality discussion under the following conditions:

(a) Spherically symmetric geometry

(b) Variable (if desired, continuously variable) ... osition along the radius, of active mat... ...rial (28 or Be or ...

13

Page Eleven

...les can be inserted in a subsequent operation, and the cards with $v = 0$ ..., corresponding to neutrons that were absorbed within the assembly) as well those with $i = N+1$ (i.e., corresponding to neutrons that escaped from the assembly), may be sorted out.

The manner in which this material can then be used for all kinds of neutron statistic investigations is obvious.

I append a tentative "computing sheet" for the calculation above. It is, of course, neither an actual "computing sheet" for a (human) computer group, nor a set-up for the ENIAC, but I think that it is well suited to serve as a basis for either. It should give a reasonably immediate idea of the amount of work that is involved in the procedure in question.

I cannot assert this with certainty yet, but it seems to me very likely that the instructions given on this "computing sheet" do not exceed the 'logical' capacity of the ENIAC. I doubt that the processing of 100 'neutrons' will take much longer than the reading, punching and (once) sorting time of 100 cards; i.e., about 3 minutes. Hence, taking 100 'neutrons' through 100 of these stages should take about 300 minutes; i.e., 5 hours.

Please let me know what you and Stan think of these things. Does the approach and the formulation and generality of the criticality problem seem reasonable to you, or would you prefer some other variant?

Would you consider coming East some time to discuss matters further? When could this be?

With best regards;

Very truly yours,

John von Neumann

TENTATIVE COMPUTING SHEET

Data:

(1) $r_i, x_i, y_i, z_i$
as functions of $i = 1, \ldots, N$ 1).

(2) $\Sigma_{aA}(v), \Sigma_{aT}(v), \Sigma_{aS}(v), \Sigma_{sA}(v), \Sigma_{sT}(v), \Sigma_{sS}$ ... $\Sigma^{(2)}_{fA}(v), \Sigma^{(3)}_{fA}(v), \Sigma^{(4)}_{fA}(v)$ $(v_0 = 0.)$
as functions of $v \geq 0, \leq v_0$ 2).

(3) $v_0$.

(4) $\varphi_A(v), \varphi_T(v), \varphi_S(v)$
as functions of $v \geq 0, \leq$ ... 2).

(5) $-10 \log \lambda$
as function of $\lambda \geq 0, \leq 1$ 2).

Card C:

$C_1$ $i$
$C_2$ $r$
$C_3$ $s$
$C_4$ $v$
$C_5$ $t$

1) Tabulated. (Discrete domain.)

2) Tabulated, to be int... or approxi...

15

Page 11

$R_5$ $\rho''$
$R_6$ $\rho'''$
$R_7$ $\rho''''$

| | Instructions: Calculation: | Explanations: |
|---|---|---|
| | 1 $r$ of $C_1 - 1$, see (1) | $r_{i-1}$ |
| | 2 $r$ of $C_1$, see (1) | $r_i$ |
| | 3 $(C_3)^2$ | $s^2$ |
| | 4 $(C_2)^2$ | $r^2$ |
| | 5 $3 - 4$ | $s^2 - r^2$ |
| | 6 $C_3 \{ \geq 0 :: A;\ < 0 :: B$ | $s \{ \geq 0 :: A;\ < 0 :: B$ |
| Only for $B$: | 7 $(1)^2$ | $r_{i-1}^2$ |
| Only for $B$: | 8 $5 + 7$ | $r_{i-1}^2 + s^2 - r^2$ |
| Only for $B$: | 9 $8 \{ \leq 0 :: B';\ > 0 :: B''$ | $r_{i-1}^2 + s^2 - r^2 \{ \leq 0;\ > 0$ |
| | 10 $A$ or $B' :: 2$; $B'' :: 1$ | $A$ or $B' :: r_i$; $B'' :: r_{i-1}$ |
| | 11 $A$ or $B' :: +1$; $B'' :: -1$ | $A$ or $B' :: +1$; $B'' :: -1$ |
| | 12 $(10)^2$ | $r^{*2}$ |
| | 13 $5 + 12$ | $r^{*2} + s^2 - r^2$ |
| | 14 $11(\text{sign}) \times \sqrt{13}$ | $s^*$ |
| | 15 $14 - C_3$ | $d$ |
| | 16 $x$ of $C_1$, see (1) | $x_i$ |
| | 17 $y$ of $C_1$, see (1) | $y_i$ |
| | 18 $z$ of $C_1$, see (1) | $z_i$ |
| | 19 $\Sigma_{aA}$ of $C_4$, see (2) | $\Sigma_{aA}(v)$ |

**Fig. 1.** The first and last pages of von Neumann's remarkable letter to Robert Richtmyer are shown above, as well as a portion of his tentative computing sheet. The last illustrates how extensivly von Neumann had applied himself to the details of a neutron-diffusion calculation.

urations). This outline was the first formulation of a Monte Carlo computation for an electronic computing machine.

In his formulation von Neumann used a spherically symmetric geometry in which the various materials of interest varied only with the radius. He assumed that the neutrons were generated isotropically and had a known velocity spectrum and that the absorption, scattering, and fission cross-sections in the fissionable material and any surrounding materials (such as neutron moderators or reflectors) could be described as a function of neutron velocity. Finally, he assumed an appropriate accounting of the statistical character of the number of fission neutrons with probabilities specified for the generation of 2, 3, or 4 neutrons in each fission process.

The idea then was to trace out the history of a given neutron, using random digits to select the outcomes of the various interactions along the way. For example, von Neumann suggested that in the compution "each neutron is represented by [an 80-entry punched computer] card . . . which carries its characteristics," that is, such things as the zone of material the neutron was in, its radial position, whether it was moving inward or outward, its velocity, and the time. The card also carried "the necessary random values" that were used to determine at the next step in the history such things as path length and direction, type of collision, velocity after scattering—up to seven variables in all. A "new" neutron was started (by assigning values to a new card) whenever the neutron under consideration was scattered or whenever it passed into another shell; cards were started for several neutrons if the original neutron initiated a fission. One of the main quantities of interest, of course, was the neutron multiplication rate—for each of the 100 neutrons started, how many would be present after, say, $10^{-8}$ second?

At the end of the letter, von Neumann attached a tentative "computing sheet" that he felt would serve as a basis for setting up this calculation on the ENIAC. He went on to say that "it seems to me very likely that the instructions given on this 'computing sheet' do not exceed the 'logical' capacity of the ENIAC." He estimated that if a problem of the type he had just outlined required "following 100 primary neutrons through 100 collisions [each]. . . of the primary neutron or its descendants," then the calculations would "take about 5 hours." He further stated, somewhat optimistically, that "in changing over from one problem of this category to another one, only a few numerical constants will have to be set anew on one of the 'function table' organs of the ENIAC."

His treatment did not allow "for the displacements, and hence changes of material distribution, caused by hydrodynamics," which, of course, would have to be taken into account for an explosive device. But he stated that "I think that I know how to set up this problem, too: One has to follow, say 100 neutrons through a short time interval $\Delta t$; get their momentum and energy transfer and generation in the ambient matter; calculate from this the displacement of matter; recalculate the history of the 100 neutrons by assuming that matter is in the middle position between its original (unperturbed) state and the above displaced (perturbed) state;. . . iterating in this manner until a "self-consistent" system of neutron history and displacement of matter is reached. This is the treatment of the first time interval $\Delta t$. When it is completed, it will serve as a basis for a similar treatment of the second time interval. . . etc., etc."

Von Neumann also discussed the treatment of the radiation that is generated during fission. "The photons, too, may have to be treated 'individually' and statistically, on the same footing as the neutrons. This is, of course, a non-trivial complication, but it can hardly consume much more time and instructions than the corresponding neutronic part. It seems to me, therefore, that this approach will gradually lead to a completely satisfactory theory of efficiency, and ultimately permit prediction of the behavior of all possible arrangements, the simple ones as well as the sophisticated ones."

And so it has. At Los Alamos in 1947, the method was quickly brought to bear on problems pertaining to thermonuclear as well as fission devices, and, in 1948, Stan was able to report to the Atomic Energy Commission about the applicability of the method for such things as cosmic ray showers and the study of the Hamilton Jacobi partial differential equation. Essentially all the ensuing work on Monte Carlo neutron-transport codes for weapons development and other applications has been directed at implementing the details of what von Neumann outlined so presciently in his 1947 letter (see "Monte Carlo at Work").

In von Neumann's formulation of the neutron diffusion problem, each neutron history is analogous to a single game of solitare, and the use of random numbers to make the choices along the way is analogous to the random turn of the card. Thus, to carry out a Monte Carlo calculation, one needs a source of random numbers, and many techniques have been developed that pick random numbers that are *uniformly* distributed on the unit interval (see "Random-Number Generators"). What is really needed, however, are *nonuniform* distributions that simulate probability distribution functions specific to each particular type of decision. In other words, how does one ensure that in random flights of a neutron, on the average, a fraction $e^{-x/\lambda}$ travel a distance $x/\lambda$ mean free paths or farther without colliding? (For a more mathematical discussion of random variables, probability distribution functions, and Monte Carlo, see pages 68–73 of "A Tutorial on Probability, Measure, and the Laws of Large Numbers.")

The history of each neutron is gener-

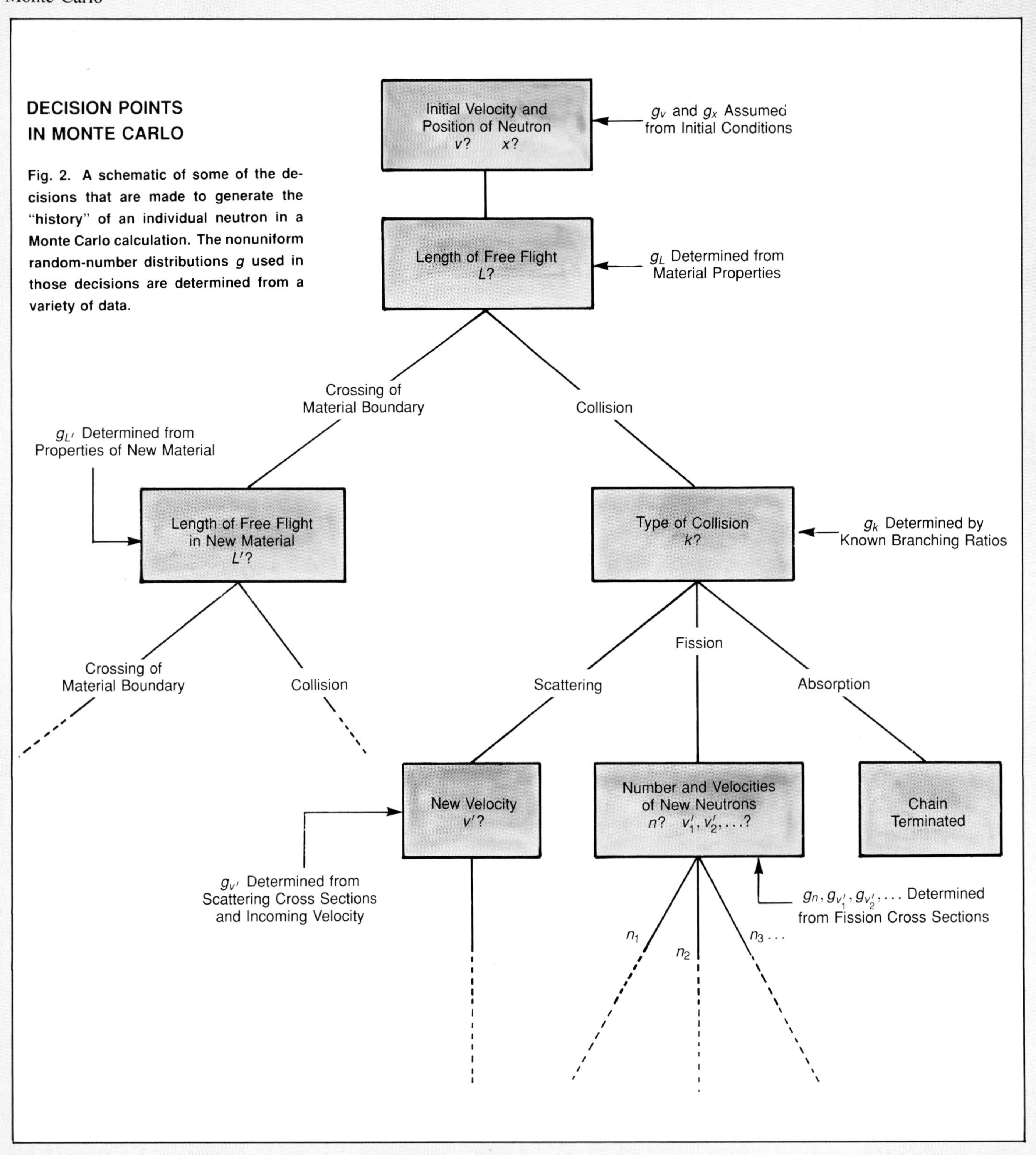


**DECISION POINTS IN MONTE CARLO**

**Fig. 2. A schematic of some of the decisions that are made to generate the "history" of an individual neutron in a Monte Carlo calculation. The nonuniform random-number distributions $g$ used in those decisions are determined from a variety of data.**

ated by making various decisions about the physical events that occur as the neutron goes along (Fig. 2). Associated with each of these decision points is a known, and usually nonuniform, distribution of random numbers $g$ that mirrors the probabilities for the outcomes possible for the event in question. For instance, returning to the example above, the distribution of random numbers $g_L$ used to determine the distance that a neutron travels before interacting with a nucleus is exponentially decreasing, making the selection of shorter distances more probable than longer distances. Such a distribution simulates the observed exponential falloff of neutron path lengths. Similarly, the distribution of random numbers $g_k$ used to select between a scattering, a fission, and an absorption must reflect the known probabilities for these different outcomes. The idea is to divide the unit interval $(0, 1)$ into three subintervals in such a way that the probability of a uniform random number being in a given subinterval equals the probability of the outcome assigned to that set.

In another 1947 letter, this time to Stan Ulam, von Neumann discussed two techniques for using uniform distributions of random numbers to generate the desired nonuniform distributions $g$ (Fig. 3). The first technique, which had already been

## ANOTHER VON NEUMANN LETTER

**Fig. 3. In this 1947 letter to Stan Ulam, von Neumann discusses two methods for generating the nonuniform distributions of random numbers needed in the Monte Carlo method. The second paragraph summarizes the inverse-function approach in which ($x^i$) represents the uniform distribution and ($\xi^i$) the desired nonuniform distribution. The rest of the letter describes an alternative approach based on *two* uniform and independent distributions: ($x^i$) and ($y^i$). In this latter approach a value $x^i$ from the first set is accepted when a value $y^i$ from the second set satisfies the condition $y^i \leq f(x^i)$, where $f(\xi^i)\,d\xi$ is the density of the desired distribution function. (It should be noted that in von Neumann's example for forming the random pairs $\xi = \sin x$ and $\eta = \cos x$, he probably meant to say that $x$ is equidistributed between 0 and 360 degrees (rather than "300"). Also, his notation for the tangent function is "tg," so that the second set of equations for $\xi$ and $\eta$ are just half-angle ($y = x/2$) trigonometric identities.)**

THE INSTITUTE FOR ADVANCED STUDY
SCHOOL OF MATHEMATICS
PRINCETON, NEW JERSEY

Mr. Stan Ulam
Post Office Box 1663
Santa Fe
New Mexico

May 21, 1947

Dear Stan:

Thanks for your letter of the 19th. I need not tell you that Klari and I are looking forward to the trip and visit at Los Alamos this Summer. I have already received the necessary papers from Carson Mark. I filled out and returned mine yesterday; Klari's will follow today.

I am very glad that preparations for the random numbers work are to begin soon. In this connection, I would like to mention this: Assume that you have several random number distributions, each equidistributed in $0, 1$ : $(x^i), (y^i), (z^i), \ldots$. Assume that you want one with the distribution function (density) $f(\xi)\,d\xi : (\xi^i)$. One way to form it is to form the cumulative distribution function: $g(\xi) = \int^{\xi} f(\xi)\,d\xi$ to invert it $h(x) = \xi \rightleftarrows x = g(\xi)$, and to form $\xi^i = h(x^i)$ with this $h(x)$, or some approximant polynomial. This is, as I see, the method that you have in mind.

An alternative, which works if $\xi$ and all values of $f(\xi)$ lie in $0, 1$, is this: Scan pairs $x^i, y^i$ and use or reject $x^i, y^i$ according to whether $y^i \leqq f(x^i)$ or not. In the first case, put $\xi^j = x^i$; in the second case form no $\xi^j$ at that step.

The second method may occasionally be better than the first one. In some cases combinations of both may be best; e.g., form random pairs $\xi = \sin x$, $\eta = \cos x$ with $x$ equidistributed between $0°$ and $300°$. The obvious way consists of using the sin - cos - tables (with interpolation). This is clearly closely related to the first method. This is an alternative procedure:

Put $\xi = \frac{2t}{1+t^2}$, $\eta = \frac{1-t^2}{1+t^2}$, $t = tg\, y$,

with y (which is $\frac{x}{2}$) equidistributed between $0°$ and $180°$. Restrict y to $0°$ to $45°$. Then the $\xi, \eta$ will have to be replaced randomly by $\eta, \xi$ and again by $\pm\xi, \pm\eta$. This can be done by using random digits 0, . . . , 7. It is also feasible with random digits 0, . . . , 9:

| | | |
|---|---|---|
| 0 | Replace $\xi, \eta$ by | $\xi, \eta$ |
| 1 | " | $-\xi, \eta$ |
| 2 | " | $\xi, -\eta$ |
| 3 | " | $-\xi, -\eta$ |
| 4 | " | $\eta, \xi$ |
| 5 | " | $\eta, -\xi$ |
| 6 | " | $-\eta, \xi$ |
| 7 | " | $-\eta, -\xi$ |
| 8 | Reject this digit | |
| 9 | " " " | |

Now $t = tg\, y$, $0° \leqq y \leqq 45°$, lies between 0 and 1, and its distribution function is $\frac{dt}{1+t^2}$. Hence one may pick pairs of numbers $t, s$ both (independently) equidistributed between 0 and 1, and then

use $t$ for $(1+t^2)\,s \leqq 1$

reject $t, s$ and form no $t$ at this step for $(1+t^2)\,s > 1$

Of course, the first part requires a divider, but the method may still be worth keeping in mind, especially when the ENIAC is available.

* * *

With best regards from house to house.

Yours, as ever,

John

John von Neumann

JvN:MW

proposed by Stan, uses the inverse of the desired function $f = g^{-1}$. For example, to get the exponentially decreasing distribution of random numbers on the interval $(0, \infty)$ needed for path lengths, one applies the inverse function $f(x) = -\ln x$ to a uniform distribution of random numbers on the open interval (0, 1).

What if it is difficult or computationally expensive to form the inverse function, which is frequently true when the desired function is empirical? The rest of von Neumann's letter describes an alternative technique that will work for such cases. In this approach *two* uniform and independent distributions $(x^i)$ and $(y^i)$ are used. A value $x^i$ from the first set is accepted when a value $y^i$ from the second set satisfies the condition $y^i \leq f(x^i)$, where $f(\xi^i)d\xi$ is the density of the desired distribution function (that is, $g(x) = \int f(x)dx$).

This acceptance-rejection technique of von Neumann's can best be illustrated graphically (Fig. 4). If the two numbers $x^i$ and $y^i$ are selected randomly from the domain and range, respectively, of the function $f$, then each pair of numbers represents a point in the function's coordinate plane $(x^i, y^i)$. When $y^i > f(x^i)$ the point lies above the curve for $f(x)$, and $x^i$ is rejected; when $y^i \leq f(x^i)$ the point lies on or below the curve, and $x^i$ is accepted. Thus, the fraction of accepted points is equal to the fraction of the area below the curve. In fact, the proportion of points selected that fall in a small interval along the $x$-axis will be proportional to the average height of the curve in that interval, ensuring generation of random numbers that mirror the desired distribution.

## THE ACCEPTANCE-REJECTION METHOD

Fig. 4. If two independent sets of random numbers are used, one of which $(x^i)$ extends uniformly over the range of the distribution function $f$ and the other $(y^i)$ extends over the domain of $f$, then an acceptance-rejection technique based on whether or not $y^i \leq f(x^i)$ will generate a distribution for $(x^i)$ whose density is $f(x^i)\,dx^i$.

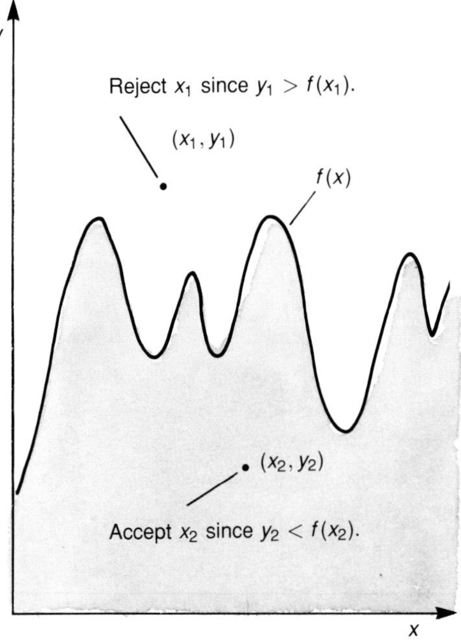


After a series of "games" have been played, how does one extract meaningful information? For each of thousands of neutrons, the variables describing the chain of events are stored, and this collection constitutes a numerical model of the process being studied. The collection of variables is analyzed using statistical methods identical to those used to analyze experimental observations of physical processes. One can thus extract information about any variable that was accounted for in the process. For example, the average energy of the neutrons at a particular time is calculated by simply taking the average of all the values generated by the chains at that time. This value has an uncertainty proportional to $\sqrt{V/(N-1)}$, where $V$ is the variance of, in this case, the energy and $N$ is the number of trials, or chains, followed.

It is, of course, desirable to reduce statistical uncertainty. *Any* modification to the stochastic calculational process that generates the same expectation values but smaller variances is called a variance-reduction technique. Such techniques frequently reflect the addition of known physics to the problem, and they reduce the variance by effectively increasing the number of data points pertinent to the variable of interest.

An example is dealing with neutron absorption by weighted sampling. In this technique, each neutron is assigned a unit "weight" at the start of its path. The weight is then decreased, bit by bit at each collision, in proportion to the absorption cross section divided by the total collision cross section. After each collision an outcome *other* than absorption is selected by random sampling and the path is continued. This technique reduces the variance by replacing the sudden, one-time process of neutron absorption by a gradual elimination of the neutron.

Another example of variance reduction is a technique that deals with outcomes that terminate a chain. Say that at each collision *one* of the alternative outcomes terminates the chain and associated with this outcome is a particular value $x_t$ for the variable of interest (an example is $x_t$ being a path length long enough for the neutron to escape). Instead of collecting these values *only* when the chain terminates, one can generate considerably more data about this particular outcome by making an extra calculation at *each* decision point. In this calculation the know value $x_t$ for termination is multiplied by the probability that that outcome will occur. Then *random* values are selected to continue the chain in the usual manner. By the end of the calculation, the "weighted values" for the terminating outcome have been summed over all decision points. This variance-reduction technique is especially useful if the probablity of the alternative in question is low. For example, shielding calculations typically predict that only one in many thousands of neutrons actually get through the shielding. Instead of accumulating those rare paths, the small probabilities that a neutron will get through the shield on its very next free flight are accumulated after each collision.

The Monte Carlo method has proven to be a powerful and useful tool. In fact, "solitaire games" now range from the neutron- and photon-transport codes through the evaluation of multi-dimensional integrals, the exploration of the properties of high-temperature plasmas, and into the quantum mechanics of systems too complex for other methods.

Quite a handful. ■

# Random-Number Generators

*by Tony Warnock*

Random numbers have applications in many areas: simulation, game-playing, cryptography, statistical sampling, evaluation of multiple integrals, particle-transport calculations, and computations in statistical physics, to name a few. Since each application involves slightly different criteria for judging the "worthiness" of the random numbers generated, a variety of generators have been developed, each with its own set of advantages and disadvantages.

Depending on the application, three types of number sequences might prove adequate as the "random numbers." From a purist point of view, of course, a series of numbers generated by a truly random process is most desirable. This type of sequence is called a *random-number sequence*, and one of the key problems is deciding whether or not the generating process is, in fact, random. A more practical sequence is the *pseudo-random sequence*, a series of numbers generated by a deterministic process that is intended merely to imitate a random sequence but which, of course, does not rigorously obey such things as the laws of large numbers (see page 69). Finally, a *quasi-random sequence* is a series of numbers that makes no pretense at being random but that has important predefined statistical properties shared with random sequences.

## Physical Random-Number Generators

Games of chance are the classic examples of random processes, and the first inclination would be to use traditional gambling devices as random-number generators. Unfortunately, these devices are rather slow, especially since the typical computer application may require millions of numbers per second. Also, the numbers obtained from such devices are not always truly random: cards may be imperfectly shuffled, dice may not be true, wheels may not be balanced, and so forth. However, in the early 1950s the Rand Corporation constructed a million-digit table of random numbers using an electrical "roulette wheel." (The device had 32 slots, of which 12 were ignored; the others were numbered from 0 to 9 twice.)

Classical gambling devices appear random only because of our ignorance of initial conditions; in principle, these devices follow deterministic Newtonian physics. Another possibility for generating truly random numbers is to take advantage of the Heisenberg uncertainty principle and quantum effects, say by counting decays of a radioactive source or by tapping into electrical noise. Both of these methods have been used to generate random numbers for computers, but both suffer the defects of slowness and ill-defined distributions (however, on a different but better order of magnitude than gambling devices).

For instance, although each decay in a radioactive source may occur randomly and independently of other decays, it is not necessarily true that successive counts in the detector are independent of each other. The time it takes to reset the counter, for example, might depend on the previous count. Furthermore, the source itself constantly changes in time as the number of remaining radioactive particles decreases exponentially. Also, voltage drifts can introduce bias into the noise of electrical devices.

There are, of course, various tricks to overcome some of these disadvantages. One can partially compensate for the counter-reset problem by replacing the string of bits that represents a given count with a new number in which all of the original 1-1 and 0-0 pairs have been discarded and all of the original 0-1 and 1-0 pairs have been changed to 0 and 1, respectively. This trick reduces the bias caused when the probability of a 0 is different from that of a 1 but does not completely eliminate nonindependence of successive counts.

A shortcoming of *any* physical generator is the lack of reproducibility. Reproducibility is needed for debugging codes that use the random numbers and for making correlated or anti-correlated computations. Of course, if one wants random numbers for a cryptographic one-time pad, reproducibility is the last attribute desired, and time can be traded for security. A radioactive source used with the bias-removal technique described above is probably sufficient.

## Arithmetical Pseudo-Random Generators

The most common method of generating pseudo-random numbers on the computer uses a recursive technique called the linear-congruential, or Lehmer, generator. The sequence is defined on the set of integers by the recursion formula

$$x_{n+1} = Ax_n + C \pmod{M},$$

where $x_n$ is the $n$th member of the sequence, and $A$, $C$, and $M$ are parameters that can be adjusted for convenience and to ensure the pseudo-random nature of the sequence. For example, $M$, the modulus, is frequently taken to be the word size on the computer, and $A$, the multiplier, is chosen to yield both a long period for the sequence and good statistical properties.

When $M$ is a power of 2, it has been shown that a suitable sequence can be generated if, among other things, $C$ is odd and $A$ satisfies $A = 5 \pmod 8$ (that is, $A - 5$ is a multiple of 8). A simple example of the generation of a 5-bit number sequence using these conditions would be to set $M = 32$ (5 bits), $A = 21$, $C = 1$, and $x_0 = 13$. This yields the sequence

$$13, 18, 27, 24, 25, 14, 7, 20, 5, 10 \ldots,$$

or, in binary,

$$01101, 10010, 11011, 11000, 11001, 01110, 00111, 10100, 00101, 01010, \ldots. \quad (1)$$

This type of generator has the interesting (or useful, or disastrous) property, illustrated by Seq. 1, that the least significant bit always has the alternating pattern 101010.... Further, the next bit has a pattern with period 4 (0110 above), the third bit has period 8, and so forth. Ultimately, the most significant bit has period $M$, which becomes the period of the sequence itself. Our example uses a short 5-bit word, which generates a sequence with a period of only 32. It is not unusual in many computer applications, however, to use many more bits (for example, to use a 32-bit word to generate a sequence with period $M = 2^{32}$).

One must be careful not to use such sequences in a problem with structures having powers of 2 in their dimensions. For example, a sequence with period $2^{32}$ would be a poor choice if the problem involved, say, a 3-dimensional lattice with sides of 128 ($= 2^7$) because the structure of the sequence can then interact unfavorably with the structure of the problem. Furthermore, there would be only $2^{32}/(2^7)^3 = 2048$ possible states. The usual assumption in Monte Carlo computations is that one has used a "representative" sample of the total number of possible computations—a condition that is certainly not true for this example.

One method of improving a pseudo-random-number generator is to combine two or more unrelated generators. The length of the hybrid will be the least common multiple of the lengths of the constituent sequences. For example, we can use the theory of *normal numbers* to construct a sequence that has all the statistical features of a "truly random" sequence and then combine it with a linear-congruential sequence. This technique yields a hybrid possessing the strengths of both sequences—for example, one that retains the statistical features of the normal-number sequence.

We first construct a normal number, that is, a number in base $b$ for which each block of $K$ digits has limiting frequency $(1/b)^K$. A simple example in base 2 can be constructed by concatenating the sequence of integers

$$1, 10, 11, 100, 101, 110, 111, 1000, 1001, 1010, 1011, 1100, 1101, 1110, 1111, \ldots$$

to form the normal number

$$1101110010111011110001001101010111100110111101111\ldots.$$

If the number is blocked into 5-digit sets

$$11011, 10010, 11101, 11100, 01001, 10101, 01111, 00110, 11110, 11111, \ldots, \quad (2)$$

it becomes a sequence of numbers in base 2 that satisfy all linear statistical conditions for randomness. For example, the frequency of a specific 5-bit number is $(1/2)^5$.

Sequences of this type do not "appear" random when examined; it is easy for a person to guess the rule of formation. However, we can further disguise the sequence by combining it with the linear-congruence sequence generated earlier (Seq. 1). We do this by performing an *exclusive-or* (XOR) operation on the two sequences:

$$01101, 10010, 11011, 11000, 11001, 01110, 00111, 10100, 00101, 01010, \ldots \quad (1)$$

and

$$11011, 10010, 11101, 11100, 01001, 10101, 01111, 00110, 11110, 11111, \ldots \quad (2)$$

yield

$$10110, 00000, 00110, 00100, 10000, 11011, 01000, 10010, 11011, 10101, \ldots . \quad (3)$$

Of course, if Seq. 3 is carried out to many places, a pattern in it will also become apparent. To eliminate the new pattern, the sequence can be XOR'ed with a third pseudo-random sequence of another type, and so on.

This type of hybrid sequence is easy to generate on a binary computer. Although for most computations one does not have to go to such pains, the technique is especially attractive for constructing "canonical" generators of apparently random numbers.

A key idea here is to take the notion of randomness to mean simply that the sequence can pass a given set of statistical tests. In a sequence based on normal numbers, each term will depend nonlinearly on the previous terms. As a result, there are nonlinear statistical tests that can show the sequence not to be random. In particular, a test based on the transformations used to construct the sequence itself will fail. But, the sequence will pass all *linear* statistical tests, and, on that level, it can be considered to be random.

What types of linear statistical tests are applied to pseudo-random numbers? Traditionally, sequences are tested for uniformity of distribution of single elements, pairs, triples, and so forth. Other tests may be performed depending on the type of problem for which the sequence will be used. For example, just as the correlation between two sequences can be tested, the auto-correlation of a single sequence can be tested after displacing the original sequence by various amounts. Or the number of different types of "runs" can be checked against the known statistics for runs. An increasing run, for example, consists of a sequential string of increasing numbers from the generator (such as, 0.08, 0.21, 0.55, 0.58, 0.73, ...). The waiting times for various events (such as the generation of a number in each of the five intervals $(0, 0.2), (0.2, 0.4), \ldots, (0.8, 1)$) may be tallied and, again, checked against the known statistics for random-number sequences.

If a generator of pseudo-random numbers passes these tests, it is deemed to be a "good" generator, otherwise it is "bad." Calling these criteria "tests of randomness" is misleading because one is testing a hypothesis known to be false. The usefulness of the tests lies in their similarity to the problems that need to be solved using the stream of pseudo-random numbers. If the generator fails one of the simple tests, it will surely not perform reliably for the real problem. (Passing all such tests may not, however, be enough to make a generator work for a given problem, but it makes the programmers setting up the generator feel better.)

## Quasi-Random Numbers

For some applications, such as evaluating integrals numerically, the use of quasi-random sequences is much more efficient than the use of either random or pseudo-random sequences. Although quasi-random sequences do not necessarily mimic a random sequence, they can be tailored to satisfy the equi-distribution criteria needed for the integration. By this I mean, roughly speaking, that the numbers are spread throughout the region of interest in a much more uniform manner than a random or pseudo-random sequence.

For example, say one needs to find the average of the quantity $f(x)$ over the set of coordinates $x$, knowing the distribution of coordinate values $\rho(x)$ for the system being considered. Ordinarily, the average is given by the expression

$$\langle f \rangle = \frac{\int \rho(x) f(x)\, dx}{\int \rho(x)\, dx}.$$

Rather than evaluating this integral, however, one can evaluate $f(x)$ at a series of random points. If the probability of picking a particular point $x$ is proportional to the statistical weight $\rho(x)$, then $\langle f \rangle$ is given by the expression

$$\langle f \rangle = \sum_{i=1}^{N} f(x_i)/N,$$

where $N$ is the total number of points chosen. This idea is the basis of the Metropolis technique of evaluating integrals by the Monte Carlo method.

Now if the points are taken from a random or a psuedo-random sequence, the statistical uncertainty will be proportional to $1/\sqrt{N}$. However, if a quasi-random sequence is used, the points will occupy the coordinate space with the correct distribution but in a more uniform manner, and the statistical uncertainty will be proportional to $1/N$. In other words, the uncertainty will decrease much faster with a quasi-random sequence than with a random or pseudo-random sequence.

How are quasi-random sequences generated? One type of sequence with a very uniform distribution is based on the radical-inverse function. The radical-inverse function $\phi(N, b)$ of a number $N$ with base $b$ is constructed by

1. writing the number in base $b$ (for example, 14 in base 3 is 112);
2. reversing the digits (112 becomes 211); and
3. writing the result as a fraction less than 1 in base $b$ (211 becomes 211/1000 in base 3 and, thus, $\phi(14, 3) = .211$).

A sequence based on the radical-inverse function is generated by choosing a prime number as the base $b$ and finding $\phi(1, b), \phi(2, b), \phi(3, b), \phi(4, b), \ldots$. For a problem with $k$ dimensions, the first $k$ primes are used, and $(\phi(N, b_1), \phi(N, b_2), \ldots \phi(N, b_k))$ becomes the $N$th point of the $k$-dimensional sequence. This sequence has a very uniform distribution and is useful in mutiple integration or multi-dimensional sampling.

There are many other types of random, pseudo-random, or quasi-random sequences than the ones I have discussed here, and there is much research aimed at generating sequences with the properties appropriate to the desired application. However, the examples I have discussed should illustrate both the approaches being taken and the obstacles that must be overcome in the quest of suitable "random" numbers. ■

# Monte Carlo at Work

*by Gary D. Doolen and John Hendricks*

Every second nearly 10,000,000,000 "random" numbers are being generated on computers around the world for Monte Carlo solutions to problems that Stan Ulam first dreamed of solving forty years ago. A major industry now exists that has spawned hundreds of full-time careers invested in the fine art of generating Monte Carlo solutions—a livelihood that often consists of extracting an answer out of a noisy background. Here we focus on two of the extensively used Monte Carlo solvers: MCNP, an internationally used neutron and photon transport code maintained at Los Alamos; and the "Metropolis" method, a popular and efficient procedure for computing equilibrium properties of solids, liquids, gases, and plasmas.

## MCNP

In the fifties, shortly after the work on the Monte Carlo method by Ulam, von Neumann, Fermi, Metropolis, Richtmyer, and others, a series of Monte Carlo transport codes began emerging from Los Alamos. The concepts on which these codes were based were those outlined by von Neumann (see "Stan Ulam, John von Neumann, and the Monte Carlo Method"), but a great deal of detailed work was needed to incorporate the appropriate physics and to develop shorter routes to statistically valid solutions.

From the beginning the neutron transport codes used a general treatment of the geometry, but successive versions added such features as cross-section libraries, variance-reduction techniques (essentially clever ways to bias the random numbers so that the guesses will cluster around the correct solution), and a free-gas model treating thermalization of the energetic fission neutrons. Also, various photon transport codes were developed that dealt with photon energies from as low as 1 kilo-electron-volt to the high energies of gamma rays. Then, in 1973, the neutron transport and the photon transport codes were merged into one. In 1977 the first version of MCNP appeared in which photon cross sections were added to account for production of gamma rays by neutron interactions. Since then the code has been distributed to over two hundred institutions worldwide.*

The Monte Carlo techniques and data now in the MCNP code represent over three hundred person-years of effort and have been used to calculate many tens of thousands of practical problems by scientists throughout the world. The types of problems include the design of nuclear reactors and nuclear safeguard systems, criticality analyses, oil well logging, health-physics problems, determinations of radiological doses, spacecraft radiation modeling, and radiation damage studies. Research on magnetic fusion has used MCNP heavily.

The MCNP code features a general three-dimensional geometry, continuous energy or multigroup physics packages, and sophisticated variance reduction techniques. Even very complex geometry and particle transport can be modeled almost exactly. In fact, the complexity of the geometry that can be represented is limited only by the dedication of the user.

---

*The MCNP code and manual can be obtained from the Radiation Shielding Information Center (RSIC), P.O. Box X, Oak Ridge, TN 37831.

**PAIR-DISTRIBUTION FUNCTION**

This plot gives the probability of pairs of charged particles in a plasma being separated by a certain distance. The probabilities are plotted as a function of the distance between the pair of particles (increasing from left to right) and temperature (decreasing from front to back). At the left edge, both the distance and the probability are zero; at the right edge, the probability has become constant in value. Red indicates probabilities below this constant value, yellow and green above. As the temperature of the plasma decreases, lattice-like peaks begin to form in the pair-distribution function. The probabilites, generated with the Metropolis method described in the text, have been used for precise tests of many theoretical approximations for plasma models.

## The Metropolis Method

The problem of finding the energy and configuration of the lowest energy state of a system of many particles is conceptually simple. One calculates the energy of the system, randomly moves each particle a small distance, and recalculates the energy. If the energy has decreased, the new position is accepted, and the procedure continues until the energy no longer changes.

The question of how to calculate equilibrium properties of a finite system at a given temperature is more difficult, but it was answered in a 1953 *Journal of Chemical Physics* article by Metropolis, Rosenbluth, Rosenbluth, Teller, and Teller, who decided that the calculation should follow the same steps for finding the minimum energy but with one important change. When a move results in an *increased* energy, one accepts the new position with probability $e^{-\Delta E/T}$, where $\Delta E$ is the change in energy and $T$ is the temperature. This procedure gives the equilibrium solution for any physical system. In fact, a system with many particles can be solved with only a few lines of code and a fast computer.

Although calculations for short-range forces are much easier than for long-range forces (such as the Coulomb force), the Metropolis technique has been used for most physical systems in which the forces between particles are known. Wayne Slattery, Hugh DeWitt, and one of the authors (GD) applied the technique to a neutral Coulomb plasma consisting of thousands of particles in a periodic box. The purpose was to calculate such physical properties as the temperature at which this type of plasma freezes and the pair distribution function, which is the probability of finding one particle at a given distance from another (see accompanying figure). Because the uncertainty in a Monte Carlo result is proportional to $1/\sqrt{N}$, where $N$ is the number of moves of a single particle, several million moves requiring several hundred Cray hours were needed to obtain accurate results for the plasma at many temperatures.

As computers become faster and their memories increase, larger and more complicated systems are being calculated far more accurately than even Stan Ulam probably expected. ■

# Early Work in Numerical Hydrodynamics

*by Francis H. Harlow*

I met Stan Ulam shortly after coming to Los Alamos in 1953. As an eager youngster chasing new dreams, I was inspired and encouraged (and sometimes properly chastised) by the older resident scientists and Laboratory consultants. Several stand out especially for their powerful encouragement; one of these is Stan.

Some of my associates, especially during the first six years, didn't like many of my wild ideas about fluid dynamics and the techniques for solving such problems by high-speed computers. Stan continually took the time to see what was going on and had the faith (not always justified) to tell others positive things about my explorations. I shall always be grateful for Stan's, as well as Conrad Longmire's, crucial influence in establishing our fluid dynamics group in the Theoretical Division.

Stan and I had many talks, especially on the stochastic behavior of complex systems. He seemed to *feel* how these systems worked: their collective properties were very real to him. He was intrigued by the almost-cyclic properties they sometimes could exhibit and participated in pioneering numerical experiments on fluid-like, many-particle dynamics.

His early work with John Pasta* created the grandaddy of the free-Lagrangian method of modeling turbulence and, in the sixties, led ultimately to the Particle-and-Force technique for the calculation of shock formation and interaction problems. Although couched in terms of hydrodynamics, the pioneering work has had significant impact on many branches of numerical analysis, especially in terms of the interpretation and meaning of results. The main thrust of their thinking is captured in the following excerpts.

"Our approach to the problem of dynamics of continua can be called perhaps "kinetic"—the continuum is treated, in an approximation, as a collection of a finite number of elements of "points;" these "points" can represent actual points of the fluid, or centers of mass of zones, i.e., globules of the fluid, or, more abstractly, *coefficients* of functions, representing the fluid, developed into series."

One of the motivations behind the free-Lagrangian approach was the computational difficulties for fluid flow with large internal shears in which elements that were initially close later found themselves widely separated.

"It was found impractical to use a "classical" method of calculation for this hydrodynamical problem, involving two independent spatial variables in an essential way .... This "classical" procedure, correct for infinitesimal steps in time and space, breaks down for any reasonable (i.e., practical) finite length of step in time. The reason is, of course, that the computation ... assumes that "neighboring" points, determining a "small" area—stay as neighbors for a considerable number of cycles. It is clear that in problems which involve mixing specifically this is not true ... the classical way of computing by referring to initial (at time $t = 0$) ordering of points becomes meaningless."

The next point is one that Stan emphasized repeatedly, illustrating what he felt to be a potential power of their approach.

"The meaningful results of the calculations are not so much the precise positions of our elements themselves as the behavior, in time, of a *few functionals* of the motion of the continuum.

"Thus in the problem relating to the *mixing* of two fluids, it is not the exact position of each globule that is of interest but quantities such as the *degree* of mixing (suitably defined); in problems of turbulence, not the shapes of each portion of the fluid, but the *overall* rate at which energy goes from simple *modes* of motion to higher frequencies."

As it turned out, the behavior, in time, of the functionals of the motion that they calculated was very smooth despite the complicated, turbulent nature of the fluid's motion. Thus, an important perspective on the modeling of complex phenomenon had been established. Indeed, turbulence transport theory, the subject of the following article, *depends* upon the strong tendencies in nature towards universal behavior that are the basis for the observed smoothness in their functionals. This theory is an excellent example of Stan's idea that wonderful numerical results can emerge from averaging discrete-representations over a set of possible scale sizes. But the theory goes further in providing an analytic formulation of turbulence transport.

Stan lived to see the realization of some of his ideas—others are still being investigated—but I always had confidence that if Stan had a *feeling* for something, it was sure to be signifcant. He was a friend I shall long remember. ■

---

*John Pasta and Stan Ulam. 1953. Heuristic studies in problems of mathematical physics on high speed computing machines. Los Alamos report LA-1557.

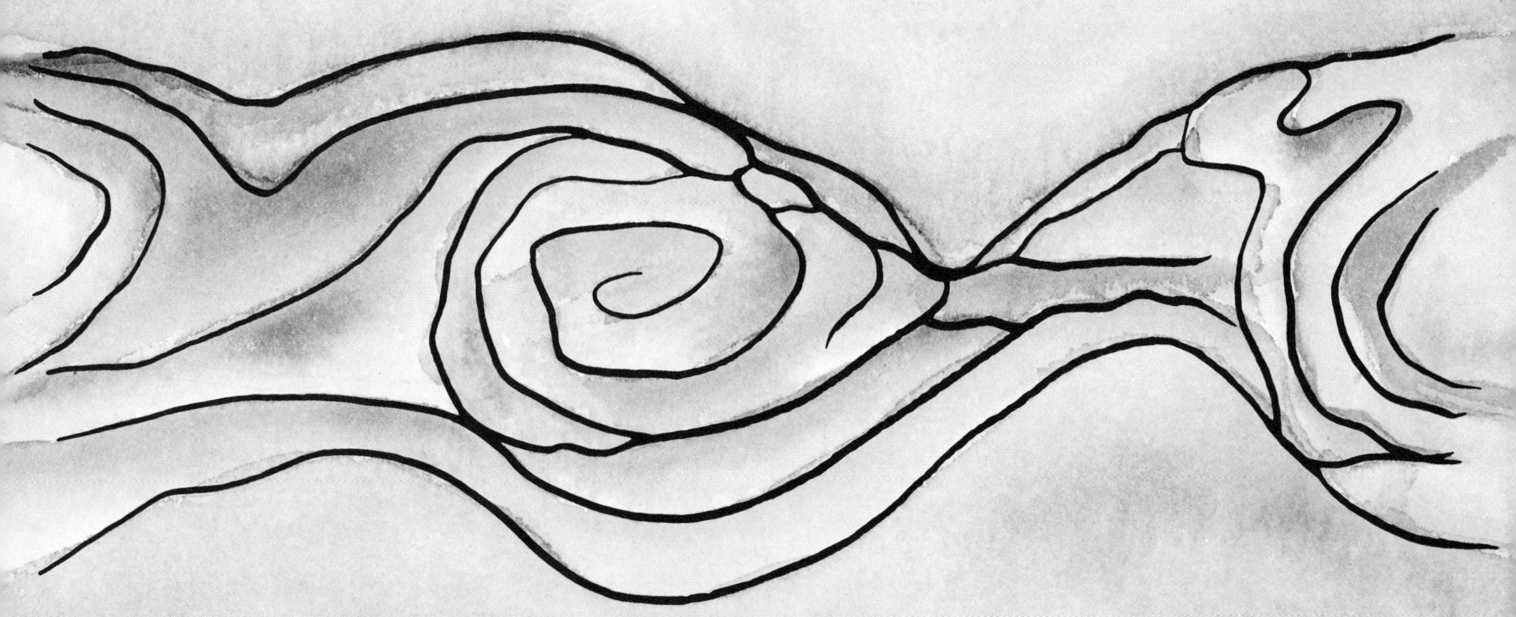

# Instabilities & Turbulence

*by Didier Besnard,*
*Francis H. Harlow,*
*Norman L. Johnson,*
*Rick Rauenzahn,*
*and Jonathan Wolfe*

When the interface between two materials experiences strong accelerative or shearing forces, the inevitable results are instability, turbulence, and the mixing of materials, momentum, and energy. One of the most important and exciting breakthroughs in our understanding of these disruptive processes has been the recent discovery that the features of the processes often are *independent of the initial interface perturbations.* This discovery is so important that scientists at Los Alamos National Laboratory, the California Institute of Technology, the Atomic Weapons Research Establishment in Great Britain, Lawrence Livermore National Laboratory, as well as scientists in France, and no doubt in the Soviet Union, are working hard to confirm and extend this new understanding experimentally.

Theoretical analyses are likewise showing a firm basis for this astonishing discovery. Two types of theory are being employed, gradually combined, and even proved essentially equivalent. These are the *multifield-interpenetration approach* and the *single-field turbulence approach.* Even brute-force hydrodynamics calculations are demonstrating this same property of independence from initial perturbation.

The consequences for developments in such main-line Laboratory projects as inertial-confinement fusion are profound. Our entire view of material mixing, turbulence shear impedance, and energy transport has undergone a revolutionary shift to qualitatively different directions.

What is the physical essence of this new way of thinking? No matter how

**TURBULENCE EFFECTS**

**Fig. 1. The effects of turbulence include increased mixing of initially separated materials, an increase in shear impedance of fluid near rough boundaries due to the turbulent viscosity, and increased transport of heat into surrounding cooler regions.**

Increased Mixing of Materials

Increased Shear Impedance

Increased Heat Diffusion

carefully we attempt to achieve smoothness and homogeneity, any sufficiently strong destabilizing influence at a material discontinuity will *inevitably* be disruptive. Indeed, the disruptive effects will be manifested in essentially the same manner as if there were a considerable roughness or inhomogeneity at or near the interface. Add to this the effects of any long-wavelength asymmetries, and we have an immutable inevitability for major instabilities in virtually every experimental circumstance of accelerative or shearing dynamics of interest to the Laboratory. Reliable predictability of new weapons designs in a comprehensive test ban, the design of any locally intense energy source, the development of workable concepts in Strategic Defense, the achievement of successful inertial-confinement fusion devices, and the success of many other Laboratory programs will depend crucially on our ability to model these instability and turbulence effects realistically.

## What Is Turbulence?

To describe the techniques we are using to model these effects, we must first consider in more detail the properties of turbulence itself. Turbulence is the random fluctuation in fluid motion that often is superimposed on the average course of the flow. The effects of turbulence can be highly significant (Fig. 1), increasing the fluid's effective viscosity and enhancing the mixing of initially separated materials, such as the mixing of dust into air or bubbles into a liquid. Turbulence is a significant factor in the wind resistance of a vehicle, in the dispersal of fuel droplets in an internal combustion engine, in mixing and transporting materials in chemical plants, indeed in virtually every circumstance of high-speed fluid flow.

It is easy to be deceived into thinking that turbulence is rare, because it often is not directly visible to the casual observer. Although water flowing rapidly through a transparent pipe may look completely smooth, touching the pipe can reveal large vibrations and the injection of dye through a tiny hole in the wall can demonstrate rapid downstream mixing. Both effects are a direct result of intense turbulent fluctuations.

Turbulence in air can be demonstrated—even in a relatively calm room—by holding one end of a long thread and watching its fluctuating response to air currents. Sunshine streaming over the top of a hot radiator creates shadow patterns on a nearby wall that dance restlessly in the never-ending turbulence that accompanies the upward flow of air.

Why is nature discontent with the smooth and peaceful flow of liquids and gases, especially at high flow speeds? What are the processes that feed energy into turbulent fluctuations? The answers lie in the behavior of energy. In contrast to momentum, energy has the peculiar ability to assume numerous and varied configurations. Momentum constraints, while restrictive, are helpless to prevent seemingly capricious energy rearrangements. In any real fluid flow, these rearrangements are triggered by inevitable perturbations that can be fed from the reservoir of mean-flow energy.

It is helpful at this point to compare turbulence with the random motion of simple gas molecules in a box because the approaches to both of these problems include much that is similar. However, the analogy becomes seriously misleading if pushed too far.

**Molecular Systems.** In a box of molecules the dynamics of each individual can be described quite accurately by Newton's laws. Yet we seldom try to analyze the complex interactions of all the trajectories, which are seemingly capable of very chaotic behavior. Instead, we appeal to the remarkably organized *mean* properties of the motion, identifying such useful variables as density, pressure, temperature, and fluid velocity.

We cannot ignore the departure of the individual from the behavior of the mean; indeed, some of the most interesting properties of the gas are directly associated with these departures. Diffusion of heat energy, for example, represents transport of kinetic energy by fluctuations; pressure in a "stationary" gas is the result of continual bombardment of molecules against objects immersed in the gas (Fig. 2(a)); viscous drag between two opposing streams of gas (Fig. 2(b)) arises because of fluctuations from the mean-flow velocity that cause molecules to migrate from one stream to the other.

**MOMENTUM FLUX**

Fig. 2. (a) The pressure on a wall is the result of the transfer of momentum during collisions between individual molecules and the wall. (b) Viscous drag between two opposing streams of gas is a result of individual departures from the mean-flow velocity in each stream. More precisely, pressure and viscous drag represent the normal flux through any imaginary surface of the normal and the tangential components of momentum, respectively.

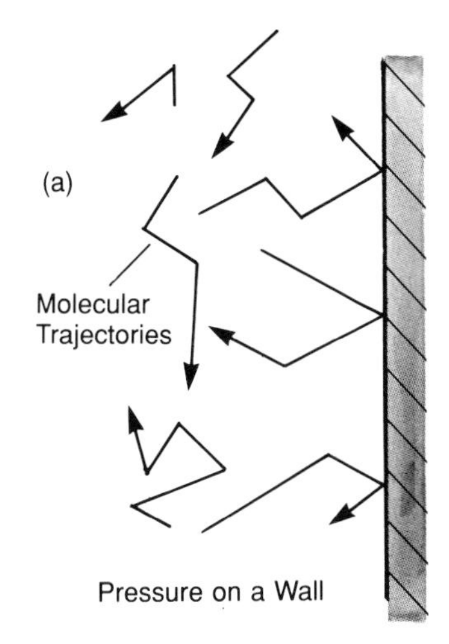


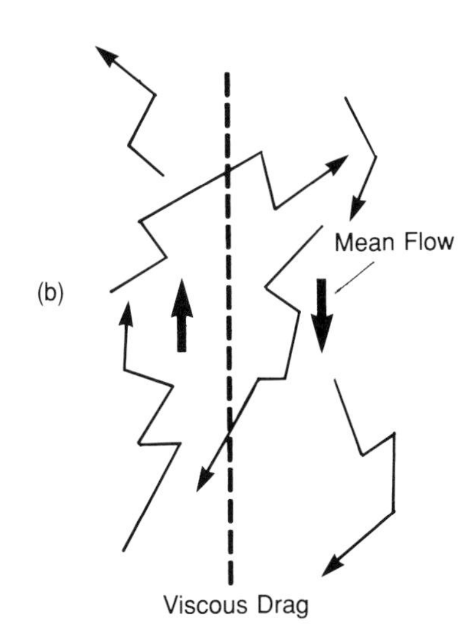


**Turbulent Eddies and Mean Flow.** Turbulent eddies in a fluid superficially resemble individual molecules in a gas. They likewise bounce around in random fashion, carrying kinetic energy in their fluctuational velocities. (Such turbulence kinetic energy is typically as much as 10 per cent of the mean-flow kinetic energy, or even more in regions where the mean flow stagnates at a solid surface.) Eddies also diffuse momentum (plus heat and any imbedded materials), exerting pressure through momentum transport and bombardment against walls.

But the concept of a turbulent eddy is nebulous at best. Gas molecules have an easily identifiable shape, size, mean separation, and mean free path between collisions. Turbulent eddies, in contrast, have a spectrum of sizes; they overlap each other; the constraint on their motion through the fluid by the immediate presence of neighboring fluid precludes the simple concept of a mean free path.

Moreover, identification of what part of the dynamics is turbulence and what part is mean flow is arbitrary. For molecules the distinction is essentially unique; in most circumstances, individual molecular fluctuations take place on a scale that is orders of magnitude smaller than the scale of collective, fluid-like motion. For turbulent eddies the fluctuational scale may be an appreciable fraction of the mean-flow scale. More to the point, the observer's experimental configuration itself establishes the distinction between turbulence and mean flow.

To put the matter succinctly, mean flow is that part of the dynamics directly associated with the macroscopic conditions established or measured by the observer, whereas turbulence is the more capricious part of the flow associated with finer-scaled perturbations not controlled by the observer but inevitably present in any real flow.

As an example, consider air flow around a parked automobile on a gusty day. With suitable instruments an observer can record variations in the approaching wind velocity. These measurements describe the source of the mean flow, and the macroscopic features of the car constitute the boundary conditions. Mean-flow patterns in the wake on the downwind side of the vehicle can be observed either with a ribbon that stretches out with the average air velocity at each place it is held or with an upstream smoke generator emitting a thin filament of smoke that can be photographed as it passes over the car.

Both the ribbon and the filament have an *average* direction to their motion that varies on the same time scale as that of the monitored gusts of wind; the relationship between these two features is the correlation that our investigator is seeking. In addition,

**TANGENTIAL DRAG**

Fig. 3. Fluid moving in a circular trough loses mean-flow kinetic energy because of tangential drag on the walls. Although this entire loss in energy will eventually appear as heat, a significant fraction may first appear as the kinetic energy of turbulence.

however, the ribbon flutters rapidly about that average (at the rate of many fluctuations per second), and the smoke filament diffuses in contorted kinks into the surrounding air. This capricious variation around the time-varying average is what our observer calls turbulence.

A second observer standing nearby, but paying no attention to the detailed observations of the first, feels buffeted by the gusts and, likewise, would agree that there is much turbulence. However, this observer can legitimately disagree as to which part of the air flow is mean flow and which part is turbulence, seeing an average southwesterly wind with turbulent variations that last several seconds. Meanwhile, an earth-orbiting satellite reveals that the southwesterly wind is simply a momentary fluctuation (of a half hour or so) from the general westerlies crossing the continent that day.

This example has three different fluctuational scales, all properly identified as turbulence on the basis of the observer's chosen viewpoint. The difference, however, is not merely one of semantics, and we discuss below the consequences of this multiple viewpoint to mathematical modeling of the flow processes. Important guidance is furnished by a careful consideration of interactions among the various dynamical scales.

There is thus a seemingly random nature to both molecular dynamics and turbulence. The detailed flow field of a group of molecules or eddies can vary by large amounts as a result of minor initial perturbations on a microscopic scale. But the remarkable feature of these dynamical systems is that the overall stochastic behavior is essentially independent of the manner in which the fluctuations are introduced.

However, not every fluid flow is sensitive to minor perturbations. Viscous or *slowly* moving fluids travel in a purely laminar fashion, responding negligibly to fine-scale perturbations. Why does flow remain stable for some conditions and exhibit turbulence for others? The answer lies in the ways in which energy is drawn from the mean flow as the motion gradually decays to quiescence.

## Turbulence Energy: Sources and Sinks

The statements of mass, momentum, and energy conservation lie at the foundations of fluid dynamics. In particular, fluid flow implies the presence of energy, which can exist in any of various forms: kinetic, heat, turbulence, potential, chemical. For the moment we are concerned only with the first three. By kinetic energy we mean the motion energy carried by the *mean flow*; heat energy refers to the kinetic energy of *molecular* fluctuations. Turbulence energy is at a scale between these first two: it is the kinetic energy of fluctuations that are large compared with the individual molecular scale but small compared with the mean-flow scale.

As we said earlier, in contrast to mass and momentum, which are highly constrained by their conservation laws, energy behaves very capriciously. Although *total* energy is rigorously conserved, transitions among the many manifestations of energy occur continuously. It is a remarkable fact of nature that, as a result of such transitions, any system devoid of remedial influences inevitably tends to move from order to disorder. An egg hitting the floor turns to a mess as ordered kinetic energy is converted into splat. Cars break down, rust, and eventually end up as nondescript piles of metallic and organic compounds blowing in the wind or leached by groundwater into a progressively wider and less ordered distribution. Fluids in a nicely ordered state of mean flow likewise

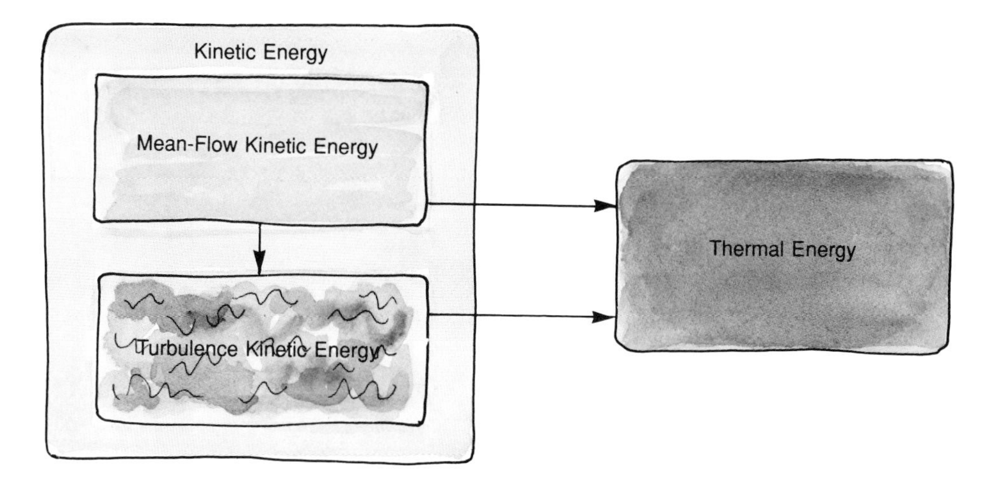


## ENERGY DEGRADATION

**Fig. 4. The mean-flow kinetic energy of a moving fluid inevitably degrades to thermal energy. Frequently, however, part of that kinetic energy is first transformed to kinetic energy of turbulence.**

tend toward (mass- and momentum-conserving) states of disordered energy in which the only residuum is heat—and even that leaks off to the less ordered state of wide dispersal as a result of conduction and radiation. In thermodynamics, the trend from order to disorder is called the Second Law; its profound scientific and philosophical implications have been discussed and debated for many decades; its validity is beyond doubt.

Consider a fluid that has been set into smooth and uniform motion in a circular trough (Fig. 3). It has zero total (vector) momentum: as much is moving east as is moving west at every instant. Tangential shearing drag on the walls slows the motion so that mean-flow kinetic energy is lost. Where does the energy go—to turbulence or to heat? The competition is fierce, and heat always wins in the end, but fluids yield themselves to the inevitable only grudgingly. If at all possible, they transform at least part of their kinetic energy to turbulence as an intermediate step along the way (Fig. 4).

Let's replace this animistic description with physics. The conversion of mean-flow kinetic energy directly to heat is limited by the viscosity of the fluid and by the steepness of the mean-flow velocity gradients. For example, consider fluid flow between two plates moving in parallel but opposite directions (Fig. 5). Although a variety of flow-velocity profiles could have been depicted, the one shown has the smallest fluid kinetic energy of any flow profile with that same momentum between the moving plates. This profile is thus the flow distribution to which all others inevitably tend.

Suppose we now examine a flow profile at the opposite extreme—one in which the gradient at the midpoint between the plates is very sharp (Fig. 6(a)). Both this distribution and the stable one in Fig. 5 have the same total fluid momentum (namely zero); however, in the distribution in Fig. 6(a), every fluid element has the same speed ($u_0$), whereas in the stable distribution, most elements are moving slower than $u_0$. Thus the fluid in Fig. 6(a) possesses an excess of kinetic energy compared to the fluid with the stable profile and will lose part of this energy as it transforms toward the stable configuration. Will turbulence be an intermediate state in this evolution? To answer this question we must dig deeply into the competitive processes of dissipation and instability.

## STABLE FLOW BETWEEN MOVING PLATES

**Fig. 5. When fluid is trapped between two plates moving at speed $u_0$ in parallel but opposite directions, a gradient in fluid velocity is established. The linear flow-velocity profile shown here has the smallest kinetic energy of any profile at that same total momentum for the given boundary conditions.**

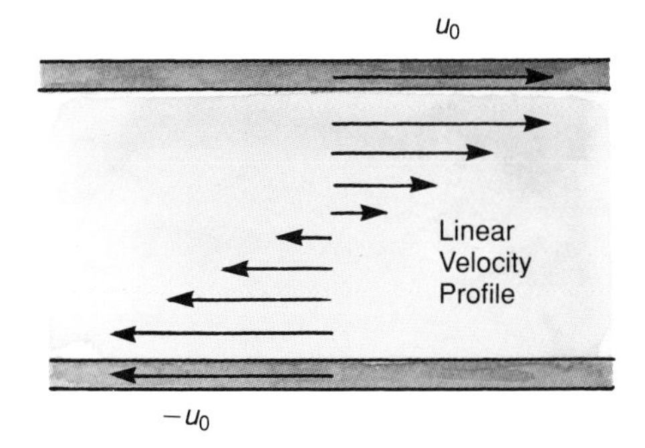

## UNSTABLE FLOW

Fig. 6. A discontinuity in the velocity profile between two oppositely moving fluid regions can lead to a Kelvin-Helmholtz instability at that interface, resulting in turbulence. For example, if, as in (b), the interface experiences a sinusoidal perturbation of wavelength $\lambda$ and amplitude $A$, such a perturbation will act effectively as a series of Venturi nozzles (c) that alter the mean-flow velocities and pressure $p$. These pressure variations, in turn, further increase the distortion.

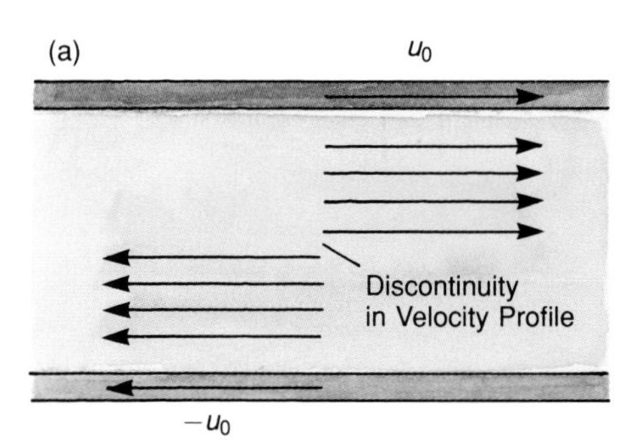


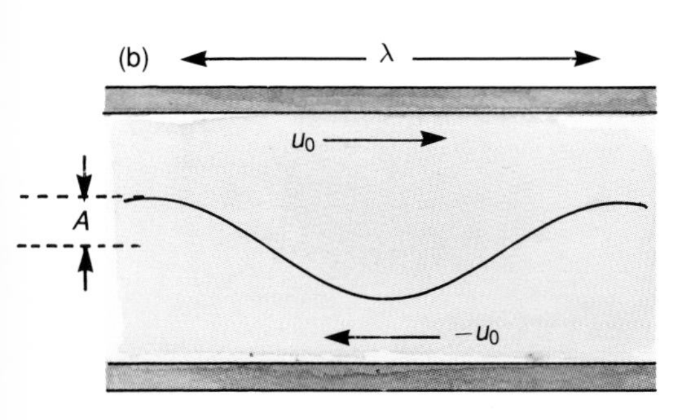


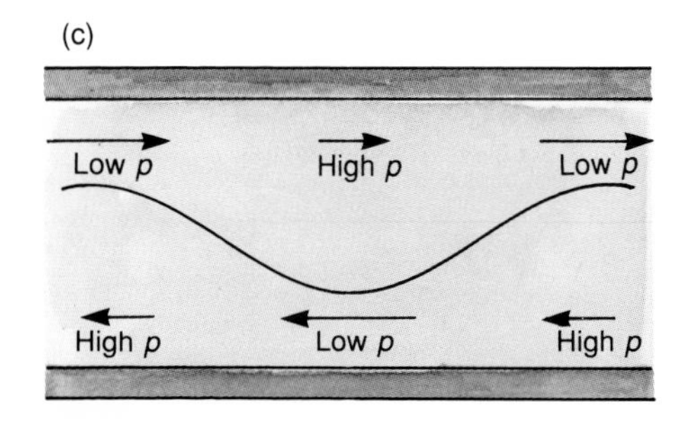


Consider first the dissipation of mean-flow kinetic energy into heat. Let $H$ be heat energy per unit volume and $du/dy$ describe some measure of the mean-flow velocity gradient in a fluid with molecular viscosity $\mu_{\rm m}$. Then the rate at which heat is generated is given by

$$\frac{dH}{dt} = \mu_{\rm m}\left(\frac{du}{dy}\right)^2. \tag{1}$$

To estimate the rate at which turbulence energy is generated, we return to the flow described in Fig. 6(a), which is susceptible to a destabilizing process called the Kelvin-Helmholtz instability. The presence of such an instability is easily demonstrated for an incompressible fluid if we arbitrarily assume that the slip interface between the upper and lower halves of the flow profile is distorted by a sinusoidal wave of wavelength $\lambda$ and amplitude $A$ (Fig. 6(b)). Because the fluid is incompressible, wherever the flow area is constricted the fluid has to move faster than average, and wherever the flow area is expanded the fluid has to move more slowly (Fig. 6(c)). What is the associated behavior of the pressure? Each cycle in the perturbation is like a Venturi nozzle, for which Bernoulli's law says the pressure is less in the constricted region where fluid speed is higher and is greater in the expanded region where fluid speed is lower. Thus, there is a pressure difference across the perturbed slip plane, acting in *exactly the right direction to enhance the perturbation amplitude*.

More formally, we can associate an appropriate inertia with the material being accelerated (the acceleration of the perturbation in the slip plane is $d^2A/dt^2$), and we can use Bernoulli's law to calculate the pressure difference (the driving force for enhancing the perturbation), which is proportional to the square of the fluid speed $u_0^2$. Newton's second law then leads to the following formula for the behavior of the perturbed slip plane:

$$\frac{d^2A}{dt^2} = \left(\frac{2\pi u_0}{\lambda}\right)^2 A. \tag{2}$$

The growth in amplitude is $dA/dt$, so the kinetic energy per unit volume involved in this turbulent-like motion is

$$K \approx \frac{1}{2}\rho\left(\frac{dA}{dt}\right)^2, \tag{3}$$

where $\rho$ is the density of the fluid. Differentiating Eq. 3 and substituting Eq. 2, we see that turbulence energy $K$, in turn, grows as

$$\frac{dK}{dt} \approx \left(\frac{2\pi u_0}{\lambda}\right)^2 \rho A \frac{dA}{dt}. \tag{4}$$

With $\omega \equiv (2\pi u_0/\lambda)$, a solution of Eq. 2 is $A = A_0 e^{\omega t}$, and

$$\frac{dK}{dt} \approx \omega^3 \rho A_0{}^2 e^{2\omega t}, \tag{5}$$

or

$$\frac{dK}{dt} \approx 2\omega K. \tag{6}$$

The essence of these results is that $dK/dt$ *increases* with time ($\omega$ is positive), whereas, because viscosity smears out the sharp velocity transition, thus decreasing $du/dy$, $dH/dt$ *decreases* with time (Eq. 1). Whenever the amplitude (scale) of the disturbance is large enough, turbulence creation will dominate.

The important dimensionless quantities involved in the competition between turbulence energy creation and heat dissipation can be illustrated by taking the ratio of the growth rates of turbulence energy and heat energy at $t = 0$. Using Eqs. 1 and 6 and setting $du/dy \approx 2u_0/A$ at $t = 0$ we find

$$\left(\frac{dK/dt}{dH/dt}\right)_{t=0} \approx 2\pi^3 \underbrace{\frac{u_0 \rho A_0}{\mu_m}}_{\substack{\text{Local}\\ \text{Reynolds}\\ \text{Number}}} \underbrace{\left(\frac{A_0}{\lambda}\right)^3}_{\substack{\text{Initial}\\ \text{Pertur-}\\ \text{bation}}}, \tag{7}$$

where $A_0/\lambda$ can be thought of as a measure of the extent of the initial perturbation. The appearance in this equation of the local Reynolds number is not surprising, given that the number is a measure of the competition between the inertial and viscous effects in any flow (see "Reynolds Number").

As long as $dH/dt$ dominates, the mean-flow kinetic energy dissipates to heat, and the intermediate turbulent stage is bypassed; we say that the mean flow is stable. If $dK/dt$ dominates, then the mechanism driving the instability draws the excess kinetic energy into turbulence. We can thus formulate a stability criterion, based on the Reynolds number, in which molecular viscosity plays a central role. For large viscosity, $dH/dt$ is able to exactly balance the loss rate for mean-flow kinetic energy. Decreasing the viscosity eventually drops $dH/dt$ below the mean-flow loss rate, and the flow becomes unstable.

As a corollary, note that conservation of *total* energy raises an interesting question about mean-flow dynamics. What mechanism accounts for destruction of mean-flow kinetic energy at exactly the required rate to ensure conservation? The answer is viscosity—molecular viscosity and *turbulence* viscosity.

For the case of *only* direct viscous dissipation to heat, viscous drag between the opposing currents causes each to slow down, and the corresponding loss rate for kinetic energy exactly accounts for the dissipative heating. For the case of transfer of mean-flow kinetic energy to turbulence, a directly analogous process occurs in which turbulence viscosity produces drag. More precisely, the presence of turbulence induces a fluid shear stress, the Reynolds stress, that is independent of the molecular viscosity of the fluid. Expression of the components of the Reynolds stress tensor in terms of readily measured flow quantities (such as pressure and mean-flow velocity) lies at the heart of our theoretical work and is discussed in detail in the next section.

Analogous to molecular viscosity, turbulence viscosity depletes mean-flow kinetic energy at precisely the same rate that turbulence energy is growing. A direct consequence is that turbulence contributes to the effective viscosity of the fluid, enhancing the rate of momentum diffusion from one part of the fluid to another as it simultaneously destroys the excess mean-flow kinetic energy. As we shall see, turbulence diffuses *anything* imbedded in the fluid—momentum, heat, dye, dust particles, dissolved salts.

# Reynolds Number

To design and test proposed large-scale equipment, such as airfoils or entire aircraft, it is often much more practical to experiment with scaled-down versions. If such tests are to be successful, however, dynamic similitude must exist between model and field equipment, which, in turn, implies that geometric, inertial, and kinematic similitude must exist.

The Navier-Stokes equations (Eqs. 9 and 10 in the main text) are a good starting point for deriving the relationships needed to establish dynamic similitude. First, we look at the case of laminar flow. Ignoring body force and pressure effects, we examine the momentum conservation relationship for steady, laminar, incompressible, two-dimensional flow, equating just the advection and diffusion terms in the $x$-direction:

$$\underbrace{\frac{\partial uu}{\partial x} + \frac{\partial vu}{\partial y}}_{\text{Advection}} = \nu_{\mathrm{m}} \underbrace{\left[\frac{d^2 u}{dx^2} + \frac{d^2 u}{dy^2}\right]}_{\text{Diffusion}} . \quad (1)$$

Here $u$ and $v$ are the $x$ and $y$ components of the velocity and $\nu_{\mathrm{m}}$ is the molecular kinematic viscosity (the ratio of fluid viscosity to fluid density $\mu_{\mathrm{m}}/\rho$). Advection has to do with kinematic effects, that is, the transport of fluid properties by the motion of the fluid, and thus accounts for momentum transport *along* streamlines; the diffusion terms represent viscous effects that cause momentum to diffuse *between* streamlines, thereby tending to diminish any sharp velocity gradients.

We can write Eq. 1 in dimensionless form by introducing a length scale $L$ and a fluid velocity in the free stream $u_0$. The result is

$$u_0 \left[\frac{\partial \mathbf{uu}}{\partial \mathbf{x}} + \frac{\partial \mathbf{vu}}{\partial \mathbf{y}}\right] = \frac{\nu_{\mathrm{m}}}{L} \left[\frac{d^2 \mathbf{u}}{d\mathbf{x}^2} + \frac{d^2 \mathbf{u}}{d\mathbf{y}^2}\right], \quad (2)$$

where the highlighted variables are dimensionless. This portion of the momentum equation can thus be uniquely characterized by the ratio of the coefficients multiplying the dimensionless advection and diffusion terms. The ratio, called the Reynolds number

$$R = \frac{u_0 L}{\nu_{\mathrm{m}}}, \quad (3)$$

can be thought of as a comparative measure of inertial and viscous (diffusive) effects within the flow field. To achieve dynamic similitude in two different laminar-flow situations, the Reynolds numbers for both must be identical.

What happens if we increase the flow speed to the point that viscous dissipation can no longer stabilize the flow, and the macroscopic balance between mean-flow inertia and viscous effects breaks down? At this point there is a transition from purely laminar flow to turbulence. In similar flows, the transition occurs at a specific Reynolds number characteristic of the flow geometry. For instance, any fluid traveling inside a circular pipe—regardless of the specific fluid or conduit being used—experiences the onset of turbulence at $R \cong 2000$.

At or near this "critical" Reynolds number, inertial contributions to mean-flow momentum that cannot be dissipated by viscous stresses must be absorbed by newly formed turbulent eddies. The presence of turbulence energy is often described in terms of an effective turbulence viscosity $\nu_{\mathrm{t}}$, defined as the ratio of the turbulence-shear, or Reynolds, stress to the mean-flow strain rate. With this in mind, an effective *turbulence* Reynolds number—one that includes molecular viscous effects—is

$$R_{\mathrm{eff}} = \frac{u_0 L}{\nu_{\mathrm{t}} + \nu_{\mathrm{m}}}. \quad (4)$$

Molecular viscous effects are overwhelmed if $\nu_{\mathrm{t}} \gg \nu_{\mathrm{m}}$. In those instances the exact value of the kinematic viscosity $\nu_{\mathrm{m}}$ is immaterial, and flow behavior is dominated by turbulence effects.

Although a turbulence Reynolds number may be entirely adequate for research on macroscopic flows, the analysis of turbulence *substructure* requires a third Reynolds number, a local turbulence Reynolds number based not on $L$ and $u_0$ but on representative eddy size $s$ and eddy velocity $u'$:

$$R_{\mathrm{s}} = \frac{u's}{\nu_{\mathrm{m}}}. \quad (5)$$

Note that the molecular kinematic viscosity $\nu_{\mathrm{m}}$ is retained in this definition. The choice of molecular viscosity to characterize the dissipative mechanisms responsible for tearing eddies apart is based on the ultimate transformation of turbulence into heat energy. Molecular processes are, in the end, dominant at the smallest scales, and $R_{\mathrm{s}}$ is a relative measure of the loss of kinetic energy from an eddy of a given size to heat. For the smallest eddies in a flow system, $R_{\mathrm{s}} \cong 1$; that is, all the energy of the eddy is dissipated into heat. ■

What then can we deduce from this example about the features necessary for the creation of turbulence?

- A mean-flow profile richer in kinetic energy than other momentum-conserving states to which it can transform (such as the profile in Fig. 6(a) that can transform to the one in Fig. 5).
- A viscosity low enough that dissipation to heat cannot absorb all the mean-flow energy during the transition to the low-energy profile.
- A driving mechanism for enhancement of the inevitable microscopic perturbations (such as the Kelvin-Helmholtz instability in Fig. 6).

However, the energy of turbulence frequently comes from sources (Fig. 7) other than a velocity profile rich in mean-flow kinetic energy. For example, turbulence can be fed directly from potential energy as when a Rayleigh-Taylor instability develops at the interface between, say, water overlying a less dense layer of oil or cold air overlying warm air. The latter instance, called buoyancy-driven turbulence, produces the dancing air currents that can be seen by looking across the surface of a sunlit roof on a cold day. Similarly, turbulence can be fed by accelerative forces as when a Richtmyer-Meshkov instability develops at the deformable interface between two materials that are perturbed by, say, a passing shock wave or the sudden acceleration of the entire system.

Droplets, particles, or bubbles projected through a liquid or gaseous fluid with some relative velocity likewise can serve as a good source of turbulence energy. The momentum-conserving transition induced by drag tends always to bring such entities and the fluid to the same velocity. Competition for the center-of-mass kinetic energy results in a partition into both heat and turbulence—the winner again depending on the level of viscosity.

Likewise, if a quiescent suspension is subjected to a pressure gradient or shock, a differential acceleration occurs that is in proportion to the difference in densities between the suspended entities and the surrounding medium. Turbulence often gleans a significant share of the resulting interpenetrational energy.

**Turbulence Sinks.** So far we have been discussing only *sources* for turbulence and the manner in which the turbulence *decays*. Here we must return to what constitutes turbulence and, in particular, reaffirm that the existence of turbulence depends on the observer's point of view. Mean flow is that part of the dynamics whose structure is comparable in size to the region being measured; it is capable of being reproducibly duplicated or monitored—at least in some statistical sense. Finer dynamical scales of a capricious nature arising from random initial, boundary, or bulk perturbations constitute the fluid's turbulence. But the mean flow for one observer may simply be the larger scales of a turbulence spectrum for an observer whose field of view encompasses a somewhat larger domain. Thus, the *source* of turbulence seen by one observer becomes the energy *sink* for the decay of turbulence at the larger scales of another observer.

This principle and its generalizations have powerful consequences for our mathematical modeling of turbulence dynamics, leading to the concept of a *turbulence cascade*. In this process turbulence energy is transferred to progressively smaller and smaller fluctuational scales with the source of energy for each scale coming from the mean-flow velocity contortions of the next larger scale (Fig. 8). At each stage, there is competition for the energy, part going into heat and part going into even smaller

## SOURCES OF TURBULENCE

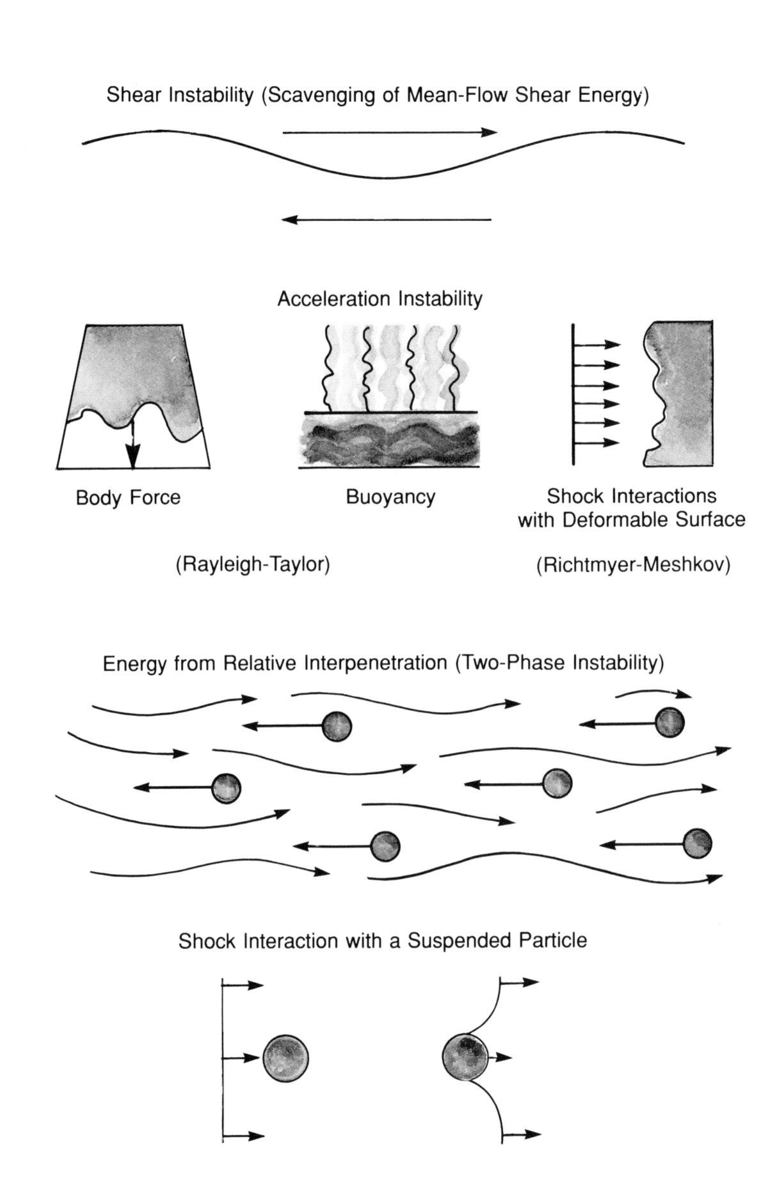


**Fig. 7. Although we have so far dealt only with shear instabilities (Figs. 5 and 6), there are many other sources of turbulence, ranging from the instability of one fluid overlying a less dense one, through the interpenetration of two distinct phases, to the interaction of a shock wave with particles or surfaces.**

turbulent fluctuations. However, as the scale decreases, the characteristic length of the eddies decreases, and the velocity gradients in the eddies become steeper and steeper. In other words, $dH/dt$ eventually wins, and, at the smallest of turbulence scales, energy goes directly to heat.

Thus, cascading of turbulence is consistent with nature's universal law dictating that ordered motion must become progressively more disordered until the energy in a flow degrades to heat. The direction and magnitude of energy flow within the cascade guides us in mathematically describing the decay of turbulence, not only into heat from very small-scale eddies but also from large scales to smaller scales. Because the transfer of energy through the cascade is, in some sense, equal at all steps, we can easily describe the energy decay rate in a manner independent of molecular processes. We will describe this approach more extensively when we consider detailed modeling in the next section.

In an idealized steady-state approximation of turbulence, exactly as much energy enters the fluctuational spectrum of motion at the largest scale as leaves it to become heat at the smallest scale. More accurately, there is *some* loss of energy to heat at every scale, but the loss at the smallest scale is dominant (Fig. 9). Although these ideas have been exploited to derive interesting properties of the small-scale spectrum of turbulence energy, our principal concern here is with the largest scales. It is these scales that contain most of the energy and thus exert the dominant effects on mean-flow dynamics.

## Transport Modeling of Turbulence

There are numerous theoretical approaches to turbulence: some reach to the conceptual heart of the matter, others are directed toward the solution of practical problems, and a few attempt to cover the entire range. Despite its present shortcomings, *turbulence transport theory*, which fits into the last category, already shows promise of considerable success in both illuminating the fundamental dynamical processes and serving as a vehicle for the solution of practical problems.

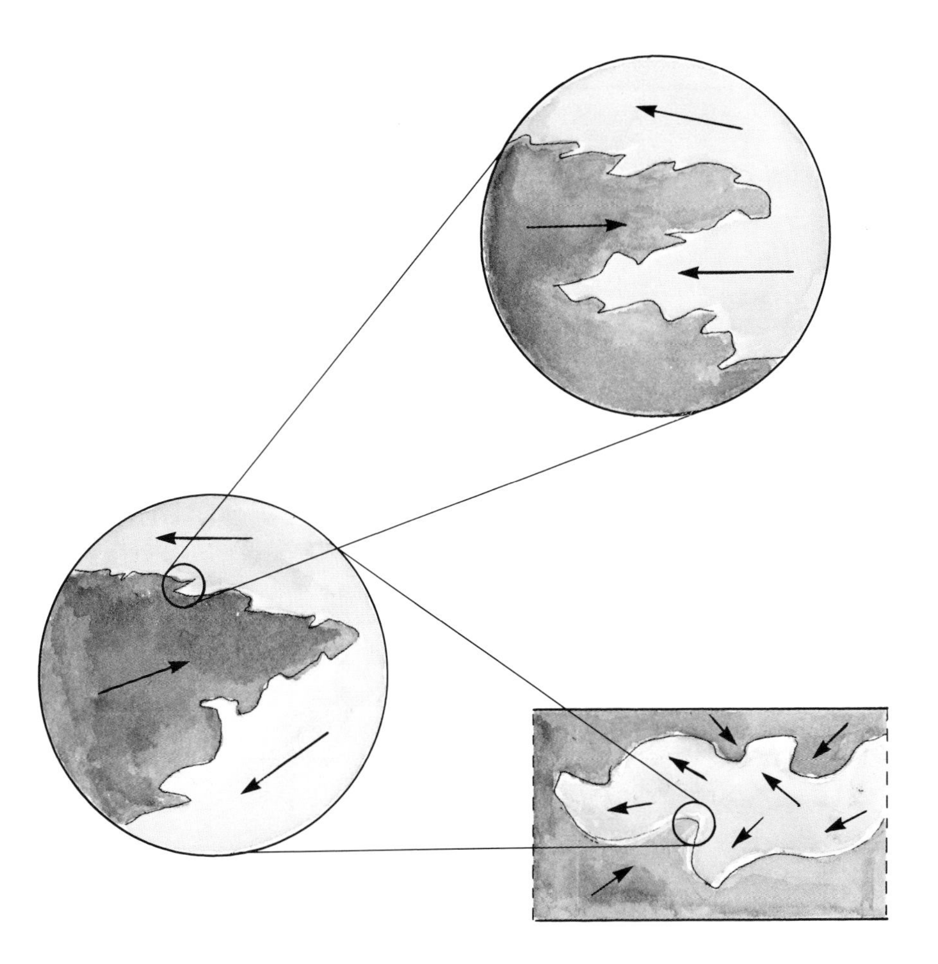

**TURBULENCE CASCADE**

Fig. 8. With each reduction in scale, turbulent motion of the larger scale becomes mean-flow motion of the smaller scale (arrows). Because each reduction in scale has approximately the same change in mean-flow velocity occurring over a much smaller distance, velocity gradients become steeper, and a larger fraction of the turbulence energy goes directly into heat.

## TURBULENCE ENERGY FLOW

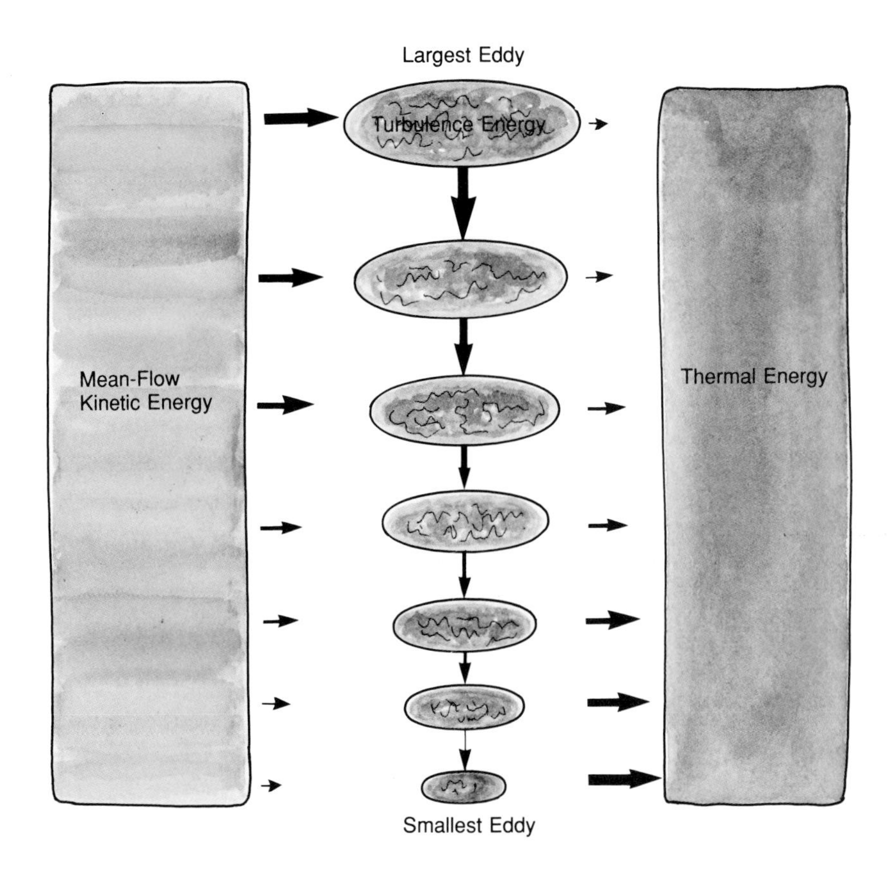


Fig. 9. Mean-flow kinetic energy transforms into turbulence energy and thence to heat energy, but the relative amounts (indicated by the sizes of the arrows) transported by each mechanism changes as the scale of the turbulence changes. For example, in the highly turbulent system being illustrated here, much of the mean-flow kinetic energy feeds into large-scale turbulence, whereas thermal energy receives much of its energy from small-scale turbulence.

Even for a single fluid with constant, uniform density the relevant mathematical formulations are lengthy, and there are significant difficulties yet to be resolved. Nevertheless, we can capture in a relatively simple manner much of the flavor of turbulence modeling by starting with the Navier-Stokes fluid-dynamics equations for an incompressible fluid—that is, a fluid of constant density and viscosity everywhere and for all time. One of our fundamental assumptions is that these familiar and deceptively simple equations describe everything we need to understand about the turbulence of such a fluid, including every "microscopic" detail in every fluctuating part of the turbulent flow.

The Navier-Stokes equations describe the variations of pressure and velocity in the fluid. Using Cartesian index notation with the summation convention, we can write the first equation, which is an expression of the conservation of mass, as

$$\frac{\partial u_i}{\partial x_i} = 0, \tag{9}$$

and the second, which is an expression of the conservation of momentum, as

$$\underset{\text{Rate of Change}}{\frac{\partial u_i}{\partial t}} + \underset{\text{Advection}}{\frac{\partial u_i u_j}{\partial x_j}} = \underset{\text{Driving Force}}{-\frac{\partial p}{\partial x_i}} + \underset{\text{Diffusion}}{\nu_{\mathrm{m}} \frac{\partial^2 u_i}{\partial x_k^2}}. \tag{10}$$

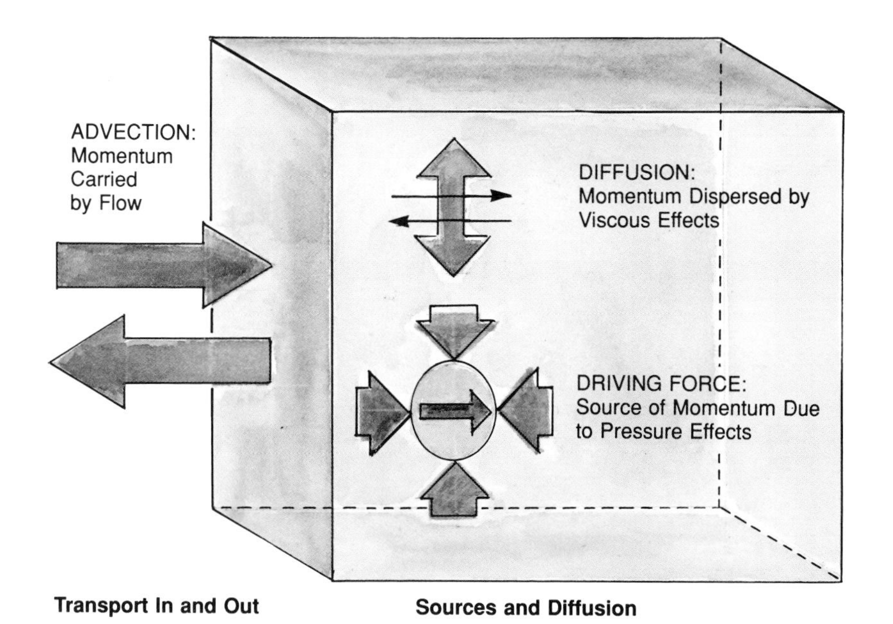


**MOMENTUM TRANSPORT**

**Fig. 10. The diffusion, driving-force, and advection terms of the Navier-Stokes momentum equation represent the ways in which momentum is locally added to or taken away from a region in the fluid.**

Here $\nu_m$ is the molecular *kinematic* viscosity coefficient (the ratio of molecular viscosity to density $\mu_m/\rho$), $u_i$ is the $i$th component of the vector velocity, and $p$ is the ratio of pressure to density. Figure 10 illustrates the effects of the various terms in the momentum equation.

So, there it is in a nutshell: the entire, mysterious world of turbulent fluid flow described by two short lines of mathematical symbols. Well, not quite the entire *real* world, because many fluids are not the idealized incompressible materials of constant viscosity and density considered here, but we shall return to that point below.

One obvious approach to modeling turbulence is to solve the Navier-Stokes equations directly (the left path in Fig. 11). However, certain difficulties limit the success of this approach. For example, even when the mean flow is one-dimensional, the equations must be solved numerically in three dimensions because turbulence is inherently three-dimensional. Only recently have computers had enough computational capability to begin meeting the task of solving three-dimensional fluid-flow problems. To describe the full spectrum of eddies, the computational mesh would have to be fine enough for the smallest eddies, yet cover a domain large enough to include the mean flow and the largest eddies. Another complication occurs if the system includes a solid boundary. Because the turbulent flow depends on the minute details of the boundary conditions (even *stochastic* quantitites depend on minute perturbations in the initial and boundary conditions, such as wall roughness), these details must be specified. Furthermore, because a particular set of minute perturbations describe only one possible representation of the boundary conditions, repeated calculations must be made with various boundary conditions and the results of the calculations averaged to give a complete description of the turbulent flow. The memory and speed requirements for the calculations would

## APPROACHES TO TURBULENCE MODELING

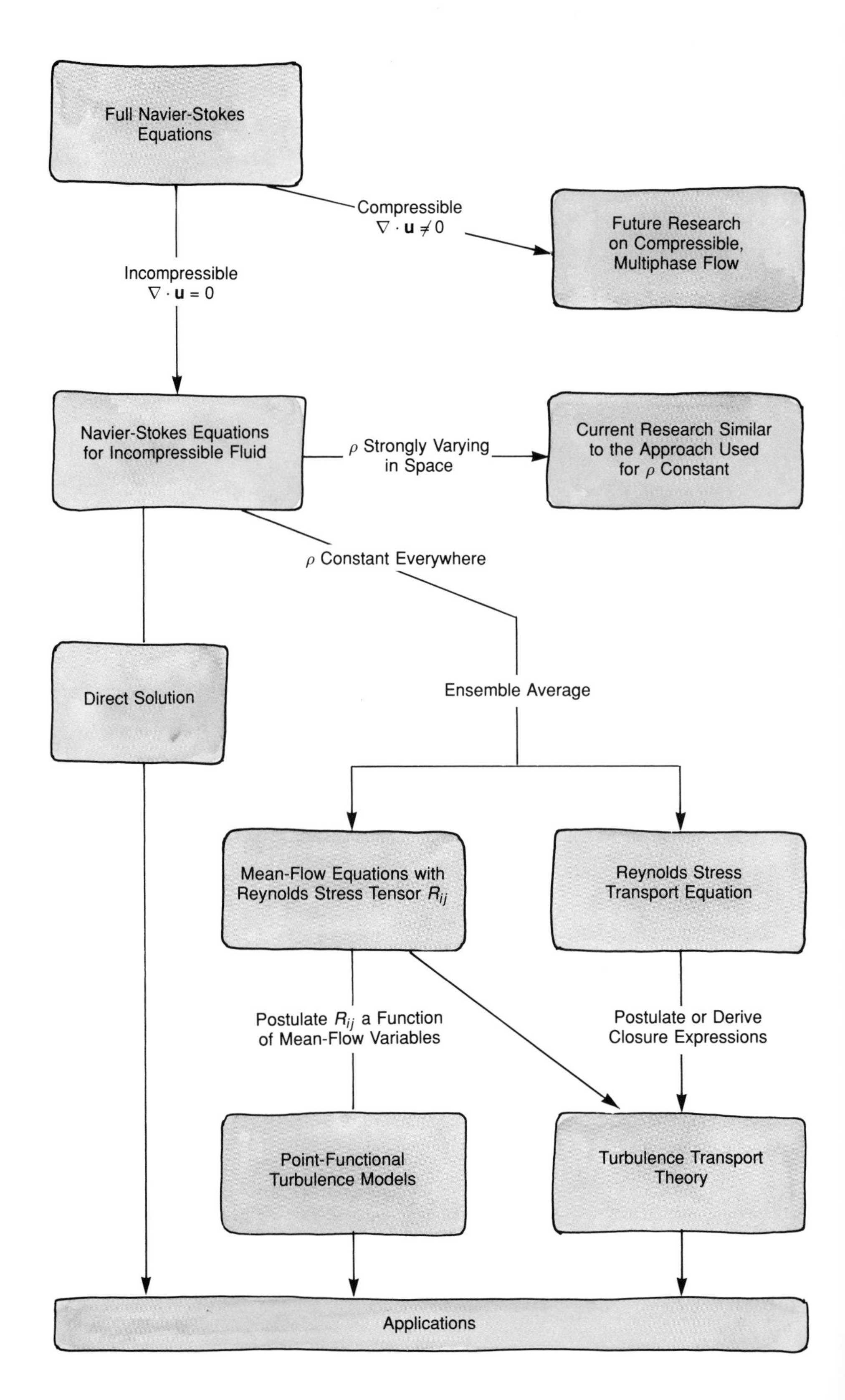


**Fig. 11. Typically, modeling of turbulence makes two simplifying assumptions with respect to the full Navier-Stokes equations: an incompressible fluid ($\nabla \cdot \mathbf{u} = 0$) and the density $\rho$ constant everywhere. Using just the first of these assumptions, one could, in principle, solve the equations directly (left path)—a difficult task. However, if one uses *both* simplifying assumptions together with ensemble averaging, the result is two sets of equations: the mean-flow equations, which include the Reynolds stress tensor $R_{ij}$, and the Reynolds stress transport equation. Turbulence transport theory (right) uses input from both sets, whereas point-functional turbulence models (middle) deal only with the mean-flow equations (by postulating that $R_{ij}$ is a function of mean-flow variables). Later in this article we describe work on multiphase flow in which the assumption of constant density has been dropped. Current research is just beginning to approach the full Navier-Stokes equations for compressible, multiphase flow.**

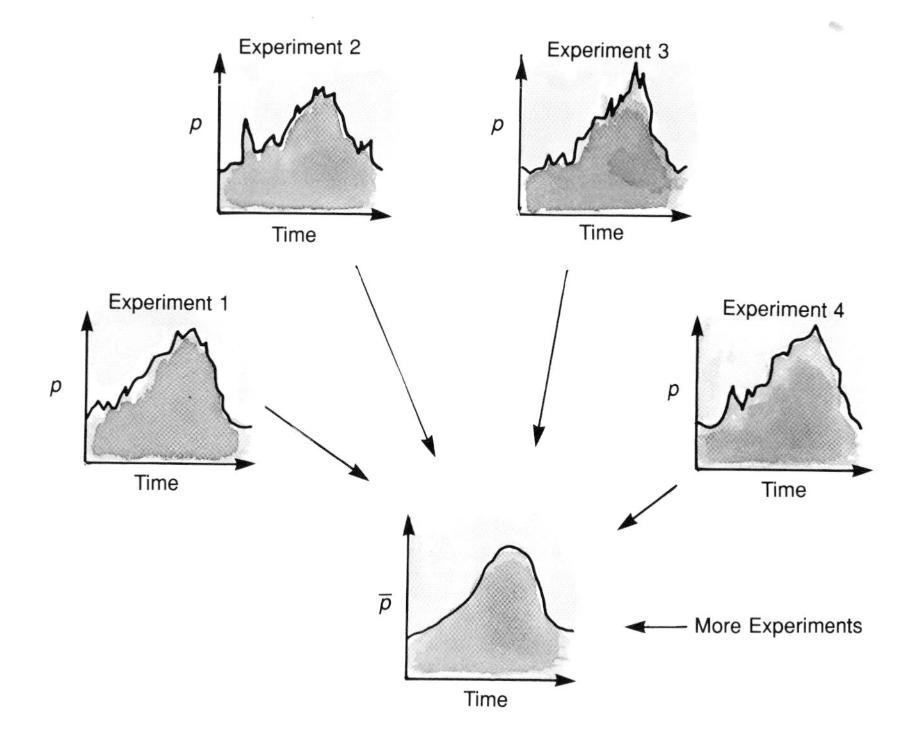


**ENSEMBLE AVERAGING**

**Fig. 12. Consider a series of experiments, each conducted with the same initial and boundary conditions. For each, we determine the pressure *p* at a particular point in space as a function of time. An average of all these experiments would represent an ensemble average.**

tax even the most powerful of our modern computers. If these calculations could be accomplished, however, the *advantage* of a direct calculation of turbulence would be that no approximations or empirical postulates are required.

**Ensemble Averages.** Largely for the reasons given above, almost all theoretical approaches to turbulence modeling use some type of averaging—either temporal, spatial, or ensemble. With the proper statistical treatment, the solution of turbulent flow problems need not resolve the full spectrum of eddies, initial and boundary conditions need not be specified in minute detail, and a flow whose mean velocity is one-dimensional can be numerically calculated in one dimension even though the resolved turbulence is three-dimensional. However, with these advantages for turbulence transport modeling come the disadvantages of assumptions and approximations needed to obtain a set of solvable equations.

What is meant by the average of any flow variable in a turbulent flow? Time averages are easy to understand. We say that fluid flow is statistically steady if the time average of many fluctuations at some point in space is independent of the averaging period chosen. Spatial averages, likewise, are easy to visualize but are relevant only when the structural scale of the turbulence is *very* small compared with that of the mean-flow fluctuations—a relatively rare condition. Here we will focus on ensemble averaging, which is the most general type of averaging with the fewest restrictions.

We can intuitively sense what an ensemble average is if we imagine a very large number of experiments, all with the same macroscopic initial and boundary conditions, but each with its own particular realization of the turbulent part of the flow (Fig 12). The ensemble average of some flow parameter at any given point and time is then the average of that parameter over all the experiments. For the condition of steady flow, time and ensemble averages are the same.

The ensemble average is most conveniently formulated in terms of *moments* of an appropriate distribution function. (Here, rather than integral nomenclature, we will simply use an overline to designate an ensemble average.) Thus $\overline{p}$ is the moment, or ensemble average, of the pressure (per unit density), and $\overline{pu_i}$ is the ensemble average of the product of pressure and the $i$th component of fluid velocity. (Note that $\overline{pu_i}$ does not necessarily equal $\overline{p}\,\overline{u_i}$.)

For each experiment in a series, detailed measurements give $p$ and $u_i$, both of which fluctuate strongly as a function of position and time. Likewise, $\overline{p}$ and $\overline{u_i}$ vary with position and time but in a much calmer fashion. The difference between the individual experimental value and the ensemble average is the fluctuating part of the variable, denoted by a prime: $p' \equiv p - \overline{p}$; $u_i' \equiv u_i - \overline{u_i}$. The ensemble average of this fluctuating part must be zero for each variable (that is, $\overline{p'} = 0$ and $\overline{u_i'} = 0$), but it does not follow that the moment, or ensemble average, of a product of fluctuational variables (such as $\overline{p'u_i'}$ or $\overline{u_i'u_j'}$) vanishes. Indeed, the essence of our turbulence modeling is contained in the behavior of such ensemble averages of fluctuational products.

**Reynolds-Stress Transport Equation.** One of these fluctuational products, the Reynolds stress tensor, is especially important; it is defined by

$$R_{ij} \equiv \overline{u_i'u_j'}. \tag{11}$$

Notice that the contraction of the Reynolds stress tensor (that is, when $i = j$) is exactly twice the turbulence kinetic energy per unit mass of fluid ($R_{ii} \equiv \overline{u_i'u_i'} = 2K$).

The importance of $R_{ij}$ in turbulence modeling can be demonstrated quite handily. First we rewrite the Navier-Stokes equations, expressing each of the variables as the sum of its mean and fluctuating parts:

$$\frac{\partial\left(\overline{u_i} + u_i'\right)}{\partial x_i} = 0$$

and

$$\frac{\partial\left(\overline{u_i} + u_i'\right)}{\partial t} + \frac{\partial}{\partial x_j}\left[\left(\overline{u_i} + u_i'\right)\left(\overline{u_j} + u_j'\right)\right] = -\frac{\partial\left(\overline{p} + p'\right)}{\partial x_i} + \nu_{\mathrm{m}}\frac{\partial^2\left(\overline{u_i} + u_i'\right)}{\partial x_k^2}. \tag{12}$$

Then we take the ensemble average of these equations (commuting averages and derivatives where necessary and remembering that the average of a single fluctuating variable is zero) and obtain the *mean-flow equations*:

$$\frac{\partial \overline{u_i}}{\partial x_i} = 0$$

and

$$\frac{\partial \overline{u_i}}{\partial t} + \frac{\partial}{\partial x_j}\left(\overline{u_i}\,\overline{u_j}\right) = -\frac{\partial \overline{p}}{\partial x_i} - \frac{\partial R_{ij}}{\partial x_j} + \nu_{\mathrm{m}}\frac{\partial^2 \overline{u_i}}{\partial x_k^2}. \tag{13}$$

A single term involving the Reynolds stress has emerged, and we see that the *only* effect of turbulence on the mean flow is through the addition of that term to the equations. We note in passing that Eqs. 13 form the basis of *point-functional turbulence models* (the middle branch in Fig. 11) and will return to this point shortly.

The mean-flow equations (Eqs. 13) can be subtracted from the full equations (Eqs. 12) to show that the fluctuating parts of the variables obey the equations

$$\frac{\partial u_i'}{\partial x_i} = 0 \tag{14a}$$

and

$$\frac{\partial u_i'}{\partial t} + \frac{\partial}{\partial x_j}\left(\overline{u_i}u_j' + u_i'\overline{u_j} + u_i'u_j' - \overline{u_i'u_j'}\right) = -\frac{\partial p'}{\partial x_i} + \nu_{\mathrm{m}}\frac{\partial^2 u_i'}{\partial x_k^2}. \tag{14b}$$

We need Eqs. 14 to derive the *Reynolds-stress transport equation*, that is, a description of the behavior of the Reynolds stress itself (the right branch of Fig. 11). This derivation is straightforward but tedious. We merely note that the following steps are involved:

1. multiply Eq. 14b by $u_j'$ to obtain Eq. 14c,
2. interchange $i$ and $j$ in Eq. 14c to obtain Eq. 14d,
3. add Eqs. 14c and 14d, and
4. take the ensemble average.

With some rearrangement of terms and the identification of $R_{ij}$ in several places, the result is

$$\underbrace{\frac{\partial R_{ij}}{\partial t}}_{\text{Rate of Change}} + \underbrace{\frac{\partial \overline{u_k}R_{ij}}{\partial x_k}}_{\text{Advection}} + \underbrace{R_{ik}\frac{\partial \overline{u_j}}{\partial x_k} + R_{jk}\frac{\partial \overline{u_i}}{\partial x_k}}_{\text{Mean-Flow Source and Rotation}} + \underbrace{\frac{\partial \overline{u_i'u_j'u_k'}}{\partial x_k}}_{\text{Triple Correlation}} =$$

$$\underbrace{-\overline{u_i'\frac{\partial p'}{\partial x_j}} - \overline{u_j'\frac{\partial p'}{\partial x_i}}}_{\text{Driving Force}} + \nu_{\mathrm{m}}\Bigg(\underbrace{\frac{\partial^2 R_{ij}}{\partial x_k^2}}_{\text{Diffusion}} - \underbrace{4D_{ij}}_{\text{Decay}}\Bigg), \tag{15}$$

in which

$$D_{ij} = \frac{1}{2}\overline{\frac{\partial u_i'}{\partial x_k}\frac{\partial u_j'}{\partial x_k}}. \tag{16}$$

We now can distinguish the turbulence-transport theories and their predecessors, the point-functional turbulence theories. As we remarked earlier, point-functional turbulence theories use Eqs. 13 by postulating a form for the Reynolds stress $R_{ij}$ that is a function of the mean-flow variables themselves. As a result, such theories are called "point-functional" because the description of the turbulence at some point in the flow depends only on the current value of the mean-flow variables. Point-functional theories have the advantage of being as easy to solve as the original Navier-Stokes equations but have the shortcoming that the theories are largely empirical and have limited regions of applicability.

In contrast to point-functional theories are the *history-dependent*, or *turbulence-transport*, theories. These theories, the focus of our interest here, include a set of one or more *auxiliary* equations that describe the history, or transport, of the variables associated with turbulence and that are solved in conjunction with the mean-flow equations (Eqs. 13). The auxiliary equations can range from empirical postulations to some form of the Reynolds transport equation (Eq. 15).

Because our starting point was the Navier-Stokes equations, turbulence-transport theory based on Eqs. 13 and 15 should, in principle, contain all the necessary information to describe the mean properties of turbulent flow. However, in practice it is necessary to introduce additional constraints or empirical information to yield a solvable set of equations. This procedure of "closing" the set of governing equations is called *closure modeling* and plays a central role in turbulence-transport theory.

The development of a solvable set of equations is beyond the scope of this article (although, in the following section we do so for a simple treatment of turbulence). We can nevertheless capture much of the flavor of the necessary developments by considering the significance of the terms in the Reynolds transport equation (Fig. 13 graphically illustrates the nature of each) and by considering the difficulties of describing their properties in terms of the macroscopically accessible mean-field quantities.

**Advection, Mean-Flow Source, and Rotation.** The advection and the mean-flow source and rotation terms of Eq. 15 contain only the unknown tensor $R_{ij}$ and the mean-flow velocities; no reference to the detailed turbulence structure occurs. These terms constitute a bulwark of settled mathematical structure for which there are essentially no uncertainties or controversies about the physics. In essence they describe the manner

**TURBULENCE TRANSPORT**

**Fig. 13. Just as Fig. 10 illustrates the various terms of the Navier-Stokes momentum equation, this figure illustrates the various terms of the Reynolds-stress transport equation. The driving force and the diffusion terms appear twice because each can be decomposed into a contribution to the transport of turbulence and a contribution to the generation or diffusion of turbulence.**

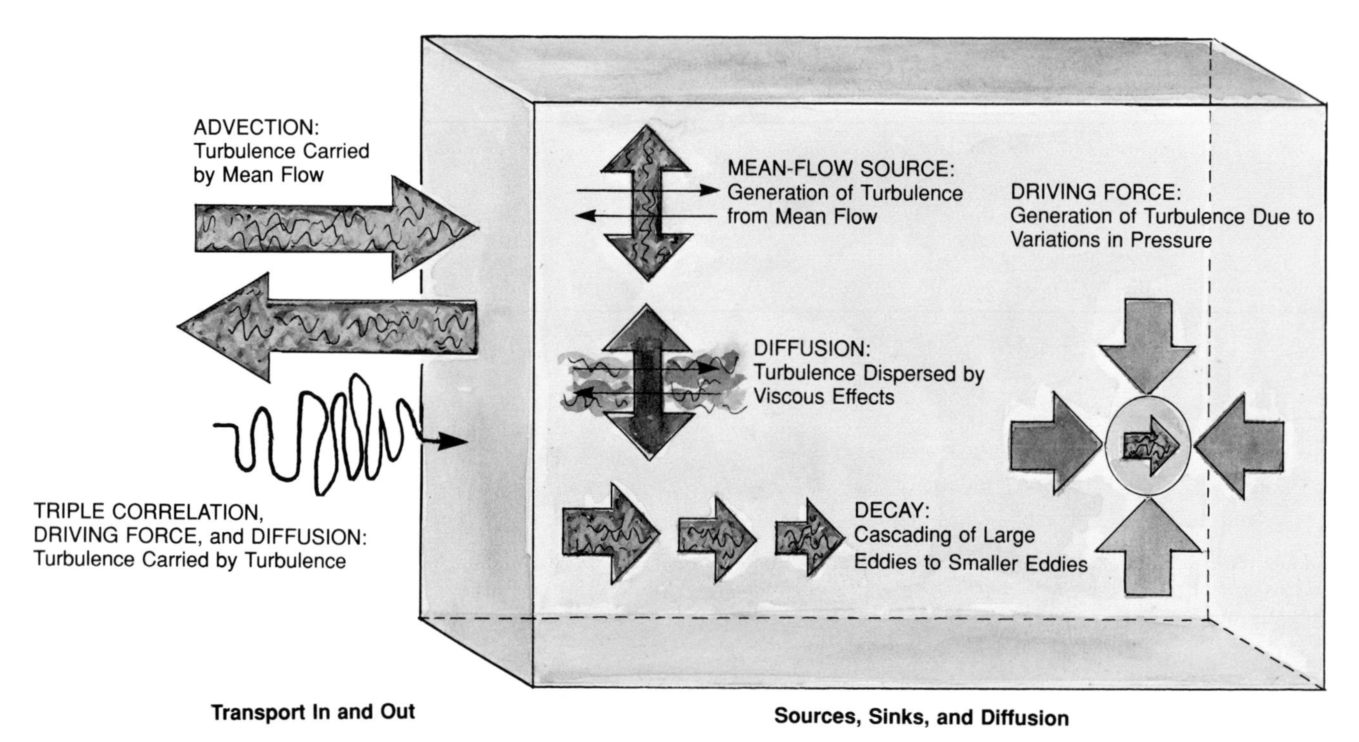

in which the mean flow moves turbulence from one place to another by translation, rotation, and stretching or contraction of the fluid.

**Triple Correlation.** The triple-correlation tensor $\overline{u_i'u_j'u_k'}$ that appears in the next term in Eq. 15 is usually interpreted as a diffusive flux of the Reynolds stress generated by the action of the stress itself. Thus, this term can be called the turbulence self-diffusion term because it describes the turbulent diffusion of turbulence.

We can show in more detail how this identification is made and, at the same time, illustrate what is meant by closure modeling. If $Q$ represents some quantity (such as the concentration of a dissolved, neutrally-buoyant substance) that is purely advected by the incompressible fluid, its transport equation is simply

$$\frac{\partial Q}{\partial t} + \frac{\partial (Qu_i)}{\partial x_i} = 0. \tag{17}$$

Decomposing the variables into mean and fluctuating parts and taking the ensemble average (as we did before with Eqs. 12 and 13), we find that

$$\frac{\partial \overline{Q}}{\partial t} + \frac{\partial (\overline{Q}\,\overline{u_i})}{\partial x_i} = -\frac{\partial \overline{Q'u_i'}}{\partial x_i}. \tag{18}$$

Since the right side describes the diffusion of $Q$ due to the effects of turbulence, we directly identify $\overline{Q'u_i'}$ as a diffusive flux. Just as the flux of a chemical species is proportional to its concentration gradient (Fick's law), the diffusive flux is proportional to the gradient of $Q$ itself:

$$\overline{Q'u_i'} \propto \frac{\partial \overline{Q}}{\partial x_i}. \tag{19}$$

The proportionality constant is a function of the turbulence intensity; indeed, more detailed considerations indicate that

$$\overline{Q'u_i'} \approx -\left(\frac{s}{\sqrt{2K}}\right) R_{il} \frac{\partial \overline{Q}}{\partial x_l}, \tag{20}$$

in which $s$ is the length scale of the turbulence. It follows that

$$\overline{u_i'u_j'u_k'} \equiv \overline{R_{ij}'u_k'} \approx -\left[\frac{s}{(2K)^{1/2}}\right] R_{kl} \frac{\partial R_{ij}}{\partial x_l}. \tag{21}$$

In this manner, we see what is meant by closure modeling, that is, the elimination of any residual reference to details of the turbulence. For our purposes we need not delve any deeper into this aspect of turbulence modeling; the example is sufficient to indicate some of the heuristic and empirical procedures we inevitably have been forced to employ.

**Driving Force.** The pressure-velocity correlation terms (the first two terms on the right side of Eq. 15) are especially important to the transport modeling of turbulence. They describe one of the principal driving forces by which mean-flow energy finds its way into fluctuations. Moreover, they contribute significantly to the isotropic rearrangement of anisotropic turbulence.

The unstable slip plane of Fig. 6a is excellent for visualizing the effects of these terms. Previously, we observed that on either side of the slip plane a slight increase in velocity over the mean for that side is accompanied by a slight decrease in the pressure, whereas exactly the opposite occurs in the other half of the fluid. Thus, fluctuations in pressure and velocity are strongly correlated. Because $u_x'$ and $\partial p'/\partial y$ have opposite signs (for example, an increase in $u_x'$ in the lower half results in a downward or negative pressure gradient), the ensemble average of the product of these variables is always negative. But the two $\overline{u_i' \frac{\partial p'}{\partial x_j}}$ terms in Eq. 15 are negative, so these terms are a *positive* source to $R_{xy}$ (that is, to the *anisotropic, or off- diagonal, components* of the Reynolds stress tensor).

Once $R_{xy}$ is created, it ultimately contributes to the turbulence kinetic energy $K$, which, as we noted earlier, is proportional to $R_{ii}$ (that is, to the sum of *diagonal components* of the stress tensor). That $R_{xy}$ contributes to $K$ is easily illustrated by examining the contracted form of the Reynolds stress transport equation for, say, the type of flow illustrated in Fig. 5. In this case, the mean-flow source terms contribute to the rate of change only as follows:

$$\frac{\partial R_{ii}}{\partial t} \equiv 2\frac{\partial K}{\partial t} = -2R_{xy}\frac{\partial \overline{u}_x}{\partial y}. \tag{22}$$

Hence, the anisotropic, or off-diagonal, components of $R_{ij}$, once created by the mean flow from the driving-force terms, eventually contribute to the turbulence kinetic energy through the mean-flow source terms.

**Diffusion and Decay.** Of the last two terms in Eq. 15, the first is usually negligible and represents diffusion of turbulence by molecular viscosity, which requires no further modeling. The second involves the tensor $D_{ij}$, for which the usual procedure has been to derive a horrendously complicated transport equation and attempt to solve this simultaneously with the Reynolds transport equation. Such a procedure introduces a host of additional correlation terms to be modeled, and much appeal to "intuition" is invoked in the process.

Bypassing the fascinating but tedious discussion of these derivations, we can nevertheless describe several interesting properties of this second term. First, its contraction

$$D_{ii} \equiv \frac{1}{2}\overline{\left(\frac{\partial u_i'}{\partial x_k}\right)^2} \tag{23}$$

is positive definite so that the $-4\nu_m D_{ij}$ term in Eq. 16 always describes a *decay* of the turbulence energy.

The second interesting property, deduced by extensive manipulations of the $D_{ij}$ transport equation, is that $D_{ij}$ should vary inversely as the molecular viscosity under almost all circumstances. Therefore $\nu_m D_{ij}$ is essentially independent of viscosity, which seems paradoxical. Resolution of this paradox hinges on an important property of turbulence: most of the turbulence effects and energy are associated with the largest of the eddies, which decay first by cascading to smaller eddies before converting to thermal energy (Fig. 9). Thus, an alternative to the usual modeling of the behavior of $D_{ij}$ has recently emerged. We can get the same results by treating the decay of the large-scale eddies as the energy source of the small-scale eddies. For this purpose the large-scale eddies are momentarily thought of as being "mean flow." In some complicating circumstances, such as interpenetration of particles, this alternative modeling technique has proven so far to be the only tractable approach.

## Simpler Transport Models and Examples of Their Application

Some problems do not warrant the degree of complexity and closure approximation required to numerically solve the full Reynolds-stress transport equation. A more conventional and practical approach uses the following approximation (called Boussinesq's approximation) for turbulence stresses in an incompressible fluid:

$$R_{ij} = -\nu_t \left( \frac{\partial \overline{u_i}}{\partial x_j} + \frac{\partial \overline{u_j}}{\partial x_i} \right) + \frac{2}{3} K \delta_{ij}, \quad (24)$$

in which $\nu_t$, the *turbulence* viscosity, is a measure of the increase in viscosity due to turbulence (see "Reynolds Number" and "Reynolds Number Revisited"), and $\delta_{ij}$ is the Kronecker delta function ($\delta_{ij} = 0$ if $i \neq j$, $\delta_{ij} = 1$ if $i = j$, and $\delta_{ii} = 3$). This approximation is consistent with the definition of turbulence kinetic energy in terms of the Reynolds stress: $R_{ii} = 2K$. Furthermore, the approximation bears a strong resemblance to the Stokes formulation for laminar-flow stresses $p_{ij}$, in which the stresses are related to *molecular* viscosity and fluid pressure (rather than turbulence viscosity and kinetic energy):

$$p_{ij} = \nu_m \left( \frac{\partial \overline{u_i}}{\partial x_j} + \frac{\partial \overline{u_j}}{\partial x_i} \right) - p\delta_{ij}. \quad (25)$$

The chief advantage of using Boussinesq's approximation is that transport relationships for all individual components of $R_{ij}$ are replaced by a single expression involving an effective turbulence, or eddy, viscosity.

How does one describe $\nu_t$? The simplest imaginable description of the turbulence viscosity is that it is a constant that depends on some average mean-flow parameters. Somewhat better is a formulation that relies on a mixing length $l$, which is usually an algebraic estimate of the size of the main energy-containing eddies as a function of flow geometry. For example, one approach that has proven quite successful for boundary-layer flow and some other well-defined jet flows is to define $\nu_t$ by modifying Prandtl's mixing-length theory so that

$$\nu_t = l^2 \frac{\partial \overline{u}}{\partial n}. \quad (26)$$

In this equation, $n$ is the local distance to a rigid object or axis of symmetry and $u$ is a representative free-stream velocity. Note that Eq. 26 makes $\nu_t$ a function of mean-flow parameters only and is thus an example of point-functional modeling.

# Reynolds Number Revisited

As discussed in the earlier sidebar, the Reynolds number is a convenient and physically sound basis for comparing similar flows under different circumstances. For instance, as flow speed through a pipe increases, the drag on the fluid increases and, consequently, so also does the required applied pressure; these increases are reflected in a corresponding increase in Reynolds number. Total friction experienced by the fluid undergoing laminar flow is usually expressed in terms of the Reynolds number $R$, allowing easy comparison between widely varying tests.

Once the Reynolds number reaches a critical value, however, laminar flow in the pipe becomes turbulent, and further increases in Reynolds number no longer reflect significant changes in measured drag. At this point, the effective turbulence Reynolds number $R_{\text{eff}}$ becomes a more appropriate gauge, reflecting the ratio of inertial to turbulence momentum-dissipation effects (rather than inertial to viscous-dissipation effects).

Although the Reynolds number can, in theory, be increased without bound, the turbulence Reynolds number cannot. The value of $R_{\text{eff}}$ is not directly and uniquely set by readily measured properties and flow geometry but rather depends on eddy generation and the resulting eddy sizes within the flow field. A limiting value of $R_{\text{eff}}$ is observed in turbulent-flow experiments.

To demonstrate this behavior quantitatively, it is convenient to make some simplifying assumptions. Typically, the turbulence viscosity $\nu_t$, which is much larger than the molecular viscosity $\nu_m$, is taken to be equal to the product of eddy size $s$ (the turbulence length scale), an appropriate turbulence velocity (here taken as $K^{1/2}$, where $K$ is the specific turbulence kinetic energy), and a universal constant $1/C_\nu$. Thus

$$R_{\text{eff}} \approx C_\nu \frac{u_0 L}{s K^{1/2}}, \qquad (1)$$

where $L$ is a characteristic length for the mean flow.

If, as is usually the case, the turbulence kinetic energy is some fraction of the mean-flow kinetic energy ($K \cong \frac{1}{2} f_K u_0^2$), then

$$R_{\text{eff}} \cong C_\nu \sqrt{\frac{2}{f_K}} \left(\frac{L}{s}\right); \qquad (2)$$

that is, $R_{\text{eff}}$ is proportional to the ratio of the length scales. As turbulence gains in intensity, its *average* length scale usually decreases slightly, but not without limit. In fact, the largest eddies, those that contain the major fraction of the turbulence kinetic energy, will be some portion of the mean-flow length scale (such as pipe diameter). Therefore, since $C_\nu$ is usually about 10 and an upper bound on $L/s$ is typically 20, $R_{\text{eff}}$ will seldom exceed several hundred, even in the most intensely turbulent flows. On the other hand, $R$ can be several million or more. ■

For more complex flows the simplifications introduced through Eq. 26 are not justified, and a transport model that details evolution of $\nu_t$, or quantities related to it, is needed. Although this approach is viable only at the expense of much added complexity, it has recently been favored by investigators working with the complicated flow patterns of high-speed jets, shock-boundary-layer interactions, and two-phase flows.

To produce a simplified transport description of turbulence, we rely on flow properties already introduced to develop a dimensionally correct form of $\nu_t$. For example,

$$\nu_t \equiv \frac{C_\nu K^2}{2\nu_m D_{ii}} = C_\nu \frac{K^2}{\epsilon}, \tag{27}$$

where $C_\nu$ is a model constant, hopefully universal in its applicability, and $\epsilon = 2\nu_m D_{ii}$. The last parameter, $\epsilon$, is related to the mean rate of dissipation of turbulence kinetic energy and, as discussed in the earlier section entitled "Diffusion and Decay," is independent of molecular viscosity.

From this definition of turbulence viscosity, we can generate transport equations for $\epsilon$ and $K$. For instance, after performing the tensor contraction of the Reynolds transport equation (Eq. 15) and introducing appropriate closure expressions, we obtain the following simplified transport equation for $K$:

$$\frac{\partial K}{\partial t} + \overline{u_k}\frac{\partial K}{\partial x_k} = \frac{\partial}{\partial x_k}\left(\frac{\nu_t}{\sigma_K}\frac{\partial K}{\partial x_k}\right) + \nu_t\left(\frac{\partial \overline{u_i}}{\partial x_k} + \frac{\partial \overline{u_k}}{\partial x_i}\right)\frac{\partial \overline{u_i}}{\partial x_k} - \epsilon. \tag{28}$$

Here $\sigma_K$ is a model constant of order one, and, because of Eq. 27, $\nu_t$ is itself a function of $K$ and $\epsilon$.

Next, a treatment similar to that used earlier to split the flow variables into mean and fluctuating parts is applied to the Navier-Stokes momentum equation to create a transport equation for $D_{ij}$. Again, after contraction and closure modeling, we get the following transport equation for $\epsilon$:

$$\frac{\partial \epsilon}{\partial t} + \overline{u_k}\frac{\partial \epsilon}{\partial x_k} = \frac{\partial}{\partial x_k}\left(\frac{\nu_t}{\sigma_\epsilon}\frac{\partial \epsilon}{\partial x_k}\right) + \frac{C_1 \epsilon \nu_t}{K}\left(\frac{\partial \overline{u_i}}{\partial x_k} + \frac{\partial \overline{u_k}}{\partial x_i}\right)\frac{\partial \overline{u_i}}{\partial x_k} - \frac{C_2 \epsilon^2}{K}, \tag{29}$$

where $C_1$, $C_2$, and $\sigma_\epsilon$ are model constants, all of order one.

To show how we apply this model, we return to the problem of turbulence in a slip plane (Fig. 6). Our goal is to demonstrate numerically that turbulence is indeed generated by such a configuration and that we can follow its development throughout a two-dimensional flow field as a function of time and position. We use three different methods of accounting for the turbulence and discuss the pros and cons of each.

The first method involves direct solution of the Navier-Stokes equations by a finite-difference method as an approximation to the left path in Fig. 11. Our calculations use a two-dimensional velocity field, and turbulence below the scale of the computational grid is thus ignored. We assume that a slight, sinusoidal vertical velocity is imparted to the interface separating the oppositely flowing fluids. The maximum speed of this perturbation is only 1 per cent of the mean translational speed (and thus the kinetic energy associated with the perturbation is, at most, $10^{-4}$ times the mean-flow kinetic energy).

## THREE SIMPLIFIED TURBULENCE CALCULATIONS

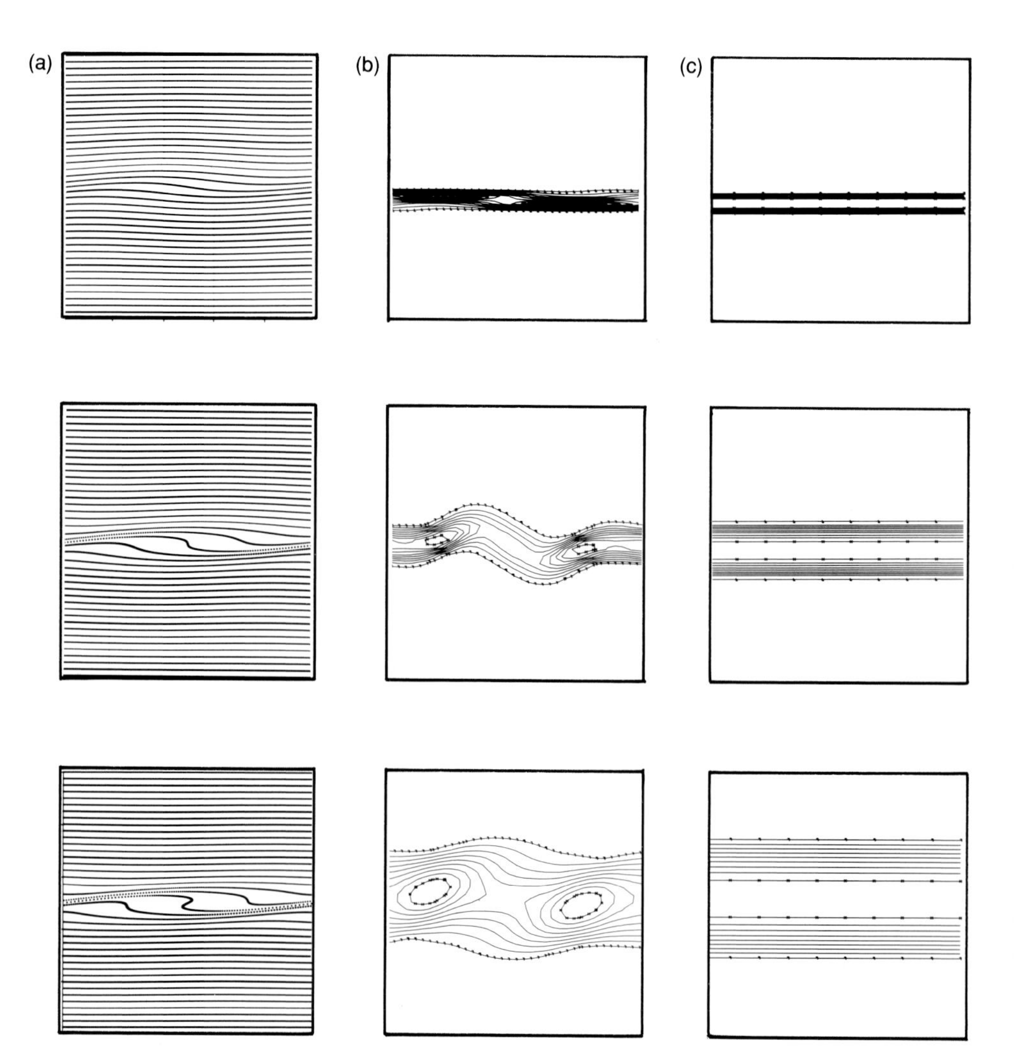


Fig. 14. The evolution in time (from top to bottom) of the turbulence in the slip-plane problem of Fig. 6, as determined by three different types of simplified calculations. In (a), a large-scale sinusoidal perturbation in the vertical direction is calculated with full equations without modeling the unresolved turbulence. The marker-particle plot (corresponding to mean-flow streaklines) in the third panel shows that a slip-plane instability is a strong source of perturbation in the velocity field. In (b), the perturbational energy of (a) has been increased 10 per cent with the addition of small-scale fluctuations. These fluctuations are accounted for with a turbulence kinetic energy $K$ and its transport equation, and the panels show contour plots of $K$. This more realistic approach reveals a faster growth in the turbulence. Finally, in (c), all the perturbational energy (both small- and large-scale motion in the vertical direction) is accounted for as turbulence kinetic energy. From this perspective, mean flow can only be horizontal and thus varies in only one (vertical) direction. The contour plots of turbulence kinetic energy show the same growth rate as in (b) for mixing between the layers of undisturbed flow.

The results of our calculations are shown in the left column of Fig. 14 as *marker-particle plots* in which the lines correspond to mean-flow streaklines (representing what you would see if you had introduced a stream of smoke). As time progresses (from top to bottom in the figure), we see that the width of the mixing layer spreads and displays wave-like structures characteristic of the Kelvin-Helmholtz instability. Thus, our calculations show that a slip-layer instability is indeed a strong source of turbulent mixing.

A more interesting and realistic approach incorporates simplified transport of turbulence in the calculations. Consider the same flow, only with additional small-scale sinusoidal perturbations superimposed on the initial large-scale perturbation. If we were to use the first method and treat these minute fluctuations as part of the resolved flow, we would need a much finer computational grid to resolve the details of the velocity field. Rather than do this, we account for the microscopic perturbations through a turbulence kinetic energy $K$ and its corresponding transport equation, then plot the results of our calculations as contour plots of $K$. This model is more realistic because the kinetic-energy variable incorporates all length scales of turbulence, as well as the three-dimensionality.

For our simplified transport calculation, we assume the same large-scale perturbation (with an energy $10^{-4}$ times the mean-flow kinetic energy) and then add a further 10 per cent (or $10^{-5}$ times the mean-flow energy) in small-scale energy. Our results (the second column of Fig. 14) show that accounting for small-scale perturbational energy causes the growth rate of the turbulence to be much larger than in the first method.

We should also note that even though turbulence is inherently three-dimensional, both the first and second methods deal with *mean flow* in two dimensions. In the second case, however, we are able to account for the third dimension in an average sense via turbulence kinetic energy and its transport.

A third approach is to treat *all* length scales as turbulence—even the large-scale perturbation, which, so far, has been treated as part of the mean flow. We can do this because the exact definition of turbulence is a relative one that depends on the observer's point of view. If we adopt this scheme, the flow becomes *one-dimensional*, that is, only vertical *changes* occur in $\overline{u}$, $K$, and $\epsilon$.

In our calculations with this method, we assume the same *total* initial perturbational energy as in the second example, but with the large- and small-scale energies lumped together. Once again, all length scales of turbulence are incorporated, and our results (the third column of Fig. 14) show that the growth rate of the layer matches that in the second example very closely. On the other hand, turbulence on the largest length scale, which corresponds to mean flow in the earlier examples, is not resolved in detail.

Thus, to effectively use turbulence modeling, one must decide which length scales will be considered mean flow and which will be considered turbulence. Once this has been decided, the power of the method allows us to describe the flow accurately without having to dedicate excessive computer resources to resolving minute flow structures in detail.

## Current Research

So far we have concentrated on turbulence in a single incompressible fluid with density perfectly constant in position and time (the downward branches of Fig. 11). Recently, our research has included additional features that are of interest to many of the new scientific and engineering directions at Los Alamos and other laboratories. These features are

- Two-phase flow interactions: the sources, sinks, and effects of turbulence in a fluid containing particles, droplets, or bubbles of another material.
- Density gradients: turbulence in an incompressible fluid for which variations of temperature or the presence of some dissolved substance cause large variations in density.
- Supersonic turbulence: the effects of high-speed processes on turbulence.

In all cases, we continue to use the basic philosophy of transport modeling, which, despite some obvious difficulties, seems at present to be by far the most promising approach for the solution of practical problems.

**Two-Phase Flow.** Particles, drops, or bubbles suspended in a fluid—whether that fluid is a liquid or a gas—can significantly alter the turbulence and its effects. Intuitively,

## TWO-PHASE INTERPENETRATION

**Fig. 15. Our transport modeling techniques are able to handle both ordered interpenetration of two phases, such as occurs in the laminar-flow transport of blood cells, and disordered interpenetration, such as occurs in a rapidly moving gas that contains suspended particles.**

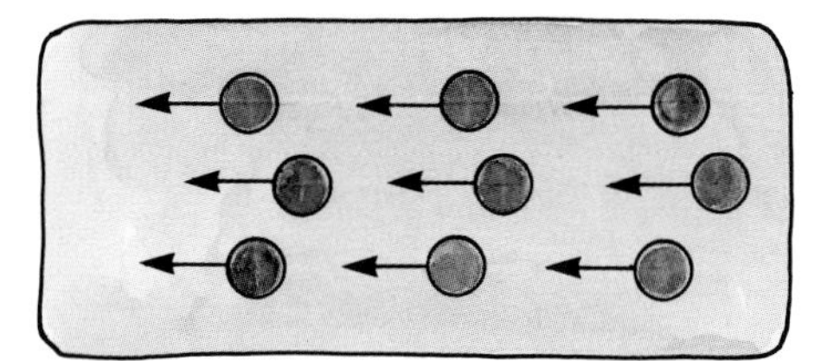

Ordered

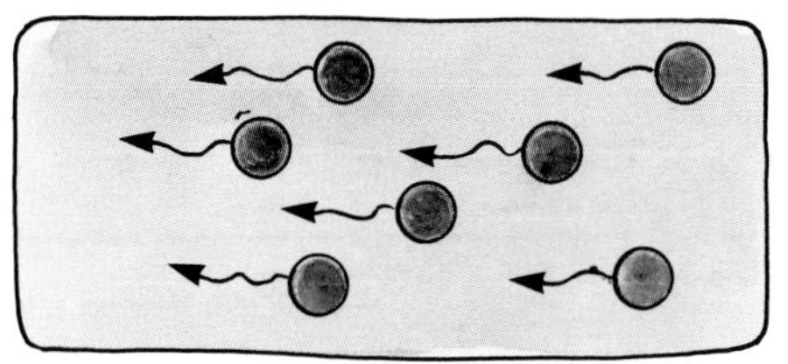

Disordered

we expect that when distinct entities interpenetrate a surrounding fluid the *creation* of turbulence is enhanced; on the other hand, we also expect the inertial properties of heavy entities to dampen turbulent fluctuations. How can we describe these effects quantitatively?

From considerations similar to those for incompressible flow of a single fluid, we know that extra turbulence is generated by pressure gradients producing *differences* in the accelerations of the particles and of the surrounding fluid. Such differential acceleration induces distortions of the fluid around the particles, thereby creating disturbances in the velocity field that would be absent if there were no particles.

For example, consider the flow field of a shock wave moving horizontally and passing a rigid particle suspended in the fluid. If no particle was present, the flow would remain completely horizontal. However, as the shock wave passes the particle, local velocity fluctuations appear, including changes in the horizontal velocity and the generation of vertical velocity. As soon as there is a velocity difference between the velocity fields of the particle and the fluid, viscous drag forces, competing with differential acceleration, begin to diminish any velocity perturbations.

In a manner analogous to that for single-phase flow, the relative contributions of acceleration and viscous drag can be compared through a *particle* Reynolds number

$$R_{\mathrm{p}} = D_{\mathrm{p}} \frac{|\overline{u_{\mathrm{f}}} - \overline{u_{\mathrm{p}}}|}{\nu_{\mathrm{m}}}, \tag{30}$$

where $D_{\mathrm{p}}$ is the particle diameter, $u_{\mathrm{p}}$ and $u_{\mathrm{f}}$ are the local velocities of the particle and fluid, respectively, and $\nu_{\mathrm{m}}$ is the molecular kinematic viscosity of the fluid.

Consider a shock moving with a high velocity through a collection of particles that are initially at rest, such that $R_{\mathrm{p}} \gg 1$. At first, the effects of differential acceleration dominate and turbulence kinetic energy is created. Then, as viscous drag causes the particles to be swept along with the fluid, the velocity difference and the particle Reynolds number decrease, corresponding to a dampening of turbulent fluctuations. Since the amount of drag depends on the volume fraction of the particles, the turbulence level that is induced will also depend on this parameter.

These effects, however, address only a small fraction of the rich spectrum of dynamic processes that can occur in multifield turbulent flows. In our approach we discard the more conventional procedure of decomposing *velocities* and volume fractions and, instead, consider *momentum* and volume fractions as the primary variables to be conserved, decomposing these into their mean and fluctuating components. Such an approach allows us to derive two limiting fluid behaviors: diffusion (in the limit of strong momentum coupling between the particle and fluid fields), and wave-like interpenetration (in the weak-coupling limit). Our model is thus strongly analogous to the interpenetration of two different molecular species: diffusive when the mean free path is short, and wave-like when little or no coupling is present and the species transport as if each were expanding into a vacuum.

In addition, our model handles both ordered and disordered interpenetration of two phases as illustrated in Fig. 15. Other technical accomplishments include the resolution of mathematical ill-posedness of the multiphase flow equations, the emergence of a new closure principle (based on the constraint, with generalized Reynolds-stress expressions, of exactly neutral stability for the mean-flow equations), and the development of practical modeling equations.

The modeling of turbulent flow with dispersed particles, droplets, or bubbles is of interest to a wide variety of scientific projects at the Laboratory. For example, to model the transport of dust and debris by volcanic eruptions, one must concentrate on the interactions between particulate and hot-gas flows. To improve the design of internal combustion engines, one needs an accurate prediction of both the combustion efficiency and the spatial distribution of heat generation, which, in turn, requires knowing the details of the mixing of fuel droplets and air. Although flow within the body's circulatory system is normally *not* turbulent, the transport of blood cells can be analyzed by using the equations for ordered two-field interpenetration. Other applications include modeling of the flow within nuclear reactors and the analysis of shock-wave motion in a gas that contains suspended particles.

**Density Gradients.** The second area we are currently striving to understand with transport modeling is turbulent mixing generated by strong density gradients that are sustained by large variations in thermal or material composition. Coupled with pressure gradients, such density gradients can lead to strongly contorted flow with intense vorticity near the steepest density variations. Again, the proper basis for deriving a generalized Reynolds stress lies in decomposing the momentum rather than the velocity.

Among the most important configurations to be studied are those for which adjacent materials—initially quiescent and of very different densities—are rapidly accelerated by a strong pressure gradient or heated by a sudden influx of radiation. The ensuing fluid instability (Richtmyer-Meshkov if the shock is going from heavy to light material, Rayleigh-Taylor for the opposite case (Fig. 7)) can act as a strong source for the turbulent mixing of the two materials.

For example, consider an experiment in which a plane shock wave progresses down a closed cylindrical tube divided into two sections by a permeable membrane with air in the first section and helium in the second. As the shock passes from the dense to the less-dense gas, the air-helium interface is accelerated. Later, the interface is repeatedly decelerated by reflections from the rigid wall at the end of the tube. Interface instabilities lead to turbulent mixing of the two gases, and the initially sharp plane separating the gases becomes smeared and indistinct. Our work allows prediction of the average concentration across any strip of fluid taken normal to the nominal streaming direction and calculation of velocity and density profiles within the turbulent mixing zone.

Instabilities driven by density gradients are important to the study of the implosion dynamics of pellets used in inertial confinement fusion (Fig. 16). Radiation from a high-power laser initiates the implosion of an outer spherical capsule, creating a strong shock wave. This shock passes over the interface between the inner surface of the capsule and the enclosed gas, is reflected from the core, and returns to the interface where it induces Rayleigh-Taylor instability. The resultant mixing of gas and capsule in the central region of the pellet can, in many cases, reduce neutron yield.

Another area of interest is the dynamics of fire plumes in the postulated circumstances of "nuclear winter." Extreme heating of the ambient atmosphere produces up to four-fold expansions, resulting in a powerful updraft with intense turbulence.

## CURRENT APPLICATIONS

Fig. 16. We are currently incorporating additional features in transport modeling so that more complex phenomena can be described adequately. An example is implosion of an inertial-confinement fusion capsule, during which two-phase turbulent interactions between the capsule and the hot fuel gases decrease the efficiency of the implosion. We also are investigating the density-driven turbulence that enhances mixing in fire plumes.

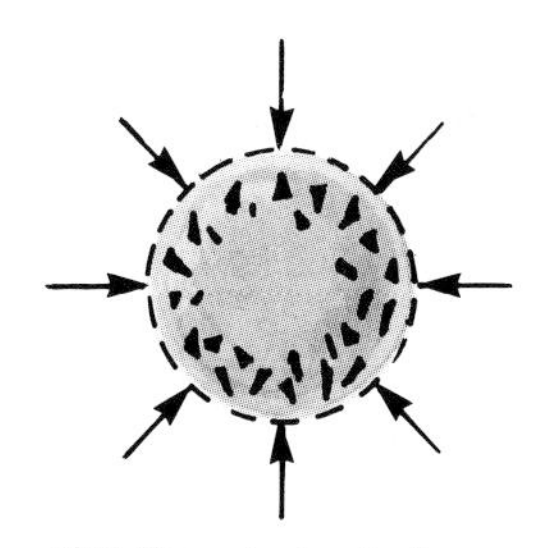

ICF Capsule Implosions

Nuclear Winter Fire Plumes

**Supersonic Turbulence.** Mach-number effects often can be ignored, but, in some cases (such as the high-Mach-number mitigation of a Kelvin-Helmholtz instability), such effects are significant. Thus, a third feature of our recent work has been to include the principal phenomena resulting from supersonic flow speeds. These effects arise across shock waves, in the shear layers behind Mach-reflection triple-shock intersections, and in the shear layers behind shock waves normal to a deformable wall.

An unexpected result of our work is the discovery that laminar instability theory (as sketched out in the section entitled "Turbulence Energy: Sources and Sinks") is applicable to the study of supersonic turbulence. Despite the seeming inconsistency, this theory is providing highly relevant guidance to our early modeling efforts.

## Concluding Remarks

A pertinent question is: What good is all this? Not only has our discussion illustrated several ways in which turbulence transport theory is heuristic or empirical, but the current large inventory of undetermined "universal" dimensionless parameters in its formulation is disturbing. Moreover, full expression of the theory is long and complicated, involving numerous coupled nonlinear partial differential equations. As a result, a transport calculation requires either costly numerical solutions or questionable approximations, or both.

What are the alternatives? There is no way to resolve turbulence in sufficient detail for numerical calculations based on turbulence transport theory to represent the effects of any but the simplest circumstances. Mixing-length theories and other point-functional approaches are hopelessly limited in their applicability. Fundamental approaches purporting to describe turbulence without empiricism are, in general, also restricted to highly idealized circumstances. Yet we are faced with the task of solving an endless variety of fluid-flow problems, a large fraction of which include significant turbulence effects. We need to supply answers to old questions and guidance for new developments in a meaningful way. At present, there seems to be no better approach to these challenging analytical tasks than that provided by turbulence transport theory.

Despite the shadows cast by these comments, the situation is actually far from gloomy. Turbulence transport theory seems to be functioning far better than we have any right to expect. There are at least four reasons for this good performance.

First, complex processes of nature often display a near universality in the collective effects that are of most interest. Just as gas molecules almost always have a nearly Maxwell-Boltzmann velocity distribution, it appears that turbulence tends toward a similar universality in its stochastic structure. The success of the few-variable (or collective, or moment) approach to turbulence modeling relies strongly on the validity of this contention. Although the extent to which universal behavior underlies most of the random processes of nature is currently a matter of intense scientific and philosophical discussion, much evidence supports the ubiquitous nature of this property. Perhaps, eventually, such universalities will help to successfully model such diverse instances as thoughts in a brain, activities of groups of organisms (such as mobs of people), and the dynamics of galaxies.

Next, turbulence transport modeling pays close attention to the binding constraints of real physics: conservation of mass, momentum, and energy, as well as rotational and translational invariance. Such modeling also accounts for history-dependent variations lacking in many other turbulence theories.

We have also paid great care to *physically* meaningful closure modeling. Auxiliary derivations (like those of laminar instability analysis) combine with new formulations of mathematical restrictions (like that of precisely neutral mean-flow stability in the presence of generalized Reynolds-stress terms) to constrain our modeling procedures in the most physically meaningful manner possible at each stage of the development.

Finally, investigators throughout the world have made numerous comparisons with experiments, leading to corrections, improvements, and ultimately to considerable confidence in the broad applicability of the results.

Future research will concentrate on several significant aspects of the theory. Closure modeling, of course, continually needs strengthening, especially by first-principle techniques that decrease our reliance on empiricism. The numerical techniques need greater stability, accuracy, and efficiency for a host of larger and more complicated problems.

But the most intriguing challenge is how to incorporate new and different physical processes into our theories. For example, with dispersed-entity flow, we have scarcely begun to understand the effects of a spectrum of entity sizes or the deformation of individual entities (including their fragmentation and coalescence) or the modifications that arise when the entities become close-packed (as they do, for example, during deposition and scouring of river-bed sand). The dispersal of turbulence energy through acoustic or electromagnetic radiation is another interesting topic that needs considerable development. Deriving, testing, and applying the appropriate models will keep many investigators busy for a long time. ■

## Further Reading

B. J. Daly and F. H. Harlow. 1970. Transport equations in turbulence. *Physics of Fluids* 13: 2634–2649.

B. E. Launder and D. B. Spalding. 1974. The numerical computation of turbulent flows. *Computer Methods in Applied Mechanics and Engineering* 3: 269–289.

C. J. Chen and C. P. Nikitopoulos. 1979. On the near field characteristics of axisymmetric turbulent buoyant jets in a uniform environment. *International Journal of Heat and Mass Transfer* 22: 245–255.

D. Besnard and F. H. Harlow. 1985. Turbulence in two-field incompressible flow. Los Alamos National Laboratory report LA–10187–MS.

F. H. Harlow, D. L. Sandoval, and H. M. Ruppel. 1986. Mathematical modeling of biological ensembles. Los Alamos National Laboratory report LA–10765–MS.

D. Besnard, F. H. Harlow, and R. M. Rauenzahn. 1987. Conservation and transport properties of turbulence with large density gradients. Los Alamos National Laboratory report LA–10911–MS.

D. Besnard and F. H. Harlow. Turbulence in multiphase flow. Submitted for publication in *International Journal of Multiphase Flow*.

Left to right: Norman L. Johnson, Rick Rauenzahn, Francis H. Harlow, Jonathan Wolfe, and Didier Besnard.

## Authors

**Didier Besnard** is a frequent visitor to Los Alamos from the Commissariat à l'Energie Atomique in France. He is the recipient of several advanced degrees in the fields of applied mathematics and engineering and has been employed since 1979 at the Centre d'Etudes de Limeil-Valenton, near Paris, where he is head of the Groupe Interaction et Turbulence. He is especially active in the analytical and numerical solution of problems in plasma physics and the high-speed dynamics of compressible materials. At Los Alamos he has worked in the Center for Nonlinear Studies and been a Visiting Scientist in the Theoretical Fluid Dynamics Group, collaborating in the development of turbulence transport theories, especially for multiphase flows and fluids with large density variations. An avid rock climber, hiker, and skier, he enjoys frequent excursions to the backwoods areas of both New Mexico and the French Alps. His wife, Anne, and daughter, Gaelle, join him in being a truly international family, spending at least a month in New Mexico each summer.

**Francis H. Harlow** came to Los Alamos in September 1953 after receiving his Ph.D. from the University of Washington and has been a physicist in the Theoretical Division during his entire employment at the Laboratory. Special interests include fluid dynamics, heat transfer, and the numerical solution of continuum dynamics problems. He was Leader of the Fluid Dynamics Group for fourteen years and became a Laboratory Fellow in 1981. His extensive publications describe a variety of new techniques for solving fluid flow problems and discuss the basic physics and the application to practical problems. Northern New Mexico has served as a strong stimulus to his collateral activities in paleontology, archeology, and painting. Writings include one book on fossil brachiopods and four on the Pueblo Indian pottery of the early historic period. His paintings have been the subject of several one-man shows and are included in hundreds of collections throught the United States.

**Norman L. Johnson** came to Los Alamos in the summer of 1981 as a graduate student working on cavity radiation and the numerical solution of highly nonlinear equations within a thermal-fluid finite element code. In the spring of 1983, he completed his Ph.D. in free Lagrangian methods and kinetic theory for viscoelastic flows at the University of Wisconsin in Madison. He returned to Los Alamos as a postdoctorate and developed an interest in modeling the complex physics of hypervelocity impact phenomena. Concurrently with his stay at Los Alamos, he held a position at the National Bureau of Standards in Boulder, Colorado, working on facilitated transport across membranes. In 1985 he became a permanent staff member in the Theoretical Fluid Dynamics Group and is currently working on the development of a local intense energy source and on numerical methods for flows with high distortions. Special interests include impact phenomena and the rheology of viscoelastic fluids and suspensions. In off hours he is currently learning how to juggle, as well as continuing his enjoyment of modern dance and the Japanese language and culture.

**Rick Rauenzahn** came to Los Alamos in 1982 as a graduate student in the Earth and Space Sciences Division to research the fundamentals of thermal spallation drilling. Prior to this, Rick had received his B.S. from Lehigh University in 1979 and his S.M. from MIT in 1980, both in chemical engineering. Two one-year stints in the chemical industry prepared him to enter MIT once again, and, in 1986, after doing thesis work at the Laboratory, he received his Ph.D. in Chemical Engineering. Since then, he has been in the Theoretical Fluid Dynamics Group, working on the fundamentals of turbulence modeling, mixing at interfaces, and two-dimensional code development.

**Jonathan Wolfe** is an undergraduate student in the Mechanical Engineering Department at the University of New Mexico, participating in the Cooperative Education Program with Los Alamos. In a cross-country pursuit of his education, Jon came to New Mexico in 1983 and to Los Alamos soon afterwards. Working for the Earth and Space Sciences Division, he contributed to the development of innovative computational techniques for the solution of large, sparse matrix systems, and worked as a field engineer ("grunt") at the Fenton Hill geothermal test site. More recently, he has been a member of the Theoretical Fluid Dynamics Group, collaborating in the numerical investigation of problems in turbulence and multiphase flow. Jon has fallen in love with the New Mexico outdoors, and spends lots of time windsurfing, motorcycling, and playing ultimate frisbee, despite the threat of impending graduation and responsibility.

## PART I
## BACKGROUND FOR LATTICE GAS AUTOMATA

*The lattice gas automaton is an approach to computing fluid dynamics that is still in its infancy. In this three-part article one of the inventors of the model presents its theoretical foundations and its promise as a general approach to solving partial differential equations and to parallel computing. Readers less theoretically inclined might begin by reading "Calculations Using Lattice Gas Techniques" at the end of Part II. This sidebar offers a summary of the model's advantages and limitations and a graphic display of two- and three-dimensional lattice gas simulations.*

The invention of a totally discrete model for natural phenomena was made by Ulam and von Neumann in the early fifties and was developed to the extent possible at the time. A few years earlier von Neumann had designed the architecture for the first serial digital computers containing stored programs and capable of making internal decisions. These machines are built of electronic logic devices that understand only packets of binary bits. Hierarchies of stored translators arrange them into virtual devices that can do ordinary or radix arithmetic at high speed. By transcribing continuum equations into discrete form, using finite difference techniques and their variants, serial digital computers can solve complex mathematical systems such as partial differential equations. Since most physical systems with large numbers of degrees of freedom can be described by such equations, serial digital machines equipped with large memories have become the standard way to simulate such phenomena.

As the architecture of serial machines developed, it became clear to both Ulam and von Neumann that such machines were not the most natural or powerful way to solve many problems. They were especially influenced by biological examples. Biological systems appear to perform computational tasks using methods that avoid both arithmetical operations and discrete approximations to continuous systems.

Though motivated by the complex information processing of biological systems, Ulam and von Neumann did not study how such systems actually solve tasks. Biological processes have been operating in hostile environments for a long time, finding the most efficient and often devious way to do something, a way that is also resistant to disturbance by noise. The crucial principles of their operation are hidden by the evolutionary process. Instead, von Neumann chose the task of simulating on a computer the least complex discrete system capable of self-reproduction. It was Ulam who suggested an abstract setting for this problem and many other totally discrete models, namely, the idea of cellular spaces. The reasoning went roughly like this.

The question is simple: Find a minimal logic structure and devise a dynamics for it that is powerful enough to simulate complex systems. Break this up into a series of sharper and more elementary pictures. We begin by setting up a collection of very simple finite-state machines with, for simplicity, binary values. Connect them so that given a state for each of them, the next state of each machine depends only on its immediate environment. In other words, the state of any machine will depend only on the states of machines in some small neighborhood around it. This builds in the constraint that we only want to consider local dynamics.

We will need rules to define how states combine in a neighborhood to uniquely fix the state of every machine, but these can be quite simple. The natural space on which to put all this is a lattice, with elementary, few-bit, finite-state machines placed at the vertices. The rules for updating this array of small machines can be done concurrently in one clock step, that is, in parallel.

One can imagine such an abstract machine in operation by thinking of a fishnet made of wires. The fishnet has some regular connection geometry, and there are lights at the nodes of the net. Each light can be on or off. Draw a disk around each node of the fishnet, and let it have a 1-node radius. On a square net there are four lights on the edge of each disk, on a triangular net six lights (Fig. 1). The next state of the light at the center of the disk depends on the state of the lights on the edge of the disk and on nothing else. Imagine all the disks in the fishnet asking their neighbors for their state at the same time and switching states according to a definite rule. At the next tick of an abstract clock, the pattern of lights on the fishnet would in general look different. This is what Ulam and von Neumann called a nearest-neighbor-connected cellular space. It is the simplest case of a parallel computing space. You can also see that it can be imaged directly in hardware, so it is also the architecture for a physical parallel computing machine.

## CELLULAR SPACES

**Fig. 1. Two examples of fishnets made of wires with lights at the nodes. The lights are either on or off. In each example a disk with a radius of 1 node is drawn around one of the lights. The next state of the light at the center depends on the states of the lights on the edge of the disk and on nothing else. Thus these are examples of nearest-neighbor-connected cellular spaces.**

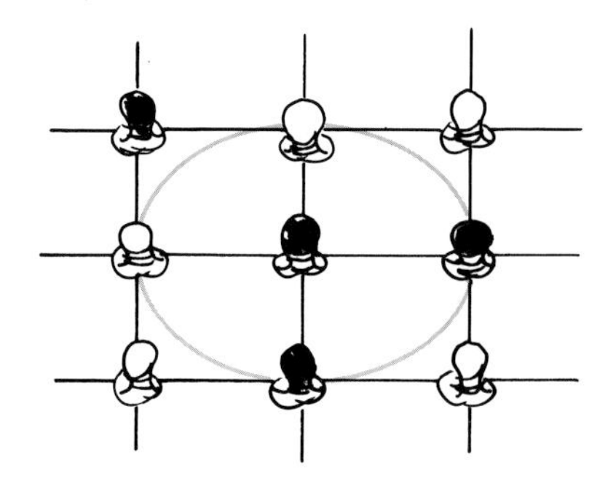

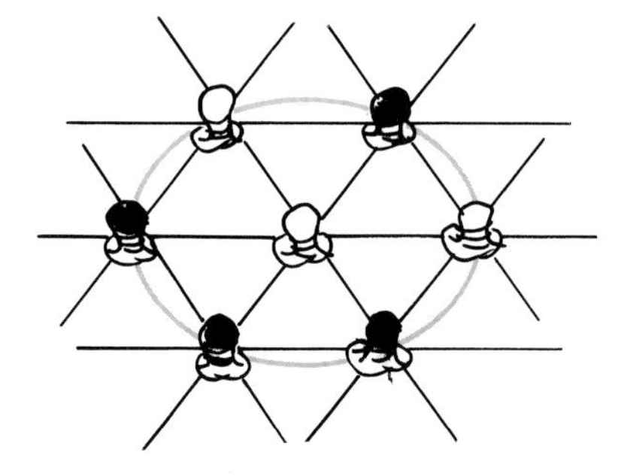

We have not shown that such a device can compute. At worst, it is an elaborate light display. Whether or not such a cellular space can compute depends on the definition of computation. The short answer is that special cases of fishnets are provably universal computers in the standard Turing machine sense; that is, they can simulate the architecture of any other sequential machine.

But there are other interpretations of computation that lie closer to the idea of simulation. For any given mathematical situation, we want to find the minimum cellular space that can do a simulation of it: At what degree of complexity does repeated iteration of the space, on which are coded both data and a solution algorithm, possess the power to come close to the solution of a complex problem? This depends on the complexity or degrees of freedom present in the problem.

An extreme case of complexity is physical systems with many degrees of freedom. These systems are ordinarily described by field theories in a continuum for which the equations of motion are highly nonlinear partial differential equations. Fluid dynamics is an example, and we will use it as a theoretical paradigm for many "large" physical systems. Because of the high degree of nonlinearity, analytic solutions to the field equations for such systems are known only in special cases. The standard way to study such models is either to perform experiments or simulate them on computers of the usual digital type.

Suppose a cellular space existed that evolved to a solution of a fluid system with given boundary conditions. Suppose also that we ask for the simplest possible such space that captured at least the qualitative and topological aspects of a solution. Later, one can worry about spaces that agree quantitatively with ordinary simulations. The problem is threefold: Find the least complex set of rules for updating the space; the simplest geometry for a neighborhood; and a method of analysis for the collective modes and time evolution of such a system.

At first sight, modeling the dynamics of large systems by cellular spaces seems far too difficult to attempt. The general problem of a so-called "inverse compiler"—given a partial differential system, find the rules and interconnection geometry that give a solution—would probably use up a non-polynomial function of computing resources and so be impractical if not impossible. Nevertheless cellular spaces have been actively studied in recent years. Their modern name is cellular automata, and specific instances of them have simulated interesting nonlinear systems. But until recently there was no example of a cellular automaton that simulated a large physical system, even in a rough, qualitative way.

Knowing that special cases of cellular automata are capable of arbitrarily complex behavior is encouraging, but not very useful to a physicist. The important phenomenon in large physical systems is not arbitrarily complex behavior, but the collective motion that develops as the system evolves, typically with a characteristic size of many elementary length scales. The problem is to simulate such phenomena and, by using simulations, to try to understand the origins of collective behavior from as many points of view as possible. Fluid dynamics is filled with examples of collective behavior—shocks, instabilities, vortices, vortex streets, vortex sheets, turbulence, to list a few. Any deterministic cellular-automaton model that attempts to describe non-equilibrium fluid dynamics must contain in it an iterative mechanism for developing collective motion. Knowing this and using some very basic physics, we will construct a cellular automaton with the appropriate geometry and updating rules for fluid behavior. It will also be the simplest such model. The methods we use to do this are very conservative from the viewpoint of recent work on cellular automata, but rather drastic compared to the approaches of standard mathematical physics. Presently there is a large gap between these two viewpoints. The simulation of fluid dynamics by cellular automata shows that there are other complementary and powerful ways to model phenomena that would normally be the exclusive domain of partial differential equations.

## The Example of Fluid Dynamics

Fluid dynamics is an especially good large system for a cellular automaton formulation because there are two rich and complementary ways to picture fluid motion. The kinetic picture (many simple atomic elements colliding rapidly with simple interactions) coincides with our intuitive picture of dynamics on a cellular space. Later we will exploit this analogy to construct a discrete model.

The other and older way of approaching flow phenomena is through the partial differential equations that describe collective motions in dissipative fluids—the Navier-Stokes equations. These can be derived without any reference to an underlying atomic picture. The derivation relies on the idea of the continuum; it is simpler to grasp than the kinetic picture and mathematically cleaner. Because the continuum argument leads to the correct functional form of the Navier-Stokes equations, we spend some time describing why it works. The continuum view of fluids will be called "coming down from above," and the microphysical view "coming up from below" (Fig. 2). In the intersection of these two very different descriptions, we can trap the essential elements of a cellular-automaton model that leads to the Navier-Stokes equations. Through this review we wish to show that cellular automaton models are a natural and evolutionary idea and not an invention come upon by accident.

## Coming down from Above—The Continuum Description

The notion of a smooth flow of some quantity arises naturally from a continuum description. A flow has physical conservation laws built-in, at least conservation of mass and momentum. With a few additional remarks one can include conservation of energy. The basic strategy for deriving the Euler and Navier-Stokes equations of fluid dynamics is to imbed these conservation laws into statements about special cases of the generalized Stokes theorem. We use the usual Gauss and Stokes theorems, depending on dimension, and apply them to small surfaces and volumes that are still large enough to ignore an underlying microworld. The equations of fluid dynamics are derived with no reference to a ball-bearing picture of an underlying atomic world, but only with a serene reliance on the idea of a smooth flow in a continuum with some of Newton's laws added to connect to the observed world. As a model (for it is not a theory), the Navier-Stokes equations are a good example of how concepts derived from the intuition of daily experience can be remarkably successful in building effective phenomenological models of very complex phenomena. It is useful to go through the continuum derivation of the Euler and Navier-Stokes equations presented in "The Continuum Argument" for several reasons: First, the reasoning is short and clear; second, the concepts introduced such as the momentum flux tensor, will appear pervasively when we pass to discrete theories of fluids; third, we learn how few ingredients are really necessary to build a fluid model and so mark out that which is essential—the role of conservation laws.

It is clear from its derivation that the Euler equation describing inviscid flows is essentially a geometrical equation. The extension to the full Navier-Stokes equations, for flows with dissipation, contains only a minimal reference to an underlying fluid microphysics, through the stress-rate of strain relation in the momentum stress tensor. So we see that continuum reasoning alone leads to nonlinear partial differential equations for large-scale physical observables that are a phenomenological description of fluid flow. This description is experimentally quite accurate but theoretically incomplete. The coupling constants that determine the strength of the nonlinear terms—that is, the transport coefficients such as viscosity—have a direct physical interpretation in a microworld picture. In the continuum approach however, these must be measured and put in as data from the outside world. If we do not use some microscopic model for the fluid, the transport coefficients cannot be derived from first principles.

**Solution Techniques—The Creation of a Microworld.** The Navier-Stokes equations are highly nonlinear; this is prototypical of field-theoretical descriptions of large physical systems. The nonlinearity allows analytic solutions only for special cases and, in general, forces one to solve the system by approximation techniques. Invariably these are some form of perturbation methods in whatever small parameters can be devised. Since there is no systematic way of applying perturbation theory to highly nonlinear partial differential systems, the analysis of the Navier-Stokes equations has been, and still remains, a patchwork of ingenious techniques that are designed to cover special parameter regimes and limited geometries.

After an approximation method is chosen, the next step toward a solution is to discretize the approximate equations in a form suitable for use on a digital computer. This discretization is equivalent to introducing an artificial microworld. Its particular form is fixed by mathematical considerations of elegance and efficiency applied to simple arithmetic operations and the particular architecture of available machines. So, even if we adopt the view that the molecular kinetics of a fluid is unimportant for describing the general features of many fluid phenomena, we are nevertheless forced to describe the sys-

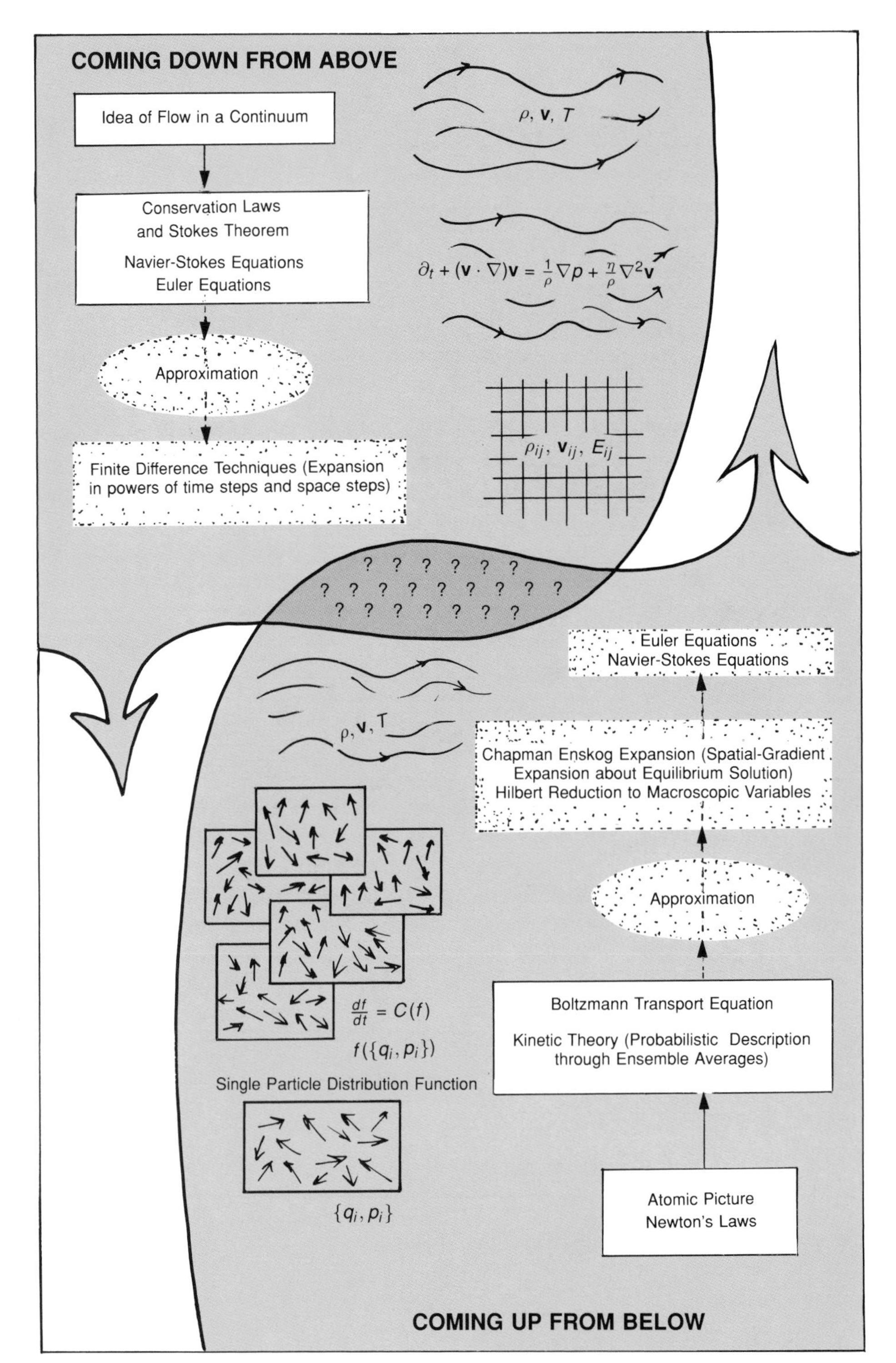


Fig. 2. Both the continuum view of fluids and the atomic picture lead to the Navier-Stokes equations but not without approximations (dashed lines). The text emphasizes how cellular-automaton models embody the essentials of both points of view.

tem by a microworld with a particular microkinetics. The idea of a partial differential equation as a physical model is tied directly to finding an analytic solution and is not particularly suited to machine computation. In a sense, the geometrically motivated continuum picture is only a clever and convenient way of encoding conservation laws into spaces with which we are comfortable.

## Coming up from Below—The Kinetic Theory Description

Kinetic theory models a fluid by using an atomic picture and imposing Newtonian mechanics on the motions of the atoms. Atomic interactions are controlled by potentials, and the number of atomic elements is assumed to be very large. This attempt at fluid realism has an imme-

*continued on page 181*

# THE CONTINUUM ARGUMENT

Let $\mathbf{v}(\mathbf{x}, t)$ be a vector-valued field referred to a fixed origin in space, which we identify with the velocity of a "macroscopic" fluid cell. The cell is not small enough to notice a particle structure for the fluid, but it is small enough to be treated as a mathematical point and still agree with physics.

To derive the properties of a flow defined by the vector field, one now invokes the generalized Stokes theorem:

$$\oint_{\partial\Sigma} \mathbf{A} = \oint_{\Sigma} d\mathbf{A},$$

where $\Sigma$ is a generalized surface or volume, $\partial\Sigma$ the boundary of $\Sigma$, $\mathbf{A}$ an $n$-differential form and $d\mathbf{A}$ an $(n+1)$-differential form. This very general theorem has two familiar forms: one is the classical Stokes theorem from one to two dimensions,

$$\oint_{\partial\Sigma} \mathbf{A} \cdot \mathbf{d}\ell = \oint_{\Sigma} \nabla \times \mathbf{A} \cdot d\mathbf{S},$$

and the other is the Gauss law from two to three dimensions,

$$\oint_{\partial\Sigma} \mathbf{A} \cdot d\mathbf{S} = \int_{\Sigma} (\nabla \cdot \mathbf{A}) dV,$$

where $\ell$ is a curve, $\mathbf{S}$ is a surface, and $V$ is a volume in three-dimensional Euclidean space $\mathbf{R}^3$.

**Conservation Laws and Euler's Equation.** First, we deal with the idea of continuity, or conservation of flow. If $\rho$ is the density, or mass per unit volume, then the mass of the fluid in volume $V$ (that is, $\Sigma$), is equal to $\int_{\Sigma} \rho dV$. A two-dimensional surface in $\mathbf{R}^3$ has an outward normal vector $\mathbf{n}$ which is defined to be positive. The total mass of fluid flowing out of a volume $\Sigma$ can be written as

$$\oint_{\partial\Sigma} \rho\mathbf{v} \cdot d\mathbf{S} = \int_{\partial\Sigma} \rho\mathbf{v} \cdot \mathbf{n} dS.$$

Continuity of the flow implies a balance between the flow through the surface and the loss of fluid from the volume. That is, the decrease in mass in the volume must equal the outflow of fluid mass through the surface of the volume, which implies by the Gauss law that

$$\oint_{\partial\Sigma} \rho\mathbf{v} \cdot d\mathbf{S} = -\partial_t \int_{\Sigma} \rho dV = \int_{\Sigma} \nabla \cdot (\rho\mathbf{v}) dV.$$

This gives the first evolution equation for a fluid, the continuity, or mass-conservation, equation:

$$\partial_t \rho + \nabla \cdot (\rho\mathbf{v}) = 0. \tag{1}$$

Now we introduce the idea of pressure $p$ as the force exerted by the fluid on a unit surface area of an enclosed volume and use Newton's second law, $\mathbf{F} = ma$. The total force acting on a volume of fluid due to the remainder of the fluid is given by $-\oint_{\partial\Sigma} p d\mathbf{S}$. Using Stokes theorem we can write

$$-\oint_{\partial\Sigma} p d\mathbf{S} = -\oint_{\Sigma} \nabla p dV.$$

The translation of $\mathbf{F} = m\mathbf{a}$ to a continuous medium is

$$-\nabla p = \rho d\mathbf{v}/dt,$$

where $d\mathbf{v}/dt$ is a total derivative. The chain rule on $d\mathbf{v}(\mathbf{x}, t)/dt$ gives

$$\rho d\mathbf{v}/dt = \rho\{\partial_t \mathbf{v} + \mathbf{v} \cdot \nabla\mathbf{v}\}$$

Substituting this result into the equation for $-\nabla p$ yields Euler's equation for an ideal, dissipation-free fluid:

$$\partial_t \mathbf{v} = -(\mathbf{v} \cdot \nabla)\mathbf{v} - \frac{1}{\rho}\nabla p. \tag{2}$$

## THE CONTINUUM ARGUMENT (continued)

One can generalize Euler's equation to a form more useful for a dissipative fluid. For this we look at the flux of momentum through a fluid volume. The momentum of fluid passing through an element $dV$ is $\rho\mathbf{v}$, and its time rate of change expressed in components is

$$\partial_t(\rho \mathrm{v}_i) = (\partial_t \rho)\mathrm{v}_i + \rho(\partial_t \mathrm{v}_i).$$

We can rewrite $\partial_t \rho$ and $\partial_t \mathrm{v}_i$ as spatial derivatives by using Eqs. 1 and 2. Then

$$\partial_t(\rho \mathrm{v}_i) = -\partial_k \Pi_{ik}, \tag{3}$$

where the momentum flux tensor $\Pi_{ik} \equiv p\delta_{ik} + \rho \mathrm{v}_i \mathrm{v}_k$.

The meaning of the momentum flux tensor can be seen immediately by integrating Eq. 3 and applying Stokes theorem.

$$\partial_t \int_\Sigma \rho \mathrm{v}_i d\Sigma = -\int_\Sigma \partial_k \Pi_{ik} d\Sigma = -\oint_{\partial\Sigma} \Pi_{ik} n_k dS.$$

So

$$\partial_t \int_\Sigma \rho \mathrm{v}_i dV = -\oint_{\partial\Sigma} \Pi_{ik} n_k dS, .$$

where the left-hand side is the rate of change of the $i$th component of momentum $\rho v_i$ in the volume and $\Pi_{ik} n_k d\Sigma$ is the $i$th component of momentum flowing through $dS$. Therefore, $\Pi_{ik}$ is the $i$th component of momentum flowing in the $k$th direction. This is more easily seen by writing

$$\Pi_{ik} n_k = p\delta_{ik} n_k + \rho \mathrm{v}_i \mathrm{v}_k n_k = p\mathbf{n} + \rho\mathbf{v}(\mathbf{v} \cdot \mathbf{n}).$$

Equations 1, 2, and 3 are the basic formalism for classical Newtonian ideal fluids (fluids with no dissipation) and are also true for flows in general.

**Classical Dissipative Fluids—The Navier-Stokes Equations.** The general Euler's equation is $\partial_t(\rho \mathrm{v}_i) = -\partial_k \Pi_{ik}$, where $\Pi_{ik}$ is now the momentum stress tensor. The form of this tensor changes if the fluid is dissipative, for example, if viscous forces convert the energy in the flow into heat. Traditionally, $\Pi_{ik}$ is modified in the following way. Take $\Pi_{ik} = p\delta_{ik} + \rho v_i v_k$ and introduce an unknown tensor $\sigma'_{ik}$ that describes the effects of viscious stress. Then rewrite the momentum stress tensor as

$$\Pi_{ik} = p\delta_{ik} + \rho \mathrm{v}_i \mathrm{v}_k - \rho\sigma'_{ik} \equiv \sigma_{ik} + \rho \mathrm{v}_i \mathrm{v}_k,$$

where $\sigma_{ik} = p\delta_{ik} - \rho\sigma'_{ik}$ is called the stress tensor and $\sigma'_{ik}$ the viscosity stress tensor.

The form of $\sigma'_{ik}$ can be deduced on general grounds. First we assume that the gradient of the velocity changes slowly so $\sigma_{ik}$ is linear in $\partial_k v_i$. Moreover, $\sigma'_{ik}$ is zero for $\mathbf{v} = 0$, and under rotation it must vanish since uniform rotation produces no overall transport of momentum. The unique form that has these properties is

$$\sigma'_{ik} = a(\partial_k \mathrm{v}_i + \partial_i \mathrm{v}_k) + b\delta_{ik}\partial_j \mathrm{v}_j,$$

where $a$ and $b$ are unknown coefficients. It is usually written in the form

$$\sigma'_{ik} = \nu(\partial_k \mathrm{v}_i - \partial_i \mathrm{v}_k - 2/3\delta_{ik}\partial_j \mathrm{v}_j) + \zeta\delta_{ik}\partial_j \mathrm{v}_j,$$

where $\nu$ is the kinematic shear viscosity and $\zeta$ is the kinematic bulk viscosity.

For an incompressible fluid (the density is constant so $\rho = \rho_0$) this tensor simplifies, and Euler's equation goes over to the incompressible Navier-Stokes equations:

$$\partial_t \mathbf{v} + (\mathbf{v} \cdot \nabla)\mathbf{v} = -\frac{1}{\rho}\nabla p + \mathbf{v}\nabla^2 \mathbf{v} \quad \text{and} \quad \nabla \cdot \mathbf{v} = 0.$$

In tensor notation we have

$$\partial_t \mathrm{v}_i + (\mathrm{v}_j \partial_j)\mathrm{v}_i = -\frac{1}{\rho}\partial_i p + \nu\frac{\partial^2}{\partial_k \partial_k}\mathrm{v}_i \quad \text{and} \quad \partial_k \mathrm{v}_k = 0.$$

In Part II we end the theoretical discussion of the lattice gas by giving the incompressible limit of the lattice gas Navier-Stokes equations. ■

*continued from page 178*

diate difficulty. We are unable to specify completely the initial state of the system or to follow its microdynamics. It follows that we cannot use a microdynamics that is this detailed. The obvious strategy is to make a smoothened model that reduces the number of degrees of freedom in the system to just a few. This reduction assumes maximum ignorance of the details of the system below some time and distance scale and replaces exact data on events by probabilistic outcomes. Measurements are assumed to be average values of quantities over large ensembles of representative systems. The assumption is that after a sufficiently long time these average observables are a close description of the fluid.

This approach seems very familiar and obvious from elementary courses in statistical mechanics. But it is unclear how to go from a statistical-mechanical description of an atomic system to the prediction of the details of collective motions that come from the evolution of that system. Fidelity to the atomic picture brings with it considerable mathematical difficulties. As we will see below and in "The Hilbert Contraction," the success of the derivation of the Navier-Stokes equations from the kinetic theory picture—that one derives the Navier-Stokes equations with the correct coefficients and not some other macrodynamics—is justified after the fact.

**Kinetic Theory and the Boltzmann Transport Equation.** Complete information on the statistical description of a fluid or gas at, or near, thermal equilibrium is assumed to be contained in the one-particle phase-space distribution function $f(t, \mathbf{r}, \Gamma)$ for the atomic constituents of the system. The variables $t$ and $\mathbf{r}$ are the time and space coordinates of the atoms and $\Gamma$ stands for all other phase-space coordinates (for example, momenta). In this rapid overview of kinetic transport theory, we will not dwell on the many and difficult questions raised by this description but keep to a level of precision consistent with a general understanding of the basic ideas.

The distribution function $f$ is basically a weighting function that is used to define the mean values of physical observables. The relation

$$N(t, \mathbf{r}) \equiv \int f(t, \mathbf{r}, \Gamma)\, d\Gamma \qquad (1)$$

defines the density function $N(t, \mathbf{r})$ for the particles in the system over all space. Therefore $NdV$ is the mean number of particles in the volume $dV$. Here $dV$ is a physical volume $\propto L^3$ whose characteristic length $L$ is much larger than $l_m$, the mean free path of a particle, and much smaller than $L_g$, some global length, such as the edge of a container for the whole gas. Thus $l_m \ll L \ll L_g$.

The basic equation of kinetic theory is the evolution equation for $f(t, \mathbf{r}, \Gamma)$ in the presence of gas collisions. Imagine first that the system has no collisions. Conservation of phase-space volumes, or Liouville's theorem, tells us that

$$\frac{df}{dt} = 0, \qquad (2)$$

where $d/dt$ is a total derivative. In an isolated system with no external fields, we can expand the total derivative as

$$\frac{df}{dt} = \partial_t f + \mathbf{v} \cdot \nabla f \equiv \partial_t f + \mathrm{v}_i \partial_i f. \quad (3)$$

(We use the convention that repeated indices are summed over.) Equation 3 defines the free-streaming operator, which represents the local change in $f$ per unit time caused by the independent motion of particles alone.

Now imagine a simple isolated gas with collisions. If $C(f)$ is a function that models the rate of change of the distribution function $f$ caused by collisions, then $C(f)\, dVd\Gamma$ is the rate of change per unit time of the number of molecules in the phase-space volume element $dVd\Gamma$. The Liouville statement now is modified to become the transport equation:

$$\frac{df}{dt} = C(f), \qquad (4)$$

where $C(f)$ is in general a highly nonlinear function of $f$.

Boltzmann first gave a simple approximation for the collision operator, which can be thought of as a gain-minus-loss $(\mathcal{G} - \mathcal{L})$ operator. A straightforward physical argument defining its general structure is presented below and is due to Landau.

**The Boltzmann Form of the Collision Term.** Let the particles in a two-body collision process have incoming distribution functions $g_1$ and $g_2$ and outgoing distribution functions $\bar{g}_1$ and $\bar{g}_2$. Fixing attention on particle 1, assume that before colliding it occupies a phase-space region $d\Gamma_1$, and after collision it occupies $d\bar{\Gamma}_1$; similarly, particle 2 occupies $d\Gamma_2$ before colliding and $d\bar{\Gamma}_2$ afterwards. If particle 1 undergoes a collision, $d\bar{\Gamma}_1$ will not in general be in $d\Gamma_1$, and particle 1 is said to be lost from $d\Gamma_1$. From these considerations we can compute the functional structure of the general loss term for a binary collision.

The probability of loss will be proportional to the product of four terms: (1) the number of particles of type 1 already in the volume, namely $g_1$; (2) the number of type-2 particles that enter the volume from some phase-space range $d\Gamma_2$, namely, $g_2 d\Gamma_2$; (3) the total volume of allowed outgoing phase space, $d\bar{\Gamma}_1 d\bar{\Gamma}_2$; and finally (4) a probability for the collision process $P_g\{\Gamma\}$. Now we sum over all possible allowed volumes of phase space. So the total number of losses $\mathcal{L}$ in the volume $dV$ and from $d\Gamma$ due to binary collision processes is

$$\mathcal{L} = dV\, d\Gamma \int P_g\{\Gamma\} g_1 g_2\, d\Gamma_2 d\bar{\Gamma}_1 d\bar{\Gamma}_2.$$

Similarly, particle gain into the phase

space volume $d\Gamma$ can only come from reversed channel processes $\bar{g}_1, \bar{g}_2 \rightarrow g_1, g_2$, with fixed $\Gamma_1$, and summed over all of $\bar{\Gamma}_1$, $\bar{\Gamma}_2$, and $\Gamma_2$, so

$$\mathcal{G} = dV\,d\Gamma \int P_g\{\Gamma\}\bar{g}_1\bar{g}_2\,d\Gamma_2 d\bar{\Gamma}_1 d\bar{\Gamma}_2.$$

The Boltzmann form for $C(f)$ is the net flow into the region, which is $\mathcal{G} - \mathcal{L}$. Using this form, we get the Boltzmann transport equation, a highly nonlinear integro-differential equation:

$$\frac{df}{dt} = \partial_t f + \mathrm{v}_i \partial_i f = \mathcal{G} - \mathcal{L}. \qquad (5)$$

In Part II we will use the same reasoning to construct the Boltzmann equation for the discrete lattice gas. The explicit form of the lattice gas collision operator is much simpler than in standard kinetic models.

Note that the Boltzmann form for the $(\mathcal{G} - \mathcal{L})$ collision term implicitly assumes only two-body collisions. It also assumes the collisions are pairwise statistically independent events occurring at a single point with detailed, or at most semi-detailed, balance symmetry for collision probabilities.

**Solutions to the Boltzmann Transport Equation.** Even though the Boltzmann equation is intractable in general, by using entropy arguments (Boltzmann's H theorem), the following can be stated about possible functional forms for $f$, the one-particle distribution function. If the system is uniform in space, any form for $f$ will relax monotonically to the global Maxwell-Boltzmann form:

$$f_{\text{global}} \sim \rho e^{-E(\rho,\mathrm{v})/T},$$

in which the macroscopic variables $\rho$, v, and $T$ (density, macrovelocity, and temperature) are independent of position, or global. In the non-equilibrium case, with a soft space dependence, any distribution function will relax monotonically in velocity space to a *local* Maxwell-Boltzmann form. This means that $\rho$, v, and $T$ will depend on space as well as time. These local distribution functions are solutions to the Boltzmann transport equation. For the non-uniform case, one gets a picture of the full solution as an ensemble of local Maxwell-Boltzmann distributions covering the description space of the fluid, with some gluing conditions providing the consistency of the patching.

## RANGE OF THE BOLTZMANN TRANSPORT EQUATION

The rigorous range of physical parameters in which the Boltzmann transport equation is mathematically meaningful is

$$N \rightarrow \infty \text{ such that } mN \rightarrow K$$

$$N \rightarrow \infty,\ \sigma \rightarrow 0 \text{ such that } (N\sigma^2) \rightarrow l_m^{-1}$$

where $N$ is the number of particles,
$m$ is the mass of each particle,
$\sigma$ is the range of the force or the effective interaction ball,
$l_m$ is the mean free path,
$K$ is a constant.

These conditions imply a dilute gas, binary collisions, and slowly varying spatial dependence (that is, slow space gradients). As an additional axiom we require that there be no long-range forces in the sense of photon excitations, etc.

**Recovering Macrodynamics–The Euler Equations.** If we assume a simple fluid and neglect all dissipative processes (viscosity, heat transfer, etc.), we can quickly derive the Euler equations (presented in "The Continuum Argument") from the Boltzmann transport equation. But first we need the notion of average quantities and some observations about collisions in a dissipation-free system.

As before, let $\rho(t,r) = \int f(t,\mathbf{r},\Gamma)\,d\Gamma$ be the density field of the gas. Then a mean gas velocity $\mathbf{v} = \frac{1}{\rho}\int \mathbf{v}' f(t,\mathbf{r},\Gamma)\,d\Gamma$, where $\mathbf{v}'$ is a microvelocity. We will use $\mathbf{v}$ as a macroscopic variable that characterizes cells whose length $L$ in any direction is much, much greater than the mean free path in the gas, $l_m$; that is, $L \gg l_m$.

Since, by assumption, collisions preserve conservation laws exactly, the moments of $C(f)$, in particular the integrals $\int C(f)\,d\Gamma$ and $\int \mathbf{v}\ C(f)\,d\Gamma$, are equal to zero (similarly for any conserved quantity). We use this fact by integrating the Boltzmann equation in two ways: $\int$(B.E.)$\,d\Gamma$ and $\int \mathbf{v}$(B.E.)$\,d\Gamma$ (where B.E. stands for the Boltzmann equation). The first integral gives the continuity equation:

$$\partial_t \rho + \partial_i(\rho \mathrm{v}_i) = 0. \qquad (6)$$

The second integral gives the momentum tensor equation:

$$\partial_t(\rho \mathrm{v}_i) + \partial_k \Pi_{ik} = 0, \qquad (7)$$

where the momentum flux tensor $\Pi_{ik}$ is given by

$$\Pi_{ik} \equiv \int \mathrm{v}_i \mathrm{v}_k f\,d\Gamma.$$

In order to derive the Euler equation for ideal gases with the usual form for the momentum flux tensor, we need to assume that each region in the gas has a local Maxwell-Boltzmann distribution. With this assumption one can show that the momentum flux tensor in Eq. 7 has the following form:

$$\Pi_{ik} = \rho \mathrm{v}_i \mathrm{v}_k + \delta_{ik} p,$$

where $p$ is the pressure. This form of $\Pi_{ik}$ gives the same Euler equation that we found by general continuum arguments. (We will see in Part II that the form of $\Pi_{ik}$ for the totally discrete fluid is not so simple but depends upon the geometry of the underlying lattice. Again by assuming a form for the local distribution function (the appropriate form will turn out to be Fermi-Dirac rather than Boltzmann), $\Pi_{ik}$ will reduce to a form that gives the lattice Euler equation.)

**Recovering the Navier-Stokes Equation.** The derivation of the Navier-Stokes equation from the kinetic theory picture is more involved and requires us to face the full Boltzmann equation. Hilbert accomplished this through a beautiful argument that relies on a spatial-gradient perturbation expansion around some single-particle distribution function $f_L$ assumed to be given at $t_0$. In "The Hilbert Contraction" we discuss the main outline of his argument emphasizing the assumptions involved and their limitations. Here we will summarize his argument. Hilbert was able to show that the evolution of $f$ for times $t > t_0$ is given in terms of its initial data at $t_0$ by the first three moments of $f$, namely the familiar macroscopic variables $\rho$ (density), $\mathbf{v}$ (mean velocity), and $T$ (temperature). In other words, he was able to contract this many-degree-of-freedom system down to a low-dimensional descriptive space whose variables are the same as those used in the usual hydrodynamical description. The beauty of Hilbert's proof is that it is constructive. It explicitly displays a recursive closed tower of constraint relations on the moments of $f$ that come directly from the

# THE HILBERT CONTRACTION

The Boltzmann equation is a microscopic equation for colliding-gas evolution valid in a very tight regime. It is first order in time and so requires a complete description of the one-particle distribution function at one time, say $t = 0$, after which its functional form is completely fixed by the Boltzmann transport equation.

Describing the one-particle distribution function completely is a hopeless procedure, since the amount of information is too large. However, one wants to recover hydrodynamics, which is essentially a partial differential equation for a macroscopic description of the fluid at long times and distances compared to molecular scales. So there must exist a contraction mechanism that reduces the number of degrees of freedom required to describe the solution to the Boltzmann transport equation at such long times and distances. It is not obvious how that can happen, but Hilbert gave a proof that is central to understanding that it must happen and in a rather surprising way. We will call this process the Hilbert contraction. All analyses of the Boltzmann equation are based on this contraction. We would like to give it in detail because it is a beautiful argument, but space forbids this, so we outline how Hilbert reasoned.

Since we don't know what else to do when faced with such a highly nonlinear system, we construct a perturbation expansion in a small variable around some distribution function $f$, assumed to be given to us at $t_0$. Under some very mild assumptions, and assuming the existence of such a general perturbation expansion in some parameter $\delta$, Hilbert was able to show that the evolution of $f$ for $t > t_0$ is given in terms of its initial data at $t_0$ by the first three moments of $f$, namely $\rho, \mathbf{v}$, and $T$. The system has contracted down to a low-dimensional descriptive manifold whose coordinates are the same variables used by the hydrodynamic description. The beauty of Hilbert's proof is that it is constructive. It explicitly displays a recursive closed tower of constraint relations on the moments of $f$ that come directly from the Boltzmann equation. The proof also shows that such a contracted description is unique—a very powerful result.

It must be pointed out that Hilbert's construction is on the time-evolved solution to the Boltzmann transport equation, not on the equation itself, which still requires a complete specification of $f$. It amounts to a hard mathematical statement on an effective field-theory description for times much greater than elementary collision times, but with space gradients still smooth enough to entertain a serious gradient perturbation expansion. As such, it says nothing about the turbulent regime, for example, where all these assumptions fail.

In standard physics texts one can read all kinds of plausibility arguments as to why this contraction process should exist, but they lack force, for, by arguing tightly, one can make the conclusion go the other way. This is why the Hilbert contraction is important. It is really a powerful and mathematically unexpected result about a highly nonlinear integro-differential equation of very special form. Beyond Hilbert's theorem and within the Boltzmann transport picture, we can say nothing more about the contraction of descriptions.

The construction of towers of moment constraints, coupled to a perturbation expansion that Hilbert developed for his proof of contraction, was used in a somewhat different form by Chapman and Enskog. Their main purpose was to devise a perturbation expansion with side constraints in such a way as to pick off the values of the coupling constants—which are called transport coefficients in standard terminology—for increasingly more sophisticated forms of macrodynamical equations.

One makes the usual kinetic assump-

## THE HILBERT CONTRACTION (continued)

tions: The gas reaches local equilibrium in a collision time or so; the one-particle distribution function has a local Maxwell-Boltzmann form (or whatever form is appropriate), call it $f_L$; a second time scale is assumed where space gradients are still small, but collective modes develop at large distances and at times much greater than molecular collision times. Then one assumes a general functional perturbation expansion exists of the form

$$f = f_L(1 + \xi^{(1)} + \xi^{(2)} + \cdots),$$

which turns out to be explicitly a spatial gradient expansion:

$$f = f_L(1 + c_1(\mathrm{v})(\lambda\nabla) + c_2(\mathrm{v})(\lambda\nabla)^2 + \cdots)$$

where $\lambda$ is the mean free path in the system and v is the macrovelocity.

The perturbation expansion is set up so that at $n$th order, the correction to $f_L$ obeys an integral equation of the form $f_L C(\xi^{(n)}) = L_n$, where $C$ is the Boltzmann collision operator and $L_n$ is an operator that depends only on lower order spatial derivatives. This generates a recursive tower of relations $\xi^{(n)}$ whose solubility conditions at order $n$ are the $(n-1)$th-order hydrodynamical equations.

For example, assume

$$f = f_L\left(1 + \xi^{(1)}\right);$$

that is, we keep only 1st order in $\xi$. Then in the Boltzmann collision term keep consistently only order $\xi^{(1)}$ and in the streaming operator put $\xi^{(0)} \equiv f_L$. So we get

$$(\partial_t + \mathrm{v}_\alpha\partial_\alpha + a_\alpha\frac{\partial}{\partial \mathrm{v}_\alpha})f_L = f_L C(\xi^{(1)}),$$

which is of the form

$$f_L C(\xi^{(1)}) = L_1.$$

The solubility conditions for this are that $L_1$ must be orthogonal to the five zero eigenmodes of $C(\xi^{(1)}) = 0$ (the solutions are $1, \mathbf{v}$, and $\mathbf{v}^2$). These solubility conditions are the Euler equation for $\rho$, v, and $T$ and the ideal gas equation of state. In this way one derives a sequence of hydrodynamical equations with explicit forms for the transport coefficients. Order 0 gives the Euler equation, order 1 gives the Navier-Stokes equations, order 2 and greater give the generalized hydrodynamical equations, which have some validity only in special situations. The expansion is an asymptotic functional expansion, so going beyond Navier-Stokes takes one away from ordinary fluids rather than closer to them. Solving explicitly for the various $\xi^{(n)}$ gives a way to evaluate the transport quantities (viscosity, etc.).

There are many other ways to do the same thing—multiple time expansions, dispersion methods, etc. We have developed everything so far within the conceptual frame of the Boltzmann transport equation. Within that framework the problem of deriving macrodynamical equations and associated transport coefficients reduces to tedious but straightforward linear algebra that has absorbed the best efforts of excellent technical people since the turn of the century. It is a problem best suited to a computer but only recently have algebraic processors of sufficient power been available.

This asymptotic perturbation expansion is a way to compute measurable quantities from microdynamical properties, but the physical insight one gains from doing it is small. The other methods mentioned, especially correlation-function techniques, are much more revealing. All of these comments and approaches carry over directly to the discrete case of the lattice gas. Nothing conceptually new arises in the totally discrete case, but explicit calculations are a great deal easier. ■

Boltzmann equation. The zero-order relation gives the Euler equations and the second-order relation gives the Navier-Stokes equations. However, Hilbert's method is an asymptotic functional expansion, so that the higher order terms take one away from ordinary fluids rather than closer to them. Nevertheless, solving explicitly for the terms in the functional expansion provides a way of evaluating transport coefficients such as viscosity. (See the "Hilbert Contraction" for more discussion.)

**Summary of the Kinetic Theory Picture.** Our review of the kinetic theory description of fluids introduced a number of important concepts: the idea of local thermal equilibrium; the characterization of an equilibrium state by a few macroscopic observables; the Boltzmann transport equation for systems of many identical objects (with ordinary statistics) in collision; and the fact that a solution to the Boltzmann transport equation is an ensemble of equilibrium states. In "The Hilbert Contraction" we introduced the linear approximation to the Boltzmann equation with which one can derive the Navier-Stokes equations for systems not too far (in an appropriate sense) from equilibrium in terms of these same macroscopic observables (density, pressure, temperature, etc.). We then outlined a method for calculating the coupling constants in the Navier-Stokes system—that is, the strengths of the nonlinear terms—as a function of any particular microdynamics.

This review was intended to give a flavor for the chain of reasoning involved. We will use this chain again in the totally discrete lattice world. However, just as important as understanding the kinetic theory viewpoint is keeping in mind its limitations. In particular, notice that perturbation theory was the main tool used for going from the exact Boltzmann transport equation to the Navier-Stokes equations. We did not discover more powerful techniques for finding solutions to the Navier-Stokes equations than we had before. To go from the Boltzmann to the Navier-Stokes description, we made many smoothness assumptions in various probabilistic disguises; in other words, we recreated an approximation to the continuum. It is true one could compute (at least for relatively simple systems) the transport coefficients, but in a sense these coefficients are a property of microkinetics, not macrodynamics.

We are at a point where we can ask some questions about the emergence of macrodynamics from microscopic physics. It is clear by now that microscopic conservation laws, those of mass, momentum, and energy are crucial in fixing the form of large-scale dynamics. These are in a sense sacred. But one can question the importance of the description of individual collisions. How detailed must micromechanics be to generate the qualitative behavior predicted by the Navier-Stokes equations? Can it be done with simple collisions and very few classes of them? There exists a whole collection of equations whose functional form is very nearly that of the Navier-Stokes equations. What microworlds generate these? Do we have to be exactly at the Navier-Stokes equations to generate the qualitative behavior and numerical values that we derive from the Navier-Stokes equations or from real fluid experiments? Is it possible to design a collection of synthetic microworlds that could be considered local-interaction board games, all having Navier-Stokes macrodynamics? In other words, does the detailed microphysics of fluids get washed out of the macrodynamical picture under very rapid iteration of the deterministic system? If the microgame is simple enough to update it deterministically on a parallel machine, is the density of states required to see everything we see in ordinary Navier-Stokes simulations much smaller than the density of atoms in real physical fluids? If so, these synthetic

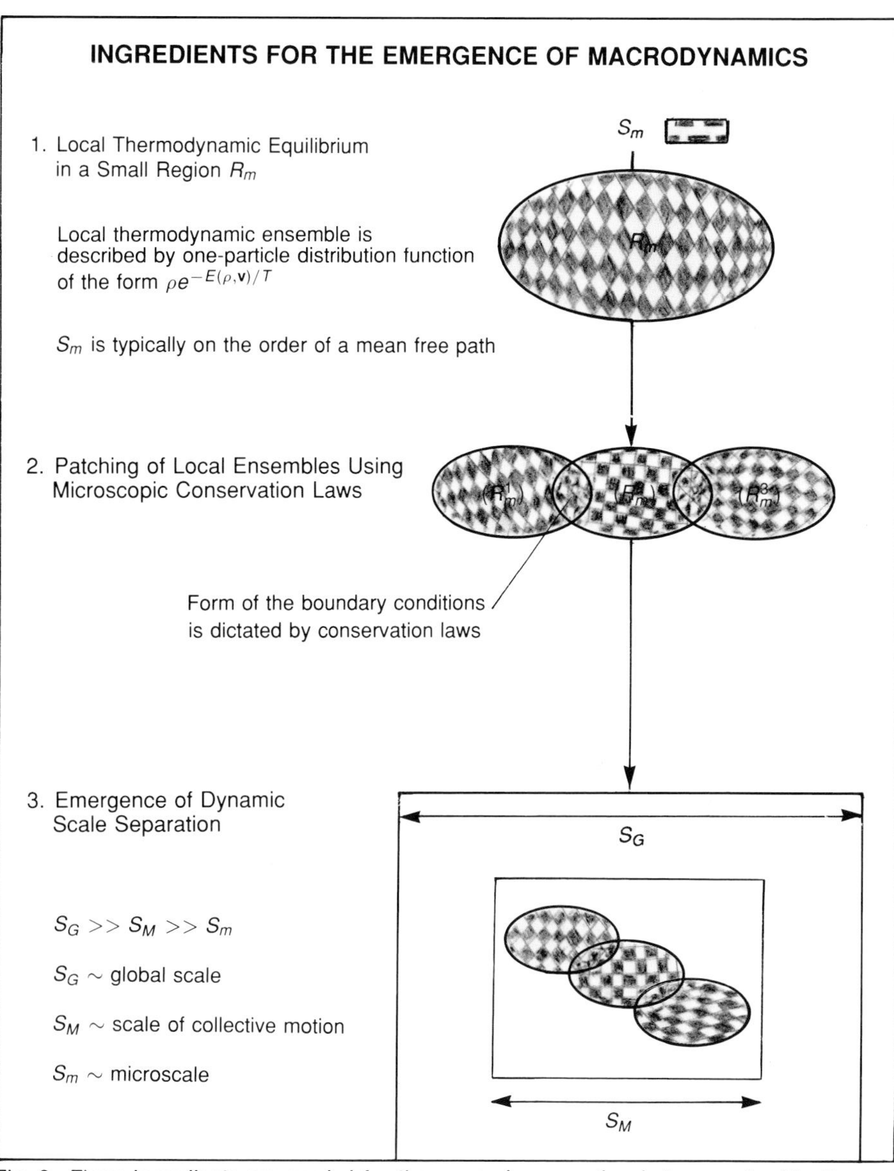


**Fig. 3. Three ingredients are needed for the emergence of macrodynamics: local thermodynamic equilibrium, conservation laws, and scale separation between microkinetics and collective motion.**

microworlds become a potentially powerful analytic tool.

Our approach in building a cellular space is to move away from the idea of a fluid state and focus instead on the idea of the macrodynamics of a many-element system. In abstract terms, we want to devise the simplest deterministic local game made of a collection of few-bit, finite-state machines that has the Navier-Stokes equations as its macrodynamical description. From our brief look at kinetic transport theory, we can abstract the essential features of such a game (Fig. 3). The many-element system must be capable of supporting a notion of local thermodynamic equilibrium and must also include local microscopic conservation laws. The state of a real fluid can be imagined as a collection of equilibrium distribution functions whose macroscopic parameters are unconstrained. These distribution functions have a Maxwell-Boltzmann form, $e^{-E(\rho,\mathrm{v})/T}$. If these distribution functions are made to deviate slightly from equilibrium, then local conservation laws impose consistency conditions among their parameters, which become constrained variables. These consistency conditions are the macrodynamical equations necessary to put a consistent equilibrium function description onto the many-element system. In physical fluids they are the Navier-Stokes equations. This is the general setup that will guide us in creating a lattice model.

## Evolution of Discrete Fluid Models

**Continuous Network Models.** The Navier-Stokes equations, however derived, are analytically intractable, except in a few special cases for especially clean geometries. Fortunately, one can avoid them altogether for many problems, such as shocks in certain geometries. The strategy is to rephrase the problem in a very simple phase space and solve the Boltzmann transport equation directly. If a single type of particle is constrained to move continuously only along a regular grid, the Boltzmann equation is so tightly constrained that it has simple analytic solutions. In the early 1960s Broadwell and others applied this simplified method of analysis to the dynamics of shock problems. Their numerical results agreed closely with much more elaborate computer modeling from the Navier-Stokes equations. However, there was no real insight into why such a calculation in such a simplified microworld should give such accurate answers. The accuracy of the limited phase-space approach was considered an anomaly.

**Discrete Skeletal Models.** The next development in discrete fluid theory was a discrete modification of the continuous-speed network models of the Broadwell class. By forming a loose analogy to the structure of the Ising model (spins on a lattice), Hardy, de Pazzis, and Pomeau created the first minimalist fluid model on a two-dimensional square lattice. It was a simple, binary-valued, nearest-neighbor gas with a single species of molecule, limited to binary collisions. The new feature was a totally discrete velocity and state space for the gas. Particles hopped from one site to the next without a notion of continuous movement between sites. Particles were confined to the vertices of the network, and the velocity vector of each particle could point in only one of four directions. Since there was no natural way to deal with bound states, these authors imposed the arbitrary rule that the maximum number of particles occupying any vertex be four.

This simple model possessed remarkable properties including local thermodynamic equilibrium and the emergence of a scale separation; that is, the typical collective motion scale $L$ is much greater than the microscopic mean free path $l_m; L \gg l_m$. However, the macrodynamics that emerged was not that of the Navier-Stokes equations but a more complex one with unphysical features. The square model was the first example of rich dynamics emerging wholly on a cellular space. It had all the right ingredients except one: isotropy under the rotation group of the lattice. The momentum flux tensor must reduce to a scalar for isotropy, but this is impossible with a square lattice. In two dimensions the neighborhood that has the minimal required symmetry and tiles the plane is a hexagonal neighborhood. In Part II we will present the simple hexagonal model, analyze it mathematically, and describe the simulations of fluid phenomena that have been done so far.

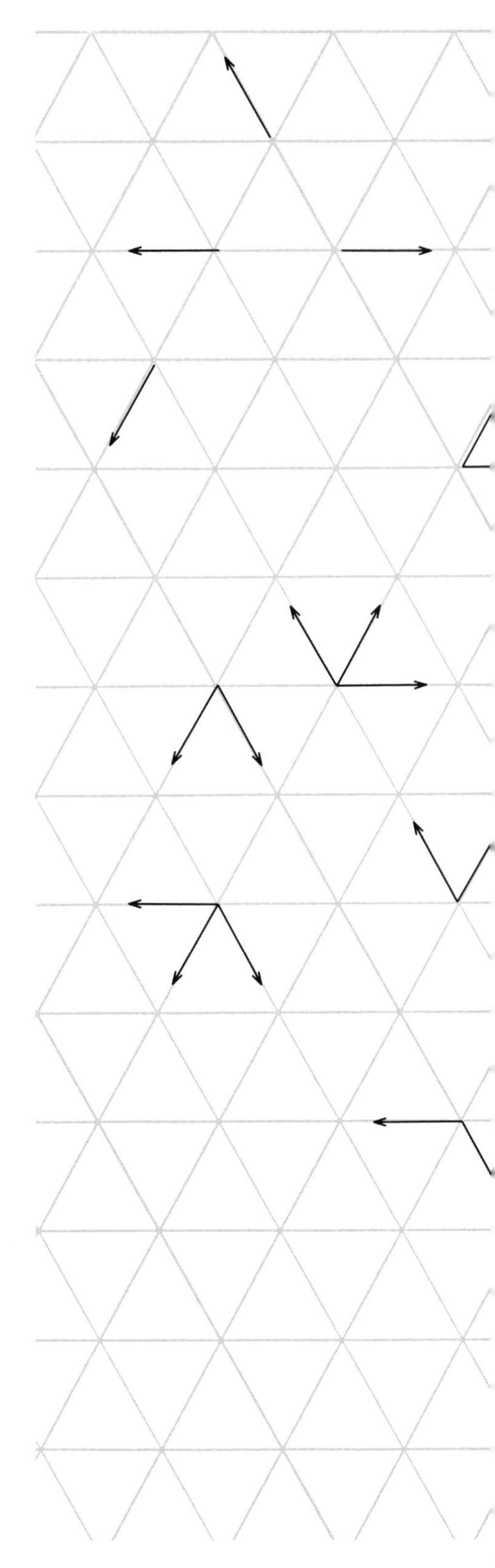

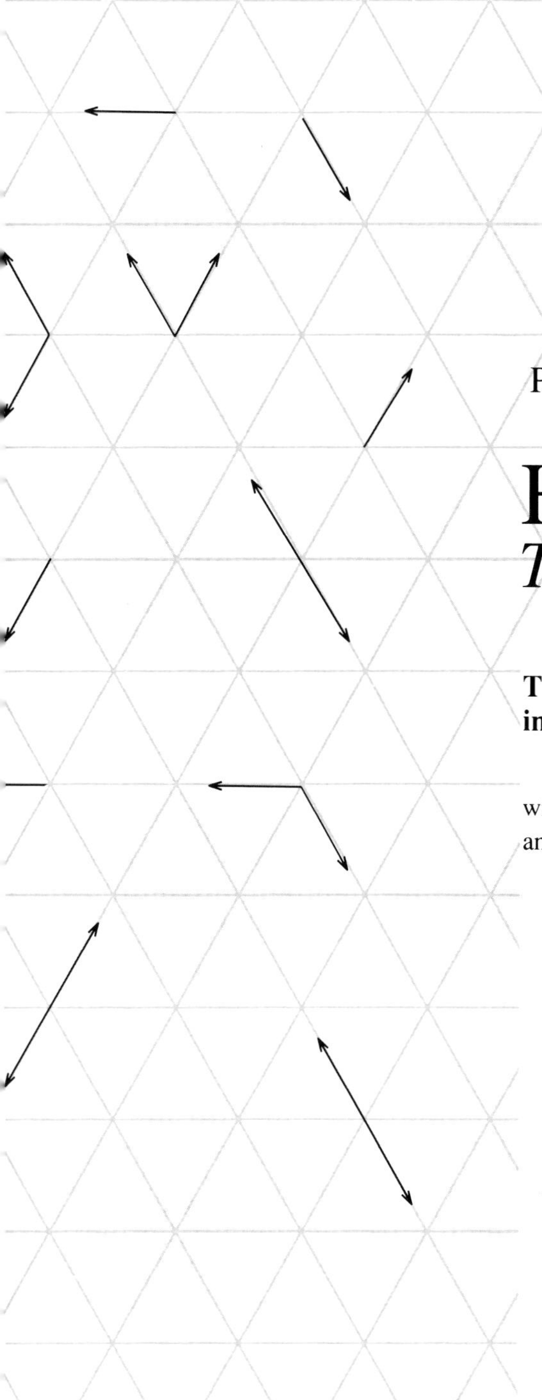

PART II

# THE SIMPLE HEXAGONAL MODEL

## *Theory and Simulation*

### The Minimal Totally Discrete Model of Navier-Stokes in Two Dimensions

We can now list the ingredients we need to build the simplest cellular-space world with a dynamics that reproduces the collective behavior predicted by the compressible and incompressible Navier-Stokes equations:

1. A population of identical particles, each with unit mass and moving with the same average speed $c$.
2. A totally discrete phase space (discrete values of $x, y$ and discrete particle-velocity directions) and discrete time $t$. Discrete time means that the particles hop from site to site.
3. A lattice on which the particles reside only at the vertices. In the simplest case the lattice is regular and has a hexagonal neighborhood to guarantee an isotropic momentum flux tensor. We use a triangular lattice for convenience.
4. A minimum set of collision rules that define symmetric binary and triple collisions such that momentum and particle number are conserved (Fig. 4).
5. An exclusion principle so that at each vertex no two particles can have identical velocities. This limits the maximum number of particles at a vertex to six, each one having a velocity that points in one of the six directions defined by the hexagonal neighborhood.

The only way to make this hexagonal lattice gas simpler is to lower the rotation symmetry of the lattice, remove collision rules, or break a conservation law. In a two-dimensional universe with boundaries, any such modification will not give Navier-Stokes dynamics. Left as it is, the model will. Adding attributes to the model, such as different types of particles, different speeds, enlarged neighborhoods, or weighted collision rules, will give Navier-Stokes behavior with different equations of state and different adjustable parameters such as the Reynolds number (see the discussion in Part III). The hexagonal model defined by the five ingredients listed above is the simplest model that gives Navier-Stokes behavior in a sharply defined parameter regime.

At this point it is instructive to look at the complete table of allowed states for the model (Fig. 5). The states and collision rules can be expressed by Boolean logic

**Fig 4. SCATTERING RULES FOR SIMPLE HEXAGONAL MODEL**

**Scattering Rules**

**Results of Scattering Plus Transport**

Two-Body Scattering Rules

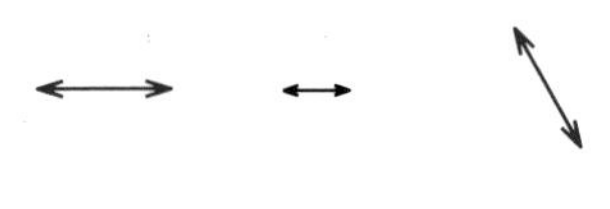

or

Only a head-on collision of two particles causes scattering, that is, the particles change direction by ±60°. The particles then continue to move at constant speed (one node per time stop) in the new direction.

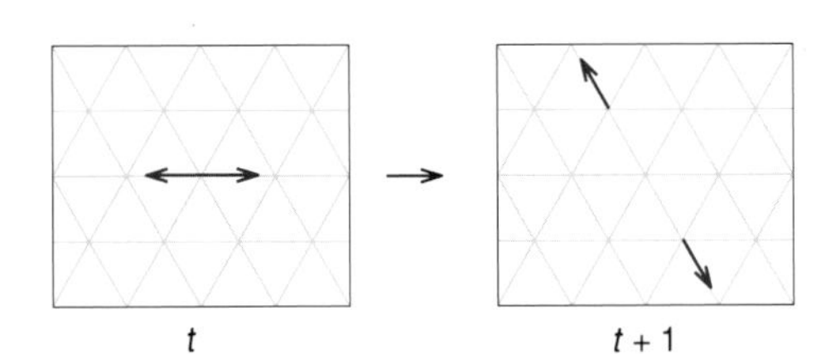


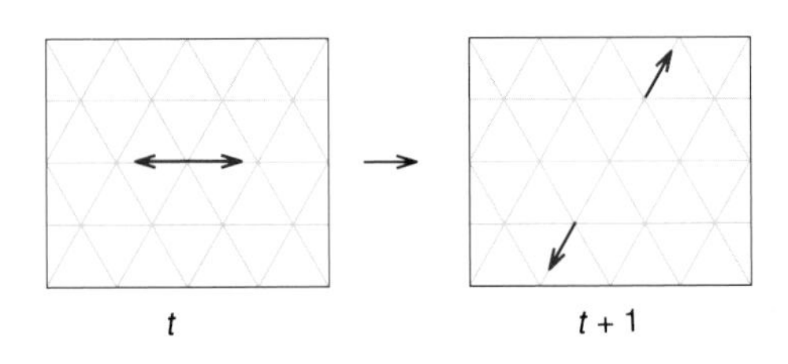


Three-Body Scattering Rules

Three particles colliding at 120° angles to each other change directions by 60° in the scattering process. All other configurations of these particles do not affect particle direction.

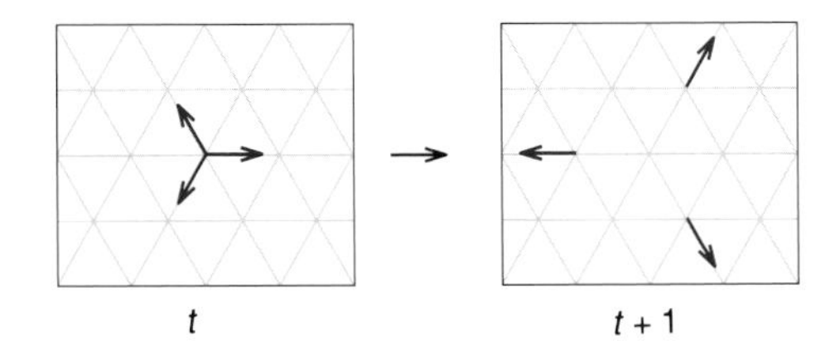


Other Configurations Don't Scatter

For Example

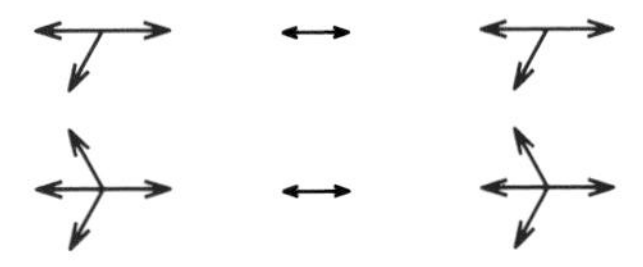

In most configurations particles do not scatter, that is, they do not change direction but are simply transported at constant speed.

Pure Transport

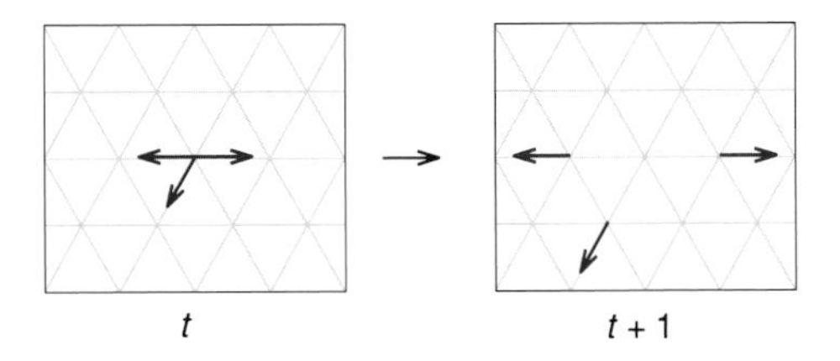

operations with the two allowed values taken as 0 and 1. From this organization scheme we see that the hexagonal lattice gas can be seen as a Boolean parallel computer. In fact, a large parallel machine can be constructed to implement part or all of the state table locally with Boolean operations alone. Our simulations were done this way and provide the first example of the programming of a cellular-automaton, or cellular-space, machine that evolves the dynamics of a many-degrees-of-freedom, nonlinear physical system.

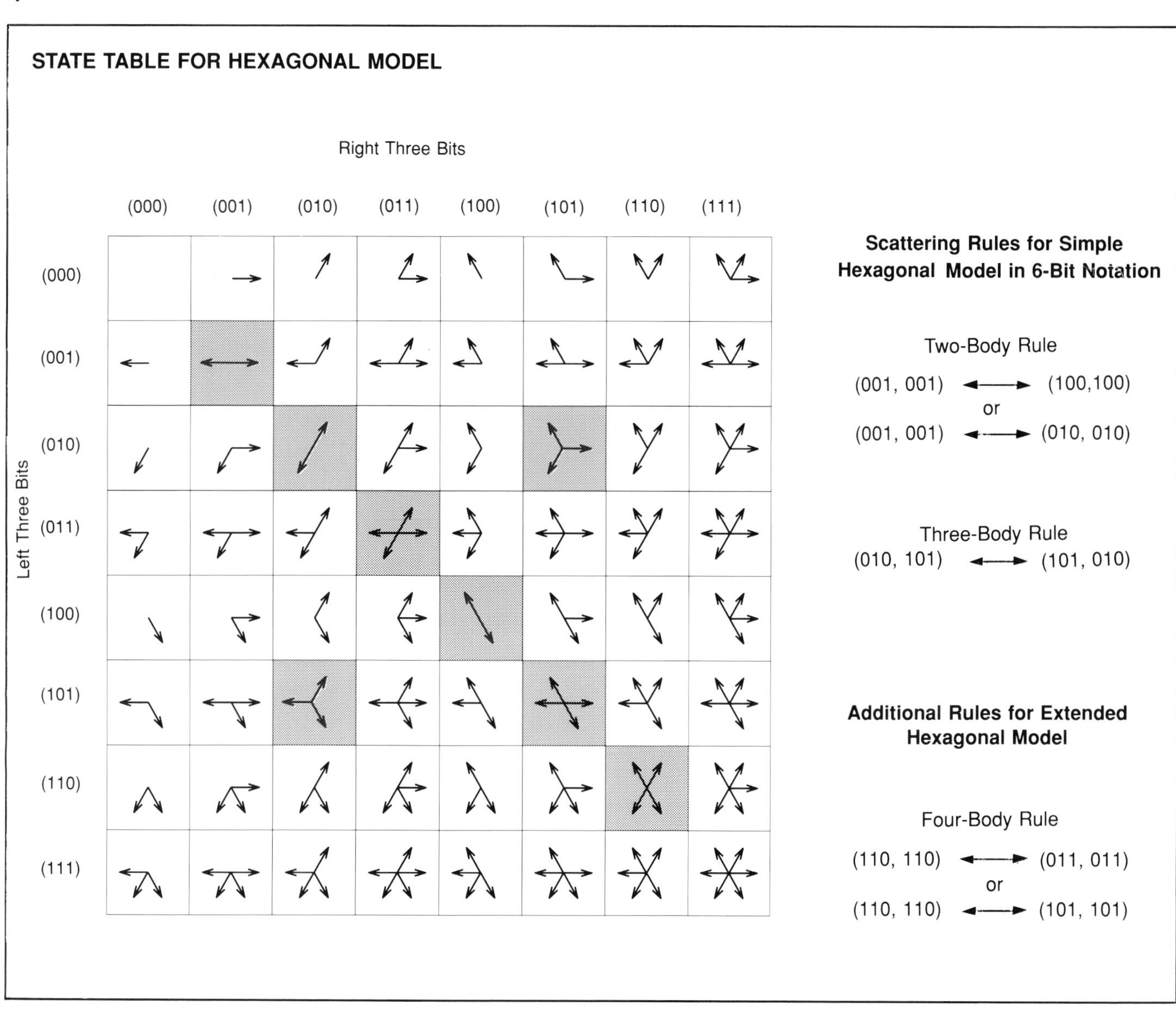


Fig. 5. All possible states of the hexagonal lattice gas are shown in the state table. Each state can be expressed in 6-bit notation (a combination of 3 right bits and 3 left bits). For example, the empty state is written (000,000) and the maximally occupied state shown in the lower right hand corner of the table is written (111, 111). Collision states for the simplest hexagonal model are shown in red and shaded in gray. The scattering rules for these states are written beside the table. All other states do not result in scattering. The extended hexagonal model includes scattering rules for four-body states (shaded in gray). The extended model lowers the viscosity of the lattice gas.

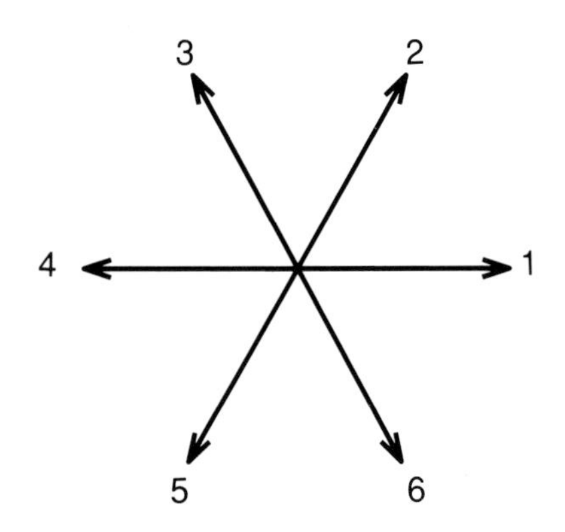


**PARTICLE DIRECTIONS IN THE HEXAGONAL MODEL**

Fig. 6. The velocity vector of each particle can point in one of six possible directions. All particles have the same speed $c$.

## Theoretical Analysis of the Discrete Lattice Gas

Before presenting the results of simulations with the lattice-gas automaton, we will analyze its behavior theoretically. The setup we work on is a regular triangular grid with hexagonal neighborhood. The natural explicit coordinate system for a single-speed, six-directional world (Fig. 6) is the set of unit vectors:

$$\hat{\mathbf{i}}_\beta = \left\{ \cos\left(\frac{2\pi\beta}{6}\right), \ \sin\left(\frac{2\pi\beta}{6}\right) \right\}, \quad \beta = 1, \cdots, 6. \tag{8}$$

One never requires this much detail except to work out explicit tensor structures and scalar products particular to the hexagonal model case, but the index conventions are important to avoid disorientation. From now on the Greek indices $\alpha, \beta, \cdots$ label lattice direction indices; $\hat{\mathbf{i}}, \hat{\mathbf{j}}, \hat{\mathbf{k}}, \ldots$ are lattice unit vectors and $i, j, k$ label space indices $(x_1, x_2, \ldots)$; on a square lattice we have $\mathbf{r} = (x_1, x_2) = (x, y)$.

The first thing we will look at is pure transport on the lattice with no collisions. Because the basic space is a discrete lattice with a fundamental lattice spacing, rather than a continuum, a shadow of the lattice is induced into the coupling constant of the theory, namely the viscosity. This lattice effect is not obvious, but we will make it so by looking at transport on the lattice in detail. As a corollary we will derive the usual Euler equations for the "macroscopic" flow of the lattice gas.

To do a quick analysis on lattice models we lift the restriction of a deterministic gas and pass to a probabilistic description familiar from kinetic theory; then we can use familiar stochastic and kinetic theory tools outlined in Part I of this article. In going from a continuous to a discrete probabilistic formalism we introduce the lattice form of the single-particle distribution function by making the identifications

$$f(\mathbf{r},\Gamma,t) \rightarrow f_\beta(\mathbf{r},t)$$

$$\int f(\mathbf{r},\Gamma,t)d\Gamma \rightarrow \Sigma_\beta f_\beta(\mathbf{r},t) \equiv \rho, .$$

and

$$\rho\mathbf{v} \rightarrow \Sigma_\beta \hat{\mathbf{i}}_\beta f_\beta$$

To begin we write the master equation for $f_\beta$ in the absence of collisions. The master equation expresses conservation of probability. For simplicity we write it for a square lattice with the following conventions: $n_\beta(\mathbf{r}+\hat{\mathbf{i}}_\beta, t)$ = number of particles in the direction $\beta$ at the node $\mathbf{r}+\hat{\mathbf{i}}_\beta$ at time $t$. The master equation for the system, neglecting collisions and written in a continuum notation for convenience, is

$$f_\beta(\mathbf{r}+h, t+k) - f_\beta(\mathbf{r},t) = 0, \ \text{with} \ h = \hat{\mathbf{i}}_\beta d_x, k = d_t,$$

where $d_t, d_x \ll 1$.

If we expand the first term in the master equation out to $O^2(h,k)$ using the Taylor series expansion $f(x_0+h, y_0+k) = \sum_{\lambda=0}^{m-1} \frac{1}{\lambda!}(h\partial_x + k\partial_y)^\lambda f(x_0,y_0) + R_m$, we obtain

$$0 = d_t\partial_t f_\beta + d_x\hat{\mathbf{i}}_\beta \cdot \nabla f_\beta + \frac{1}{2}d_t^2\partial_t^2 f_\beta + \frac{1}{2}d_x^2(\hat{\mathbf{i}}_\beta \cdot \nabla)^2 f_\beta + d_x d_t(\hat{\mathbf{i}}_\beta \cdot \nabla)\partial_t f_\beta.$$

To lowest order in $h$ and $k$, we have

$$\partial_t f_\beta + \hat{\mathbf{i}}_\beta \cdot \nabla f_\beta = 0, \tag{9}$$

which has the standard form of the kinetic theory transport equation in the absence of collisions. If we include collisions, the full Boltzmann transport equations schematically become

$$\partial_t f_\beta + \hat{\mathbf{i}}_\beta \cdot \nabla f_\beta = C_\beta(f) \tag{10}$$

where $C_\beta(f)$ is the collision operator on the lattice. The form of the lattice collision operator will tell us a great deal about how the model works, but for the moment we just look at the general structure of the "macroscopic" equations for the lattice gas to the lowest order in the lattice expansion parameters.

As in standard kinetic theory, the usual zero integrals of the motion hold, since the lattice model is assumed to have some kind of detailed balance (that is, microscopic reversibility of reaction pathways). Accordingly $\sum_\beta C_\beta(f) = 0$ and $\sum_\beta \hat{\mathbf{i}}_\beta C_\beta(f) = 0$ for a skeletal gas. Following the kinetic theory procedure, we write the continuity and momentum equations that follow from these conditions as:

$$\partial_t \rho + \partial_i(\rho \mathrm{v}_i) = 0 \tag{11}$$

and

$$\partial_t(\rho \mathrm{v}_i) + \partial_j \Pi_{ij} = 0, \tag{12}$$

where the tensor $\Pi_{ij}$ is defined as

$$\Pi_{ij} = \sum_\beta (\hat{i}_\beta)_i (\hat{i}_\beta)_j f_\beta. \tag{13}$$

So far we have kept only the leading terms of the Taylor series expansion in the scaling factors that relate to the discreteness of the lattice. It's easy to show that keeping quadratic terms in this lattice-size expansion leaves the continuity equation invariant but alters the momentum equation by introducing a free-streaming correction to the measured viscosity. This rather elegant way of viewing this correction was first developed by D. Levermore. The correction comes from breaking the form of a Galilean covariant derivative and is a geometrical effect. Specifically, to second order in the lattice size expansion, the momentum equation does not decompose simply into factors of these covariant derivatives but instead the expansion introduces a nonvanishing covariant-breaking term:

$$\text{Noncovariant term} = \sum_\beta \{(\hat{i}_\beta)_i \partial_i \partial_t + \left((\hat{i}_\beta)_i \partial_i\right)^2\} \hat{\mathbf{i}}_\beta f_\beta. \tag{14}$$

This term is of the same order as those terms that contribute to the viscosity. Later we will show how to use the Chapman-Enskog expansion to compute an explicit form for the lattice-gas viscosity.

**The Chapman-Enskog Expansion and the Direct Expansion.** The form of $\Pi_{ij}$ depends on the form of $f$, the solution to the full lattice Boltzmann transport equation.

By Hilbert's construction we know that an efficient expansion can be developed in terms of the collision invariants of the model up to powers of terms linear in the gradient of the macroscopic velocity. In whatever perturbation expansion of $f$ we choose, the coefficients in the expansion are fixed by solving for them under the Lagrange multiplier constraints of mass and momentum conservation: $\rho = \sum_\beta f_\beta$ and $(\rho\mathbf{v}) = \sum_\beta \hat{\mathbf{i}}_\beta f_\beta$. In the simple hexagonal model there is no explicit mechanism provided for storing energy in internal state space, so there is no independent energy equation.

For the lattice case, the Chapman-Enskog version of Hilbert's expansion reduces to an expansion in all available scalar products using the vectors $\hat{\mathbf{i}}_\beta, \mathbf{v}$ and the vector operator $\overrightarrow{\partial}$. The expansion is made around the global equilibrium solution for $\mathbf{v} = 0$, which we will call $N_{\mathrm{eq}}^{\mathbf{v}=0}$ and terms are kept up to those linear in $\overrightarrow{\partial}$. The relevant scalar products are

$$(\hat{\mathbf{i}}_\beta \cdot \hat{\mathbf{i}}_\beta),\ (\hat{\mathbf{i}}_\beta \cdot \mathbf{v}),\ (\hat{\mathbf{i}}_\beta \cdot \overrightarrow{\partial})(\hat{\mathbf{i}}_\beta \cdot \mathbf{v}),\ (\overrightarrow{\partial} \cdot \mathbf{v}),\ (\hat{\mathbf{i}}_\beta \cdot \mathbf{v})^2,\ (\mathbf{v} \cdot \mathbf{v}) + O(\mathbf{v}^3).$$

The systematic expansion becomes

$$f_\beta = N_{\mathrm{eq}}^{\mathbf{v}=0} \left\{ 1 + \alpha \hat{\mathbf{i}}_\beta \cdot \mathbf{v} + \beta \left[ (\hat{\mathbf{i}}_\beta \cdot \mathbf{v})^2 - \frac{1}{2}|\mathbf{v}|^2 \right] + \beta_1 \left[ (\hat{\mathbf{i}}_\beta \overrightarrow{\partial})(\hat{\mathbf{i}}_\beta \mathbf{v}) - \frac{1}{2} \overrightarrow{\partial} \cdot \mathbf{v} \right] + O(\mathbf{v}^3) \ldots \right\}. \tag{15}$$

In the usual kinetic theory approach the coefficients $\alpha$ and $\beta$ can be found by neglecting collisions and $\beta_1$, the gradient term, can be determined only by an explicit solution to the full Boltzmann equation including collision terms. In this way one obtains the viscosity in terms of $\beta_1$. For the discrete lattice, however, both $\beta$ and $\beta_1$ depend on the explicit form of the solution to the full Boltzmann equation with collisions. We also need that form to recover the correction to the raw viscosity that, as mentioned in the last section, comes from pure translation effects on the lattice.

Given that we have to use the full solution to the Boltzmann transport equation almost immediately, we now derive its structure, find the general and equilibrium solution, and then use a direct expansion to fix both $\beta$ and $\beta_1$. In the process we will recover the Euler equations for inviscid flow and the Navier-Stokes equations for the flow with dissipation.

**The Lattice Collision Operator and the Solution to the Lattice Boltzmann Transport Equation.** We will write down the discrete form of the Boltzmann equation, especially noting the collision operator, for a number of reasons. First, writing the explicit form of the collision kernel builds up an intuition of how the heart of the model works; second, we can show in a few lines that the Fermi-Dirac distribution satisfies the lattice gas Boltzmann equation; third, knowing this, we can quickly compute the lattice form of the Euler equations; fourth, we can see that many properties of the lattice-gas model are independent of the types of collisions involved and come only from the form of the Fermi-Dirac distribution.

Collision operators for lattice gases with continuous speeds were derived by Broadwell, Harris, and other early workers on continuum lattice-gas systems. For totally discrete lattice gases with an exclusion principle, we must be careful to apply this principle correctly. It is similar to the case of quasi-particles in quantum Fermi liquids.

The construction reduces to following definitions of collision operators introduced in the section on classical kinetic theory and counting properly.

Taking any hexagonal neighborhood, let $i$ be one of the six directions and use the convention $i, i+1, i+2, \ldots \equiv \hat{\mathbf{i}}_{\beta=i}, \hat{\mathbf{i}}_{\beta=i+1}, \hat{\mathbf{i}}_{\beta=i+2}, \ldots$ for convenience. (Later we will return to our original notation.) First consider binary collisions alone, and assume detailed balance, which implies microscopic reversibility of a collision at each vertex. One need not use detailed balance, but other balancing schemes are algebraically tedious and conceptually similar extensions of this basic case. Given a vertex at $(\mathbf{r}, t)$ we compute the gain and loss of particles into a neighborhood along a fixed direction, say $i$. This is, by definition, the collision kernel for binary processes. First compute the number of particles thrown in a collision into a phase-space region along the direction $i$. Let $n_i(\mathbf{r}, t)$ be the probability that a particle is at the node $(\mathbf{r}, t)$ and has a velocity in the $i$th direction.

If a particle scatters into a vector direction $i$, it must have come from binary processes along directions ($i+1$ and $i+4$) or ($i+2$ and $i+5$) (see the two-body scattering rules in Fig. 4). Interpreted as probabilities for the two events to happen, the probability for gain in the $i$ direction due to binary processes alone is

$$P_i^{\text{binary}} = n_{i+1}n_{i+4}\bar{n}_i\bar{n}_{i+2}\bar{n}_{i+3}\bar{n}_{i+5} + n_{i+2}n_{i+5}\bar{n}_i\bar{n}_{i+1}\bar{n}_{i+3}\bar{n}_{i+4},$$

where $\bar{n}_k \equiv (1-n_k)$. The $\bar{n}_k$'s impose the exclusion rule in the output channel, namely, that a particle cannot scatter there if one is already present.

Loss of a particle from direction $i$ can occur only by the binary collision $(i+3, i)$, and this can happen for each of the two choices of gain collisions separately. So we have $(-2n_i n_{i+3}\bar{n}_{i+1}\bar{n}_{i+2}\bar{n}_{i+4}\bar{n}_{i+5})$ as the probability for loss in the $i$ direction due to binary collisions alone. Note that these products can be compactly expressed as $\hat{n}_i\hat{n}_{i+3}\Pi_{i=0}^{5}(1-n_i)$ where $\hat{n}_i \equiv \frac{n_i}{1-n_i}$.

The three-body gain-loss term can be written down by inspection in the same way as the binary term. The complete two- and three-body collision term for the $i$th direction, in compact notation, is

$$C_i^{2+3} = [(\hat{n}_{i+1}\hat{n}_{i+4} + \hat{n}_{i+2}\hat{n}_{i+5} - 2\hat{n}_i\hat{n}_{i+3}) + (\hat{n}_{i+1}\hat{n}_{i+3}\hat{n}_{i+5} - \hat{n}_i\hat{n}_{i+2}\hat{n}_{i+4})]\,\Pi_{i=0}^{5}(1-n_i).$$

For extensive calculations more compact notations are easily devised, but this one clearly brings out the essential idea in constructing arbitrary collision schema. With some minor modifications this form for the collision operator can be reinterpreted as a master equation for a transition process, which is useful as a starting point for a detailed microkinetic analysis by stochastic methods.

Given the $C(f)$ for two- and three-body collisions in the above compact form, and given detailed balance, we show that $C(f_\beta) = 0$ for the Fermi-Dirac distribution. The proof is simple and well known from quantum Fermi-liquid theory where the same functional form for the collision operator appears but with a different interpretation.

If $n$ is a Fermi-Dirac distribution, it has the form $(1+e^E)^{-1} = n(E)$ where $E$ is expanded in collision invariants, in this case particle number and momentum. Then note that $\frac{n}{1-n} \equiv \hat{n} = e^{-E}$, the form of the Maxwell-Boltzmann distribution. This is also the form of the collision kernel, and the exponential terms just contain the sum of momenta in the collision. Since this sum is conserved, each collision term (binary, triple, etc.) vanishes separately, because of the exclusion principle. So the solution is a Fermi-Dirac distribution. This proof also shows that as long as conservation laws of any kind are embodied in the collision term, each type of collision is separately zero under the Fermi-Dirac distribution. Accordingly, the Fermi-Dirac solution is universal across collision types. This implies that one cannot alter the character of the Fermi-Dirac distribution in the lattice gas by adding collision types that respect collision invariants.

Since $f_\beta$ is now assumed to be a Fermi-Dirac distribution, we take it as

$$f_\beta = (1 + e^E)^{-1},$$

with

$$E = \alpha(\rho, \mathrm{v}) + \vec{\beta}(\rho, \mathrm{v}) \cdot \hat{\mathbf{i}}_\beta.$$

(Here we have returned to our original conventions for $\hat{\mathbf{i}}_\beta$.) The equilibrium value for $f_\beta$ at $\mathbf{v} = 0$, namely $N_{\mathrm{eq}}^{\mathrm{v}=0}$, is $\frac{\rho}{6}$ where $\rho$ is the density. Expanding the Fermi-Dirac form for $f_\beta$ about this equilibrium value gives us

$$f_\beta = \frac{\rho}{6}\left\{1 + \alpha(\rho)(\hat{\mathbf{i}}_\beta \cdot \mathbf{v}) + \beta(\rho)\left[(\hat{\mathbf{i}}_\beta \cdot \mathbf{v})^2 - \frac{1}{2}|\mathbf{v}|^2\right] + \cdots\right\}, \tag{16}$$

the same form as the Chapman-Enskog expansion (Eq. 15). To fix $\alpha$ and $\beta$ we use number and momentum conservation as constraints, so that $f_\beta$ becomes

$$f_\beta = \frac{\rho}{6}\left\{1 + 2(\hat{\mathbf{i}}_\beta \cdot \mathbf{v}) + 4g(\rho)\left[(\hat{\mathbf{i}}_\beta \cdot \mathbf{v})^2 - \frac{1}{2}|\mathbf{v}|^2\right] + \cdots\right\},$$

where we have taken the particle speed as 1 ($c = 1$). The coefficient $g(\rho)$ is

$$g(\rho) = \frac{3 - \rho}{6 - \rho}.$$

If we substitute this result for $f_\beta$ in the momentum tensor (Eq. 13) and do the sum over $\beta$, the particle directions, we have

$$\Pi_{ik} = \frac{\rho}{2}\left(1 - g(\rho)\mathbf{v}^2\right)\delta_{ik} + \rho g(\rho)\mathbf{v}_i \mathbf{v}_k.$$

The lattice Euler equation (Eq. 12) thus becomes

$$\partial_t(\rho \mathrm{v}_i) + \partial_j\left[\rho g(\rho)\mathrm{v}_i \mathrm{v}_j + \cdots\right] = -\partial_i p. \tag{17}$$

In the usual Euler equation $g(\rho) = 1$. Here $g(\rho)$ is the lattice correction to the convective term due to the explicit lattice breaking of Galilean invariance. The equation of state for Eq. 17 is

$$p = \frac{\rho}{2}\left(1 - g(\rho)\frac{\mathrm{v}^2}{2}\right).$$

For general single-speed models with particle speed $c$ and $b$ velocity vectors in $D$ dimensions, the result above generalizes to

$$f_\beta = \frac{\rho}{b} \text{when} \quad \mathbf{v} = \mathbf{0}$$

and

$$g(\rho) = \frac{D}{D+2} \frac{b - 2\rho}{b - \rho}.$$

These forms depend only on the structure of tensor products of $\hat{i}_\alpha$ in $D$ dimensions.

When we discuss the full Navier-Stokes equations, we will show how to absorb the $g(\rho)$ Galilean-invariance-breaking term in Eq. 17 into a rescaling of variables.

**Isotropy and The Momentum Tensor.** We will go on to discuss viscosity and the lattice form of the Navier-Stokes equation, but first we comment briefly on how the structure of the momentum tensor depends on the geometry of the lattice. Those interested in all the details can find them discussed from several viewpoints in Frisch, d'Humières, Hasslacher, Lallemand, and Pomeau 1987.

By definition $\Pi_{ij} \equiv \sum_\beta (\hat{i}_\beta)_i (\hat{i}_\beta)_j f_\beta$, where $f_\beta$ is determined by the Chapman-Enskog, or direct, expansion (Eq. 15). Isotropy implies invariance under rotations and reflections; tensors that are isotropic are proportional to a scalar. Define the tensors $E^{(n)} = \sum_\beta (\hat{i}_\beta)_{i_1} \ldots (\hat{i}_\beta)_{i_n}$. For $E^{(n)}$ with regular $b$-sided polygons, we can derive conditions on $b$ for $E^{(n)}$ to be isotropic. These conditions are

$$\left(E^{(2)}|b > 2\right), \; \left(E^{(3)}|b \geq 2, b \neq 3\right), \; \left(E^{(4)}|b > 2, b \neq 4\right), \; \left(E^{(5)}|b \geq 2, b \neq 3, 5\right), \ldots.$$

For $b = 4$, the case of the HPP (Hardy, de Pazzis, and Pomeau) square lattice, $E^{(4)}$ is not isotropic. For $b = 6$, the hexagonal neighborhood case, all tensors up to $n = 5$ are isotropic.

Using the Chapman-Enskog expansion for $f_\beta$ and the notation above for tensors, $\Pi_{ij}$ has the following tensor structure.

$$\Pi_{ij} \approx N_{\text{eq}}^{\mathbf{v}=\mathbf{0}} \left( E_{ij}^{(2)} + \alpha E_{ijk}^{(3)} \mathrm{v}_k + \beta \left[ E_{ijkl}^{(4)} \mathrm{v}_k \mathrm{v}_l, E_{ij}^{(2)} \mathrm{v}_k \mathrm{v}_k \right] + \beta_1 \left[ E_{ijkl}^{(4)} \partial_k \mathrm{v}_l, E_{ij}^{(2)} \partial_k \mathrm{v}_k \right] \right)$$

where we are following the discussion of Wolfram. The momentum stress tensor must be isotropic up to $E^{(4)}$ in order that the leading terms in the momentum equation (corresponding to the convective and viscous terms in the Navier-Stokes equation) be isotropic. For the square model, the original discrete-lattice model, we have nonisotropy manifested in two places through the momentum flux tensor.

$$\Pi_{11} = \rho g(\rho)(\mathrm{v}_1^2 - \mathrm{v}_2^2) + \frac{\rho}{2} + O(\mathrm{v}^4),$$
$$\Pi_{22} = \rho g(\rho)(\mathrm{v}_2^2 - \mathrm{v}_1^2) + \frac{\rho}{2} + O(\mathrm{v}^4),$$
$$\Pi_{12} = \Pi_{21} = 0,$$

where

$$g(\rho) = \frac{2 - \rho}{4 - \rho}.$$

See Frisch et al. for further discussion.

The nonisotropy implies that we do not get a Navier-Stokes type equation for the square lattice. For the hexagonal model, $\beta = 6$, isotropy is maintained through order $E^{(4)}$. By using general considerations on tensor structures for polygons and polyhedra in $D$-dimensional space, one can quickly arrive at probable models for Navier-Stokes dynamics in any dimension. The starting point is that isotropy, or the lack of it, in both convective and viscous terms (the Euler and the Navier-Stokes equations), is controlled completely by the geometry of the underlying lattice. This crucial point was missed by all earlier workers on lattice models who thought that the geometry of the underlying lattice was irrelevant.

**Viscosity for Lattice Gas Models.** In "The Continuum Argument" we saw that the general form of the compressible Navier-Stokes equation with bulk viscosity $\zeta = 0$ is

$$\partial_t(\rho \mathrm{v}_i) + \partial_j(\rho(\mathrm{v}_i \mathrm{v}_j) = -\partial_i p + \partial_j \left( \nu\rho(\partial_j \mathrm{v}_i + \partial_i \mathrm{v}_j - \frac{2}{3}\delta_{ij}\partial_k \mathrm{v}_k) \right),$$

where $\nu$ is the kniematic shear viscosity. To derive this form for the discrete model, one must solve for $\Pi_{ij}$ using both the Chapman-Enskog approximation for $f_\beta$ and the momentum-conservation equation. We noted earlier that the momentum equation contained corrections as powers of the lattice spacing but chose to ignore these at first pass. However, if we use the full Taylor expansion developed in the lattice-size scaling, we find that the contribution to the viscous term of the momentum equation is $-\frac{1}{8}\rho\nabla^2\mathbf{v}$. Note that the correction to the viscosity is a constant (see Eq. 19) that depends only on the lattice and dimension and is independent of the scattering-rule set. This extra noncovariant-derivative contribution to the viscosity must be subtracted from the bare viscosity calculated from the normal perturbation expansion to get the renormalized viscosity, which is the one actually measured in the lattice gas. In other words, the bare coupling constant of the lattice gas model gets "dressed" by this constant amount, owing to the discrete vacuum that the particle must pass through, to become the physical lattice-gas viscosity.

Viscosity is a coupling constant and can be found by any method that can isolate the $\beta_1$ term in the Chapman-Enskog expansion. The simplest methods involve solving for the eigenvalues and right eigenvectors of the linearized collision operator, which is a tedious exercise in linear algebra. Using the results of such a calculation, we can write the Navier-Stokes form of the momentum equation in which the viscosity $\nu(\rho)$ appears explicitly:

$$\partial_t(\rho \mathrm{v}_i) + \partial_j \Pi_{ij} = \partial_j S_{ij}, \tag{18}$$

where the momentum tensor $\Pi_{ij}$ and the viscosity stress tensor are

$$\Pi_{ij} = c_s^2 \rho(1 - g(\rho)\frac{\mathrm{v}^2}{c^2})\delta_{ij} + \rho g(\rho)\mathrm{v}_i \mathrm{v}_j$$

and

$$S_{ij} = \nu(\rho)\left( \partial_i(\rho \mathrm{v}_j) + \partial_j(\rho \mathrm{v}_i) - \frac{2}{D}\delta_{ij}\partial_k(\rho \mathrm{v}_k) \right).$$

The coefficient $c_s^2$ is given by

$$c_s^2 = \frac{c^2}{D},$$

and $c_s$ can be identified as the speed of sound. For the simple hexagonal model $c_s = 1/\sqrt{2}$, and the viscosity is given by

$$\nu = \frac{1}{12}\frac{1}{d(1-d)^3} - \frac{1}{8}, \tag{19}$$

where $d = \frac{\rho_0}{6}$, that is, the mass density per cell. (The $-\frac{1}{8}$ in the viscosity was mentioned above as the noncovariant correction due to the finite lattice size.)

**The Incompressible Limit.** Many features of low Mach number ($M = \mathrm{v}/c_s \ll 1$) flows in an ordinary gas can be described by the incompressible Navier-Stokes equations:

$$\partial_t \mathbf{v} + \mathbf{v} \cdot \nabla \mathbf{v} = -\nabla p + \nu \nabla^2 \mathbf{v} \tag{20}$$

and

$$\nabla \cdot \mathbf{v} = 0.$$

We end this theoretical analysis by showing under what conditions we recover these equations for lattice gases. One way is to freeze the density everywhere except in the pressure term of the momentum equation (Eq. 18). Then, in the low-velocity limit, we can write the lattice Navier-Stokes equations as

$$\rho_0 \partial_t \mathbf{v} + \rho_0 g(\rho_0) \mathbf{v} \cdot \nabla \mathbf{v} = -c_s^2 \nabla \rho' + \rho_0 \nu(\rho_0) \nabla^2 \mathbf{v}, \tag{21}$$

and

$$\nabla \cdot \mathbf{v} = 0$$

where $\rho = \rho_o + \rho'$ and we allow density fluctuations in the pressure term only. As it stands, Eq. 21 is not Galilean invariant. To make it so, we must scale away the $g(\rho_0)$ term in a consistent way. We rescale time and viscosity as follows:

$$t \longrightarrow \frac{t}{g(\rho_0)} \quad \text{and} \quad \nu \longrightarrow g(\rho_0)\nu.$$

To be more precise, we do an $\epsilon$ expansion of the momentum equation, where $\epsilon^{-1}$ is the same order as the global lattice size $L_g$ (see Frisch et al. for details), and rescale the variables as follows:

$$\mathbf{r} = \epsilon^{-1}\mathbf{r}_1, \qquad t = \frac{1}{g(\rho_0)}\epsilon^{-2}T,$$

$$\mathbf{v} = \epsilon \mathbf{V}, \qquad \rho' = \frac{\rho_0 g(\rho_0)}{c_s^2}\epsilon^2 P',$$

and

$$\nu = g(\rho_0)\nu'.$$

where $\epsilon^{-1}$ is on the order of the global lattice size $L_g$. (Note that this rescaling

## SIMULATED VELOCITY PROFILE

**Fig. 7. The predicted velocity profile was obtained in a low-velocity lattice gas simulation of two- dimensional flow in a channel with viscous boundaries (Kadanoff, McNamara, and Zanetti 1987).**

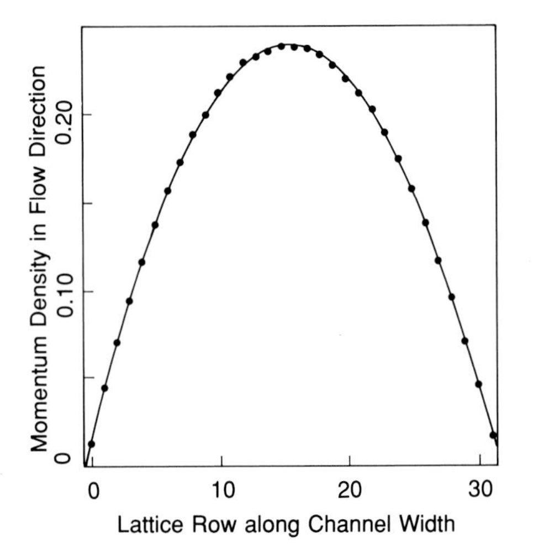


of variables keeps the Reynolds number fixed.) Now all the relevant terms in the momentum equation are of $O(\epsilon^3)$ and higher order terms are $O(\epsilon^4)$ or smaller. So to leading order (where $\nabla_1$ means $\frac{\partial}{\partial \mathbf{r}_1}$) we get

$$\partial_{\mathbf{T}}\mathbf{V} + \mathbf{V} \cdot \nabla_{\mathbf{1}}\mathbf{V} = -\nabla_1 P' + \nu' \nabla_1^2 \mathbf{V} \text{ and } \nabla_1 \cdot \mathbf{V} = 0.$$

Thus we recover the incompressible Navier-Stokes equations. To obtain this result, we have done a fixed-Reynolds-number, large-scale, low-Mach-number expansion and Galilean invariance has been restored, at least formally, by a time rescaling.

## Simulations of Fluid Dynamics with the Hexagonal Lattice Gas Automaton

In the last two years several groups in the United States and France have done simulations of fluid-dynamical phenomena using the hexagonal lattice-gas automaton. The purpose of these simulations was twofold: first, to check the internal consistency of the automaton, and second, to determine, by both qualitative and quantitative measures, whether the model behaves the same or nearly the same as the known analytic and numerical solutions of the Navier-Stokes equations.

The classes of experiments done can be grouped roughly as free flows, flow instabilities, flows past objects, and flows in channels or pipes. These simulations were run in a range of Reynolds numbers between 100 and 700 (and for relatively low mean flow velocities, so that the fluid is nearly incompressible). We first checked to see whether the automaton developed various classic instabilities when triggered by two types of mechanisms, external perturbations and internal noise. The two classic instabilities studied were the Kelvin-Helmholtz instability of two opposing shear flows and the Rayleigh-Taylor instability. We describe the Kelvin-Helmholtz instability in some detail.

In the Kelvin-Helmholtz instability one is looking for the development of a final-state vortex structure of appropriate vortex polarity. From an initial state of two opposing flows undergoing shear, the detailed development of the instability depends on the initial perturbation of the flows. Left unperturbed, except by internal noise in the automaton, at first the two opposing flows develop velocity fields that signal the development of a boundary layer, then sets of vortices develop in these boundary layers, and finally vortex interactions occur that trigger a large-scale instability and the development of large-vortex final states. The same pattern appears in standard two-dimensional numerical simulations of the Navier-Stokes equations near the incompressible regime. No pathological non-Navier-Stokes behavior was observed. These results extend over the entire range of Reynolds numbers (100–700) run with the simple hexagonal model. It is notable that the Kelvin-Helmholtz instability is self-starting due to the automaton internal noise, and the instability proceeds rapidly.

The Rayleigh-Taylor instability was simulated by a French group in a slightly compressible fluid range, where it behaves like a Navier-Stokes fluid with no anomalies.

These global topological tests check whether automaton dynamics captures the correct overall structure of fluids. In general, whenever the automaton is run in the Navier-Stokes range, it produces the expected global topological behavior and correct

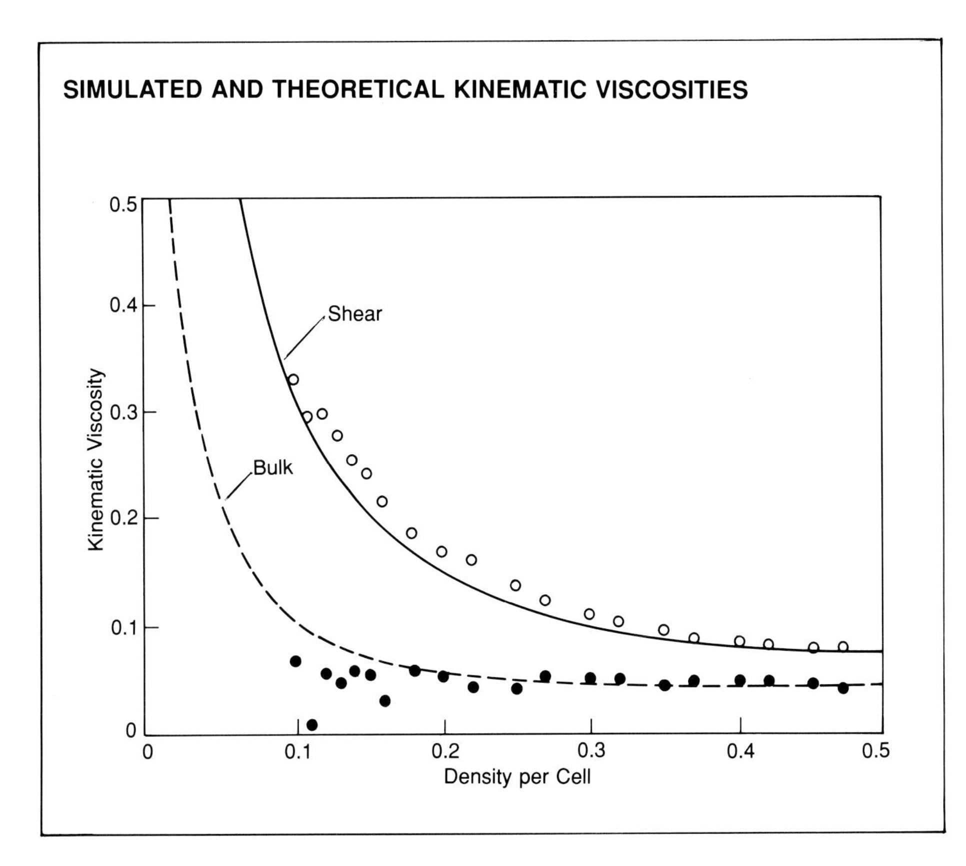


**Fig. 8. Theoretical shear (solid line) and bulk (dashed line) reduced viscosities as a function of reduced density compared with the results of hexagonal lattice gas simulation with rest particles and all possible collisions (d'Humiéres and Lallemand 1987).**

functional forms for various fluid dynamical laws. The question of quantitative accuracy of various known constants is harder to answer, and we will take it up in detail later.

The next broad class of flows studied are flows past objects. Here, we look for distinctive qualitative behavior characteristic of a fluid or gas obeying Navier-Stokes dynamics. The geometries studied, through a wide range of Reynolds numbers, were flows past flat plates placed normal to the flow, flows past plates inclined at various angles to the flow, and flows past cylinders, 60-degree wedges, and typical airfoils. The expected scenario changes as a function of increasing Reynolds number: recirculating flow behind obstacles should develop into vortices, growing couples of vortices should eventually break off to form von Karman streets with periodic oscillation of the von Karman tails; finally, and as the Reynolds number increases, the periodic oscillations should become aperiodic, and the complex phenomena characteristic of turbulent flow should appear. The lattice gas exhibits all these phenomena with no non-Navier-Stokes anomalies in the range of lattice-gas parameters that characterize near incompressibility.

The next topic is quantitative self-consistency. We used the Boltzmann transport approximation for the discrete model to calculate viscosities for the simple hexagonal automaton as well as models with additional scattering rules and rest particles. We then checked these analytic predictions against the viscosities deduced from two kinds of simulations. We ran plane-parallel Poiseuille flow in a channel, saw that it developed the expected parabolic velocity profile (Fig. 7) and then deduced the viscosity characteristic of this type of flow. We also ran an initially flat velocity distribution and deduced a viscosity from the observed velocity decay. These two simulations agree with each other to within a few percent and agree with the analytic predictions from the Boltzmann transport calculation to within 10 percent. Viscosity was also measured by observing the decay of sound waves of various frequencies (Fig. 8). The level of agreement between simulation and the computed Boltzmann viscosity is generic: we see a systematic error of approximately 10 percent. Monte Carlo calculations of viscosities computed from microscopic correlation functions improve agreement with simulations to at least 3 percent and indicate that the Boltzmann description is not as accurate an analytic tool for the automaton as are microscopic correlation techniques. One would call this type of viscosity disagreement a Boltzmann-induced error. Other consistency checks between the automaton simulation and analytic predictions display the same level of agreement.

Detailed quantitative comparisons between conventional discretizations of the Navier-Stokes equations and lattice-gas simulations have yet to be done for several reasons. The simple lattice-gas automaton has a Fermi-Dirac distribution rather than the standard Maxwell-Boltzmann distribution. This difference alone causes deviations of $O(\mathrm{v}^2)$ in the macrovelocity from standard results. For the same reason and unlike standard numerical spectral codes for fluid dynamics, the simple lattice-gas automaton has a velocity-dependent equation of state. A meaningful comparison between the two approaches requires adjusting the usual spectral codes to compute with a velocity-dependent equation of state. This rather considerable task has yet to be done. So far our simulations can be compared only to traditional two-dimensional computer simulations and analytic results derived from simple equations of state.

Some simple quantities such as the speed of sound and velocity profiles have been measured in the automaton model. The speed of sound agrees with predicted values and functional forms for channel velocity profiles and D'Arcy's law agree with calculations by standard methods. The automaton reaches local equilibrium in a few time steps and reaches global equilibrium at the maximum information-transmission speed, namely, at the speed of sound.

Simulations with the two-dimensional lattice-gas model hang together rather well as a simulator of Navier-Stokes dynamics. The method is accurate enough to test theoretical turbulent mechanisms at high Reynolds number and as a simulation tool for complex geometries, provided that velocity-dependent effects due to the Fermi nature of the automaton are correctly included. Automaton models can be designed to fit specific phenomena, and work along these lines is in progress.

Three-dimensional hydrodynamics is being simulated, both on serial and parallel machines, and early results show that we can easily simulate flows with Reynolds numbers of a few thousand. How accurately this model reproduces known instabilities and flows remains to be seen, but there is every reason to believe agreement will be good since the ingredients to evolve to Navier-Stokes dynamics are all present. We end Part II of this article with a graphical display of two- and three-dimensional simulations in "Calculations Using Lattice-Gas Techniques." My Los Alamos collaborators and I have accompanied this display with a summary of the known advantages and present limitations of lattice gas methods. (Part III begins on page 211.)

# Calculations Using Lattice Gas Techniques

by Tsutomu Shimomura,
Gary D. Doolen,
Brosl Hasslacher,
and Castor Fu

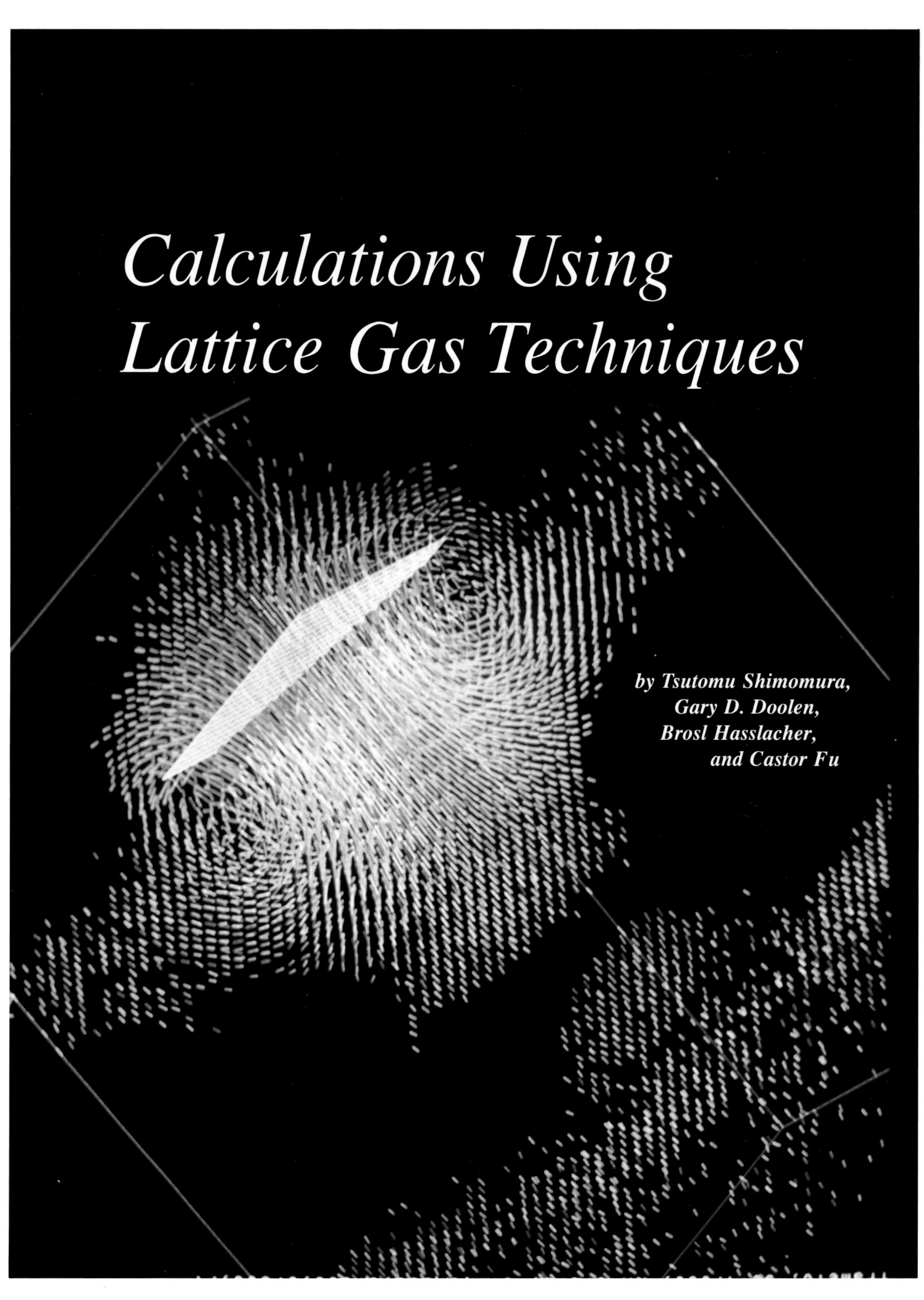

Over the last few years the tantalizing prospect of being able to perform hydrodynamic calculations orders-of-magnitude faster than present methods allow has prompted considerable interest in lattice gas techniques. A few dozen published papers have presented both advantages and disadvantages, and several groups have studied the possibilities of building computers specially designed for lattice gas calculations. Yet the hydrodynamics community remains generally skeptical toward this new approach. The question is often asked, "What calculations can be done with lattice gas techniques?" Enthusiasts respond that in principle the techniques are applicable to any calculation, adding cautiously that increased accuracy requires increased computational effort. Indeed, by adding more particle directions, more particles per site, more particle speeds, and more variety in the interparticle scattering rules, lattice gas methods can be tailored to achieve better and better accuracy. So the real problem is one of tradeoff: How much accuracy is gained by making lattice gas methods more complex, and what is the computational price of those complications? That problem has not yet been well studied. This paper and most of the research to date focus on the simplest lattice gas models in the hope that knowledge of them will give some insight into the essential issues.

We begin by examining a few of the features of the simple models. We then display results of some calculations. Finally, we conclude with a discussion of limitations of the simple models.

## Features of Simple Lattice Gas Methods

We will discuss in some depth the memory efficiency and the parallelism of lattice gas methods, but first we will touch on their simplicity, stability, and ability to model complicated boundaries.

Computer codes for lattice gas methods are enormously simpler than those for other methods. Usually the essential parts of the code are contained in only a few dozen lines of FORTRAN. And those few lines of code are much less complicated than the several hundred lines of code normally required for two- and three-dimensional hydrodynamic calculations.

There are many hydrodynamic problems that cause most standard codes (such as finite-difference codes, spectral codes, and particle-in-cell codes) to crash. That is, the code simply stops running because the algorithm becomes unstable. Stability is not a problem with the codes for lattice gas methods. In addition, such methods conserve energy and momentum exactly, with no roundoff errors.

Boundary conditions are quite easy to implement for lattice gas methods, and they do not require much computer time. One simply chooses the cells to which boundary conditions apply and updates those cells in a slightly different way. One of three boundary conditions is commonly chosen: bounce-back, in which the directions of the reflected particles are simply reversed; specular, in which mirror-like reflection is simulated; or diffusive, in which the directions of the reflected particles are chosen randomly.

We consider next the memory efficiency of the lattice gas method. When the two-dimensional hydrodynamic lattice gas algorithm is programmed on a computer with a word length of, say, 64 bits (such as the Cray X-MP), two impressive efficiencies occur. The first arises because every single bit of memory is used equally effectively. Coined "bit democracy" by von Neumann, such efficient use of memory should be contrasted with that attainable in standard calculations, where each number requires a whole 64-bit word. The lattice gas is "bit democratic" because all that one needs to know is whether or not a particle with a given velocity direction exists in a given cell. Since the number of possible velocity directions is six and no two particles in the same cell can have the same direction, only six bits of information are needed to completely specify the state of a cell. Each of those six bits corresponds to one of the six directions and is set to 1 if the cell contains a particle with that direction and to 0 otherwise. Suppose we designate the six directions by A,B,C,D, E,F as shown on the next page. We associate each bit in the 64-bit word $A$ with a different cell, say the first 64 cells in the first row. If the first cell contains (does not contain) a particle with direction A, we set the first bit in $A$ to 1 (0). Similarly, we pack information about particles in the remaining 63 cells with direction A into the remaining 63 bits of $A$. The same scheme is used for the other five directions. Consequently, all the information for the first 64 cells in the first row is contained in the six words $A$, $B$, $C$, $D$, $E$, and $F$. Note that all bits are equally important and all are fully utilized.

To appreciate the significance of such efficient use of memory, consider how many cells can be specified in the solid-state storage device presently used with the Cray X-MP/416 at Los Alamos. That device stores 512,000,000 64-bit words. Since the necessary information for $10\frac{2}{3}$ cells can be stored in each word, the device can store information for about 5,000,000,000 cells, which corresponds to a two-dimensional lattice with 100,000 cells along one axis and 50,000 cells along the other. That number of cells is a few orders of magnitude greater than the number normally treated when other methods are used. (Although such high resolution may appear to be a significant advantage of the lattice gas method, some averaging over cells is required to obtain smooth results for physical quantities such as velocity and density.)

The second efficiency is related to the

fact that lattice gas operations are bit oriented rather than floating-point-number oriented and therefore execute more naturally on a computer. Most computers can carry out logic operations bit by bit. For example, the result of the logic operation AND on the 64-bit words $A$ and $B$ is a new 64-bit word in which the $i$th bit has a value of 1 only if the $i$th bits of both $A$ and $B$ have values of 1. Hence in one clock cycle a logic operation can be performed on information for 64 cells. Since a Cray X-MP/416 includes eight logical function units, information for 8 times 64, or 512, cells can be processed during each clock cycle, which lasts about 10 nanoseconds. Thus information for 51,200,000,000 cells can be processed each second. The two-dimensional lattice gas models used so far require from about thirty to one hundred logic operations to implement the scattering rules and about another dozen to move the particles to the next cells. So the number of cells that can be updated each second by logic operations is near 500,000,000. Cells can also be updated by table-lookup methods. The authors have a table-lookup code for three-dimensional hydrodynamics that processes about 30,000,000 cells per second.

A final feature of the lattice gas method is that the algorithm is inherently parallel. The rules for scattering particles within a cell depend only on the combination of particle directions in that cell. The scattering can be done by table lookup, in which one creates and uses a table of scattering results—one for each possible cell configuration. Or it can be done by logic operations.

## Using Lattice Gas Methods To Approximate Hydrodynamics

In August 1985 Frisch, Hasslacher, and Pomeau demonstrated that one can approximate solutions to the Navier-Stokes equations by using lattice gas methods,

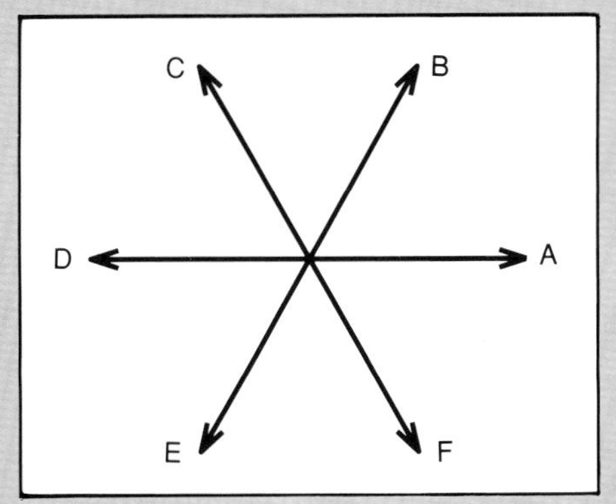


but their demonstration applied only to low-velocity incompressible flows near equilibrium. No one knew whether more interesting flows could be approximated. Consequently, computer codes were written to determine the region of validity of the lattice gas method. Results of some of the first simulations done at Los Alamos and of some later simulations are shown in Figs. 1 through 6. (Most of the early calculations were done on a Celerity computer, and the displays were done on a Sun workstation.) All the results indicate qualitatively correct fluid behavior.

Figure 1a demonstrates that a stable trailing vortex pattern develops in a two-dimensional lattice gas flowing past a plate. Figure 1b shows that without a three-particle scattering rule, which removes the spurious conservation of momentum along each line of particles, no vortex develops. (Scattering rules are described in Part II of the main text.)

Figure 2 shows that stable vortices develop in a lattice gas at the interface between fluids moving in opposite directions. The Kelvin-Helmholtz instability is known to initiate such vortices. The fact that lattice gas methods could simulate vortex evolution was reassuring and caused several scientists to begin to study the new method.

Figure 3 shows the complicated wake that develops behind a V-shaped wedge in a uniform-velocity flow.

Figure 4 shows the periodic oscillation of a low-velocity wake behind a cylinder. With a Reynolds number of 76, the flow has a stable period of oscillation that slowly grows to its asymptotic limit.

Figure 5 shows a flow with a higher Reynolds number past an ellipse. The wake here becomes chaotic and quite sensitive to details of the flow.

Figure 6 shows views of a three-dimensional flow around a square plate, which was one of the first results from Los Alamos in three-dimensional lattice gas hydrodynamic simulations.

Rivet and Frisch and other French scientists have developed a similar code that measures the kinematic shear viscosity numerically; the results compare well with theoretical predictions (see Fig. 8 in the main text).

The lattice gas calculations of a group at the University of Chicago (Kadanoff, McNamara, and Zanetti) for two-dimensional flow through a channel (Fig. 7 of the main text) agree with the known parabolic velocity profile for low-velocity channel flows.

The above calculations, and many others, have established some confidence that qualitative features of hydrodynamic flows are simulated by lattice-gas methods. Problems encountered in detailed comparisons with other types of calculations are discussed in the next section.

## Limitations of Simple Lattice Gas Models

As we discussed earlier, lattice gas methods can be made more accurate by making them more complicated—by, for example, adding more velocity directions and magnitudes. But the added complications degrade the efficiency. We mention in this section some of the difficulties (associated with limited range of speed, velocity dependence of the equation of state, and noisy results) encountered in the simplest lattice-gas models.

The limited range of flow velocities is inherent in a model that assumes a

*continued on page 210*

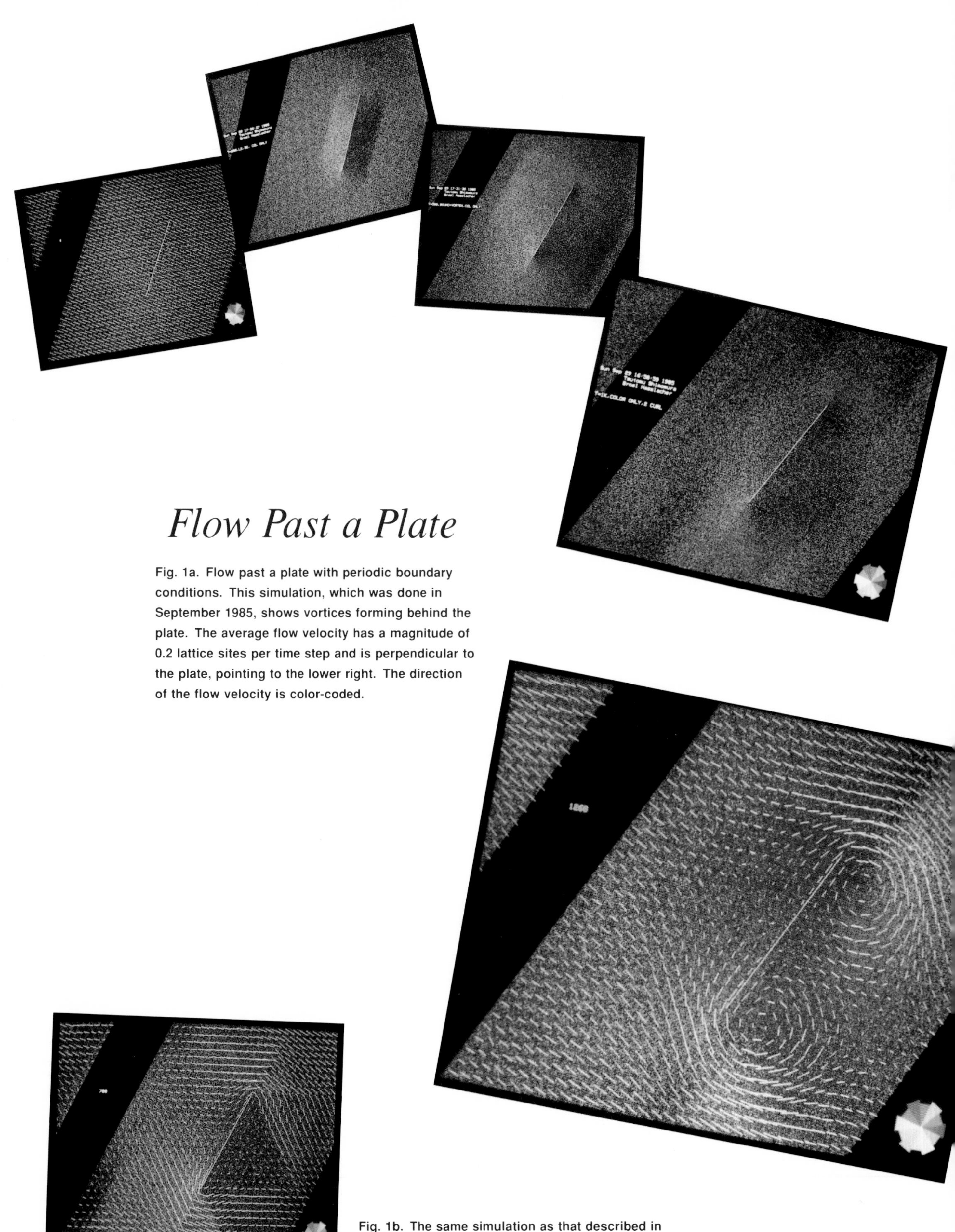

# *Flow Past a Plate*

Fig. 1a. Flow past a plate with periodic boundary conditions. This simulation, which was done in September 1985, shows vortices forming behind the plate. The average flow velocity has a magnitude of 0.2 lattice sites per time step and is perpendicular to the plate, pointing to the lower right. The direction of the flow velocity is color-coded.

Fig. 1b. The same simulation as that described in Fig. 1a but with no three-body scattering rule. As a result, spurious laws of conservation of momentum along the lines of the grid prevent the development of hydrodynamics.

## Kelvin-Helmholtz Instability

Fig. 2. A Kelvin-Helmholtz instability develops into vortices from initially opposing flows past a sinusoidal interface that is removed at $t = 0$. Periodic boundary conditions apply. For this simulation about 10,000,000 particles and 14,000,000 cells were used.

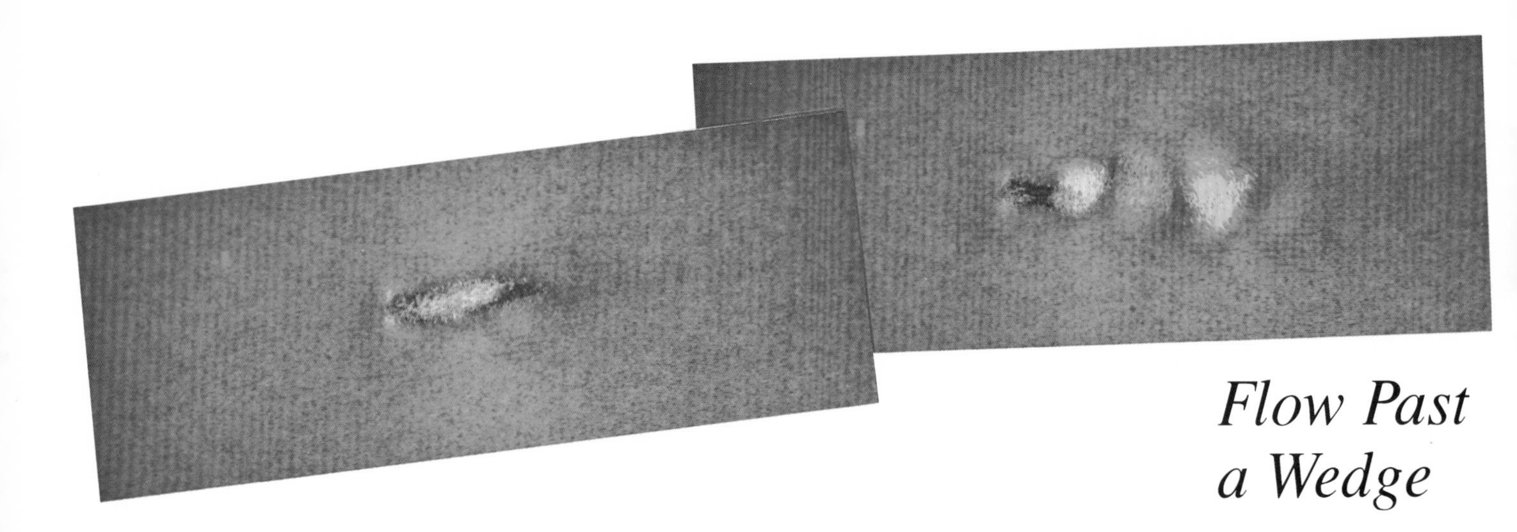

*Flow Past a Wedge*

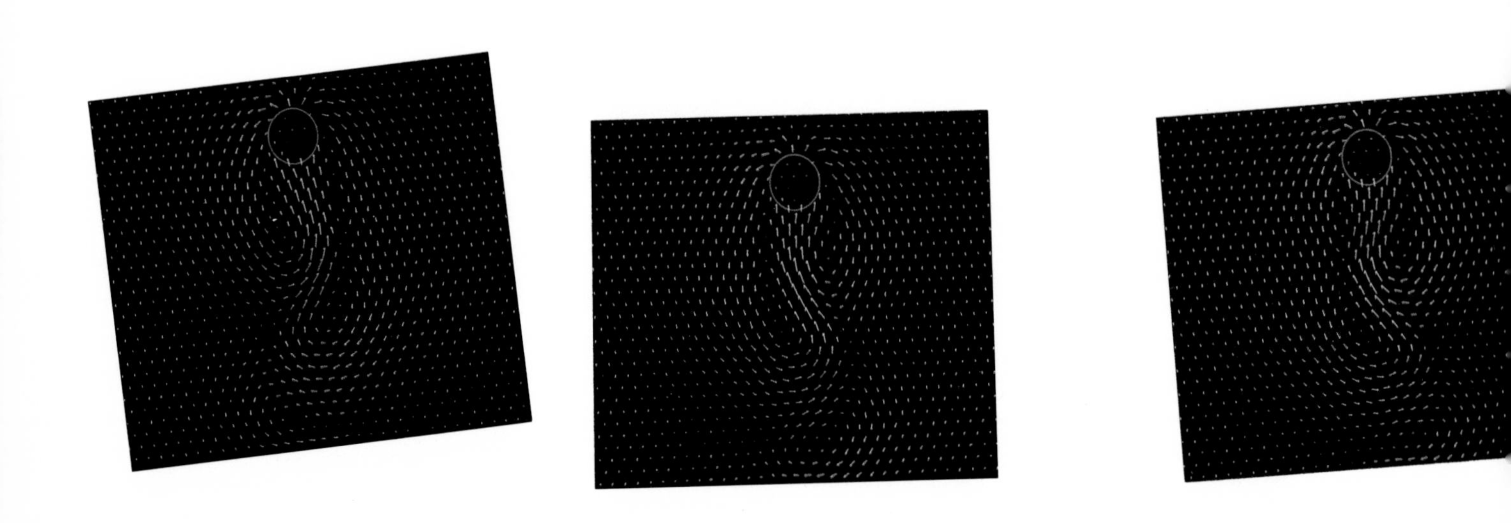

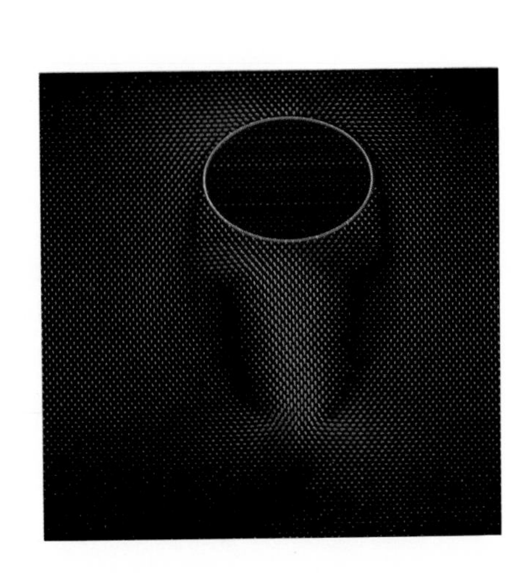

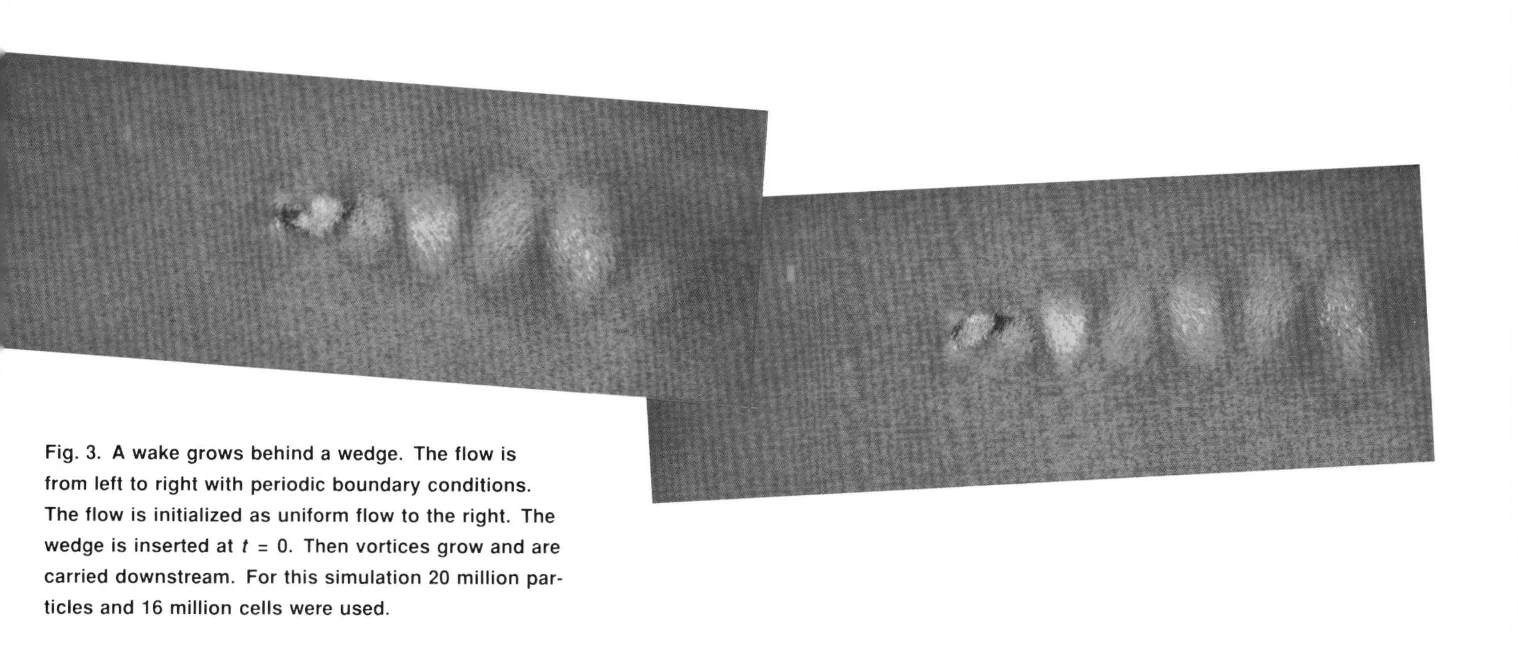

Fig. 3. A wake grows behind a wedge. The flow is from left to right with periodic boundary conditions. The flow is initialized as uniform flow to the right. The wedge is inserted at $t = 0$. Then vortices grow and are carried downstream. For this simulation 20 million particles and 16 million cells were used.

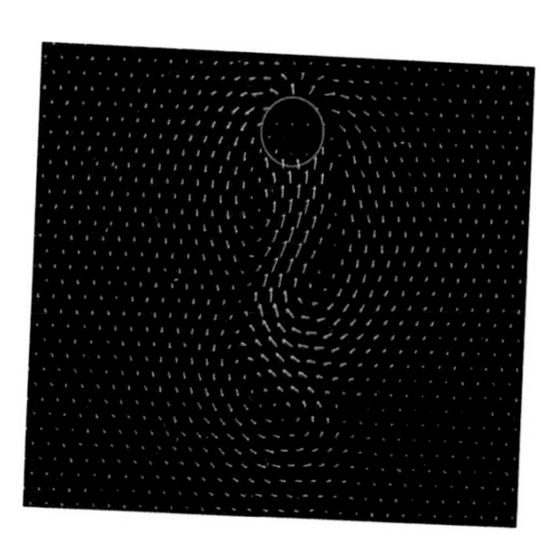

# *Flow Past a Cylinder*

**Fig. 4. Low-velocity flow (from top to bottom) past a cylinder creates a periodically oscillating wake. Four snapshots from one period of the oscillation are shown. In this simulation, which has periodic right and left boundaries, 1.4 million particles flowed through 1 million cells. The flow was initially uniform.**

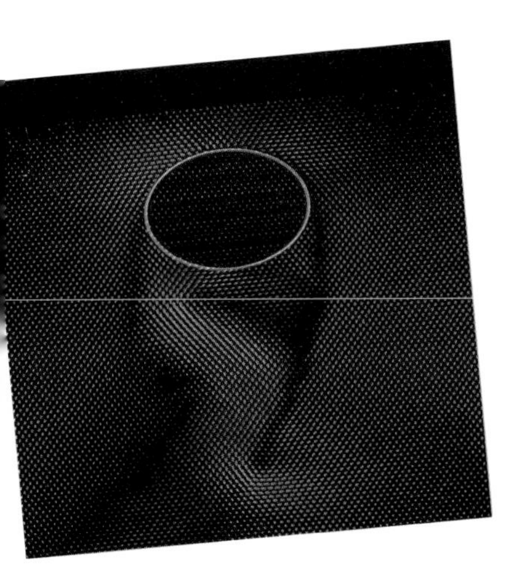

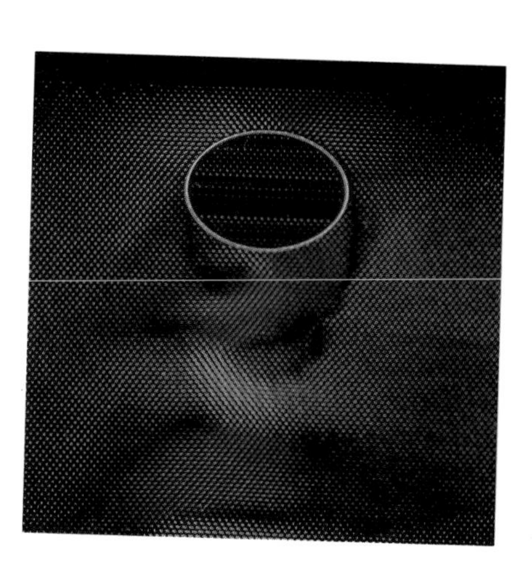

# *Turbulent Wake*

Fig. 5. A turbulent wake grows behind an ellipse being dragged through a fluid consisting of 11 million particles and 8 million cells. The ellipse is composed of about 2400 cells in which the velocity directions of the entering particles are reversed. The flow has periodic right and left boundaries. (An infinite sequence of equivalent ellipses exists to the left and right of the frame shown.) The Reynolds number in the flow is 1021.

# 3-D Flow Past a Plate

Fig. 6. Cross sections (left) and three-dimensional views show the development of vortices behind a square plate and a spherical sound wave that propagates through the system. Mean flow was subtracted to highlight the dynamics of the flow. For this simulation 80 million particles and 10 million face- centered-cubic cells were used.

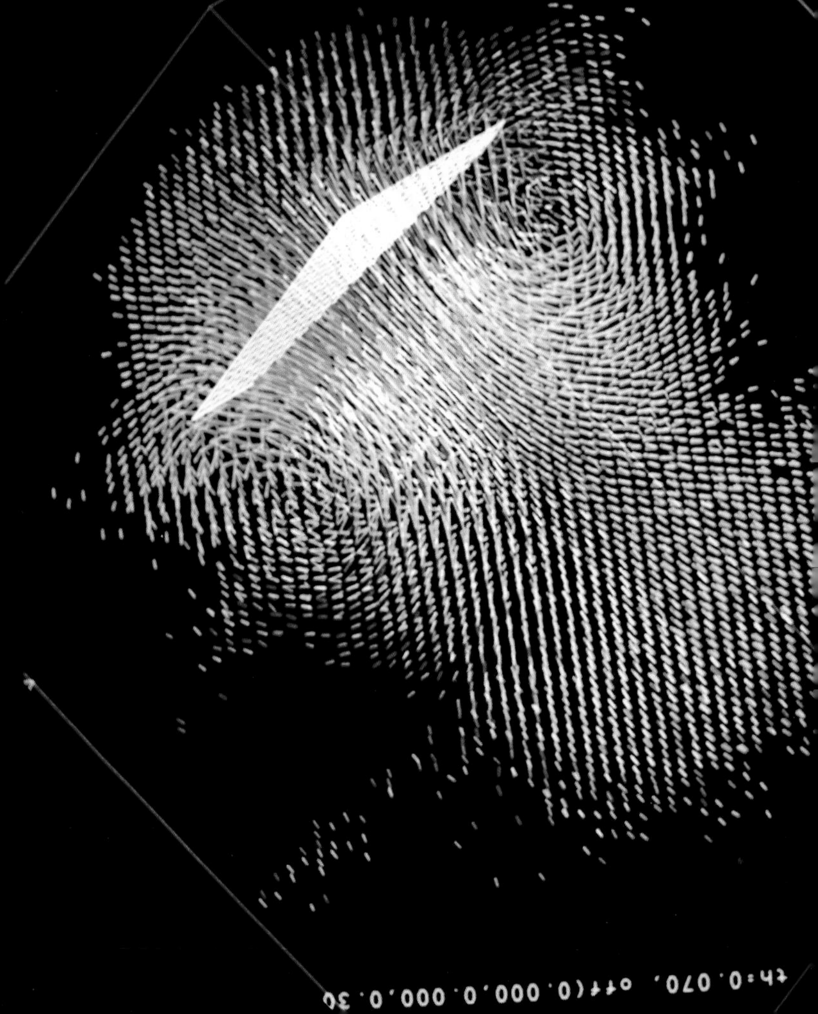

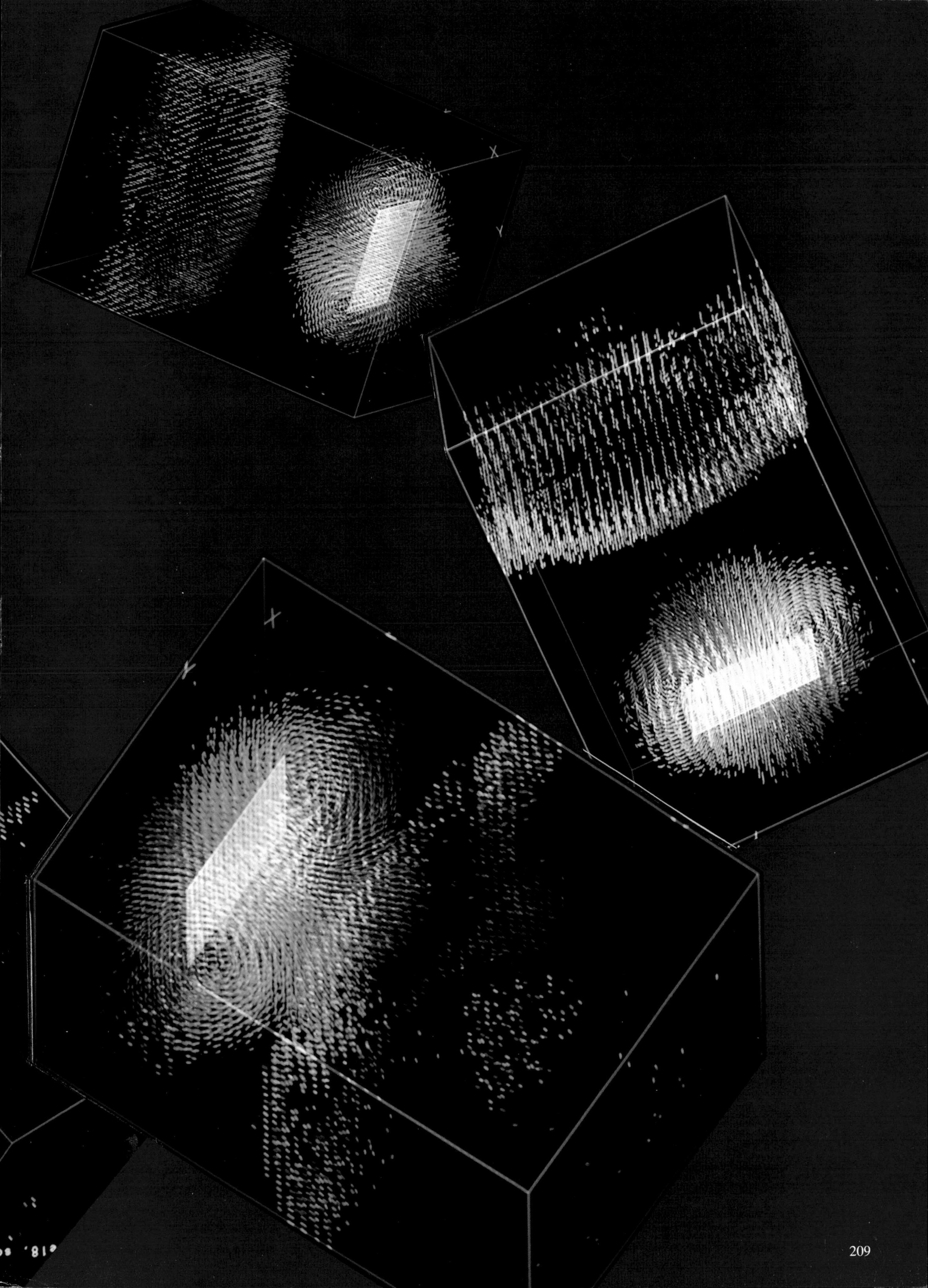

*continued from page 203*

single speed for all particles. The sound speed in such models can be shown to be about two-thirds of the particle speed. Hence flows in which the Mach number (flow speed divided by sound speed) is greater than 1.5 cannot be simulated. This difficulty is avoided by adding particles with a variety of speeds.

The limited range of velocities also restricts the allowed range of Reynolds numbers. For small Reynolds numbers (0 to 1000) the flow is smooth, for moderate Reynolds numbers (2000 to 6000) some turbulence is observed, and for high Reynolds numbers (10,000 to 10,000,000) extreme turbulence occurs. Since the effective viscosity, $\nu$, is typically about 0.2 in two-dimensional problems, the Reynolds number scales with the characteristic length, $l$, allowed by computer memory. Currently the upper bound on $l$ is of the order of 100,000.

The velocity dependence of the equation of state is unusual and is a consequence of the inherent Fermi-Dirac distribution of the lattice gas (see the section on Theoretical Analysis of the Discrete Lattice Gas in the main text). The low-velocity equation of state for a lattice gas can be written as $p = \frac{1}{2}\rho\left(1 - \frac{1}{2}\mathrm{v}^2\right)$, where $p$ is the pressure, $\rho$ is the density, and v is the flow speed. Thus, for constant-pressure flows, regions of higher velocity flows have higher densities.

The velocity dependence of the equation of state is related to the fact that lattice gas models lack Galilean invariance. The standard Navier-Stokes equation for incompressible fluids is

$$\frac{\partial \mathbf{v}}{\partial t} + \mathbf{v} \cdot \nabla \mathbf{v} = -\nabla p + \nu \nabla^2 \mathbf{v}.$$

But in the incompressible, low-velocity limit the single-speed hexagonal lattice gas follows the equation

$$\frac{\partial \mathbf{v}}{\partial t} + g(\rho)\mathbf{v} \cdot \nabla \mathbf{v} = -\nabla p + \nu \nabla^2 \mathbf{v},$$

where

$$g(\rho) = \frac{3-\rho}{6-\rho}$$

and $\rho$ is the average number of particles per cell. The extra factor $g(\rho)$ requires special treatment. The conventional way to adjust for the fact that $g(\rho)$ does not equal unity (as it does in the Navier-Stokes equation) is to simply scale the time, $t$, and the viscosity, $\nu$, by the factor $g(\rho)$ as follows: $t' = g(\rho)t$ and $\nu' = \nu/g(\rho)$. (The pressure must also be scaled.) Hence a density-dependent scaling of the time, the viscosity, and the pressure is required to bring the lattice gas model into a form that closely approximates the hydrodynamics of incompressible fluids in the low-velocity limit.

Finally, the discreteness of the lattice gas approximation introduces noise into the results. One method of smoothing the results for comparison with other methods is to average in space and time. In practice, spatial averages are taken over 64, 256, 512, or 1024 neighboring cells for time-dependent flows in two dimensions. For steady-state flows, time averaging is done. The details of noise reduction are complicated, but they must be addressed in each comparison calculation. The presence of noise is both a virtue and a defect. Noise ensures that only robust (that is, physical) singularities survive, whereas in standard codes, which are subject to less noise, mathematical artifacts can produce singularities. On the other hand, the noise in the model can trigger instabilities.

## Conclusion

In the last few years lattice gas methods have been shown to simulate the qualitative features of hydrodynamic flows in two and three dimensions. Precise comparisons with other methods of calculation remain to be done, but it is believed that the accuracy of the lattice gas method can be increased by making the models more complicated. But how complicated they have to be to obtain the desired accuracy is an unanswered question.

Calculations based on the simple models are extremely fast and can be made several orders-of-magnitude faster by using special-purpose computers, but the models must be extended to get quantitative results with an accuracy greater than 1 percent. Significant research remains to be done to determine the accuracy of a given lattice gas method for a given flow problem. ■

**Note added in proof:** Recently Kadanoff, McNamara, and Zanetti reported precise comparisons between theoretical predictions and lattice gas simulations (University of Chicago preprint, October 1987). They used a seven-bit hexagonal model on a small automaton universe to simulate forced two-dimensional channel flow for long times. Three tests were used to probe the hydrodynamic and statistical-mechanical behavior of the model. The tests determined (1) the profile of momentum density in the channel, (2) the equation of state given by the statistical mechanics of the system, and (3) the logarithmic divergence in the viscosity (a famous effect in two- dimensional hydrodynamics and a deep test of the accuracy of the model in the strong nonlinear regime).

The results were impressive. First, to within the accuracy of the simulation, there is no discrepancy between the parabolic velocity profile predicted by macroscopic theory and the lattice gas simulation data. Second, the equation of state derived from theory fits the simulation data to better than 1 percent. Finally, the measured logarithmic divergence in the viscosity as a function of channel width agrees with prediction. These results are at least one order of magnitude more accurate than any previously reported calculations.

# PART III THE PROMISE OF LATTICE GAS METHODS

In the sidebar "Calculations Using Lattice Gas Techniques" we displayed the results of generalizing the simple hexagonal model to three dimensions. Here, in the last part of the article, we will discuss numerous ways to extend and adapt the simple model. In particular, we emphasize its role as a paradigm for parallel computing.

## Adjusting the Model To Fit the Phenomenon

There are several reasons for altering the geometry and rule set of the fundamental hexagonal model. To understand the mathematical physics of lattice gases, we need to know the class of functionally equivalent models, namely those models with different geometries and rules that produce the same dynamics in the same parameter range.

To explore turbulent mechanisms in fluids, the Reynolds number must be significantly higher than for smooth flow, so models must be developed that increase the Reynolds number in some way. The most straightforward method, other than increasing the size of the simulation universe, is to lower the effective mean free path in the gas. This lowers the viscosity and the Reynolds number rises in inverse proportion. Increasing the Reynolds number is also important for practical applications. In "Reynolds Number and Lattice Gas Calculations" we discuss the computational storage and work needed to simulate high-Reynolds-number flows with cellular automata.

To apply lattice gas methods to systems such as plasmas, we need to develop models that can support widely separated time scales appropriate to, for example, both photon and hydrodynamical modes. The original hexagonal model on a single lattice cannot do so in any natural way but must be modified to include several lattices or the equivalent (see below).

Within the class of fluids, problems involving gravity on the gas, multi-component fluids, gases of varying density, and gases that undergo generalized chemical reactions require variations of the hexagonal model. Once into the subject of applications rather than fundamental statistical mechanics, there is an endless industry in devising clever gases that can simulate the dynamics of a problem effectively.

We outline some of the possible extensions to the hexagonal gas, but do so only to give an overview of this developing field. Nothing fundamental changes by making the gas more complex. This model is very much like a language. We can build compound sentences and paragraphs out of simple sentences, but it does not change the fundamental rules by which the language works.

The obvious alterations to the hexagonal model are listed below. They comprise almost a complete list of what can be done in two dimensions, since a lattice gas model contains only a few adjustable structural elements.

Indistinguishable particles can be colored to create distinguishable species in the gas, and the collision rules can be appropriately modified. Rules can be weighted to different outcomes; for example, one can create a chiral gas (left- or right-handed) by biasing collisions to make them asymmetric. In three dimensions there is an instability at any Reynolds number caused by lack of microscopic parity, so the chiral gas is an important model for simulating this instability.

At the next order of complexity, multi-speed particles can be introduced, either alone or with changes in geometry. The simplest example is a square neighborhood in two dimensions in which the collision domain is enlarged to include next-to-nearest neighbors, and a diagonal particle with speed $\sqrt{2}$ is introduced to force an isotropic lattice gas. In general, any lattice model with only two-body collisions and a single speed will contain spurious conservation laws. But if multiple speeds are allowed, models with binary collisions can maintain isotropy. In other words, models with multiple speeds are equivalent to single-speed models with a higher order rotation group and extended collision sets. Many variations are possible and each can be designed to a problem where it has a special advantage.

Finally, colored multiple-speed models are in general equivalent to single-species models operating on separate lattices. Colored collision rules couple the lattices so that information can be transferred between them at different time scales. Certain statistical-mechanical phenomena such as phase transitions can be done this way.

By altering the rule domain and adding gas species with distinct speeds, it is possible to add independent energy conservation. This allows one to tune gas models to different equations of state. Again, we gain no fundamental insight into the development of large collective models by doing so, but it is useful for applications.

In using these lattice gas variations to construct models of complex phenomena, we can proceed in two directions. The first direction is to study whether or not complex systems with several types of coupled dynamics are described by skeletal gases. Can complex chemical reactions in fluids and gases, for example, be simulated by adding collision rules operating on colored multi-speed lattice gases? Complex chemistry is set up in the gas in outline form, as a gross scheme of closed sets of interaction rules. The same idea might be used for plasmas. From a theoretical viewpoint one wants to study how much of the known dynamics of such systems is reproduced by a skeletal gas; consequently both qualitative and quantitative results are important.

**Exploring Fundamental Questions.** Models of complex gas or fluid systems, like other lattice gas descriptions, may either be a minimalist description of microphysics or simply have no relation to microphysics other than a mechanism for carrying known conservation laws and reactions. We can always consider such gas models to be pure computers, where we fit the wiring, or architecture, to the problem, in the same fashion that ordinary discretization schemes have no relation to the microphysics of the problem. However for lattice gas models, or cellular-

*continued on page 214*

# REYNOLDS NUMBER *and* Lattice Gas Calculations

The only model-dependent coupling constant in the Navier-Stokes equation is the viscosity. Its main role in lattice gas computations is its influence on the Reynolds number, an important scaling concept for flows. Given a system with a fixed intrinsic global length scale, such as the size of a pipe or box, and given a flow, then the Reynolds number can be thought of as the ratio of a typical macrodynamic time scale to a time scale set by elementary molecular processes in the kinetic model.

Reynolds numbers characterize the behavior of flows in general, irrespective of whether the system is a fluid or a gas. At high enough Reynolds numbers turbulence begins, and turbulence quickly loses all memory of molecular structure, becoming universal across liquids and gases. For this reason and because many interesting physical and mathematical phenomena happen in turbulent regimes, it is important to be able to reach these Reynolds numbers in realistic simulations without incurring a large amount of computational work or storage.

Some simple arguments based on dimensional analysis and phenomenological theories of turbulence indicate, at first glance, that any cellular automaton model has a high cost in computer resources when simulating high-Reynolds-number flows. These arguments appeared in the first paper on the subject (Frisch, Hasslacher, and Pomeau 1987) and were later elaborated on by other authors. We will go through the derivation of some of the more severe constraints on simulating high-Reynolds-number flows with cellular automata, then discuss some possible ways out, and finally estimate the seriousness of the situation for a realistic large-scale simulation.

The turbulent regime has many length scales, bounded above by the length of the simulation box and below by the scale at which turbulent dynamics degenerates into pure dissipation, the so-called dissipation scale. We focus on these extreme scales and, with a few definitions, derive a bound on the computational storage and work needed for simulating high-Reynolds-number flows with cellular automata.

The Reynolds number $R$ is usually defined not in terms of time but simply as $R = \mathrm{v}L/\nu$ where $L$ is a characteristic length, v is characteristic speed, and $\nu$ is the kinematic shear viscosity. One sees immediately why calculating viscosity functions for particular models is important. It is the only variable one can adjust in a flow problem, given a fixed flow in a fixed geometry. First, we calculate a rough upper bound on Reynolds numbers attainable with lattice models. If the speed of sound in the lattice gas is $c_s$ and the spacing between lattice nodes is $\ell$, then by definition the kinematic viscosity $\nu \geq c_s\ell$. Now viscosity estimated this way must agree with that fixed by the scale of hydrodynamic modes. Given a global length $L$ and a global velocity $V$ associated with these modes, $R = VL/\nu$ at best. In terms of the Mach number ($M = V/c_s$), the Reynolds number is equal to $ML/\ell$. But $M$ also characterizes fluid flow, and $L$ and $\ell$ are model-dependent. In a lattice gas we can relate the ratio $L/\ell$ to the number of nodes in the gas simulator, namely $n = (L/\ell)^d$, where $d$ is the space dimension of the model. Therefore, the number of nodes in a lattice model must grow at least as $n \sim (R/M)^d$. Computational work is the number of lattice nodes per time step multiplied by the number of time steps required to resolve hydrodynamical features. This is $L/\ell M$ steps (to cross the hydrodynamical feature at the given Mach number), and so we find the computational work is of order $R^{d+1}/M^{d+2}$. For a so-called normal simulation based on the usual ways of discretizing the Navier-Stokes equation, the growth in storage is roughly proportional to one power lower in the Reynolds number than the growth in storage for the lattice gas. So at first it seems that simulating high-Reynolds-number flows by lattice gas techniques is costly compared to ordinary methods.

This argument is not only approximate; it is also tricky and must be applied with great care. The normal way of simulating flows escapes power-law penalties by cutting off degrees of freedom at the turbulence-dissipation scale, which the lattice gas does not do. The gas computes within these scales and so wastes computational resources for some problems. Actually computation of these very small scales is the source of the noisy character of the gas and is responsible for its power to avoid spurious mathematical singularities. One way around this is to find an effective gas with new collision rules for which the dissipation length scales are averaged out. A possible technique uses the renormalization group, but it is useful only if the effective gas is not too complex and has the attributes that made the original gas attractive, including locality. Work is going on at present to explore this possibility, and it seems likely that some such method will be developed.

The more serious consideration is what happens in a realistic large-scale simulation, and here we will find the lattice gas does very well indeed.

First, we note that a dissipation length $l_d$ with the behavior $l_d \to \infty$ as $R \to \infty$ is actually required to guarantee the scale separation between the lattice spacing and the hydrodynamic modes that is necessary to develop hydrodynamic behavior.

The actual Reynolds number in lattice gas models is much more complex than in normal fluid models. An accurate form is $R = L\mathrm{v}g(\rho_0)/\nu(\rho_0)$, where v is an averaged velocity and the fundamental unit of distance (the lattice spacing $\ell$) and the fundamental unit of time (the speed required to traverse the lattice spacing $\ell$) have been set to 1. To remain nearly incompressible, the velocities in the model should remain small compared to the speed of sound $c_s$, but $c_s$ in lattice gases is model-dependent. So we factor the Reynolds number into model-dependent and invariant factors this way: we define $\hat{R}(\rho_0) = c_s\left(g(\rho_0)/\nu(\rho_0)\right)$ so that $R = ML\hat{R}(\rho_0)$. The value of $\hat{R}$ depends critically on the model used. In two dimensions it ranges from 0.39 to about 6 times that, depending on the amount of the state table we want to include. For the three-dimensional projection of the four-dimensional model, it is known that $\hat{R}$ is about 9.

By repeating essentially the same dimensional arguments, only more carefully, we find that the dissipation length $l_d = (M\hat{R})^{-1}R^{-1/2}$ for two dimensions and $l_d = (M\hat{R})^{-1}R^{-1/4}$ for three dimensions.

For a typical simulation in three dimensions, we take $M = 0.3$ for incompressibility, $\hat{R} \sim 9$, and $L = 10^3$, which is a large simulation, possible only on the largest Cray-class machines. Then $l_d$ is about three lattice spacings, and the simulation wastes very little computational power. The subtle point is that the highly model-dependent factor $\hat{R}$ is not of order 1, as is usually estimated. It depends critically on the complexity of the collision set, going up a factor of 20 from the elementary hexagonal model in two dimensions to the projected four-dimensional case with an optimal collision table.

There is a great deal of work to be done on the high-Reynolds-number problem, but it is clear that the situation is complicated and rich in possibilities for evading simple dimensional arguments. ■

*continued from page 212*

automaton models in general, there always seems to be a deep relation between the abstract computer embodying the gas algorithm for a physical problem and the mathematical physics of the system itself.

This duality property is an important one, and it is not well understood. One of the main aims of lattice gas theory is to make the underlying mathematics of dynamical evolution clearer by providing a new perspective on it. One would, for example, like to know the class of all lattice gas systems that evolve to a dynamics that is, in an appropriate sense, nearby the dynamics actually evolved by nature. Doing this will allow us to isolate what is common to such systems and identify universal mathematical mechanisms.

**Engineering Design Applications.** The second direction of study is highly applied. In most engineering-design situations with complicated systems, one would like to know first the general qualitative dynamical behavior taking place in some rather involved geometry and then some rough numerics. Given both, one can plot out the zoo of dynamical development within a design problem. Usually, one does not know what kinds of phenomena can occur as a parameter in the system varies. Analytic methods are either unavailable, hard to compute by traditional methods, or simply break down. Estimating phenomena by scaling or arguments depending on order-of-magnitude dimensional analysis is often inaccurate or yields insufficient information. As a result, a large amount of expensive and scarce supercomputer time is used just to scan the parameter space of a system.

Lattice gas models can perform such tasks efficiently, since they simulate at the same speed whether the geometry and system are simple or complex. Complicated geometries and boundary conditions for massively parallel lattice gas simulators involve only altering collision rules in a region. This is easily coded and can be done interactively with a little investment in expert systems. There is no question that for complex design problems, lattice gas methods can be made into powerful design tools.

## Beyond Two Dimensions

In two dimensions there exists a single-speed skeletal model for fluid dynamics with a regular lattice geometry. It relies on the existence of a complete tiling of the plane by a domain of sufficiently high symmetry to guarantee the isotropy of macroscopic modes in the model. In three dimensions this is not the case, for the minimum appropriate domain symmetry is icosahedral and such polyhedra do not tile three-space. If we are willing to introduce multiple-speed models, there may exist a model with high enough rotational symmetry, as in the square model with nearest and next-to-nearest neighbor interaction in two dimensions, but it is not easy to find and may not be efficient for simulations.

A tactic for developing an enlarged-neighborhood, three-dimensional model, which still admits a regular lattice, is to notice that the number of regular polyhedra as a function of dimension has a maximum in four dimensions. Examination of the face-centered four-dimensional hypercube shows that a single-speed model connected to each of twenty-four nearest neighbors has exactly the right invariance group to guarantee isotropy in four dimensions. So four-dimensional single-speed models exist on a regular tiling. Three-dimensional, or regular, hydrodynamics can be recovered by taking a thin one-site slice of the four-dimensional case, where the edges of the slice are identified. Projecting such a scheme into three-dimensional space generates a two-speed model with nearest and next-to-nearest neighbor interactions of the sort guaranteed to produce three-dimensional Navier-Stokes dynamics.

Such models are straightforward extensions of all the ideas present in the two-dimensional case and are being simulated presently on large Cray-class machines and the Connection Machine 2. Preliminary results show good agreement with standard computations at least for low Reynolds numbers. In particular, simulation of Taylor-Green vortices at a Reynolds number of about 100 on a $(128)^3$ universe (a three-dimensional cube with 128 cells in each direction) agrees with spectral methods to within 1 percent, the error being limited by Monte Carlo noise. The ultimate comparison is against laboratory fluid-flow experiments. As displayed at the end of Part II, three-dimensional flows around flat plates have also been done.

A more intriguing strategy is to give up the idea of a regular lattice. Physical systems are much more like a lattice with nodes laid down at random. At present, we don't know how to analyze such lattices, but an approximation can be given that is intermediate between regular and random grids. Quasi-tilings are sets of objects that completely tile space but the grids they generate are not periodic. Locally, various types of rotation symmetry can be designed into such lattices, and in three dimensions there exists such a quasi-tiling that has icosahedral symmetry everywhere. The beauty of quasi-tilings is that they can all be obtained by simple slices through hypercubes in the appropriate dimension. For three dimensions the parent hypercube is six-dimensional.

The idea is to run an automaton model containing the conservation laws with as simple a rule set as possible on the six-dimensional cube and then take an appropriately oriented three-dimensional slice out of the cube so arranged as to generate the icosahedral quasi-tiling. Since we only examine averaged quantities, it is enough to do all the averaging in six dimensions along the quasi-slice and image the results. By such a method we guarantee exact isotropy everywhere in three dimensions and avoid computing directly on the extremely complex lattices that the quasi-tiling generates. Ultimately, one would like to compute on truly random lattices, but for now there is no simple way of doing that efficiently.

The simple four-dimensional model is a good example of the limits of present super-computer power. It is just barely tolerable to run a $(1000)^3$ universe at a Reynolds number of order a few thousand on the largest existing Cray's. It is far more efficient to compute in large parallel arrays with rather inexpensive custom machines, either embedded in an existing parallel architecture or on one designed especially for this class of problems.

## Lattice Gases as Parallel Computers

Let us review the essential features of a lattice gas. The first property is the totally discrete nature of the description: The gas is updated in discrete time steps, lattice gas elements can only lie on discrete space points arranged in a space-filling network or grid, velocities can also have only discrete values and these are usually aligned with the grid directions, and the state of each lattice gas site is described by a small number of discrete bits instead of continuous values.

The second crucial property is the local deterministic rules for updating the array in space and time. The value of a site depends only on the values of a few local neighbors so there is no need for information to propagate across the lattice universe in a single step. Therefore, there is no requirement for a hardwired interconnection of distant sites on the grid.

The third element is the Boolean nature of the updating rules. The evolution of a lattice gas can be done by a series of purely Boolean operations, without ever computing with radix arithmetic.

To a computer architect, we have just described the properties of an ideal concurrent, or parallel, computer. The identical nature of particles and the locality of the rules for updating make it natural to update all sites in one operation—this is what one means by concurrent or parallel computation. Digital circuitry can perform Boolean operations naturally and quickly. Advancing the array in time is a sequence of purely local Boolean operations on a deterministic algorithm.

Most current parallel computer designs were built with non-local operations in mind. For this reason the basic architecture of present parallel machines is overlaid with a switching network that enables all sites to communicate in various degrees with all other sites. (The usual model of a switching network is a telephone exchange.) The complexity of machine architecture grows rapidly with the number of sites, usually as $n \log n$ at best with some time tradeoff and as $O(n^2)$ at worst. In a large machine, the complexity of the switching network quickly becomes greater than the complexity of the computing configuration.

In a purely local architecture switching networks are unnecessary, so two-dimensional systems can be built in a two-dimensional, or planar configuration, which is the configuration of existing large-scale integrated circuits. Such an architecture can be made physically compact by arranging the circuit boards in an accordion configuration similar to a piece of folded paper. Since the type of geometry chosen is vital to the collective behavior of the lattice gas model and no unique geometry fits all of parameter space, it would be a design mistake to hardwire a particular model into a machine architecture. Machines with programmable geometries could be designed in which the switching overhead to change geometries and rules would be minimal and the gain in flexibility large (Fig. 9).

In more than two dimensions a purely two-dimensional geometry is still efficient, using a combination of massive parallel updating in a two-dimensional plane and pipelining for the extra dimensions. As technology improves, it is easy to imagine fully three-dimensional machines, perhaps with optical pathways between planes, that have a long mean time to failure.

The basic hardware unit in conventional computers is the memory chip, since it has a large storage capacity (256 K bytes or 1 M bytes presently) and is inexpensive, reliable, and available commercially in large quantities. In fact, most modern computers have a memory-bound architecture, with a small number of controlling processors either doing local arithmetic and logical operations or using fast hashing algorithms on large look-up tables. An alternative is the local architecture described above for lattice gas simulators. In computer architecture terms it becomes very attractive to build compact, cheap, very fast simulators which are general over a large class of problems such as fluids. Such machines have a potential processing capacity much larger than the general-purpose architectures of present or foreseen vectorial and pipelined supercomputers. A number of such machines are in the process of being designed and built, and it will be quite interesting to see how these experiments in non-von Neumann architectures (more appropriately called super-von Neumann) turn out.

At present, the most interesting machine existing for lattice gas work is the Connection Machine with around 65,000 elementary processors and several gigabytes of main memory. This machine has a far more complex architecture than needed for pure lattice-gas work, but it was designed for quite a different purpose. Despite this, some excellent simulations have been done on it. The simulations at Los Alamos were done mainly on Crays with SUN workstations serving as code generators, controllers, and graphical units. The next generation of machines will see specialized lattice gas machines whether parallel, pipelined, or some combination, running either against

### ARCHITECTURE OF THE LATTICE GAS SIMULATOR

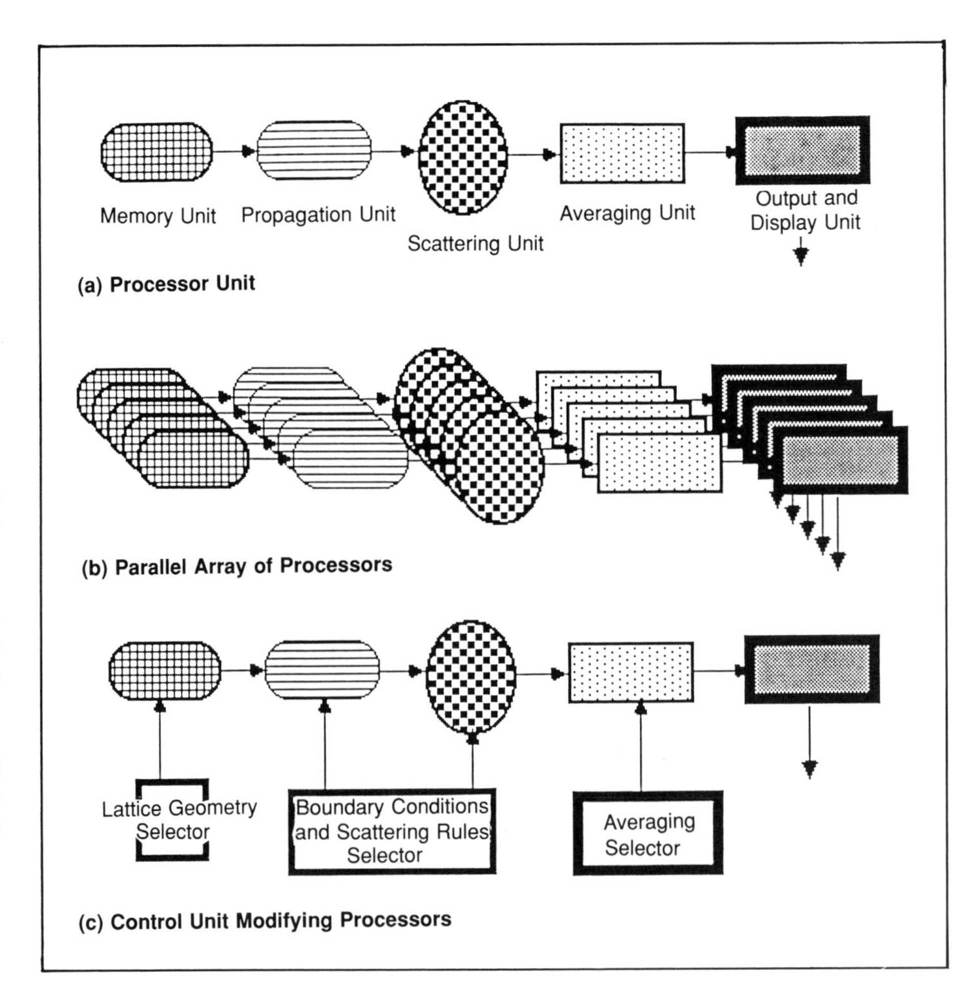


**Fig. 9. The lattice gas code is a virtual machine in the sense that the way the code works is exactly the way to build a machine.**

**(a) The basic processor unit in a lattice gas simular has five units: (1) a memory unit that stores the state at each node of the lattice grid; (2) a propagation unit that advances particles from one node to the next; (3) a scattering unit that checks the state at each node and implements the scattering rules where appropriate; (4) an averaging unit that averages velocities over a preassigned region of the lattice universe; and (5) an output and display unit.**

**(b) Processors are arranged in a parallel array. Each processor operates independently except at nodes on shared boundaries of the lattice gas universe.**

**(c) Processor units are overlaid by units that can alter the geometry of the lattice, the collision rules and boundary conditions, and the type of averaging.**

Connection Machine style architectures or using them as analyzing engines for processing data generated in lattice gas "black boxes." This will be a learning experience for everyone involved in massive simulation and provide hardware engines that will have many interesting physics and engineering applications.

Unfortunately, fast hardware alone is not enough to provide a truly useful exploration and design tool. A large amount of data is produced in a typical many degree of freedom system simulation. In three dimensions the problems of accessing, processing, storing, and visualizing such quantities of data are unsolved and are really universal problems even for standard supercomputer technology. As the systems we study become more complex, all these problems will also. It will take innovative engineering and physics approaches to overcome them.

## Conclusion

To any system naturally thought of as classes of simple elements interacting through local rules there should correspond a lattice-gas universe that can simulate it. From such skeletal gas models, one can gain a new perspective on the underlying mathematical physics of phenomena. So far we have used only the example of fluids and related systems that naturally support flows. The analysis of these systems used the principle of maximum ignorance: Even though we know the system is deterministic, we disregard that information and introduce artificial probabilistic methods. The reason is that the analytic tools for treating problems in this way are well developed, and although tedious to apply, they require no new mathematical or physical insight.

A deep problem in mathematical physics now comes up. The traditional methods of analyzing large probabilistic systems are asymptotic perturbation expansions in various disguises. These contain no information on how fast large-scale collective behavior should occur. We know from computer simulations that local equilibrium in lattice gases takes only a few time steps, global equilibrium occurs as fast as sound propagation will allow, and fully developed hydrodynamic phenomena, including complex instabilities, happen again as fast as a traverse of the geometry by a sound wave. One might say that the gas is precociously asymptotic and that this is basically due to the deterministic property that conveys information at the greatest possible speed.

Methods of analyzing the transient and invariant states of such complex multidimensional cellular spaces, using determinism as a central ingredient, are just beginning to be explored. They are non-perturbative. The problem seems as though some of the methods of dynamical systems theory should apply to it, and there is always the tempting shadow of renormalization-group ideas waiting to be applied with the right formalism. So far we have been just nibbling around the edges of the problem. It is an extraordinarily difficult one, but breaking it would provide new insight into the origin of irreversible processes in nature.

The second feature of lattice gas models, for phenomena reducible to natural skeletal worlds, is their efficiency compared to standard computational methods. Both styles of computing reduce to inventing effective microworlds, but the conventional one is dictated and constrained by a limited vocabulary of difference techniques, whereas the lattice gas method designs a virtual machine inside a real one, whose architectural structure is directly related to physics. It is not a priori clear that elegance equals efficiency. In many cases, lattice gas methods will be better at some kinds of problems, especially ones involving highly complex systems, and in others not. Its usefulness will depend on cleverness and the problem at hand. At worst the two ways of looking at the microphysics are complementary and can be used in various mixtures to create a beautiful and powerful computational tool.

We close this article with a series of conjectures. The image of the physical world as a skeletal lattice gas is essentially an abstract mathematical framework for creating algorithms whose dynamics spans the same solution spaces as many physically important nonlinear partial differential equations that have a microdynamical underpinning. There is no intrinsic reason why this point of view should not extend to those rich nonlinear systems which have no natural many-body picture. The classical Einstein-Hilbert action, phrased in the appropriate space, is no more complex than the Navier-Stokes equations. It should be possible to invent appropriate skeletal virtual computers for various gauge field theories, beginning with the Maxwell equations and proceeding to non-Abelian gauge models. Quantum mechanics can perhaps be implemented by using a variation on the stochastic quantization formulation of Nelson in an appropriate gas. When such models are invented, the physical meaning of the skeletal worlds is open to interpretation. It may be they are only a powerful mathematical device, a kind of virtual Turing machine for solving such problems. But it may also be that they will provide a new point of view on the physical origin and behavior of quantum mechanics and fundamental field-theoretic descriptions of the world. ■

## Further Reading


G. E. Uhlenbeck and G. W. Ford. 1963. *Lectures in Statistical Mechanics*. Providence, Rhode Island: American Mathematical Society.

Paul C. Martin. 1968. *Measurements and Correlation Functions*. New York: Gordon and Breach Science Publishers.

Stewart Harris. 1971. *An Introduction to the Theory of the Boltzmann Equation*. New York: Holt, Rinehart, and Winston.

E. M. Lifshitz and L. P. Pitaevskii. 1981. *Physical Kinetics*. Oxford: Pergamon Press.

Stephen Wolfram, editor. 1986. *Theory and Applications of Cellular Automata*. Singapore: World Scientific Publishing Company.

L. D. Landau and E. M. Lifshitz. 1987. *Fluid Mechanics*, second edition. Oxford: Pergamon Press.

*Complex Systems*, volume 1, number 4. The entirety of this journal issue consists of papers presented at the Workshop on Modern Approaches to Large Nonlinear Systems, Santa Fe, New Mexico, October 27–29, 1986.

U. Frisch, B. Hasslacher, and Y. Pomeau. 1986. Lattice-gas automata for the Navier-Stokes equation. *Physical Review Letters* 56: 1505–1508.

Stephen Wolfram. 1986. Cellular automaton fluids 1: Basic theory. *Journal of Statistical Physics* 45: 471–526.

Victor Yakhot and Stephen A. Orszag. 1986. *Physical Review Letters* 57: 1722.

U. Frisch, D. d'Humières, B. Hasslacher, P. Lallemand, Y. Pomeau, and J. P. Rivet. 1987. *Complex Systems* 1: 649–707.

Leo P. Kadanoff, Guy R. McNamara, and Gianluigi Zanetti. 1987. A Poiseuille viscometer for lattice gas automata. *Complex Systems* 1: 791–803.

Dominique d'Humières and Pierre Lallemand. 1987. Numerical simulations of hydrodynamics with lattice gas automata in two dimensions. *Complex Systems* 1: 599–632.

**Brosl Hasslacher** was born and grew up in Manhattan. He earned his Ph.D. in theoretical physics (quantum field theory) under Daniel Friedman, the inventor of supergravity, at State University of New York, Stonybrook. Most of his physics career has been spent at the Institute for Advanced Study in Princeton and at Caltech, with stops at the University of Illinois, École Normale Supérieure in Paris, and CERN. He is currently a staff physicist at the Laboratory. Says Brosl, "I spend essentially all my time thinking about physics. Presently I'm working on nonlinear field theories, chaotic systems, lattice gases, the architectures of very fast parallel computers, and the theory of computation and complexity."

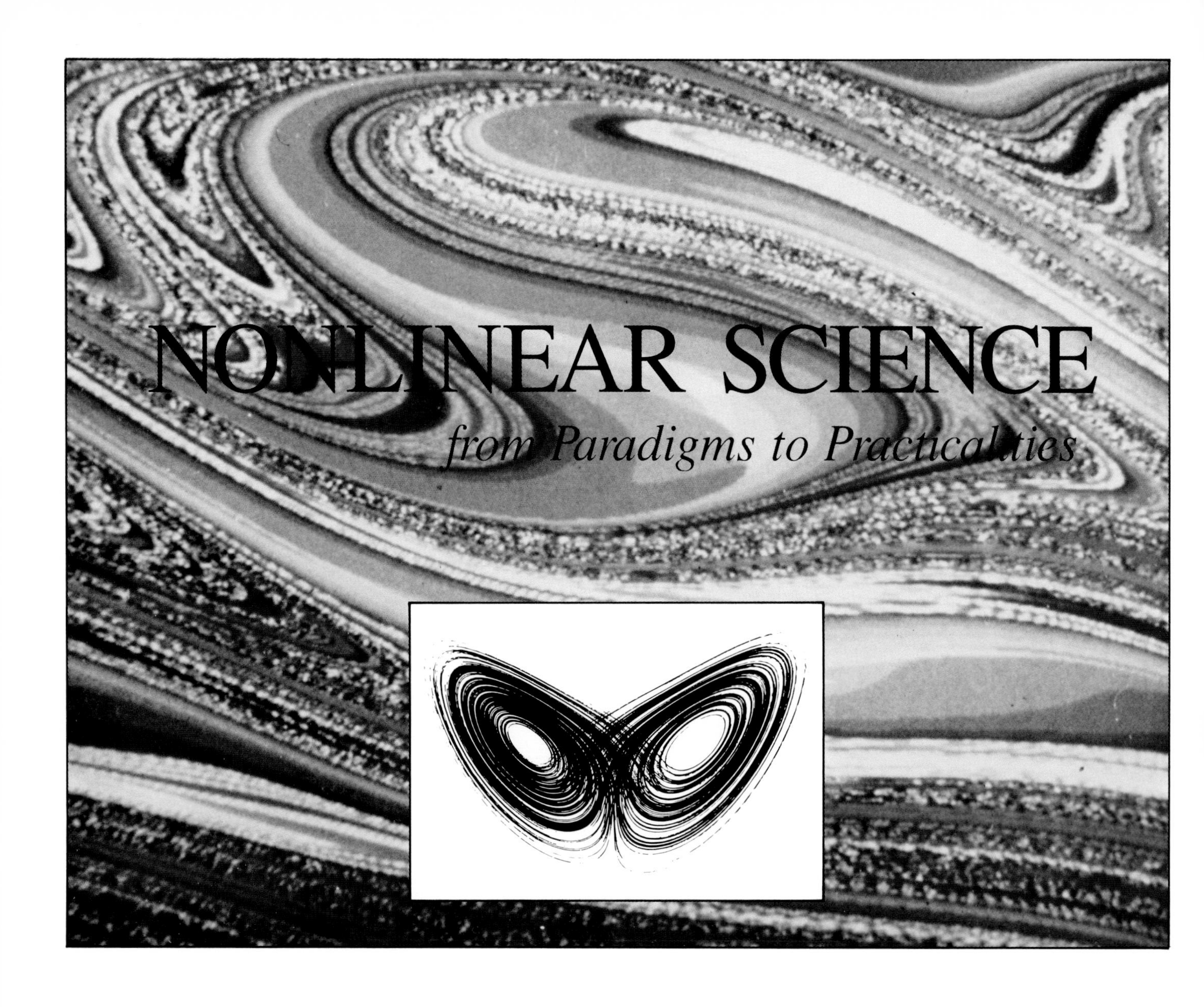

# NONLINEAR SCIENCE

## *from Paradigms to Practicalities*

***by David K. Campbell***

No tribute to the legacy of Stan Ulam would be complete without a discussion of "nonlinear science," a growing collection of interdisciplinary studies that in the past two decades has excited and challenged researchers from nearly every discipline of the natural sciences, engineering, and mathematics. Through his own research Stan played a major role in founding what we now call nonlinear science, and through his encouragement of the work of others, he guided its development. In this survey article I will try to weave the thread of Stan's contributions into the pattern of recent successes and current challenges of nonlinear science. At the same time I hope to capture some of the excitement of research in this area.

### Introduction

Let me start from a very simple, albeit circular, definition: nonlinear science is the study of those mathematical systems and natural phenomena that are *not* linear. Ever attuned to the possibility of *bons mots*, Stan once remarked that this was "like defining the bulk of zoology by calling it the study of 'non-elephant animals'." His point, clearly, was that the vast majority of mathematical equations and natural phenomena are nonlinear, with linearity being the exceptional, but important, case.

**Linear versus Nonlinear.** Mathematically, the essential difference between linear and nonlinear equations is clear. Any two solutions of a linear equation can be added together to form a new solution; this is the *superposition principle*. In fact, a moment of serious thought allows one to recognize that superposition is responsible for the systematic methods used to solve, independent of other complexities, essentially *any* linear problem. Fourier and Laplace transform methods, for example, depend on being able to superpose solutions. Putting it naively, one breaks the problem into many small pieces, then adds the separate solutions to get the solution to the whole problem.

In contrast, two solutions of a nonlinear equation *cannot* be added together to form another solution. Superposition fails. Thus, one must consider a nonlinear problem *in toto*; one cannot—at least not obviously—break the problem into small subproblems and add their solutions. It is therefore perhaps not surprising that no general analytic approach exists for solving typical nonlinear equations. In fact, as we shall discuss, certain nonlinear equations describing chaotic physical motions have *no* useful analytic solutions.

Physically, the distinction between linear and nonlinear behavior is best abstracted from examples. For instance, when water flows through a pipe at low velocity, its motion is *laminar* and is characteristic of linear behavior: regular, predictable, and describable in simple analytic mathematical terms. However, when the velocity exceeds a critical value, the motion becomes *turbulent*, with localized eddies moving in a complicated, irregular, and erratic way that typifies nonlinear behavior. By reflecting on this and other examples, we can isolate at least three characteristics that distinguish linear and nonlinear physical phenomena.

First, the motion itself is qualitatively different. Linear systems typically show smooth, regular motion in space and time that can be described in terms of well-behaved functions. Nonlinear systems, however, often show transitions from smooth motion to chaotic, erratic, or, as we will see later, even apparently random behavior. The quantitative description of *chaos* is one of the triumphs of nonlinear science.

Second, the response of a linear system to small changes in its parameters or to external stimulation is usually smooth and in direct proportion to the stimulation. But for nonlinear systems, a small change in the parameters can produce an enormous qualitative difference in the motion. Further, the response to an external stimulation can be different from the stimulation itself: for example, a periodically driven nonlinear system may exhibit oscillations at, say, one-half, one-quarter, or twice the period of the stimulation.

Third, a localized "lump," or pulse, in a linear system will normally decay by spreading out as time progresses. This phenomenon, known as dispersion, causes waves in linear systems to lose their identity and die out, such as when low-amplitude water waves disappear as they move away from the original disturbance. In contrast, nonlinear systems can have highly coherent, stable localized structures—such as the eddies in turbulent flow—that persist either for long times or, in some idealized mathematical models, for all time. The remarkable order reflected by these persistent coherent structures stands in sharp contrast to the irregular, erratic motion that they themselves can undergo.

To go beyond these qualitative distinctions, let me start with a very simple physical system—the plane pendulum—that is a classic example in at least two senses. First, it is a problem that all beginning students solve; second, it is a classic illustration of how we mislead our students about the prevalence and importance of nonlinearity.

Applying Newton's law of motion to the plane pendulum shown in Fig. 1 yields an ordinary second-order differential equation describing the time evolution:

$$\frac{d^2\theta(t)}{dt^2} + \frac{g}{l}\sin\theta(t) = 0, \tag{1}$$

where $\theta$ is the angular displacement of the pendulum from the vertical, $l$ is the length of the arm, and $g$ is the acceleration due to gravity. Equation 1 is obviously nonlinear

**THE SIMPLE PENDULUM**

**Fig. 1. It can be seen that a nonlinear equation describes the motion of the simple, plane pendulum when, in accordance with Newton's force law, the component of the gravitational force in the angular direction, $-mg\sin\theta(t)$, is set equal to the rate of change of the momentum, $ml\,d^2\theta(t)/dt^2$, in that direction.**

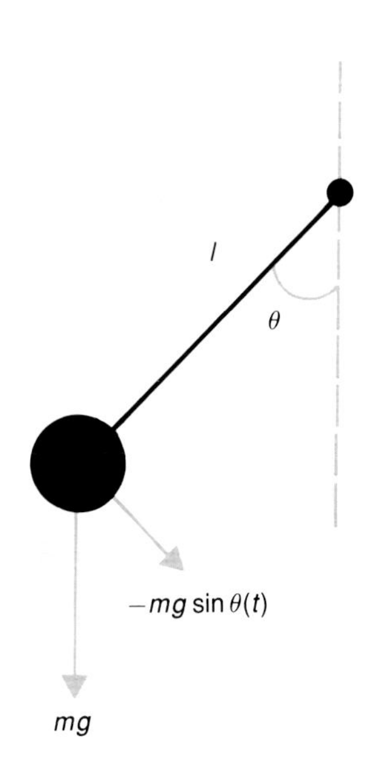

because $\sin(\theta_1 + \theta_2) \neq \sin\theta_1 + \sin\theta_2$.

What happens, however, if we go to the regime of small displacements? The Taylor expansion of $\sin\theta$ ($\approx \theta - \frac{\theta^3}{6} + \cdots$) tells us that for small $\theta$ the equation is approximately linear:

$$\frac{d^2\theta(t)}{dt^2} + \frac{g}{l}\theta(t) \approx 0. \tag{2}$$

The general solution to the linear equation is the superposition of two terms,

$$\theta(t) = \frac{1}{\omega}\left(\frac{d\theta}{dt}\right)_0 \sin\omega t + \theta_0 \cos\omega t, \tag{3}$$

where $\theta_0$ and $(d\theta/dt)_0$ are the angle and angular velocity at the initial time and the frequency $\omega$ is a constant given by $\omega = \sqrt{g/l}$.

Equation 3 is the mathematical embodiment of Galileo's famous observation that the frequency of a pendulum is independent of its amplitude. But in fact the result is a consequence of the linear approximation, valid only for small oscillations. If the pendulum undergoes very large displacements from the vertical, its motion enters the nonlinear regime, and one finds that the frequency depends on amplitude, larger excursions having longer periods (see "The Simple But Nonlinear Pendulum"). Of course, grandfather clocks would keep terrible time if the linear equation were not a good approximation; nonetheless, it remains an approximation, valid only for small-amplitude motion.

The distinction between the full nonlinear model of the pendulum and its linear approximation becomes substantially more striking when one studies the pendulum's *response* to an external stimulus. With both effects of friction and a periodic driving force added, the pendulum equation (Eq. 1) becomes

$$\frac{d^2\theta}{dt^2} + \alpha\frac{d\theta}{dt} + \frac{g}{l}\sin\theta = \Gamma\cos\Omega t, \tag{4}$$

where $\alpha$ is a measure of the frictional force and $\Gamma$ and $\Omega$ are the amplitude and frequency, respectively, of the driving force. In the regime of small displacements, this reduces to the linear equation

$$\frac{d^2\theta}{dt^2} + \alpha\frac{d\theta}{dt} + \frac{g}{l}\theta \approx \Gamma\cos\Omega t. \tag{5}$$

A closed-form solution to the linear equation can still be obtained, and the motion can be described analytically for all time. For certain values of $\alpha$, $\Gamma$, and $\Omega$, the solution to even the nonlinear equation is periodic and quite similar to that of the linear model. For other values, however, the solution behaves in a complex, seemingly random, unpredictable manner. In this chaotic regime, as we shall later see, the motion of this very simple nonlinear system defies analytic description and can indeed be as random as a coin toss.

**Dynamical Systems: From Simple to Complex.** Both the free pendulum and its damped, driven counterpart are particular examples of *dynamical systems.* The free pendulum is a *conservative* dynamical system—energy is constant in time—whereas the damped, driven pendulum is a *dissipative* system—energy is not conserved. Loosely speaking, a dynamical system can be thought of as anything that evolves in time according to a well-defined rule. More specifically, the variables in a dynamical system, such as $q$ and $p$, the canonical position and momentum, respectively, have a rate of change at a given time that is a function of the values of the variables themselves at that time: $\dot{q}(t) = f\left(q(t), p(t)\right)$ and $\dot{p}(t) = g\left(q(t), p(t)\right)$ (where a dot signifies differentiation with respect to time). The abstract "space" defined by these variables is called the *phase*

# The Simple but NONLINEAR PENDULUM

Elementary physics texts typically treat the simple plane pendulum by solving the equation of motion only in the linear approximation and then presenting the general solution as a superposition of sines and cosines (as in Eq. 3 of the main text). However, the full nonlinear equation can also be solved analytically in closed form, and a brief discussion of this solution allows us to illustrate explicitly several aspects of nonlinear systems.

It is most instructive to start our analysis using the Hamiltonian for the simple pendulum, which, in terms of the angle $\theta$ (a generalized coordinate) and the corresponding (generalized) momentum $p_\theta \equiv ml^2 \frac{d\theta}{dt}$, has the form

$$H(p_\theta, \theta) = \frac{p_\theta^2}{2ml^2} - mgl\cos\theta. \quad (1)$$

Using the Hamiltonian equations

$$\frac{\partial H}{\partial p_\theta} = \frac{d\theta}{dt},$$

and

$$\frac{\partial H}{\partial \theta} = -\frac{dp_\theta}{dt},$$

we obtain (after substituting for $p_\theta$) an equation solely in terms of the angle $\theta$ and its derivative:

$$\frac{d^2\theta}{dt^2} + \frac{g}{l}\sin\theta = 0. \quad (2)$$

Recognizing that $\frac{d\theta}{dt}\frac{d^2\theta}{dt^2} = \frac{d}{dt}(\frac{1}{2}(\frac{d\theta}{dt})^2)$ and $\frac{d\theta}{dt}\sin\theta = \frac{d}{dt}(-\cos\theta)$, we see that Eq. 2 can be converted to a perfect differential by multiplying by $d\theta/dt$:

$$\frac{d\theta}{dt}\frac{d^2\theta}{dt^2} + \frac{g}{l}\sin\theta\frac{d\theta}{dt} = \frac{d}{dt}\left(\frac{1}{2}\left(\frac{d\theta}{dt}\right)^2 - \frac{g}{l}\cos\theta\right) = 0. \quad (3)$$

Hence, we can integrate Eq. 3 immediately to obtain

$$\frac{1}{2}\left(\frac{d\theta}{dt}\right)^2 = \frac{g}{l}\cos\theta + C. \quad (4)$$

By comparing Eqs. 1 and 3 and recalling the definition of $p_\theta$, we see that

$$C = \frac{H(p_\theta, \theta)}{ml^2}.$$

That the constant $C$ is proportional to the *value* of the Hamiltonian, of course, is just an expression of the familiar conservation of energy and shows that the value of the conserved energy determines the nature of the pendulum's motion.

Restricting our considerations to librations—that is, motions in which the pendulum oscillates back and forth without swinging over the top of its pivot point—we can evaluate $C$ in terms of $\theta$ by using the condition $d\theta/dt = 0$ when $\theta = \theta_{\max}$, which yields

$$C = -\frac{g}{l}\cos\theta_{\max}.$$

This, in turn, means that

$$\frac{d\theta}{dt} = \sqrt{\frac{2g}{l}(\cos\theta - \cos\theta_{\max})}. \quad (5)$$

The full period of the motion $T$ is then the definite integral

$$T = 4\int_0^{\theta_{\max}} \frac{d\theta}{\sqrt{\frac{2g}{l}(\cos\theta - \cos\theta_{\max})}}. \quad (6)$$

This last integral can be converted, via trigonometric identities and redefinitions of variables, to an elliptic integral of the first kind. Although not as familiar as the sines and cosines that arise in the linear approximation, the elliptic integral is tabulated and can be readily evaluated. Thus, the full equation of motion for the nonlinear pendulum can be solved in closed form for arbitrary initial conditions.

An elegant method for depicting the solutions for the one-degree-of-freedom system is the "phase plane." If we examine such a plot (see Fig. 2 in the main text), we see that the origin ($\theta = 0$, $d\theta/dt = 0$)—and, of course, its periodic equivalents at $\theta = \pm 2n\pi$, $d\theta/dt = 0$—represent *stable fixed points* with the pendulum at rest and the bob pointing down. The point at $\theta = \pi$, $d\theta/dt = 0$—and, again, its periodic equivalents at $\theta = \pm(2n+1)\pi$, $d\theta/dt = 0$—represent *unstable fixed points* with the pendulum at rest but the bob inverted; the slightest perturbation causes the pendulum to move away from these points. The closed curves near the horizontal axis ($d\theta/dt = 0$) represent librations, or *periodic oscillations*. The open, "wavy" lines away from the horizontal axis (large $|d\theta/dt|$) correspond to *unbounded motions* in the sense that $\theta$ increases or decreases forever as the pendulum rotates around its pivot point in either a clockwise ($d\theta/dt < 0$) or a counterclockwise ($d\theta/dt > 0$) sense.

What about other systems? A dynamical system that can be described by $2N$ generalized position and momentum coordinates is said to have $N$ degrees of freedom. Hamiltonian systems that, like the pendulum, have only one degree of freedom can *always* be integrated completely with the techniques used for Eqs. 2–6. More generally, however, systems with $N$ degrees of freedom are *not* completely integrable; Hamiltonian systems with $N$ degrees of freedom that *are* completely integrable form a very restricted but extremely important subset of all $N$-degree-of-freedom systems.

As suggested by the one-degree-of-freedom case, complete integrability of a system with $N$ degrees of freedom requires that the system have $N$ constants of motion—that is, $N$ integrals analogous to Eq. 4—*and* that these constants be consistent with each other. Technically, this last condition is equivalent to saying that when the constants, or integrals of motion, are expressed in terms of the dynamical variables (as $C$ is in Eq. 4), the expressions must be "in involution," meaning that the Poisson brackets must vanish identically for all possible pairs of integrals of motion. Remarkably, one can find nontrivial examples of completely integrable systems, not only for $N$-degree-of-freedom systems but also for the "infinite"-degree-of-freedom systems described by partial differential equations. The sine-Gordon equation, discussed extensively in the main text, is a famous example.

In spite of any nonlinearities, systems that are completely integrable possess remarkable regularity, exhibiting smooth motion in all regions of phase space. This fact is in stark contrast to nonintegrable systems. With as few as one-and-a-half degrees of freedom (such as the damped, driven system with three generalized coordinates represented by Eq. 4 in the main text), a nonintegrable system can exhibit deterministic chaos and motion as random as a coin toss. ■

*space*, and its dimension is clearly related to the number of variables in the dynamical system.

In the case of the free pendulum, the angular position and velocity at any instant determine the subsequent motion. Hence, as discussed in "The Simple But Nonlinear Pendulum," the pendulum's behavior can be described by the motion of a point in the two-dimensional phase space with coordinates $\theta$ and $\dot{\theta}$ (Fig. 2). In the traditional parlance of mechanics, the free pendulum is a Hamiltonian system having "one degree

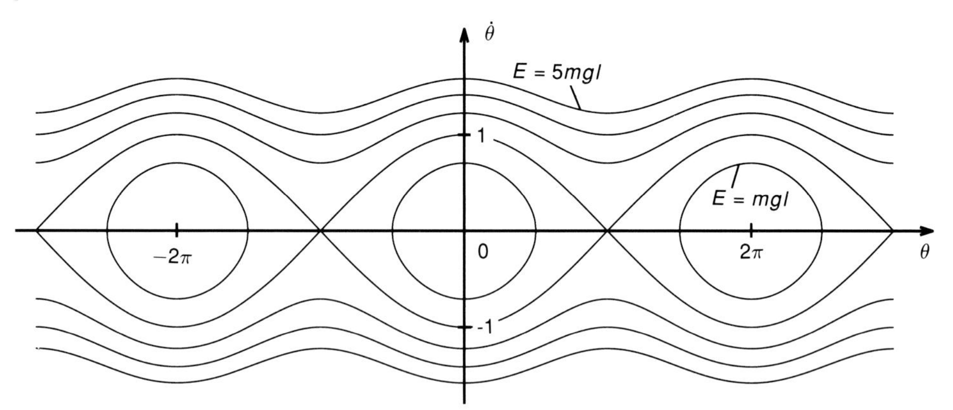


## HARMONIC-OSCILLATOR PHASE SPACE

**Fig. 2. The behavior of the simple pendulum is here represented by constant-energy contours in $\theta$-$\dot{\theta}$ (roughly, position-momentum) phase space. The closed curves around the origin ($E < 2mgl$) represent librations, or periodic oscillations, whereas the open, "wavy" lines for large magnitudes of $\dot{\theta}$ ($E > 2mgl$) correspond to motions in which the pendulum moves completely around its pivot in either a clockwise ($\dot{\theta} < 0$) or counterclockwise ($\dot{\theta} > 0$) sense, causing $\theta$ to increase in magnitude beyond $2\pi$. (Figure courtesy of Roger Eckhardt, Los Alamos National Laboratory.)**

of freedom," since it has only one spatial variable ($\theta$) and one generalized momentum (roughly, $\dot{\theta}$). Further, as discussed in the sidebar, this system is *completely integrable*, which in effect means that its motion for all time can be solved for analytically in terms of the initial values of the variables.

More typically, dynamical systems involve many degrees of freedom and thus have high-dimensional phase spaces. Further, they are in general *not* completely integrable. An example of a many-degree-of-freedom system particularly pertinent to our current discussion is the one first studied by Enrico Fermi, John Pasta, and Stan Ulam in the mid-fifties: a group of particles coupled together by nonlinear springs and constrained to move only in one dimension. Now celebrated as the "FPU problem," the model for the system consists of a large set of coupled, ordinary differential equations for the time evolution of the particles (see "The Fermi, Pasta, and Ulam Problem: Excerpts from 'Studies of Nonlinear Problems'"). Specifically, one particular version of the FPU problem has 64 particles obeying the equations

$$\ddot{x}_i = (x_{i+1} + x_{i-1} - 2x_i) + \alpha\big((x_{i+1} - x_i)^2 - (x_i - x_{i-1})^2\big) \qquad \text{for } i = 1, 2, \ldots, 64, \tag{6}$$

where $\alpha$ is the measure of the strength of the nonlinear interaction between neighboring particles. Thus there are 64 degrees of freedom and, consequently, a 128-dimensional phase space.

Still more complicated, at least *a priori*, are continuous nonlinear dynamical systems, such as fluids. Here one must define dynamical variables—such as the density $\rho(x, t)$—at every point in space. Hence the number of degrees of freedom, and accordingly the phase-space dimension, becomes infinite; further, the resulting equations of motion become nonlinear *partial* differential equations. Note that one can view these continuous dynamical systems as the limits of large discrete systems and understand their partial differential equations as the limits of many coupled ordinary differential equations.

We can illustrate this approach using a continuous nonlinear dynamical system that will be important in our later discussion. Hopefully, this example will pique the reader's interest, for it also indicates how elegantly perverse nonlinearity can be. The system is represented by the so-called sine-Gordon equation

$$\frac{\partial^2 \theta}{\partial t^2} - \frac{\partial^2 \theta}{\partial x^2} + \sin\theta = 0, \tag{7}$$

where the dependent variable $\theta = \theta(x, t)$ is a measure of the response of the system at position $x$ and time $t$.

Computationally, one natural way to deal with this system is to introduce a discrete spatial grid with spacing $\Delta x$ such that the position at the $n$th point in the grid is given by $x_n = n\,\Delta x$ and define $\theta_n(t) \equiv \theta(x_n, t)$ for $n = 1, 2, \ldots N$. Using a finite difference approximation for the second derivative,

$$\frac{\partial^2 \theta}{\partial x^2} \simeq \frac{1}{(\Delta x)^2}\big(\theta_{n+1}(t) - 2\theta_n(t) + \theta_{n-1}(t)\big) + \mathcal{O}(\Delta x), \tag{8}$$

leads to a set of $N$ coupled ordinary differential equations

$$\frac{d^2\theta_n}{dt^2} = \frac{1}{(\Delta x)^2}\big(\theta_{n+1}(t) - 2\theta_n(t) + \theta_{n-1}(t)\big) + \sin\theta_n(t) \quad n = 1, 2, \ldots N. \tag{9}$$

This is a *finite* degree-of-freedom dynamical system, like the FPU problem. In particular, it is just a set of simple plane pendula, coupled together by the discretized spatial derivative. Of course, the continuous sine-Gordon equation is recovered in the limit that $N \to \infty$ (and thus $\Delta x \to 0$). The perverseness of nonlinearity is that whereas the Hamiltonian dynamical system described by a finite number $N$ of coupled ordinary differential equations is *not* completely integrable, the infinite-dimensional Hamiltonian system described by the continuum sine-Gordon equation *is*! Further, as we shall later demonstrate, the latter system possesses localized "lump" solutions—the famed *solitons*—that persist for all time.

Hopefully, this digression on dynamical systems has made the subtlety of nonlinear phenomena quite apparent: very simple nonlinear systems—such as the damped, driven pendulum—can exhibit chaos involving extremely complex, apparently random motions, while very complicated systems—such as the one described by the sine-Gordon equation—can exhibit remarkable manifestations of order. The challenge to researchers in this field is to determine which to expect and when.

**Paradigms of Nonlinearity.** Before examining in some detail how this challenge is being confronted, we need to respond to some obvious but important questions. First, why study nonlinear *science*, rather than nonlinear *chemistry*, or nonlinear *physics*, or nonlinear *biology*? Nonlinear science sounds impossibly broad, too interdisciplinary, or "the study of everything." However, the absence of a systematic mathematical framework and the complexity of natural nonlinear phenomena suggest that nonlinear behavior is best comprehended by classifying its various manifestations in many different systems and by identifying and studying their common features. Indeed, both the interest and the power of nonlinear science arise precisely because common concepts are being discovered about systems in very different areas of mathematics and natural sciences. These common concepts, or *paradigms*, give insight into nonlinear problems in a large number of disciplines at once. By understanding these paradigms, one can hope to understand the essence of nonlinearity as well as its consequences in many fields.

Second, since it has long been known that most systems are inherently nonlinear, why has there been a sudden blossoming of interest in this field in the past twenty years or so? Why weren't many of these fundamental problems solved a century ago? On reflection, one can identify three recent developments whose synergistic blending has made possible revolutionary progress.

The first, and perhaps most crucial, development has been that of high-speed electronic computers, which permit quantitative numerical simulations of nonlinear systems. Indeed, the term *experimental mathematics* has been coined to describe computer-based investigations into problems inaccessible to analytic methods. Rather than simply confirming quantitatively results already anticipated by qualitative analysis, experimental mathematics uses the computer to generate qualitative insight where none has existed before. As the visionary of this development, John von Neumann, wrote (in a 1946 article called "On the principles of large scale computing machines"):

"Our present analytic methods seem unsuitable for the solution of the important problems rising in connection with nonlinear partial differential equations and, in fact, with virtually all types of problems in pure mathematics. ... really efficient high-speed computing devices may, in the field of nonlinear partial differential equations as well as in many other fields which are now difficult or entirely denied of access, provide us with those heuristic hints which are needed in all parts of mathematics for genuine progress."

Stan Ulam, together with many of his Los Alamos colleagues, was one of the very first to make this vision a reality. Among Stan's pioneering experimental mathematical investigations was the seminal study of the FPU problem mentioned above. Another example was his early numerical work on nonlinear mappings, carried out in collaboration with Paul Stein (see "Iteration of Maps, Strange Attractors, and Number Theory—An Ulamian Potpourri"). Both of these studies will figure in our later discussion.

The second crucial development has been the experimental observation of "universal" nonlinear characteristics in natural systems that range from chicken hearts and chemical reactors to fluids and plasmas. In the past decade these experiments have reached previously inaccessible levels of precision, so that one can measure *quantitative* similarities in, for example, the route to chaotic behavior among an enormous variety of nonlinear systems.

The third and final development has been in the area of novel analytical mathematical methods. For instance, the invention of the *inverse spectral transform* has led to a systematic method for the explicit solution of a large number of nonlinear partial differential equations. Similarly, new methods based on the theory of Hamiltonian systems allow the analysis of nonlinear stability of a wide range of physically relevant mathematical models.

As we shall shortly see, the methodology based on these three developments has been remarkably successful in solving many nonlinear problems long considered intractable. Moreover, the common characteristics of nonlinear phenomena in very distinct fields has allowed progress in one discipline to transfer rapidly to others and confirms the inherently interdisciplinary nature of nonlinear science. Despite this progress, however, we do not have an entirely systematic approach to nonlinear problems. For the *general* nonlinear equation there is simply no analog of a Fourier transform. We do, however, have an increasing number of well-defined *paradigms* that both reflect typical qualitative features and permit quantitative analysis of a wide range of nonlinear systems. In the ensuing three sections I will focus on three such paradigms: *coherent structures and solitons*, *deterministic chaos and fractals*, and *complex configurations and patterns*. Of these the first two are well developed and amply exemplified, whereas the third is still emerging. Appropriately, these paradigms reflect different aspects of nonlinearity: coherent structures reveal a surprising orderliness, deterministic chaos illustrates an exquisite disorder, and complex configurations represent the titanic struggle between opposing aspects of order and chaos.

If we were to follow the biblical sequence we would start with chaos, but because it is frankly a rather counterintuitive concept, we shall start with solitons or, more generally and accurately, coherent structures.

## Coherent Structures and Solitons

From the Red Spot of Jupiter through clumps of electromagnetic radiation in turbulent plasmas to microscopic charge-density waves on the atomic scale, spatially localized, long-lived, wave-like excitations abound in nonlinear systems. These nonlinear waves and structures reflect a surprising orderliness in the midst of complex behavior. Their ubiquitous role in both natural nonlinear phenomena and the corresponding mathematical models has caused coherent structures and solitons to emerge as one of the central paradigms of nonlinear science. Coherent structures typically represent the natural "modes" for understanding the time-evolution of the nonlinear system and often dominate the long-time behavior of the motion.

To illustrate this, let me begin with one of the most familiar (and beautiful!) examples in nature, namely, the giant Red Spot (Fig. 3a). This feature, first observed

## COHERENT STRUCTURES IN NATURE

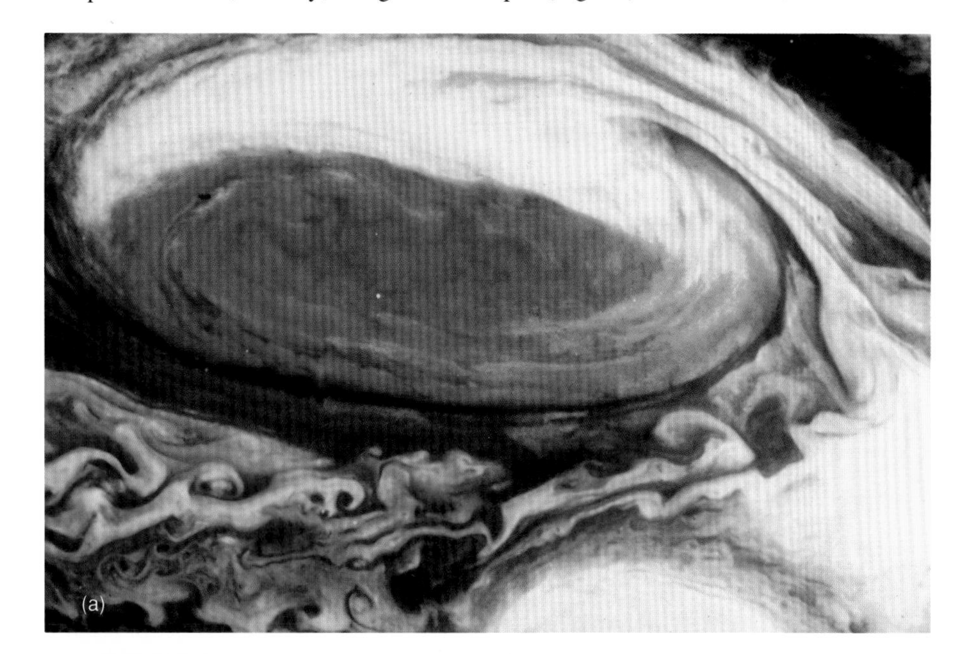


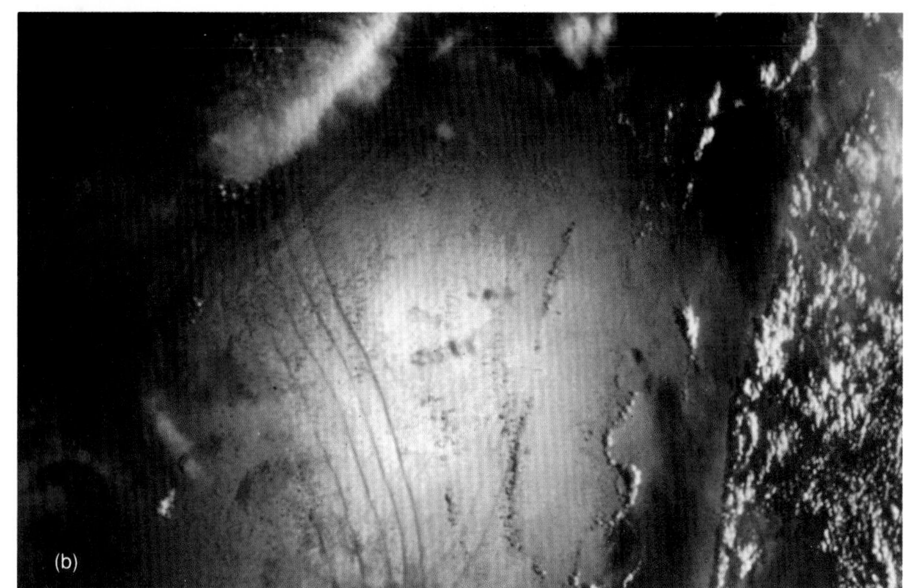


**Fig. 3.** (a) A closeup of the Red Spot of Jupiter, taken from the Voyager spacecraft. False color is used to enhance features of the image. In addition to the celebrated Red Spot, there are many other "coherent structures" on smaller scales on Jupiter. (Photo courtesy of NASA). (b) Nonlinear surface waves in the Andaman Sea off the coast of Thailand as photographed from an Apollo-Soyuz spacecraft. (Photo courtesy of NASA.)

from earth in the late seventeenth century, has remained remarkably stable in the turbulent cauldron of Jupiter's atmosphere. It represents a coherent structure on a scale of about $4 \times 10^8$ meters, or roughly the distance from the earth to the moon.

To give an example at the terrestrial level, certain classes of nonlinear ocean waves form coherent structures that propagate essentially unchanged for thousands of miles. Figure 3b is a photograph taken from an Apollo-Soyuz spacecraft of a region of open ocean in the Andaman Sea near northern Sumatra. One sees clearly a packet of five nearly straight surface waves; each is approximately 150 kilometers wide, so the scale of this phenomenon is roughly $10^5$ meters. Individual waves within the

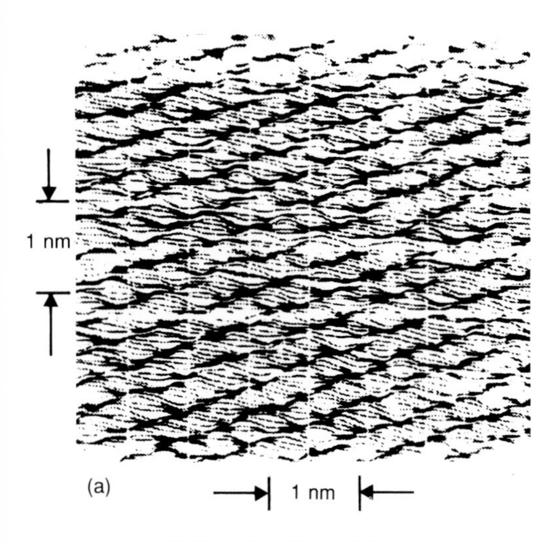


**COHERENT STRUCTURE IN CRYSTALS**

**Fig. 4. (a) An image, made by using tunneling-electron microscopy, of a cleaved surface of tantalum diselenide that shows the expected graininess around atomic sites in the crystal lattice. (b) A similar image of tantalum disulfide, showing coherent structures called charge-density waves that are *not* simply a reflection of the crystal lattice but arise from nonlinear interaction effects. (Photos courtesy of C. G. Slough, W. W. McNairy, R. V. Coleman, B. Drake, and P. K. Hansma, University of Virginia.)**

packet are separated from each other by about 10 kilometers. The waves, which are generated by tidal forces, move in the direction perpendicular to their crests at a speed of about 2 meters per second. Although the surface deflection of these waves is small—about 1.8 meters—they can here be seen from orbit because the sun is directly behind the spacecraft, causing the specular reflection to be very sensitive to variations of the surface. These visible surface waves are actually a manifestation of much larger amplitude—perhaps ten times larger—internal waves. The internal waves exist because thermal or salinity gradients lead to a stratification of the subsurface into layers. *A priori* such large internal waves could pose a threat to submarines and to off-shore structures. Indeed, the research on these waves was initiated by Exxon Corporation to assess the actual risks to the oil rigs they planned to construct in the area. Fortunately, in this context the phenomenon turned out to be more beautiful than threatening.

Our final physical illustration is drawn from solid-state physics, where the phenomenon of charge-density waves exemplifies coherent structures on the atomic scale. If one studies a crystal of tantalum diselenide using an imaging process called tunneling-electron microscopy (Fig. 4a), one finds an image that is slightly denser around the atomic sites but otherwise is uniform. Given that the experimental technique focuses on specfic electronic levels, this graininess is precisely what one would expect at the atomic level; there are no nonlinear coherent structures, no charge-density waves. In contrast, tantalum disulfide, which has nearly identical lattice parameters, exhibits much larger structures in the corresponding image (Fig. 4b); in fact, the image shows a hexagonal array of coherent structures. These charge-density waves are separated by about 3.5 normal lattice spacings, so their occurence is not simply a reflection of the natural atomic graininess. Rather, these coherent structures arise because of a nonlinear coupling between the electrons and the atomic nuclei in the lattice. Notice that now the scale is $10^{-9}$ meter.

**Solitons.** We have thus identified nonlinear coherent structures in nature on scales ranging from $10^8$ meters to $10^{-9}$ meter—seventeen orders of magnitude! Clearly this paradigm is an essential part of nonlinear science. It is therefore very gratifying that during the past twenty years we have seen a veritable revolution in the understanding of coherent structures. The crucial event that brought on this revolution was the discovery, by Norman Zabusky and Martin Kruskal in 1965, of the remarkable soliton. In a sense, solitons represent the purest form of the coherent-structure paradigm and thus are a natural place to begin our detailed analysis. Further, the history of this discovery shows the intricate interweaving of the various threads of Stan Ulam's legacy to nonlinear science.

To define a soliton precisely, we consider the motion of a wave described by an equation that, in general, will be nonlinear. A *traveling wave* solution to such an equation is one that depends on the space $x$ and time $t$ variables only through the combination $\xi = x - vt$, where $v$ is the constant velocity of the wave. The traveling wave moves through space without changing its shape and in particular without spreading out or dispersing. If the traveling wave is a localized single pulse, it is called a solitary wave. A soliton is a solitary wave with the crucial additional property that it preserves its form exactly when it interacts with other solitary waves.

The study that led Kruskal and Zabusky to the soliton had its origin in the famous FPU problem, indeed in precisely the form shown in Eq. 6. Experimental mathematical studies of those equations showed, instead of the equipartition of energy expected on general grounds from statistical mechanics, a puzzling series of recurrences of the initial state (see "The Ergodic Hypothesis: A Complicated Problem of Mathematics and Physics"). Through a series of asymptotic approximations, Kruskal and Zabusky related the recurrence question for the system of oscillators in the FPU problem to the nonlinear partial differential equation

$$\frac{\partial u}{\partial t} + u\frac{\partial u}{\partial x} + \frac{\partial^3 u}{\partial x^3} = 0. \qquad (10)$$

Equation 10, called the Korteweg-deVries or KdV equation, had first been derived in 1895 as an approximate description of water waves moving in a shallow, narrow channel. Indeed, the surface waves in the Andaman Sea, which move essentially in one direction and therefore can be modeled by an equation having only one spatial variable, are described quite accurately by Eq. 10. That this same equation should also appear as a limiting case in the study of a discrete lattice of nonlinear oscillators is an illustration of the generic nature of nonlinear phenomena.

To look analytically for a coherent structure in Eq. 10, one seeks a localized solution $u_s(\xi)$ that depends only on $\xi = x - vt$, thereby reducing the partial differential equation to an ordinary differential equation in $\xi$. The result can be integrated explicitly and, for solutions that vanish at infinity, yields

$$u_s(x, t) = 3v \operatorname{sech}^2 \frac{\sqrt{v}}{2}(x - vt). \tag{11}$$

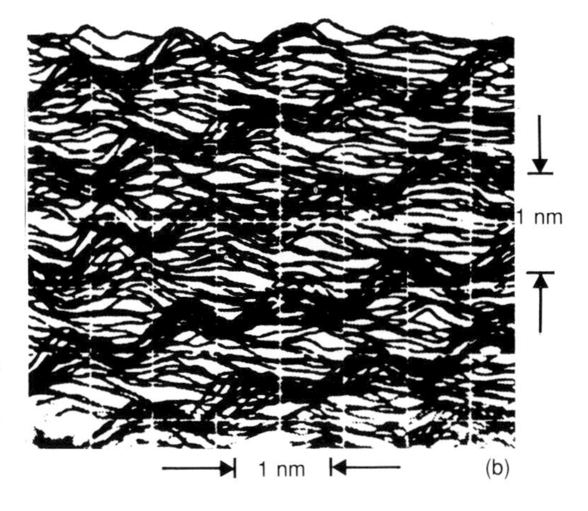


This solution describes a solitary wave moving with constant velocity $v$. Moreover, the amplitude of the wave is proportional to $v$, and its width is inversely proportional to $\sqrt{v}$. The faster the wave goes, the narrower it gets. This relation between the shape and velocity of the wave reflects the nonlinearity of the KdV equation.

Intuitively, we can understand the existence of this solitary wave as a result of a delicate balance in the KdV equation between the linear dispersive term $\frac{\partial^3 u}{\partial x^3}$, which tends to cause an initially localized pulse to spread out and change shape as it moves, and the nonlinear *convective derivative* term $u\frac{\partial u}{\partial x}$, which tends to increase the pulse where it is already large and hence to bunch up the disturbance. (For a more precise technical analysis of these competing effects in another important nonlinear equation, see "Solitons in the Sine-Gordon Equation.")

Although the solution represented by Eq. 11 is, by inspection, a coherent structure, is it a soliton? In other words, does it preserve its form when it collides with another solitary wave? Since the analytic methods of the 1960s could not answer this question, Zabusky and Kruskal followed another of Ulam's leads and adopted an experimental mathematics approach by performing computer simulations of the collision of two solitary waves with different velocities. Their expectation was that the nonlinear nature of the interaction would break up the waves, causing them to change their properties dramatically and perhaps to disappear entirely. When the computer gave the startling result that the coherent structures emerged from the interaction unaffected in shape, amplitude, and velocity, Zabusky and Kruskal coined the term "soliton," a name reflecting the particle-like attributes of this nonlinear wave and patterned after the names physicists traditionally give to atomic and subatomic particles.

In the years since 1965 research has revealed the existence of solitons in a host of other nonlinear equations, primarily but not exclusively in one spatial dimension. Significantly, the insights gained from the early experimental mathematical studies have had profound impact on many areas of more conventional mathematics, including infinite-dimensional analysis, algebraic geometry, partial differential equations, and dynamical systems theory. To be more specific, the results of Kruskal and Zabusky led directly to the invention of a novel analytic method, now known as the "inverse spectral transform," that permits the explicit and systematic solution of soliton-bearing equations by a series of effectively linear operations. Further, viewed as nonlinear dynamical systems, the soliton equations have been shown to correspond to infinite-degree-of-freedom Hamiltonian systems that possess an infinite number of independent conservation laws and are thus completely integrable. Indeed, the invariance of solitons under interactions can be understood as a consequence of these conservation laws.

**Applied Solitons.** From all perspectives nonlinear partial differential equations containing solitons are quite special. Nonetheless, as our examples suggest, there is a surprising mathematical diversity to these equations. This diversity is reflected in the

# SOLITONS in the SINE-GORDON Equation

To understand quantitatively how solitons can result from a delicate balance of dispersion and nonlinearity, let us begin with the *linear, dispersionless,* bidirectional wave equation

$$\frac{\partial^2 u}{\partial t^2} - c_0^2 \frac{\partial^2 u}{\partial x^2} = 0. \qquad (1)$$

By direct substitution into Eq. 1, it is easy to check that, with $\eta \equiv x - c_0 t$ and $\xi \equiv x + c_0 t$,

$$u(x,t) = f(\eta) + g(\xi)$$

is a solution for *any* functions $f$ and $g$. Thus if we take, for example, $f(\eta) = e^{-\eta^2}$ and $g(\xi) = e^{-\xi^2}$, we will have two "solitary waves," one moving to the left and one to the right. At $t \to -\infty$, the lumps are isolated, at $t = 0$ they collide, and at $t \to +\infty$ they re-emerge. Hence, by our definition these solutions to the linear, dispersionless wave equation are solitons, although trivial examples of such.

Now consider an equation, still linear, of the form

$$\frac{\partial^2 u}{\partial t^2} - c_0^2 \frac{\partial^2 u}{\partial x^2} + m^2 u = 0. \qquad (2)$$

Such equations arise naturally in descriptions of optically-active phonons in solid state physics and in relativistic field theories. An elementary (plane wave) solution of this equation has the form

$$u(x,t) = Ae^{i(\omega t + kx)}, \qquad (3)$$

where $A$ is a constant, $\omega$ is the frequency, and $k$ is the wave number. Substituting into Eq. 2 shows that this plane wave can be a solution of Eq. 2 *only if*

$$-\omega^2 + c_0^2 k^2 + m^2 = 0, \qquad (4a)$$

so that

$$\omega(k) = \pm\sqrt{m^2 + c_0^2 k^2}. \qquad (4b)$$

This relation between $\omega$ and $k$ is known technically as a *dispersion relation* and contains essential information about how individual plane waves with different $k$'s (and hence different $\omega$'s) propagate. In particular, the group velocity,

$$v_g(k_0) \equiv \left.\frac{\partial \omega}{\partial k}\right|_{k_0},$$

measures how fast a small group of waves with values of $k$ peaked around a particular value $k_0$ moves. Notice that for the dispersion relation Eq. 4b,

$$v_g(k_0) = \pm \frac{c_0^2 k_0}{\sqrt{c_0^2 k_0^2 + m^2}},$$

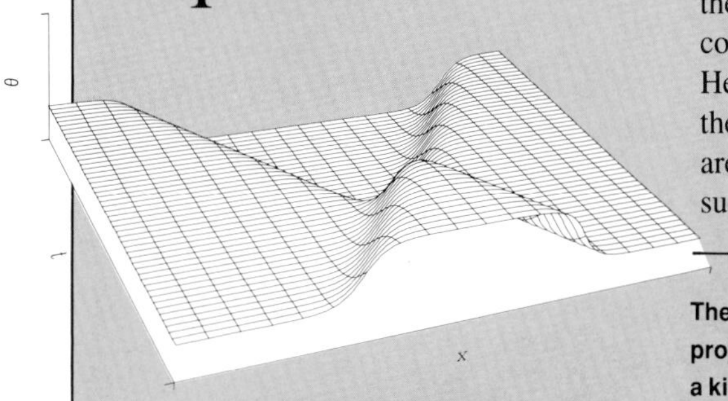

INTERACTION BETWEEN TWO SOLITONS

**The robustness of solitons is indicated in this projected space-time plot of the interaction of a kink and an antikink in the sine-Gordon equation (Eq. 5). As time develops (toward the reader), the two steplike solitary waves approach each other, interact nonlinearly, and then emerge unchanged in shape, amplitude, and velocity. The collision process is described analytically by Eq. 9. (The figure was made at the Los Alamos National Laboratory by Michel Peyrard, University of Bourgogne, France.)**

corresponding variety of real-world applications to problems in the natural sciences and engineering. In fiber optics, conducting polymers and other quasi-one-dimensional systems, Josephson transmission lines, and plasma cavitons—as well as the surface waves in the Andaman Sea!—the prevailing mathematical models are slight modifications of soliton equations. There now exist several numerical and analytic perturbation techniques for studying these "nearly" soliton equations, and one can use these to describe quite accurately the behavior of real physical systems.

One specific, decidedly practical illustration of the application of solitons concerns effective long-distance communication by means of optical fibers. Low-intensity light pulses in optical fibers propagate linearly but dispersively (as described in "Solitons in the Sine-Gordon Equation"). This dispersion tends to degrade the signal, and, as a consequence, expensive "repeaters" must be added to the fiber at regular intervals to reconstruct the pulse.

However, if the intensity of the light transmitted through the fiber is substantially increased, the propagation becomes nonlinear and solitary wave pulses are formed. In fact, these solitary waves are very well described by the solitons of the "nonlinear Schrödinger equation," another of the celebrated completely integrable nonlinear partial differential equations. In terms of the (complex) electric field amplitude $E(x,t)$, this equation can be written so that (groups of) waves with different values of $k_0$ will have different group velocities. Now consider a general solution to Eq. 2, which, by the principle of superposition, can be formed by adding together many plane waves (each with a different constant). Since the elementary components with different wave numbers will propagate at different group velocities, the general solution will change its form, or disperse, as it moves. Hence, the general solution to Eq. 2 cannot be a soliton.

Next consider adding a nonlinear term to Eq. 2. With considerable malice aforethought, we change notation for the dependent variable and choose the nonlinearity so that the full equation becomes

$$\frac{\partial^2\theta}{\partial t^2} - c_0^2\frac{\partial^2\theta}{\partial x^2} + m^2\sin\theta = 0, \quad (5)$$

the "sine-Gordon" equation. We can compare Eq. 5 to our previous Eq. 2 by noting that in the limit of small $\theta$, Eq. 5 reduces to

$$\frac{\partial^2\theta}{\partial t^2} - c_0^2\frac{\partial^2\theta}{\partial x^2} + m^2\theta - \frac{1}{6}\theta^3 + \cdots = 0, \quad (6)$$

where the remaining terms are $\mathcal{O}(\theta^5)$ and higher.

Based on remarks made in the introductory section of the main text, we see that Eq. 5 looks like a bunch of simple, plane pendulums coupled together by the spatial derivative term $\partial^2\theta/\partial x^2$. In fact, the sine-Gordon equation has many physical applications, including descriptions of chain-like magnetic compounds and transmission lines made out of arrays of Josephson junctions of superconductors. Also, the equation is one of the celebrated completely integrable, infinite-degree-of-freedom Hamiltonian systems, and the initial-value problem for the equation can be solved exactly by the analytic technique of the "inverse spectral transform." Since the details of this method are well beyond the scope of a general overview, we shall only quote the solutions relevant to our discussion. First, just as for the KdV equation (Eq. 10 in the main text), one can find directly a single solitary-wave solution:

$$\theta_k(x,t) = 4\tan^{-1} e^{\gamma(\zeta - v\tau)}, \quad (7)$$

with $\gamma = 1/\sqrt{1-v^2}$, $\zeta = mx/c_0$, and $\tau = mt$.

Since this solution approaches 0 as $x \to -\infty$ and $2\pi$ as $x \to +\infty$, it describes a glitch in the field $\theta$ localized around $\zeta = v\tau$, that is, around $x = c_0vt$. As a consequence, it is known as a "kink." Importantly, it does represent a physically truly localized excitation, because all the energy and momentum associated with this wave are exponentially centered around the kink's location. Similarly, the so-called anti-kink solution

$$\theta_{\bar{k}}(x,t) = 4\tan^{-1} e^{-\gamma(\zeta - v\tau)}, \quad (8)$$

interpolates between $2\pi$ as $x \to -\infty$ and 0 as $x \to +\infty$.

Are the kinks and anti-kinks solitons? Here we can avail ourselves of the miracle of integrability and simply write down an analytic solution that describes the scattering of a kink and an antikink. The result is

$$\theta_{k\bar{k}}(x,t) = 4\tan^{-1}\left(\frac{\sinh\frac{v\tau}{\sqrt{1-v^2}}}{v\cosh\frac{\zeta}{\sqrt{1-v^2}}}\right). \quad (9)$$

The dedicated reader can verify that as $t \to -\infty$, $\theta_{k\bar{k}}$ looks like a widely separated kink and anti-kink approaching each other at velocity $v$. For $t$ near 0 they interact nonlinearly, but as $t \to +\infty$, the kink and anti-kink emerge with their forms intact. Readers with less dedication can simply refer to the figure, in which the entire collision process is presented in a space-time plot. Note that since the equation is invariant under $\theta \to \theta + 2n\pi$, a kink that interpolates between $2\pi$ and $4\pi$ is physically equivalent to one that interpolates between 0 and $2\pi$.

In the interest of historical accuracy, we should add one final point. The analytic solution, Eq. 9, showing that the kink and anti-kink are in fact solitons, was actually known, albeit not widely, *before* the discovery of the KdV soliton. It had remained an isolated and arcane curiosity, independently rediscovered several times but without widespread impact. That such solutions could be constructed analytically in a wide range of theories was not appreciated. It took the experimental mathematics of Zabusky and Kruskal to lead to the soliton revolution. ■

$$i\frac{\partial E}{\partial t} + \frac{\partial^2 E}{\partial x^2} + |E|^2 E = 0. \quad (12)$$

The soliton corresponding to the nonlinear pulse moving with velocity $v$ through the optical fiber has the form

$$E(x,t) = \left(2\omega + \frac{v^2}{2}\right)^{\frac{1}{2}} e^{i\omega t + \frac{vx}{2}} \operatorname{sech}\left(\left(\omega + \frac{v^2}{4}\right)^{\frac{1}{2}}(x - vt)\right). \quad (13)$$

In the idealized limit of no dissipative energy loss, these solitons propagate without degradation of shape; they are indeed the natural stable, localized modes for propagation in the fiber. An intrinsically nonlinear characteristic of this soliton, shown explicitly in Eq. 13, is the relation between its amplitude (hence its energy) and its width. In real fibers, where dissipative mechanisms cause solitons to lose energy, the individual soliton pulses therefore broaden (but do not disperse). Thus, to maintain the separation between solitons necessary for the integrity of the signal, one must add optical amplifiers, based on stimulated Raman amplification, to compensate for the loss.

Theoretical numerical studies suggest that the amplification can be done very effectively. An all-optical system with amplifier spacings of 30 to 50 kilometers and

Raman pump power levels less than 100 milliwatts can use solitons of 20 picoseconds duration to send information at a bit rate of over 10 gigahertz. This is more than an order of magnitude greater than the rate anticipated for conventional (linear) systems. Although laboratory experiments have confirmed some of these results, full engineering studies have yet to be carried out. In addition, a critical and still unresolved issue is the relative cost of the repeaters supporting the linear system versus that of the amplifiers in the soliton-based approach. Nonetheless, the prospects for using optical solitons in long-distance communication are exciting and real.

**Coherent Structures.** Thus far our discussion of the coherent-structure paradigm has focused almost exclusively on solitons. Although this emphasis correctly indicates both the tremendous interest and the substantial progress to which this aspect of the paradigm has led, it obscures the much broader role that *nonsoliton* coherent structures play in nonlinear phenomena. Vortices in fluids, chemical-reaction waves and nonlinear diffusion fronts, shock waves, dislocations in metals, and bubbles and droplets can all usefully be viewed as instances of coherent structures. As in the case of the solitons, the existence of these structures results from a delicate balance of nonlinear and dispersive forces.

In contrast to solitons, however, these more general coherent structures typically interact strongly and do not necessarily maintain their form or even their separate identities for all times. Fluid vortices may merge to form a single coherent structure equivalent to a single larger vortex. Interactions among shock waves lead to diffraction patterns of incident, reflected, and transmitted shocks. Droplets and bubbles can interact through merging or splitting. Despite these nontrivial interactions, the coherent structures can be the nonlinear *modes* in which the dynamics is naturally described, and they may dominate the long-time behavior of the system. To exemplify more concretely the essential role of these general coherent structures in nonlinear systems, let me focus on two broad classes of such structures: vortices and fronts.

The importance of vortices in complicated fluid flows and turbulence has been appreciated since ancient times. The giant Red Spot (Fig. 3a) is a well-known example of a fluid vortex, as are tornados in the earth's atmosphere, large ocean circulation patterns called "modons" in the Gulf Stream current, and "rotons" in liquid helium. In terms of practical applications, the vortex pattern formed by a moving airfoil is immensely important. Not only does this pattern of vortices affect the fuel efficiency and performance of the aircraft, but it also governs the allowed spacing between planes at takeoff and landing. More generally, vortices are the coherent structures that make up the turbulent boundary layer on the surfaces of wings or other objects moving through fluids. Further, methods based on idealized point vortices provide an important approach to the numerical simulation of certain fluid flows.

The existence of fronts as coherent structures provides yet another illustration of the essential role of nonlinearity in the physical world. Linear diffusion equations cannot support wave-like solutions. In the presence of nonlinearity, however, diffusion equations can have traveling wave solutions, with the propagating wave front representing a transition from one state of the system to another. Thus, for example, chemical reaction-diffusion systems can have traveling wave fronts separating reacted and unreacted species. Often, as in flame fronts or in internal combustion engines, these traveling chemical waves are coupled with fluid modes as well. Concentration fronts arise in the leaching of minerals from ore beds. Moving fronts between infected and non-infected individuals can be identified in the epidemiology of diseases such as rabies. In advanced oil recovery processes, (unstable) fronts between the injected water and the oil trapped in the reservoir control the effectiveness of the recovery process.

Given their ubiquity and obvious importance in nonlinear phenomena, it is gratifying that recent years have witnessed remarkable progress in understanding and modeling these general coherent structures. Significantly, this progress has been achieved by precisely the synergy among computation, theory, and experiment that we have argued characterizes nonlinear science. Further, as a consequence of this progress, coherent structures and solitons have emerged as an essential paradigm of nonlinear science, providing a unifying concept and an associated methodology at the theoretical, computational, and experimental levels. The importance of this paradigm for technological applications, as well as its inherent interest for fundamental science, will guarantee its central role in all future research in this subject.

## Deterministic Chaos and Fractals

Deterministic chaos is the term applied to the aperiodic, irregular, unpredictable, random behavior that in the past two decades has been observed in an incredible variety of nonlinear systems, both mathematical and natural. Although the processes are strictly deterministic and all forces are known, the long-time behavior defies prediction and is as random as a coin toss.

That a system governed by deterministic laws can exhibit effectively random behavior runs directly counter to our normal intuition. Perhaps it is because intuition is inherently "linear;" indeed, deterministic chaos *cannot* occur in linear systems. More likely, it is because of our deeply ingrained view of a clockwork universe, a view that in the West was forcefully stated by the great French mathematician and natural philosopher Laplace. If one could know the positions and velocities of all the particles in the universe and the nature of all the forces among them, then one could chart the course of the universe for all time. In short, from exact knowledge of the initial state (and the forces) comes an exact knowledge of the final state. In Newtonian mechanics this belief is true, and to avoid any possible confusion, I stress that we are considering only dynamical systems obeying classical, Newtonian mechanics. Subsequent remarks have nothing to do with "uncertainties" caused by quantum mechanics.

However, in the real world exact knowledge of the initial state is not achievable. No matter how accurately the velocity of a particular particle is measured, one can demand that it be measured more accurately. Although we may, in general, recognize our inability to have such exact knowledge, we typically assume that if the initial conditions of two separate experiments are *almost* the same, the final conditions will be *almost* the same. For most smoothly behaved, "normal" systems this assumption is correct. But for certain nonlinear systems it is false, and deterministic chaos is the result.

At the turn of this century, Henri Poincaré, another great French mathematician and natural philosopher, understood this possibility very precisely and wrote (as translated in *Science and Method*):

> "A very small cause which escapes our notice determines a considerable effect that we cannot fail to see, and then we say that that effect is due to chance. If we knew exactly the laws of nature and the situation of the universe at the initial moment, we could predict exactly the situation of that same universe at a succeeding moment. But even if it were the case that the natural laws had no longer any secret for us, we could still only know the initial situation *approximately*. If that enabled us to predict the succeeding situation *with the same approximation*, that is all we require, and we should say that the phenomenon had been predicted, that it is governed by laws. But it is not always so; it may happen that small differences in the initial conditions produce very great ones in the final phenomena. A small error in the former will produce an enormous error in the later. Prediction becomes impossible, and we have the fortuitous phenomenon."

Despite Poincaré's remarkable insight, deterministic chaos remained virtually unexplored and unknown until the early 1960s. As the ensuing discussion will reveal, the reason for this long hiatus is that chaos defies direct analytic treatment. The seeds planted by Poincaré could only germinate when the advances in interactive computation made experimental mathematics a reality.

**The Logistic Map.** One remarkable instance of a successful experimental mathematical study occurred in a nonlinear equation simple enough to explain to an elementary school child or to analyze on a pocket calculator yet subtle enough to capture the essence of a whole class of real world phenomena. It is arguably the simplest model of a system displaying deterministic chaos, and as such has been studied by a host of distinguished researchers, including Ulam, von Neumann, Kac, Metropolis, Stein, May, and Feigenbaum (see "Iteration of Maps, Strange Attractors, and Number theory—An Ulamian Potpourri"). As we shall see, this focus of talent has been fully justified, for the simple model provides remarkable insight into a wealth of nonlinear phenomena. Thus it is a natural place to begin our quantitative study of deterministic chaos.

The model, known as the *logistic map*, is a discrete-time, dissipative, nonlinear dynamical system. The value of a variable $x_n$ at time $n$ is mapped to a new value $x_{n+1}$ at time $n + 1$ according to the nonlinear function

$$x_{n+1} = rx_n(1 - x_n), \tag{14}$$

where the *control parameter* $r$ satisfies $0 < r \leq 4$ and the allowed values—loosely speaking, the phase space—of the $x_n$ are $0 \leq x_n \leq 1$. The map is iterated as many times as desired, and one is particularly interested in the behavior as time—that is, $n$, the number of iterations—approaches infinity. Specifically, if an initial condition is picked at random in the interval $(0, 1)$ and iterated many times, what is its motion after all transients have died out?

The behavior of this nonlinear map depends critically on the control parameter and exhibits in certain regions sudden and dramatic changes in response to small variations in $r$. These changes, technically called *bifurcations*, provide a concrete example of our earlier observation that small changes in the parameters of a nonlinear system can produce enormous qualitative differences in the motion.

For $0 < r < 1$, the value of $x_n$ drops to 0 as $n$ approaches infinity no matter what its inital value. In other words, after the transients disappear, all points in the interval $(0, 1)$ are *attracted* to the *fixed point* $x^*$ at $x = 0$. This fixed point is analogous to the fixed point in Fig. 2 at ($\theta = 0$, $d\theta/dt = 0$) with the very important distinction that the fixed point in the logistic map is an *attractor*: the dissipative nature of the map causes the "volume" in phase space to collapse to a single point. Such attractors are impossible in Hamiltonian systems, since their motion preserves phase-space volumes (see "Hamiltonian Chaos and Statistical Mechanics"). The mathematical statement of this behavior then is

$$\lim_{n\to\infty} x_n = x^* = 0. \tag{15}$$

We can easily calculate the (linearized) stability of this fixed point by considering how small deviations from it behave under the map. In Eq. 14 we set $x_n = x^* + \epsilon_n$ and $x_{n+1} = x^* + \epsilon_{n+1}$ and consider only terms linear in $\epsilon_n$ and $\epsilon_{n+1}$. The resulting equation is

$$\epsilon_{n+1} = r(1 - 2x^*)\epsilon_n + \mathcal{O}(\epsilon_n^2), \tag{16}$$

so that for $x^* = 0$, the $\epsilon_n$'s will remain small for all iterations—provided $r < 1$.

This last comment suggests that something interesting happens as $r$ passes 1, and indeed for $1 < r < 3$ we find an attracting fixed point with a value that depends on $r$. This value is readily calculated, since at a fixed point $x_n = x_{n+1} = x^*$. Substituting this relation into Eq. 14, we find

$$\lim_{n\to\infty} x_n = x^*(r) = 1 - \frac{1}{r}. \tag{17}$$

Hence as the value of $r$ moves from 1 toward 3, the value of the fixed point $x^*$ moves from 0 toward $2/3$. Notice that the linear stability analysis given above shows that this $r$-dependent fixed point is stable for $1 < r < 3$. Notice also that while $x^* = 0$ is *still* a fixed point in this region, the linear stability analysis shows that it is unstable. Hence the point $x = 0$ is now analogous to the unstable fixed point in Fig. 2 at $\theta = \pi$, $d\theta/dt = 0$; the slightest perturbation will drive the solution away from $x = 0$ to the stable fixed point at $x^*(r)$.

A more interesting bifurcation occurs at $r = 3$. Suddenly, instead of approaching a single fixed point, the long-time solution oscillates between *two* values: thus the model has an *attracting limit cycle* of period 2! This limit cycle is the discrete analogue of the closed periodic oscillations shown in the phase plane of the pendulum (Fig. 2), again, of course, subject to the distinction that the logistic-map limit cycle is an attractor.

Although one can still continue analytically at this stage, it is easier to refer to the results of an experimental mathematical simulation (Fig. 5) that depicts the attracting set in the logistic map as a function of $r$. Here we see clearly the bifurcation to the period-2 limit cycle at $r = 3$. But more striking, as $r$ moves toward 3.5 and beyond, period-4 and then period-8 limit cycles occur, followed by a region in which the attracting set becomes incredibly complicated. A careful anlysis of the logistic map shows that the period-8 cycle is followed by cycles with periods 16, 32, 64, and so forth. The process continues through all values $2^n$ so that the period approaches infinity. Remarkably, all this activity occurs in the *finite* region of $r$ below the value $r_c \sim 3.57$.

Above $r_c$ the attracting set for many (but not all) values of $r$ shows no periodicity whatsoever. In fact, the set consists of a sequence of points $x_n$ that never repeats itself. For these values of $r$, the simple logistic map exhibits deterministic chaos, and the attracting set—far more complex than the fixed points and limit cycles seen below $r_c$—is called a *strange attractor*. Beyond the critical value $r_c$, the logistic map exhibits a *transition to chaos*.

Although this complicated, aperiodic behavior clearly motivates the name "chaos," does it also have the crucial feature of sensitive dependence on initial conditions that we argued was necessary for the long-time behavior to be as random as a coin toss? To study this question, one must observe how two initially nearby points separate as they are iterated by the map. Technically, this can be done by computing the *Lyapunov exponent* $\lambda$. A value of $\lambda$ greater than 0 indicates that the nearby initial points separate exponentially (at a rate determined by $\lambda$). If we plot the Lyapunov exponent as a function of the control parameter (Fig. 6), we see that the chaotic regions do have $\lambda > 0$ and, moreover, the periodic windows in Fig. 5 that exist above $r_c$ correspond to regions where $\lambda < 0$. That such a filigree of interwoven regions of periodic and chaotic motion can be produced by a simple quadratically nonlinear map is indeed remarkable.

In view of the complexity of the attracting sets above $r_c$, it is not at all surprising that this model, like the typical problem in chaotic dynamics, defied direct analytic approaches. There is, however, one elegant analytic result—made all the more relevant here by its having been discovered by Ulam and von Neumann—that further exemplifies the sensitive dependence that characterizes deterministic chaos.

For the particular value $r = 4$, if we let $x_n = \sin^2\theta_n$, the logistic map can be rewritten

$$\sin^2\theta_{n+1} = 4\sin^2\theta_n\cos^2\theta_n = (2\sin\theta_n\cos\theta_n)^2 = (\sin 2\theta_n)^2\,. \tag{18}$$

Hence the map is simply the square of the doubling formula for the sine function, and we see that the solution is $\theta_{n+1} = 2\theta_n$. In terms of the initial value, $\theta_0$, this gives

$$\theta_n = 2^n\theta_0. \tag{19}$$

This solution makes clear, first, that there is a very sensitive dependence to initial conditions and, second, that there is a very rapid exponential separation from adjacent initial conditions. For example, by writing $\theta_n$ as a binary number with a finite number of digits—as one would in any digital computer—we see that the map amounts to a simple shift operation. When this process is carried out on a real computer, round-off errors replace the right-most bit with garbage after each operation, and each time the map is iterated one bit of information is lost. If the initial condition is known to 48 bits of precision, then after only 48 iterations of the map no information about the initial condition remains. Said another way, despite the completely deterministic nature of the logistic map, the exponential separation of nearby initial conditions means that all long-time information about the motion is encoded in the initial state, whereas none (except for very short times) is encoded in the dynamics.

There is still much more that we can learn from this simple example. One question of obvious interest in nonlinear systems is the mechanism by which such systems move from regular to chaotic motion. In the logistic map, we have seen that this occurs via a *period-doubling cascade* of bifurcations: that is, by a succession of limit cycles with periods increasing as $2^n$. In a classic contribution to nonlinear science, Mitchell Feigenbaum analyzed the manner in which this cascade occurred. Among his first results was the observation that the values of the parameter $r$ at which the bifurcations occurred converged geometrically: namely, with $\delta_n$ defined by

$$\delta_n \equiv \frac{r_n - r_{n-1}}{r_{n+1} - r_n}, \tag{20}$$

he found

$$\lim_{n \to \infty} \delta_n \equiv \delta = 4.669\ldots . \tag{21}$$

More important, Feigenbaum was able to show that $\delta$ did not depend on the details of the logistic map—the function need only have a "generic" maximum, that is, one with a nonvanishing second derivative—and hence $\delta$ should be *universal* for all generic maps. Even more, he was able to argue convincingly that whenever a period-doubling cascade of bifurcations is seen in a dissipative dynamical system, the universal number $\delta$, as well as several other universal quantities, should be observed *independent of the system's phase-space dimension.*

This prediction received dramatic confirmation in an experiment carried out by Albert Libchaber and J. Maurer involving convection in liquid helium at low temperatures. Their observation of the period-doubling cascade and the subsequent extraction of $\delta$ and other universal parameters provided striking proof of universal behavior in nonlinear systems. More recently, similar confirmation has been found in experiments on nonlinear electrical circuits and semiconductor devices and in numerical simulations of the damped, driven pendulum. Further, it is now known rigorously for dissipative systems that the universal behavior of the period-doubling transition to chaos in the logistic map can occur even when the phase-space dimension becomes infinite.

It is important to emphasize that the period-doubling cascade is by no means the only way in which dissipative nonlinear systems move from regular motion to chaos (see, for example, the discussion of the indented trapezoid map on pp. 103–104). Many other routes—such as *quasiperiodic* and *intermittent*—have been identified and universality theories have been developed for some of them. But the conceptual progenitor of all these developments remains the simple logistic map.

Finally, Fig. 5b illustrates one additional obvious feature of the attracting set of Eq. 14: namely, that it contains nontrivial—and, in fact, self-similar—structure under magnification. Indeed, in the mathematical model this self-similar structure occurs on *all* smaller scales; consequently, Fig. 5b is one example of a class of complex, infinitely ramified geometrical objects called *fractals*. We shall return to this point later.

**The Damped, Driven Pendulum.** Armed with the quantitative insight gained from the logistic map, we can confront deterministic chaos in more conventional dynamical systems. We start with a very familiar example indeed: namely, the plane pendulum subjected to driving and damping. We can now make precise our earlier assertion that this simple system can behave in a seemingly random, unpredictable, chaotic manner (see "The Simple but Nonlinear Pendulum").

The motion of the damped, driven pendulum is described by Eq. 4 above. Apart from its application to the pendulum, Eq. 4 describes an electronic device called a Josephson tunneling junction in which two superconducting materials are separated

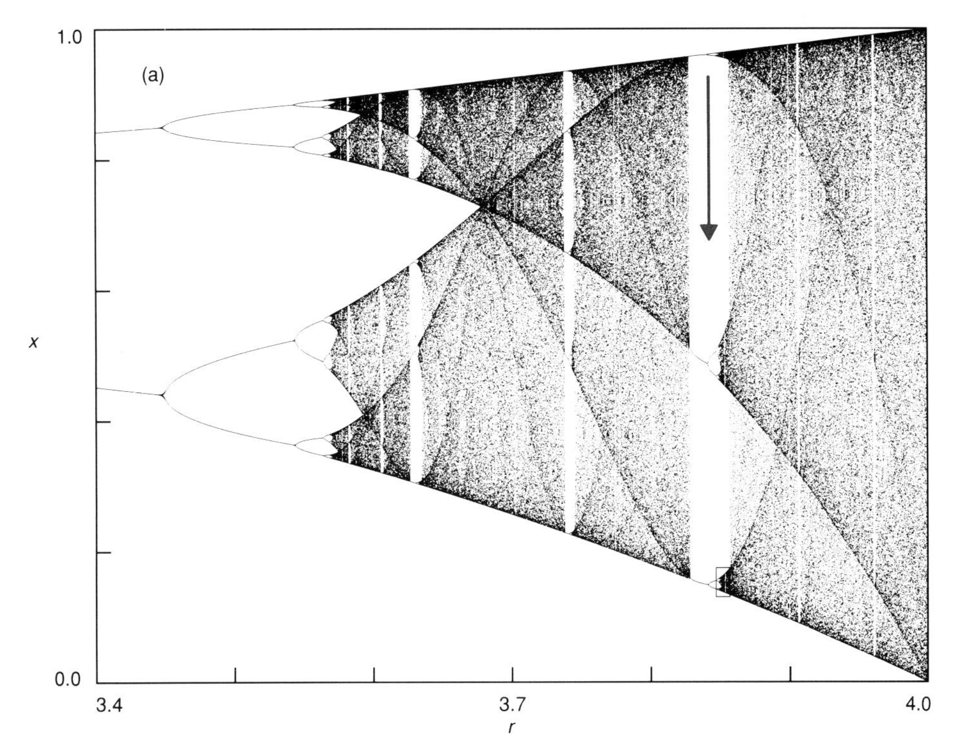


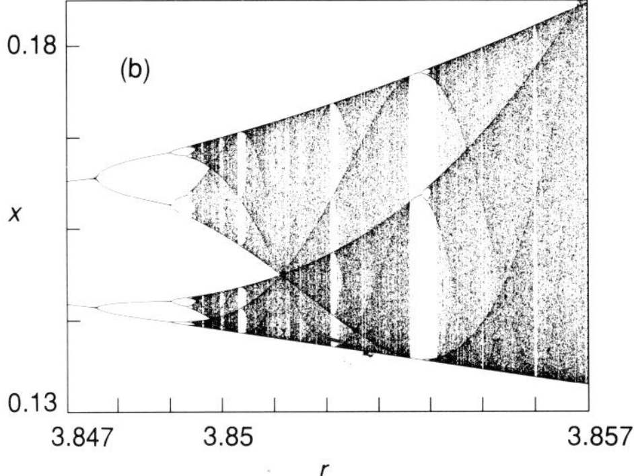


## THE LOGISTIC MAP

**Fig. 5. (a) The attracting set for the logistic map (Eq. 14 in the main text) generated by plotting 300 values of the iterated function (after the transients have died out) for each of 1150 values of the control parameter $r$. The map has a cycle of period 2 when the control parameter $r$ is at 3.4 (left edge). This cycle quickly "bifurcates" to cycles of periods 4, 8, 16, and so forth as $r$ increases, generating a period-doubling cascade. Above $r_c \approx 3.57$ the map exhibits deterministic chaos interspersed with gaps where periodic motion has returned. For example, cycles of periods 6, 5, and 3 can be seen in the three larger gaps to the right. (b) A magnified region (shown as a small rectangle in (a)) illustrates the self-similar structure that occurs at smaller scales. (Figure courtesy of Roger Eckhardt, Los Alamos National Laboratory.)**

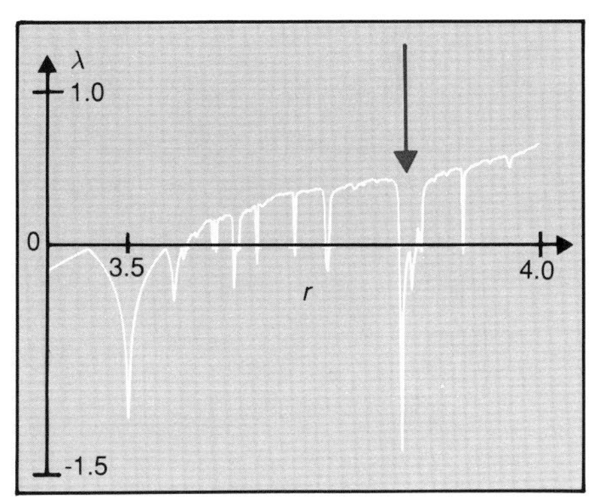


## THE LYAPUNOV EXPONENT

**Fig. 6. A positive value for the Lyapunov exponent ($\lambda > 0$) indicates that nearby initial points separate exponentially, whereas negative values ($\lambda < 0$) indicate periodic or quasiperiodic motion. Here the Lyapunov exponent is plotted as a function of the control parameter $r$ for the logistic map (Fig. 5), and it can be seen that the periodic windows of Fig. 5 correspond to regions where $\lambda < 0$. (Figure courtesy of Gottfried Mayer-Kress and Hermann Haken, Universität Stuttgart, FRG.)**

by a thin nonconducting oxide layer. Among the present practical applications of such junctions are high-precision magnetometers and voltage standards. The ability of these Josephson junctions to switch rapidly (tens of picoseconds) and with very low dissipation (less than microwatts) from one current-carrying state to another may provide microcircuit technologies for, say, supercomputers that are more efficient than those based on conventional semiconductors. Hence the nature of the dynamic response of a Josephson junction to the external driving force—the $\Gamma \cos \Omega t$ term in Eq. 4—is a matter of technological, as well as fundamental, interest.

Since analytic techniques are of limited use in the chaotic regime, we demonstrate the existence of chaos in Eq. 4 by relying on graphical results from numerical simulations. Figure 7 illustrates how the phase plane (Fig. 2) of the pendulum is modified when driving and damping forces are included and, in particular, shows how the simple structure involving fixed points and limit cycles is dramatically altered.

We note first that since there is an external time dependence in Eq. 4, the system really involves *three* first-order differential equations. In a normal dynamical system each degree of freedom results in two first-order equations, so this system is said to correspond to one-and-a-half degrees of freedom. To see this explicitly, we introduce a variable $z = \Omega t$, recall that the angular momentum $p_\theta = ml^2 d\theta/dt$, and rewrite Eq. 4, resulting in

## THE DAMPED, DRIVEN PENDULUM: A STRANGE ATTRACTOR

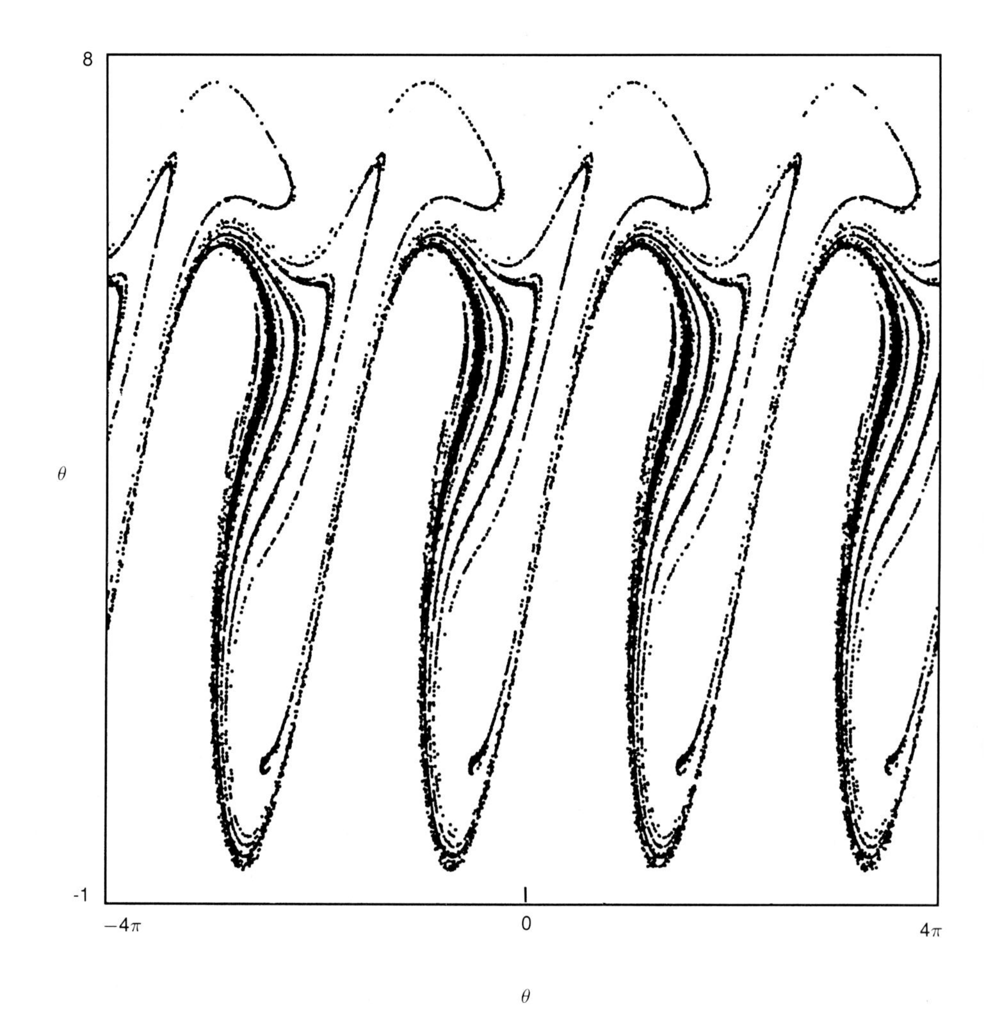


**Fig. 7. The motion of a damped, periodically driven pendulum (Eq. 4 in the main text) for certain parameter values is chaotic with the attracting set being a "strange attractor." An impression of such motion can be obtained by plotting the position $\theta$ and velocity $\dot{\theta}$ of the pendulum once every cycle of the driving force (as shown here for $\alpha$ = 0.3, $\Gamma$ = 4.5, and $\Omega$ = 0.6 in units with $g/l$ = 4). The fact that the image is repeated at higher and lower values of $\theta$ is a result of the pendulum swinging over the top of its pivot point. (Figure courtesy of James Crutchfield, University of California, Berkeley.)**

$$\begin{aligned}\frac{d\theta}{dt} &= \frac{1}{ml^2}p_\theta,\\ \frac{dp_\theta}{dt} &= -\alpha p_\theta - mgl\sin\theta + ml^2\Gamma\cos z,\\ \frac{dz}{dt} &= \Omega,\end{aligned}\tag{22}$$

which shows how the system depends on the three generalized coordinates: $\theta$, $p_\theta$, and $z$. Note further that the presence of damping implies that the system is no longer Hamiltonian but rather is dissipative and hence can have attractors.

Analysis of the damped, driven pendulum neatly illustrates two separate but related aspects of chaos: first, the existence of a strange attractor, and second, the presence of several different attracting sets and the resulting extreme sensitivity of the asymptotic motion to the precise initial conditions.

Figure 7 shows one of the attractors of Eq. 22 for the parameter values $\alpha$ = 0.3, $\Gamma$ = 4.5, and $\Omega$ = 0.6 (in units with $g/l$ = 4). As in the case of the logistic map, only the attracting set is displayed; the transients are not indicated. To obtain Fig. 7, which is a plot showing only the phase-plane variables $\theta$ and $\dot{\theta}$, one takes a "stroboscopic snapshot" of the motion once during every cycle of the driving force. The complicated attracting set shown in the figure is in fact a strange attractor and describes a never-repeating, nonperiodic motion in which the pendulum oscillates and flips over its pivot point (hence the repeated images at $2\pi$-multiples of the angle) in an irregular, chaotic manner. The sensitive dependence on initial conditions implies that nearby points on the attractor will separate exponentially in time, following totally different paths asymptotically. Enlargements of small regions of Fig. 7 show a continuation of the intricate structure on all scales; like the attracting set of the logistic map, this strange attractor is a fractal.

To visualize the motion on this attractor, it may be helpful to recall the behavior of an amusing magnetic parlor toy that has recently been quite popular. This device, for which the mathematical model is closely related to the damped, driven pendulum equation, spins first one way and then the other. At first it may seem that one can guess its behavior. But just when one expects it to spin three times to the right and then go to the left, it instead goes four, five, or perhaps six times to the right. The sequence of right and left rotations is unpredictable because the system is undergoing the aperiodic motion characteristic of a strange attractor.

Figure 8 illustrates the important point that the strange attractor of Fig. 7 is not the only attractor that exists for Eq. 22. Specifically, for $\alpha = 0.1$, $\Gamma = 7/4$, and $\Omega = 1$ (now in units of $g/l = 1$), the system is attracted to *periodic* limit cycles of clockwise or counterclockwise motion. Figure 8 demonstates this with another variant of our familiar phase-plane plot in which a color code is used to indicate the long-time behavior of all points in the plane. More precisely, this plot is a map of every initial state $(\theta, \dot{\theta})$ onto a "final state" corresponding to one of the attractors. A blue dot is plotted at a point in the plane if the solution that starts from that point at $t = 0$ is attracted asymptotically to the limit cycle corresponding to clockwise rotation of the pendulum. Similarly, a red dot is plotted for initial points for which the solution asymptotically approaches counterclockwise rotation.

In Fig. 8 we observe large regions in which all the points are colored red and, hence, whose initial conditions lead to counterclockwise rotations. Similarly, there are large blue regions leading to clockwise rotations. In between, however, are regions in which the tiniest change in initial conditions leads to alternations in the limit cycle eventually reached. In fact, if you were to magnify these regions even more, you would see further alternations of blue and red—even at the finest possible level. In other words, in these regions the final state of the pendulum—clockwise or counterclockwise motion—is an incredibly sensitive function of the exact initial point.

There is an important subtlety here that requires comment. For the red and blue regions the asymptotic state of the pendulum does *not* correspond to chaotic motion, and the two attracting sets are not strange attractors but are rather just the clockwise and counterclockwise rotations that exist as allowed motions even for the free pendulum (Fig. 2). The aspect of chaos that is reflected by the interwoven red and blue regions is the exquisite sensitivity of the final state to minute changes in the initial state. Thus, in regions speckled with intermingled red and blue dots, it is simply impossible to predict the final state because of an incomplete knowledge of initial conditions.

In addition to the dominant red and blue points and regions, Fig. 8 shows much smaller regions colored greenish white and black. These regions correspond to still other attracting limit sets, the greenish-white regions indicating oscillatory limit cycles (no rotation) and the black regions indicating points that eventually go to a strange attractor.

From the example of Fig. 8 we learn the important lesson that a nonlinear dissipative system may contain many different attractors, each with its own *basin of attraction*, or range of initial conditions asymptotically attracted to it. A subtle further consequence of deterministic chaos is that the boundaries between these basins can themselves be extraordinarily complex and, in fact, fractal. A fractal basin boundary means that qualitatively different long-time behaviors can result from nearly identical initial configurations.

**The Lorenz Attractor.** In both cases of the logistic map and the damped, driven pendulum, we have indicated that strange attractors are intimately connected with the presence of dissipative deterministic chaos. These exotic attracting sets reflect motions

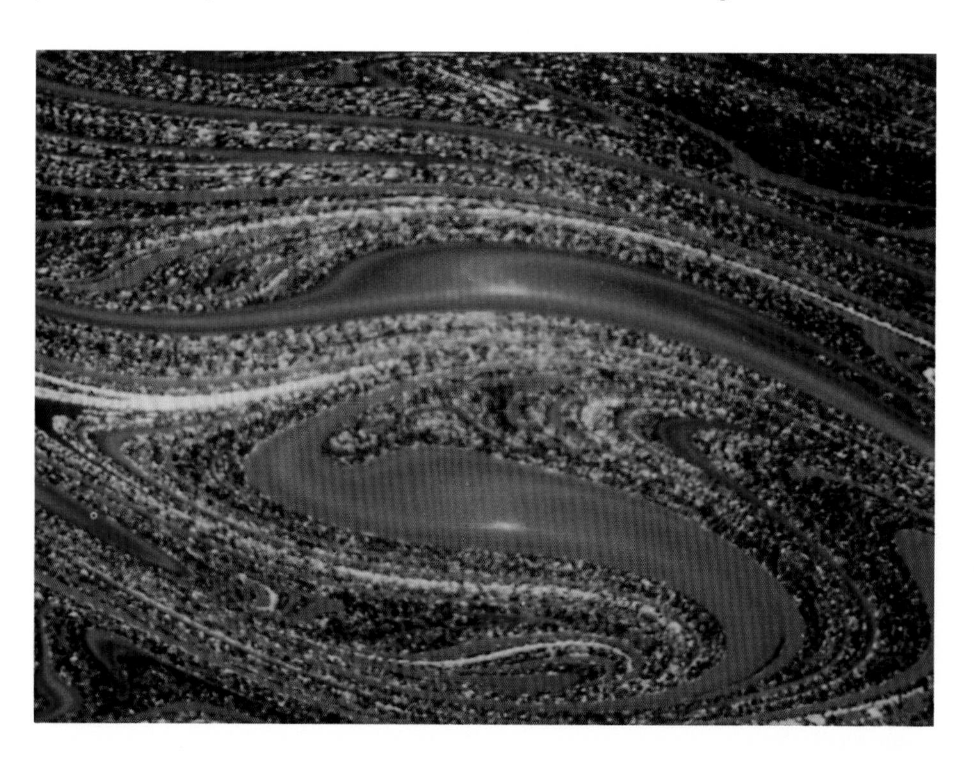

**THE DAMPED, DRIVEN PENDULUM: PERIODIC LIMIT-CYCLE ATTRACTORS**

**Fig. 8. In this variation of the phase plot for the damped, driven pendulum, a blue dot is plotted at a point in the plane if the solution that starts from that point at $t = 0$ is attracted to clockwise rotation, whereas a red dot represents an attraction to counterclockwise rotation, and a greenish-white dot represents an attraction to any oscillatory limit cycle *without* rotation. Only a portion of the phase plane is shown. The conditions used to show these limit-cycle attractors are $\alpha = 0.1$, $\Gamma = 7/4$, and $\Omega = 1$ (in units of $g/l = 1$). Despite the nonchaotic motion of the limit cycles, sensitive dependence on initial conditions is still quite evident from the presence of extensive regions of intermingled red and blue. Further, the black regions indicate initial conditions for which the limiting orbit is a strange attractor. (Figure courtesy of Celso Grebogi, Edward Ott, James Yorke, and Frank Varosi, University of Maryland.)**

of the system that, even though they may occur in a bounded region of phase space, are *not* periodic (thus never repeating), and motions originating from nearby initial points on the attractor separate exponentially in time. Further, viewed as geometric objects these attractors have an infinitely foliated form and exhibit intricate structure on all scales.

To develop a clearer understanding of these admittedly bizarre objects and the dynamical motions they depict, we turn to another simple nonlinear dynamical model. Known as the *Lorenz equations*, this model was developed in the early 1960s by Edward Lorenz, a meteorologist who was convinced that the unpredictability of weather forecasting was not due to any external noise or randomness but was in fact compatible with a completely deterministic description. In this sense, Lorenz was attempting to make precise the qualitative insight of Poincaré, who, in another prescient comment—all the more remarkable for its occurring in the paragraph immediately following our earlier quotation from *Science and Method*—observed:

> "Why have meteorologists such difficulty in predicting the weather with any certainty?... We see that great disturbances are generally produced in regions where the atmosphere is in unstable equilibrium. The meteorologists see very well that the equilibrium is unstable, that a cyclone will be formed somewhere, but exactly where they are not in a position to say; a tenth of a degree more or less at any given point, and the cyclone will burst here and not there, and extend its ravages over districts it would otherwise have spared. ... Here, again, we find the same contrast between a very trifling cause that is inappreciable to the observer, and considerable effects, that are sometimes terrible disasters."

To demonstrate this sensitive dependence, Lorenz began with a simplified model approximating fluid convection in the atmosphere. By expanding this model in (spatial) Fourier modes and by *truncating* the expansion to the three lowest modes and explicitly ignoring couplings to higher modes, Lorenz obtained a closed system of three nonlinear ordinary first-order differential equations:

$$\begin{aligned} \dot{x} &= -\sigma x + \sigma y, \\ \dot{y} &= -xz + rx - y, \text{ and} \\ \dot{z} &= xy - bz. \end{aligned} \tag{23}$$

In the application to atmospheric convection, $x$ measures the rate of convective overturning, $y$ and $z$ measure the horizontal and vertical temperature variations, respectively, $\sigma$ is the Prandtl number, $r$ is the Rayleigh number, and $b$ ($\neq 1$) reflects the fact that the horizontal and vertical temperature structures do not generally damp at the same rate.

As in the case of the damped, driven pendulum, the model describes a system with one-and-a-half degrees of freedom because it consists of three first-order equations. One set of parameters ($\sigma = 10$, $r = 28$, and $b = 8/3$) yields the celebrated *Lorenz attractor*, for which perspective views (Fig. 9) of the attracting set in the $(x, y, z)$ space reveal two "lobes" (Fig. 9a) and a thickness in the third direction (Fig. 9b) that shows the set is not planar.

Just as any initial point on a periodic orbit will eventually trace out the full orbit, so here any initial point on this strange attractor will follow a path in time that eventually traces out the full structure. Here, however, nearby initial points will diverge exponentially, reflecting the sensitive dependence on initial conditions. The two-lobed structure of the Lorenz attractor suggests a particularly useful analogy to emphasize this sensitivity. Choose two very nearby initial points and follow their evolution in time. Call each loop around the right lobe "heads" and around the left lobe "tails." Then the asymptotic sequences of heads and tails corresponding to the two points will be completely different and totally uncorrelated to each other. Of course, the nearer the initial points, the longer their motions will remain similar. But for any initial separation, there will be a finite time beyond which the motions appear totally different.

In his original study Lorenz observed this sensitive dependence in an unexpected manner, but one quite consistent with research in experimental mathematics. His own words (from p. 55 of his article in *Global Analysis*) provide a dramatic statement of the observation:

"During our computations we decided to examine one of the solutions in greater detail, and we chose some intermediate conditions which had been typed out by the computer and typed them in as new initial conditions. Upon returning to the computer an hour later, after it had simulated about two months of "weather," we found that it completely disagreed with the earlier solution. At first we expected machine trouble, which was not unusual, but we soon realized that the two solutions did not originate from identical conditions. The computations had been carried internally to about six decimal places, but the typed output contained only three, so that the new initial conditions consisted of old conditions plus small perturbations. These perturbations were amplifying quasi-exponentially, doubling in about four simulated days, so that after two months the solutions were going their separate ways."

Notice that the doubling period of the small initial perturbation corresponds directly to the binary bit shift of the logistic map at $r = 4$. Again we see the exponential loss of information about the initial state leading to totally different long-time behavior.

Let me now focus on the geometric figure that represents the strange attractor of the Lorenz equations. Figure 9 is, in fact, generated by plotting the coordinates

$x(t)$, $y(t)$, and $z(t)$ at 10,000 time steps (after transients have died out) and joining the successive points with a smooth curve. The first 5000 points are colored blue, the second 5000 green. The apparent white parts of the figure are actually blue and green lines so closely adjacent that the photographic device cannot distinguish them.

Notice how the blue and green lines interleave throughout the attractor and, in Fig. 9d, how this interleaving continues to occur on a finer scale. In fact, if the full attractor, generated by the infinite time series of points $(x(t), y(t), z(t))$, were plotted,

## LORENZ ATTRACTOR

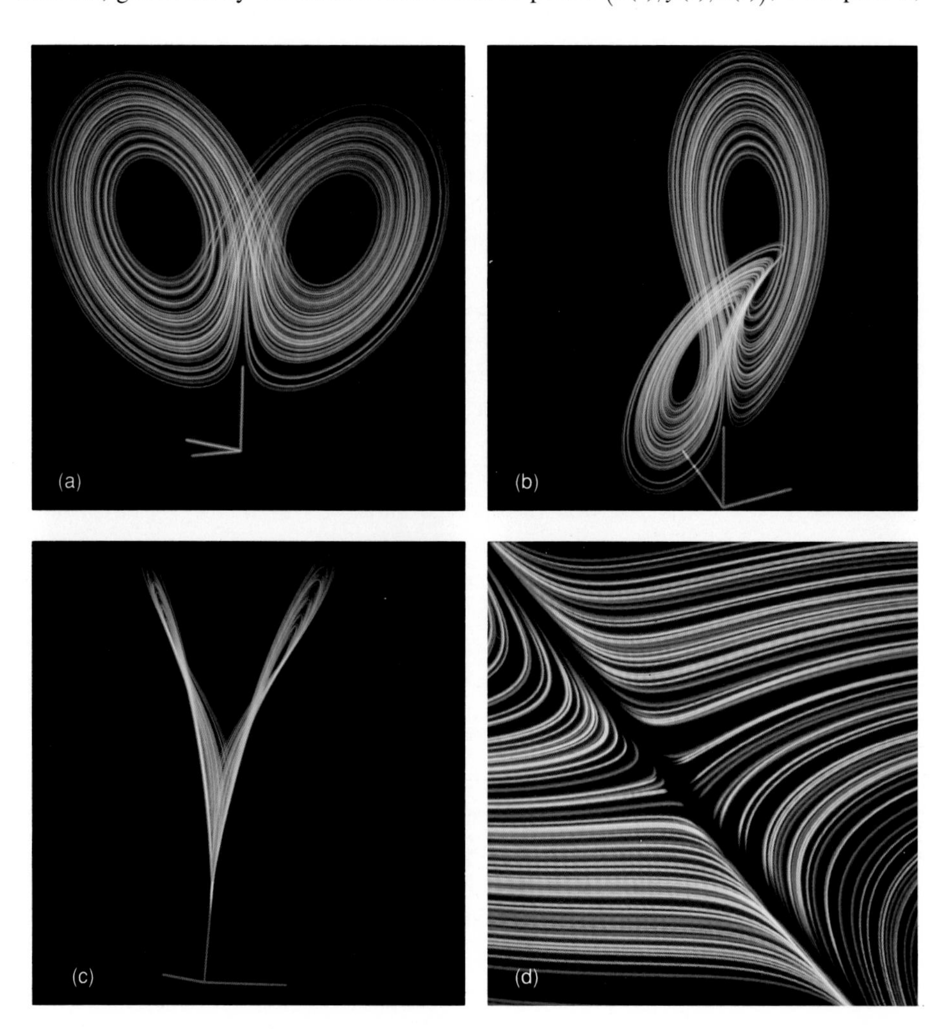


**Fig. 9. The attracting set of the Lorenz attractor (Eq. 23 in the text with $\sigma = 10$, $r = 28$, and $b = 8/3$) formed by joining 10,000 time steps of a single orbit into a smooth curve with the first 5000 points plotted in blue and the second 5000 plotted in green. (a)–(c) These perspective views reveal the two-lobed, nonplanar shape and the thickness of the attractor. The red lines indicate the direction of the coordinate axes. (d) A closeup of the interleaving of the Lorenz orbit, which, even for an infinite time series of points, would never intersect and repeat itself. The attractor has a fractal dimension of 2.04, that is, between that of an area and a volume. (Figure courtesy of Gottfried Mayer-Kress, Los Alamos National Laboratory.)**

we would see the trajectory looping around forever, never intersecting itself and hence never repeating. The exquisite filamentary structure would exist on all scales, and, even in the infinite time limit, the attractor would not form a solid volume in the $(x, y, z)$ space. In a sense that we shall make precise shortly, the attractor is a fractal object with dimension between that of an area (dimension = 2) and that of a volume (dimension = 3). Indeed, the Lorenz attractor has a *fractal dimension* of about 2.04.

**Fractals.** The term fractal was coined by Benoit Mandelbrot in 1975 to describe irregular, fragmented shapes with intricate structure on all scales. Fractals moved into the mainstream of scientific research when it became clear that these seemingly exotic geometric objects, which had previously been viewed as "a gallery of monsters," were emerging commonly in many natural contexts and, in particular, as the attracting sets of chaotic dynamical systems. In fact, Mandelbrot traced many of the core concepts related to fractals back to a number of distinguished late 19th and early 20th century mathematicians, including Cantor, Hausdorff, and Julia. But, as in the case of deterministic chaos, the flowering of these concepts came only after experimental mathematics made precise visualization of the monsters possible.

The essential feature of fractals is the existence of similar, nontrivial structure on all scales, so that small details are reminiscent of the entire object. Technically, this property is known as *scaling* and leads to a theoretical approach that allows construction of fine details of the object from crude general features. The structure need not be exactly self-similar on all scales. Indeed, much current research focuses on *self-affine* fractals, in which the structures on different scales are related by linear transformations.

One consequence of this scale invariance is that fractal objects in general have *fractional* rather than integral dimension: that is, rather than being lines, areas, or volumes, fractals lie "somewhere in between." To understand this quantitatively, we recall the example of the recursively defined Cantor set (Fig. 10). At the zeroth level, the set consists of a continuous line segment from 0 to 1. At the first level, the middle third of the segment is eliminated. At the second level, the middle third of each of the two remaining continuous segments are eliminated. At the third level, the middle third of each of these four segments is eliminated, and so forth *ad infinitum.* At each level the Cantor set becomes progressively less dense and more tenuous, so that the end product is indeed something between a point and a line. It is easy to see in Fig. 10 that at the $n$th level, the Cantor set consists of $2^n$ segments, each of length $(1/3)^n$. Thus, the "length" $l$ of the set as $n$ goes to infinity would be

$$\lim_{n\to\infty} l = \lim_{n\to\infty} 2^n \left(\frac{1}{3}\right)^n = 0. \tag{24}$$

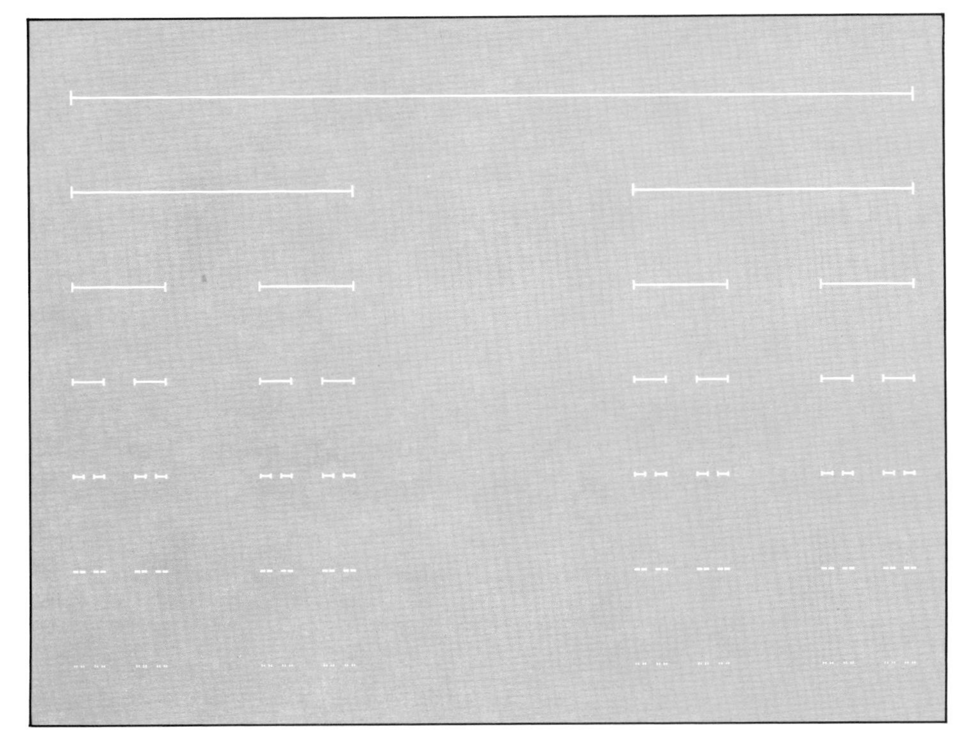

## CANTOR SET

**Fig. 10. The Cantor set is formed by starting with a line segment of unit length, removing its middle third, and, at each successive level, removing the middle third of the remaining segments. Although, the length of the remaining segments goes to zero as the number of iterations, or levels, goes to infinity, the set has a fractal dimension greater than zero, namely $\ln 2/\ln 3 \approx 0.6309$. (Figure courtesy of Roger Eckhardt, Los Alamos National Laboratory.)**

In the 1920s the mathematician Hausdorff developed a theory that can be used to study the fractional dimension of fractals such as the Cantor set. In the present simple case, this theory can be paraphrased by asking how many small intervals, $N(\epsilon)$, are required to "cover" the set at a length scale $\epsilon$. As $\epsilon \to 0$, the fractal dimension $d_f$ is defined by

$$d_f \equiv \lim_{\epsilon\to 0} \left(\frac{\ln N(\epsilon)}{\ln(\frac{1}{\epsilon})}\right). \tag{25}$$

Inverting Eq. 25, we see that

$$\lim_{\epsilon\to 0} N(\epsilon) = \left(\frac{1}{\epsilon}\right)^{d_f}. \tag{26}$$

*continued on page 246*

# HAMILTONIAN CHAOS *and* STATISTICAL MECHANICS

The specific examples of chaotic systems discussed in the main text—the logistic map, the damped, driven pendulum, and the Lorenz equations—are all dissipative. It is important to recognize that nondissipative Hamiltonian systems can also exhibit chaos; indeed, Poincaré made his prescient statement concerning sensitive dependence on initial conditions in the context of the few-body Hamiltonian problems he was studying. Here we examine briefly the many subtleties of Hamiltonian chaos and, as an illustration of its importance, discuss how it is closely tied to long-standing problems in the foundations of statistical mechanics.

We choose to introduce Hamiltonian chaos in one of its simplest incarnations, a two-dimensional discrete model called the *standard map*. Since this map preserves phase-space volume (actually *area* because there are only two dimensions) it indeed corresponds to a discrete version of a Hamiltonian system. Like the discrete logistic map for dissipative systems, this map represents an archetype for Hamiltonian chaos.

The equations defining the standard map are

$$p_{n+1} = p_n - \frac{k}{2\pi} \sin 2\pi q_n, \qquad q_{n+1} = p_{n+1} + q_n, \tag{1}$$

where, as the notation suggests, $p_n$ is the discrete analogue of the momentum, $q_n$ is the analogue of the coordinate, and the discrete index $n$ plays the role of time. Only the fractional parts of $p_n$ and $q_n$ are kept; hence the motion is on a torus, periodic in both $p$ and $q$. For any value of $k$, the map preserves the area in the $(p,q)$ plane, since the Jacobian $\partial(p_{n+1}, q_{n+1})/\partial(p_n, q_n) = 1$.

The preservation of phase-space volume for Hamiltonian systems has the very important consequence that there can be no *attractors*, that is, no subregions of lower phase-space dimension to which the motion is confined asymptotically. Any initial point $(p_0, q_0)$ will lie on some particular orbit, and the *image* of all possible initial points—that is, the unit square itself—is again the unit square. In contrast, *dissipative* systems have phase-space volumes that shrink. For example, the logistic map (Fig. 5 in the main text) at $\lambda = 3.1$ has all initial points in the interval $(0, 1)$ attracted to just *two* points.

Clearly, for $k = 0$ the standard map is trivially integrable, with $p_n = p_0$ being constant and $q_n$ increasing linearly in time ($n$) as it should for free motion. The orbits are thus just straight lines wrapping around the torus in the $q$ direction. For $k = 1.1$ the map produces the orbits shown in Figs. 1a–d. The most immediately striking feature of this set of figures is the existence of nontrivial structure on all scales. Thus, like dissipative systems, Hamiltonian chaos generates strange fractal sets (albeit "fat" fractals, as discussed below). On all scales one observes "islands," analogues in this discrete case of the periodic orbits in the phase plane of the simple pendulum (Fig. 2 in the main text). In addition, however, and again on all scales, there are swarms of dots coming from individual chaotic orbits that undergo nonperiodic motion and eventually fill a finite region in phase space. In these chaotic regions the motion is "sensitively dependent on initial conditions."

Figure 2 shows, in the full phase space, a plot of a *single* chaotic orbit followed through 100 million iterations (again, for $k = 1.1$). This object differs from the strange sets seen in dissipative systems in that it occupies a *finite* fraction of the full phase space: specifically, the orbit shown takes up 56 per cent of the unit area that represents the full phase space of the map. Hence the "dimension" of the orbit is the same as that of the full phase space, and calculating the fractal dimension by the standard method gives $d_f = 2$. However, the orbit differs from a conventional area in that it contains holes on all scales. As a consequence, the measured value of the area occupied by the orbit depends on the resolution with which this area is measured—for example, the size of the boxes in the box-counting method—and the approach to the finite value at infinitely fine resolution has definite scaling properties. This set is thus appropriately called a "fat fractal." For our later discussion it is important to note that the holes—representing periodic, nonchaotic motion—also occupy a finite fraction of the phase space.

To develop a more intuitive feel for fat fractals, note that a very simple example can be constructed by using a slight modification of the Cantor-set technique

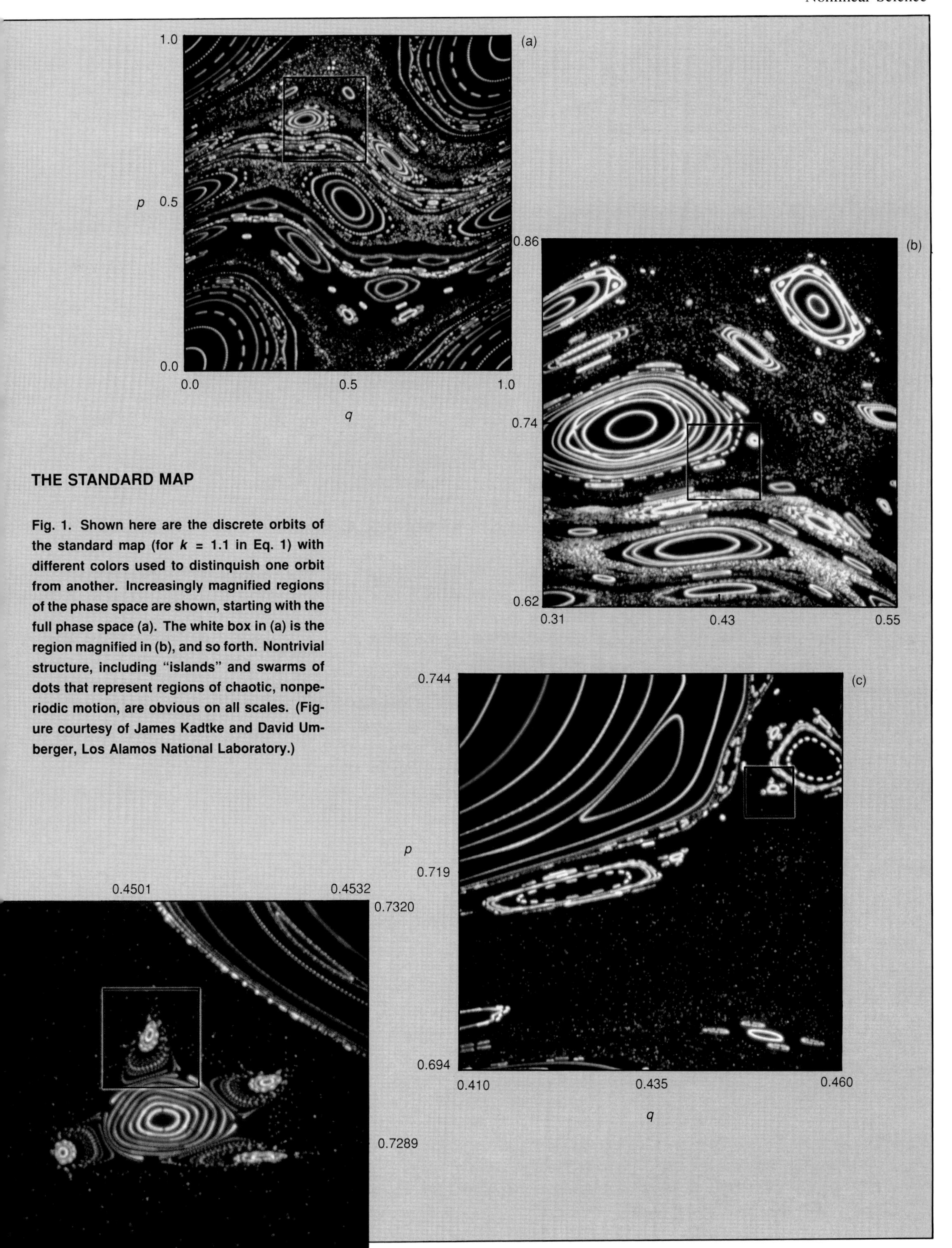


## THE STANDARD MAP

**Fig. 1. Shown here are the discrete orbits of the standard map (for $k$ = 1.1 in Eq. 1) with different colors used to distinquish one orbit from another. Increasingly magnified regions of the phase space are shown, starting with the full phase space (a). The white box in (a) is the region magnified in (b), and so forth. Nontrivial structure, including "islands" and swarms of dots that represent regions of chaotic, nonperiodic motion, are obvious on all scales. (Figure courtesy of James Kadtke and David Umberger, Los Alamos National Laboratory.)**

described in the main text. Instead of deleting the middle one-third of each interval at every scale, one deletes the middle $(1/3)^n$ at level $n$. Although the resulting set is topologically the same as the original Cantor set, a calculation of its dimension yields $d_f = 1$; it has the same dimension as the full unit interval. Further, this fat Cantor set occupies a finite fraction—amusingly but accidentally also about 56 per cent—of the unit interval, with the remainder occupied by the "holes" in the set.

To what extent does chaos exist in the more conventional Hamiltonian systems described by differential equations? A full answer to this question would require a highly technical summary of more than eight decades of investigations by mathematical physicists. Thus we will have to be content with a superficial overview that captures, at best, the flavor of these investigations.

To begin, we note that completely integrable systems can *never* exhibit chaos, independent of the number of degrees of freedom $N$. In these systems all bounded motions are quasiperiodic and occur on hypertori, with the $N$ frequencies (possibly all distinct) determined by the values of the conservation laws. Thus there cannot be any aperiodic motion. Further, since all Hamiltonian systems with $N = 1$ are completely integrable, chaos cannot occur for one-degree-of-freedom problems.

For $N = 2$, *non-integrable* systems can exhibit chaos; however, it is not trivial to determine in which systems chaos can occur; that is, it is in general not obvious whether a given system is integrable or not. Consider, for example, two very similar $N = 2$ nonlinear Hamiltonian systems with equation of motion given by:

$$\frac{d^2x}{dt^2} = -x - 2xy, \qquad \frac{d^2y}{dt^2} = -y + y^2 - x^2, \tag{2}$$

and

$$\frac{d^2x}{dt^2} = -x - 2xy, \qquad \frac{d^2y}{dt^2} = -y - y^2 - x^2, \tag{3}$$

Equation 2 describes the famous Henon-Heiles system, which is non-integrable and has become a classic example of a simple (astro-) physically relevant Hamiltonian system exhibiting chaos. On the other hand, Eq. 3 can be separated into two independent $N = 1$ systems (by a change of variables to $\zeta = x - y$ and $\eta = x + y$) and hence is completely integrable.

Although there exist explicit calculational methods for testing for integrability, these are highly technical and generally difficult to apply for large $N$. Fortunately, two theorems provide general guidance. First, Siegel's Theorem considers the space of Hamiltonians analytic in their variables: non-integrable Hamiltonians are *dense* in this space, whereas integrable Hamiltonians are not. Second, Nekhoroshev's Theorem leads to the fact that all non-integrable systems have a phase space that contains chaotic regions.

Our observations concerning the standard map immediately suggest an essential question: What is the extent of the chaotic regions and can they, under some circumstances, cover the whole phase space? The best way to answer this question is to search for *non*chaotic regions. Consider, for example, a completely integrable $N$-degree-of-freedom Hamiltonian system disturbed by a generic *non-integrable perturbation.* The famous KAM (for Kolmogorov, Arnold, and Moser) theorem shows that, for this case, there are regions of finite measure in phase space that retain the smoothness associated with motion on the hypertori of the integrable system. These regions are the analogues of the "holes" in the standard map. Hence, for a typical Hamiltonian system with $N$ degrees of freedom, the chaotic regions do *not* fill all of phase space: a finite fraction is occupied by "invariant KAM tori."

At a conceptual level, then, the KAM theorem explains the nonchaotic behavior and recurrences that so puzzled Fermi, Pasta, and Ulam (see "The Fermi, Pasta, and Ulam Problem: Excerpts from 'Studies of Nonlinear Problems'"). Although the FPU chain had many (64) nonlinearly coupled degrees of freedom, it was close enough (for the parameter ranges studied) to an integrable system that the invariant KAM tori and resulting pseudo-integrable properties dominated the behavior over the times of measurement.

There is yet another level of subtlety to chaos in Hamiltonian systems: namely, the structure of the phase space. For non-integrable systems, within every regular KAM region there are chaotic regions. Within these chaotic regions there are, in turn, regular regions, and so forth. For all non-integrable systems with $N > 3$, an orbit can move (albeit on very long time scales) among the various chaotic regions via a process known as "Arnold diffusion." Thus, in general, phase space is permeated by an Arnold web that links together the chaotic regions on all scales.

Intuitively, these observations concerning Hamiltonian chaos hint strongly at a connection to statistical mechanics. As Fig. 1 illustrates, the chaotic orbits in Hamiltonian systems form very complicated "Cantor dusts," which are nonperiodic, never-repeating motions that wander through volumes of the phase space, apparently constrained only by conservation of total energy. In addition, in these regions the sensitive dependence implies a rapid loss of information about the initial conditions and hence an effective irreversibility of the motion. Clearly, such wandering motion and effective irreversibility suggest a possible approach to the following fundamental question of statistical mechanics: How can one derive the irreversible, ergodic, thermal-equilibrium motion assumed in statistical mechanics from a reversible, Hamiltonian microscopic dynamics?

Historically, the fundamental assumption that has linked dynamics and statistical mechanics is the *ergodic hypothesis*, which asserts that time averages over actual dynamical motions are equal to ensemble averages over many different but equivalent systems. Loosely speaking, this hypothesis assumes that all regions of phase space allowed by energy conservation are equally accessed by almost all dynamical motions.

What evidence do we have that the ergodic hypothesis actually holds for realistic Hamiltonian systems? For systems with finite degrees of freedom, the KAM theorem shows that, in addition to chaotic regions of phase space, there are nonchaotic regions of finite measure. These invariant tori imply that ergodicity does *not* hold for most finite-dimensional Hamiltonian sytems. Importantly, the few Hamiltonian systems for which the KAM theorem does not apply, and for which one can prove ergodicity and the approach to thermal equilibrium, involve "hard spheres" and consequently contain non-analytic interactions that are not realistic from a physicist's perspective.

For many years, most researchers believed that these subtleties become irrelevant in the thermodynamic limit, that is, the limit in which the number of degrees of freedom ($N$) and the energy ($E$) go to infinity in such a way that $E/N$ remains a nonzero constant. For instance, the KAM regions of invariant tori may approach zero measure in this limit. However, recent evidence suggests that nontrivial counterexamples to this belief may exist. Given the increasing sophistication of our analytic understanding of Hamiltonian chaos and the growing ability to simulate systems with large $N$ numerically, the time seems ripe for quantitative investigations that can establish (or disprove!) this belief. (For additional discussion of this topic, see "The Ergodic Hypothesis: A Complicated Problem of Mathematics and Physics.")

Among the specific issues that should be addressed in a variety of physically realistic models are the following.

- How does the measure of phase space occupied by KAM tori depend on $N$? Is there a class of models with realistic interactions for which this measure goes to 0? Are there non-integrable models for which a finite measure is retained by the KAM regions? If so, what are the characteristics that cause this behavior?
- How does the rate of Arnold diffusion depend on $N$ in a broad class of models? What is the structure of important features—such as the Arnold web—in the phase space as $N$ approaches infinity?
- If there is an approach to equilibrium, how does the time-scale for this approach depend on $N$? Is it less than the age of the universe?
- Is ergodicity necessary (or merely sufficient) for most of the features we associate with statistical mechanics? Can a less stringent requirement, consistent with the behaviour observed in analytic Hamiltonian systems, be formulated?

Clearly, these are some of the most challenging, and profound, questions currently confronting nonlinear scientists. ■

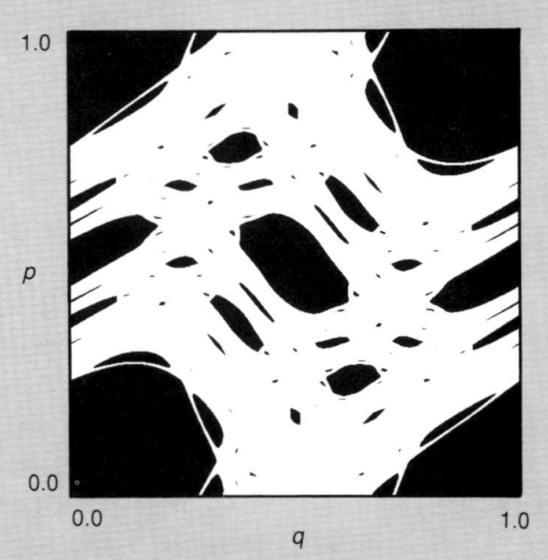


**A "FAT" FRACTAL**

**Fig. 2. A single chaotic orbit of the standard map for $k = 1.1$. The picture was made by dividing the energy surface into a 512 by 512 grid and iterating the initial condition $10^8$ times. The squares visited by this orbit are shown in black. Gaps in the phase space represent portions of the energy surface unavailable to the chaotic orbit because of various quasiperiodic orbits confined to tori, as seen in Fig. 1. (Figure courtesy of J. Doyne Farmer and David Umberger, Los Alamos National Laboratory.)**

For the Cantor set, if we look at the $n$th level and use the small interval of length $\epsilon = (1/3)^n$, we know that $N(\epsilon) = 2^n$. Since as $n \to \infty$, $\epsilon \to 0$, we can use

$$d_f = \lim_{\epsilon \to 0} \left( \frac{\ln N(\epsilon)}{\ln \frac{1}{2}} \right) = \lim_{n \to \infty} \frac{\ln 2^n}{\ln 3^n} = \frac{\ln 2}{\ln 3}. \tag{27}$$

*continued from page 241*

The simple Cantor set has, in effect, only a single scale because the factor of 1/3 is always used in constructing successive levels of the set. In contrast, fractals that arise in chaotic dynamical systems have a range of scales and, typically, different scalings apply to different parts of the set; as a consequence, these more complex sets are sometimes termed multifractals. In his original work on the logistic map, Feigenbaum defined and discussed a *scaling function* that characterized these differences. Recently, a related analytic technique—called the $f(\alpha)$ approach—has been used to provide a detailed understanding of the many different scalings occurring in a variety of chaotic dynamical systems.

Although these constructions and techniques may seem to be just mathematical manipulations, nature abounds with structures that repeat themselves on many different scales and hence have approximate fractal structure. Familiar examples include clouds, lightning bolts, ferns, and, as shown in Fig. 11, snowflakes. Less familiar but technologically significant examples include the growth of dendritic crystals, dielectric breakdown in gas-filled cells, and "viscous fingering" in certain two-fluid flows.

A laboratory experiment illustrating this last phenomenon (Fig. 12) consists of a flat, effectively two-dimensional, cylindrical cell containing a high-viscosity fluid. An inlet in the center of the cell permits the injection under pressure of a second, less viscous fluid (in this case, water). Instead of smoothly and uniformly replacing the viscous fluid in the cell, the water splits into the highly branched, coral-like fractal object shown in Fig. 12. Using a box-counting technique similar to that used to measure the dimension of the Cantor set, one finds that the fractal dimension of the viscous finger is $1.70 \pm 0.05$. Hence, although it is composed of many thin but highly branched segments, the viscous finger possesses a fractional dimension closer to that of an *area* ($d = 2$) than that of a line ($d = 1$).

To understand the processes that create such structures, one can use experimental mathematics to study specific physical models. One such study (Fig. 13) depicts the development of a fractal pattern on a triangular lattice. The model underlying the pattern depends primarily on the local pressure gradients driving the "fluid," but it also incorporates the effects of fluctuations (via a noise parameter) and of anisotropy. The study shows clearly that all the patterns grow primarily at the tips; almost no growth occurs in the "fjord" regions.

Figures 13a and 13b are examples of the fractal structures found when the noise parameter is held constant but the anisotropy $k$ is considerably decreased. Notice the striking qualitative similarity between Figs. 11 and 13a. Interestingly, the fractal dimension of both Figs. 13a and 13b is about 1.5; it is *independent* of $k$. In Figs. 13c and 13d the anisotropy is held fixed but the noise is decreased. Here the fractal dimension of both is about 1.7.

Figure 13 makes clear that $d_f$ alone is not sufficient to characterize a fractal, for although both Figs. 13a and 13b have $d_f = 1.5$, there are obvious visual differences. Mandelbrot has defined a number of higher order geometric properties—for example, *lacunarity*, a measure of the typical size of the holes in the fractal—that can be used to characterize fractals more precisely. Lacunarity and other higher-order features are, in effect, geometric restatements of our earlier remarks that multifractals generated by chaotic dynamical systems have a range of scalings and that $f(\alpha)$ and related analytic techniques can be used to study these scalings. A generally unsolved challenge in this area is the *fractal inverse problem*: given $f(\alpha)$ or related quantities, to what extent can one reconstruct the actual fractal set, including perhaps the order in which the points of the set are generated dynamically?

**Practicalities.** The impacts of deterministic chaos and fractals are only now beginning to be felt throughout science. The concepts that even simple systems can exhibit incredibly complicated behavior, that simple rules can lead to enormously intricate geometric objects, and that this behavior and these objects can be quantified are now all widely appreciated and are being applied in many fields.

The fractal viscous-fingering phenomenon (Fig. 12) is of enormous technological interest, for it represents a major barrier to the development of efficient advanced oil-recovery techniques. Nearly half the oil deposited in limestone or other porous media is typically unrecovered because it remains stuck in the pores. To force out this oil, water is injected into a second well nearby. Viscous fingering limits the effectiveness of this technique, because when one of the thin fingers of water breaks through from the injector to the recovery well, only injected water rather than oil is thereafter recovered. Clearly a full understanding of this fractal phenomenon and ways to control it are of considerable economic importance.

Similarly, a direct application of fractals occurs in the design of the toughened ceramics used as engine parts. These special ceramics are designed to tolerate flaws, such as voids and cracks, without breaking into pieces. The flaws arise primarily from voids that develop during the sintering process and fractures that arise chiefly from the use of hard materials when machining the ceramics. By adding secondary constituents to the ceramics, propagating cracks can be forced to move through the ceramic along tortuous, convoluted routes, causing more energy to be expended than if the route were smooth and regular. Hence, for a given impulse, an irregular crack does not propagate as far through the ceramic and does less overall damage. Convoluted routes should lead to cracks in the form of complex fractal patterns. Indeed, microscopic studies of high performance ceramics have revealed such patterns and established that the higher the fractal dimension of the cracks, the tougher the ceramics.

The results of deterministic chaos are also being applied across a broad range of disciplines. Experimentally, high-precision measurements of chaotic dynamics in many types of fluid flows, current and voltage responses of semiconductors and other solid-state electronic devices, and cardiac arhythymias have established the importance of dissipative chaos in fluid dynamics, condensed-matter physics, and medicine. Indeed, recent medical experiments have suggested that many physiological parameters vary chaotically in the *healthy* individual and that greater regularity can indicate a pathological condition; for example, normally chaotic oscillations of the densities of red and white blood cells become periodic in some anemias and leukemias. Hamiltonian chaos finds a direct application in accelerator design, where the potential loss of an apparently stable beam due to subtle long-time phenomena such as "Arnold diffusion" (see "Hamiltonian Chaos and Statistical Mechanics") is a vital issue of technology.

The central theoretical challenge in "applied chaos" is to develop deterministic chaotic models to explain these diverse phenomena. Rather than focusing on the details of specific applications, let me describe two broader problem areas of current research.

First, although we have stressed the randomness and unpredictability of the *long-time* behavior of chaotic systems, it nonetheless remains true that these systems are deterministic, following laws that involve *no* external randomness or uncertainty. Thus, it is possible to predict the behavior for *short* times, if the equations of motion are known. The analytic solution of the logistic map for $r = 4$ is a clear illustration; given two initial conditions known to, say, 10-bit accuracy, one can predict the relative positions—albeit with exponentially decreasing accuracy—for 10 iterations of the map. The subtler problem, currently under intense investigation, occurs when one observes that a system is deterministically chaotic but does not know the form of the underlying equations: can one nonetheless use the basic determinism to make *some* prediction? In view of the clear value of such predictive techniques—consider the stock market—

**SNOWFLAKES**

**Fig. 11. The snowflake is an example of a fractal structure in nature. (Photos reprinted from *Snow Crystals* by W. A. Bentley and W. J. Humpreys with permission of Dover Publications.)**

**VISCOUS FINGERING**

**Fig. 12. A fractal structure formed by injecting water under pressure into a high-viscosity fluid. The fractal dimension of this object has been calculated to be $d_f = 1.70 \pm 0.05$. (Figure courtesy of Gérard Daccord and Johann Nittmann, Etudes et Fabrication Dowell Schlumberger, France, and H. Eugene Stanley, Boston University.)**

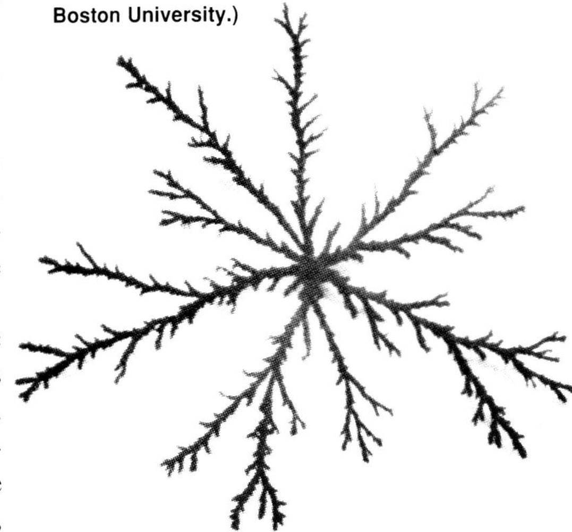

substantial efforts are being focused on this question.

Second, and at a still broader level, are the related issues of universality and mode reduction in chaos, both of which we mentioned previously. Universality implies that for certain chaotic phenomena—such as the period-doubling cascade—the details of the system and the equations describing it are *irrelevant*: the observed complex behavior develops in a similar manner in every context, be it fluid dynamics, condensed-matter physics, or biology. Indeed, the term universality is borrowed from the statistical me-

### A FRACTAL SIMULATION

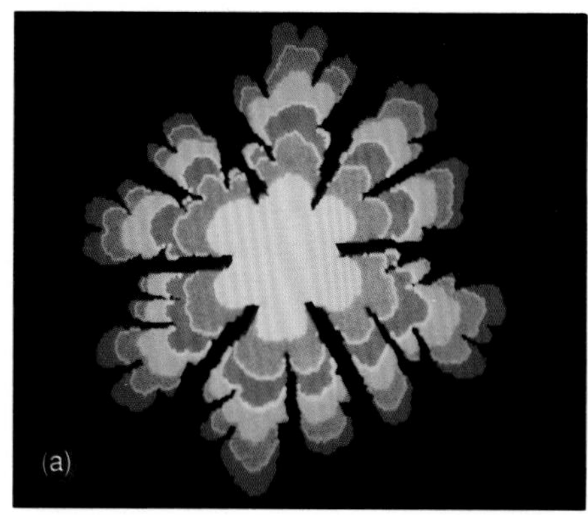


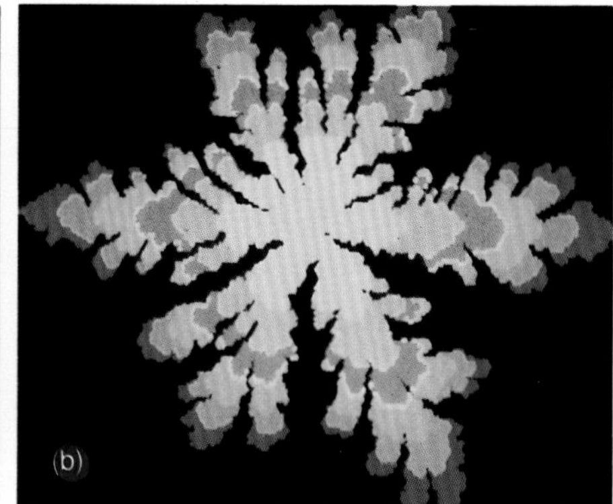


**Fig. 13. The model used in the simulation to form these fractal patterns uses local pressure gradients to "drive the fluid" across a triangular lattice. The growth patterns of the objects are indicated by the color coding; the first one-sixth of the sites to be occupied are white, the next one-sixth blue, then magenta, yellow, green, and finally red. The model also incorporates the effects of fluctuations via the noise parameter $\sigma$ and of anisotropy via the parameter $k$. The patterns in (a) and (b) have the same noise ($\sigma$ = 0.02) but different anisotropy ($k$ = 11 in (a) and $k$ = 1.3 in (b)). The patterns in (c) and (d) have the same isotropy ($k$ = 1) but the noise changes (from $\sigma$ = 0.5 in (c) to $\sigma$ = 0.005 in (d)). (Photos courtesy of Johann Nittmann, Etudes et Fabrication Dowell Schlumberger, France, and H. Eugene Stanley, Boston University.)**

chanics of phase transitions, where it has been shown that the details of the microscopic interactions are irrelevant for most of the important properties of the transitions. In the context of chaos, universality also lends tremendous power to analyses of certain phenomena; in essence, the simplest example—for instance, the logistic map for period doubling—contains the critical features of the entire effect.

The central idea of mode reduction can most easily be visualized in fluid flows. In any given fluid motion not all the (infinitely!) many possible modes are "active," so the *effective* phase-space dimension is much smaller than the full dimension of the equations. The case of laminar flow in which fluid moves *en bloc* is a trivial illustration. A more interesting and much less obvious example is observed in experiments on Couette-Taylor flows, in which fluid is contained between two concentric rotating cylinders. As the speed of relative rotation is increased, the flow forms bands of Taylor vortices. Further increases in the relative rotation causes the bands to develop "wobbling" instabilites and finally to be replaced by fully developed turbulence. In these experiments clever techniques (sometimes referred to as "geometry from a time series" and related to rigorous mathematical embedding theorems) have been used to extract phase-space information directly from a time series of measurements on a single dynamical variable. Such techniques have revealed strange attractors with effective phase-space dimensions on the order of five. In such experiments there are, in one sense, on the order of only five active modes. Mode reduction reduces the number of degrees of freedom being modeled to the minimum necessary to capture the essence of the motion.

Several important aspects of the general problem of mode reduction should be clarified. First, techniques such as "geometry from a time series" offer no immediate information about the nature of the reduced modes nor about the effective equations governing their interactions. In general, obtaining such information remains an important open problem.

Second, mode reduction is distinct from mode *truncation*. Specifically, we noted that the Lorenz equations were obtained by simply truncating the Fourier expansion of the full equations, hence ignoring certain demonstrably nonzero couplings. Ideally, the process of mode reduction should be *deductive, controlled, and constuctive*; that is, one should be able to derive the equations governing the reduced modes, to bound the error made in the reduction, and to "construct" the actual modes themselves. This, too, remains an elusive goal, despite substantial recent progress.

Third, if one is able to obtain a true mode reduction, the benefits are substantial. For instance, the parameters of the mode-reduced equations can easily be forced in a time-dependent manner, and the reduced equations themselves can be damped and driven. In this manner it may be possible to predict the behavior of the full original system, where the effects of such forcing may be difficult to determine. A specific instance of this technique has been carried out recently by Rabinowitz in the Soviet

Union: using numerical experiments on mode-reduced equations as a guide, he was able to forestall the onset of turbulence in a nozzle flow by applying a periodic stress.

Fourth, rigorous mathematical results on mode reduction have been obtained for a class of nonlinear reaction-diffusion equations that describe unstable chemical reacting fronts, such as flames. One important example is the so-called Kuramoto-Sivashinsky equation, which can be written in the form

$$\frac{\partial\phi(x,t)}{\partial t}+\frac{\partial^2\phi(x,t)}{\partial x^2}+\frac{\partial^4\phi(x,t)}{\partial x^4}+\left(\frac{\partial\phi(x,t)}{\partial x}\right)^2=0, \tag{28}$$

where $\phi(x,t)$ is the amplitude, as a function of position and time, of the phenomenon being described. Although this equation represents, in dynamical-systems parlance, an infinite-degree-of-freedom system, it is nonetheless rigorously true that in a box of finite length $L$ a finite number of modes proportional to $L$ are sufficient to capture the long-time dynamics arising from essentially any initial condition. Although the link is not yet fully constructive, the nature of these modes can be determined, and they are related to coherent structures observed for this equation. This general connection between mode reduction in chaotic systems and coherent structures in spatially extended dynamical systems will be a central issue in our discussion of complex configurations and pattern selection.

Finally, the problem of mode reduction lies at the core of attempts to understand the relation between chaos and fully developed turbulence in fluids and plasmas. Chaos, as we have stressed, involves temporal disorder and unpredictability in dynamical systems with low effective phase-space dimension. Fully developed turbulence, in addition to the temporal disorder, involves disordered, random spatial structure on all scales (at least apparently). Further, different spatial regions of the turbulent system act independently, and spatial correlation functions are short-ranged. Thus the phase-space dimension of any attractor in fully developed turbulence appears, a priori, very high.

For example, a recent numerical simulation of turbulent Poiseuille flow at a Reynolds number of 2800 suggests that the turbulent solutions to the Navier-Stokes equations for the flow do lie on a strange attractor, but one that has fractal dimension of about 400! Although it is comforting to know that the turbulence observed in this case can be described qualitatively by deterministic chaos, it is obviously disconcerting to contemplate trying either to analyze such flows experimentally or to model them theoretically in terms of a dynamical system with a 400-dimensional phase space. For higher Reynolds numbers this situation will become even worse. In the next section we will illustrate how mode reduction, coupled with a hierarchy of approximate equations, may make this situation more tractable.

In sum, the remarkable insights of the past twenty-five years have led to the emergence of deterministic chaos and fractals as a second central paradigm of nonlinear science. The impact of this paradigm on our basic view of complexity in the world, as well as on technologies affecting our daily lives, will continue to be profound for the foreseeable future.

## Complex Configurations and Patterns

When a spatially extended nonlinear system is driven far from equilibrium, the many localized coherent structures that typically arise can organize themselves into a bewildering array of spatial *patterns*, regular or random. Perhaps the most familiar example is turbulent fluid flow, in which the temporal behavior is chaotic yet one frequently observes patterns of coherent structures: recall the complex configuration of vortices surrounding the Red Spot in Fig. 3a. The process of pattern formation and selection occurs throughout nature, in nonlinear phenomena ranging from electromagnetic waves in the ionosphere through mesocale textures in metallurgy to markings on seashells and stripes on tigers. Thus, complex configurations and patterns represents a third paradigm of nonlinear science. Although somewhat less developed than solitons or chaos, the paradigm already promises to provide the basis for a unified understanding of nonlinear phenomena in many fields.

Our previous discussion of dynamical systems provides a useful conceptual framework in which to approach the general problem of patterns. A typical extended nonlinear, nonequilibrium system will have many possible configurations or patterns; some of these will be stable, others unstable, and the vast majority metastable. Highly symmetric patterns may be accessible analytically, but general, anisotropic configurations must first be studied via experimental mathematics. In dissipative extended systems these patterns are loosely analogous to the attractors of simple dynamical systems—with the important proviso that they do not correspond to true asymptotic attractors because most are, in fact, merely *metastable*. Nonetheless, the multiple-attractor analogy correctly suggests that an extended nonlinear system has many basins of temporary attraction. In view of our results on the damped, driven pendulum, we expect the basin boundaries to be complicated, perhaps fractal, objects. As a result, the study of the *dynamics* of complex configurations and of the sequence of patterns explored, as well as of the pattern ultimately selected (if any), represents one of the most daunting challenges facing nonlinear science.

At present this challenge is still being confronted primarily at the experimental level, both in actual physical systems and via numerical simulations, rather than analytically. Hence we rely here chiefly on visual results from these experiments to indicate important aspects of the paradigm.

**Experiments and Numerical Simulations.** Consider, as a first illustration, a generalization of a familiar example: the sine-Gordon equation, only now damped, driven, and with two spatial dimensions. This equation, which models certain planar magnetic materials and large-area Josephson junction arrays, has the form

$$\frac{\partial^2\theta}{\partial t^2} - \frac{\partial^2\theta}{\partial y^2} - \frac{\partial^2\theta}{\partial x^2} + \alpha\frac{\partial\theta}{\partial t} + \sin\theta = \Gamma\cos\omega t. \tag{29}$$

We can anticipate from our earlier discussion that this model will contain coherent structures (although not solitons, because the two-dimensional sine-Gordon equation is not completely integrable). We can also expect the model to contain chaos because of the driving and damping forces.

Four snapshots of the temporal development of the system are shown in Fig. 14 for $\alpha = 0.1$, $\Gamma = 1.6$, and $\omega = 0.6$. Although it may seem obvious, the use of color coding as a means of enhancing the visual interpretation should be mentioned; color graphics, especially in a high-speed, interactive mode, are *not* a frivolous luxury but, in fact, are among the most powerful tools of experimental mathematics. Here, for

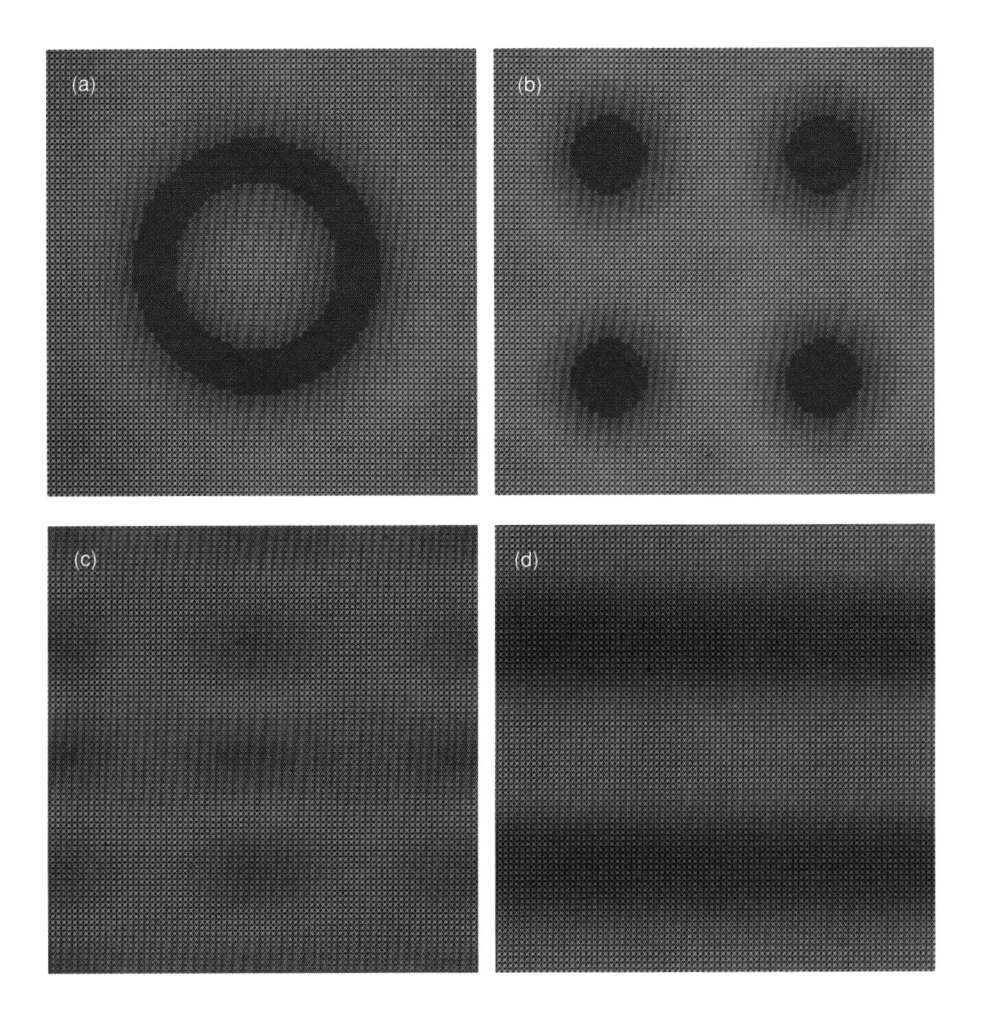


**SINE-GORDON EQUATION FOR TWO SPATIAL DIMENSIONS**

**Fig. 14. Four snapshots of the temporal behavior of the two-dimensional sine-Gordon equation. Red indicates values of $\theta$ near $2\pi$, blue indicates values near 0, and colors in the spectrum between red and blue indicate intermediate values. (a) The initial structure is annular. (b) After a time equal to approximately 100 units of the fundamental oscillation frequency of the system, the initial ring breaks into a symmetric, two-by-two pattern of four structures. (c) This last pattern is metastable and gradually slides off center, leading, at $t \sim 200$, to a pattern that is clearly beginning to "smear" in the *x*-direction. (d) Eventually, for $t \sim 300$, the smearing has led to a striped, stable configuration. The parameters used in Eq. 29 to generate these pictures are $\alpha = 0.1$, $\Gamma = 1.6$, and $\omega = 0.6$. (Figures courtesy of Peter Lomdahl, Los Alamos National Laboratory.)**

example, the color indicates the approximate value of $\theta$. The initial annular structure of this system (Fig. 14a) eventually forms other patterns that are, in fact, oscillatory in time. Because the boundary conditions are periodic in both $x$ and $y$, the system retains a high degree of symmetry as it evolves into four structures in a two-by-two pattern (Fig. 14b). Eventually, however, a "smearing" is detected parallel to the $x$ axis (Fig. 14c) that leads to the striped configuration of Fig. 14d. No further qualitative change occurs after that.

Because of the original symmetry of the problem, the emergence of a final pattern striped in the $x$ direction rather than the $y$ direction must depend on a slight asymmetry external to the equations themselves. Possibilities are a slight difference in the initial conditions for $x$ and $y$ due to computer round-off or an asymmetry in the solution algorithm. Such asymmetries can be viewed as external noise that leads to a configuration that breaks the symmetry of the equations. The extreme sensitivity of certain pattern selection processes to external noise and to minor asymmetries has already been indicated in the fractal growth models of Fig. 13 and is observed experimentally in a wide variety of contexts, including the growth of dendrites such as the snowflakes of Fig. 11.

The emergence and evolution of configurations related to those seen in the numerical simulations has been the focus of many recent experiments involving Rayleigh-Benard convection. By using shadowgraph techniques that clearly distinquish ascending and descending streams of fluid, convection-roll structures are observed in silicone oil heated from below (Figs. 15 and 16). The asymmetric pattern of Fig. 15a is typical of configurations that last for only a few minutes. On the other hand, the more symmetric pattern of Fig. 15b is more stable, maintaining its form for ten minutes or

**RAYLEIGH-BERNARD CONVECTION PATTERNS**

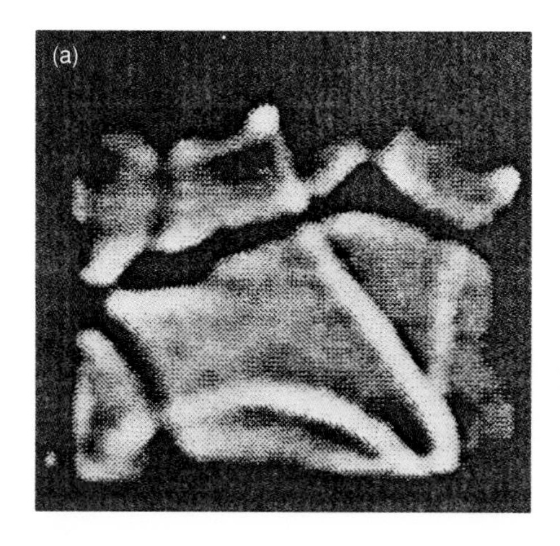


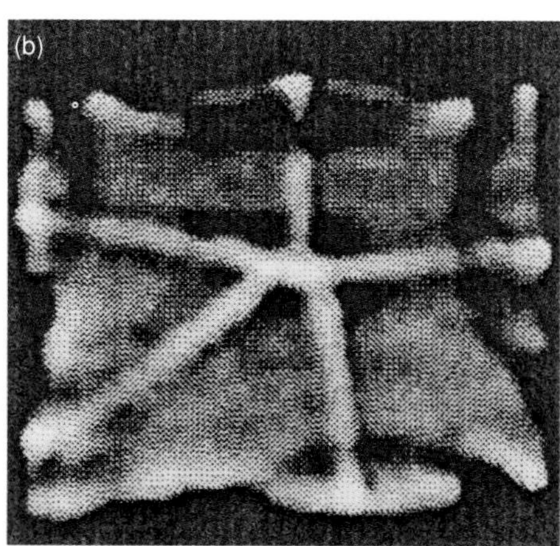


**Fig. 15. Patterns of convection-roll streaming are created here using shadowgraph techniques in an experiment in which silicone oil is heated from below. The dark lines correspond to ascending streams of fluid, the bright lines to descending streams. (Photos courtesy of Pierre Bergé, Commissariat a L'Énergie Atomique, France.)**

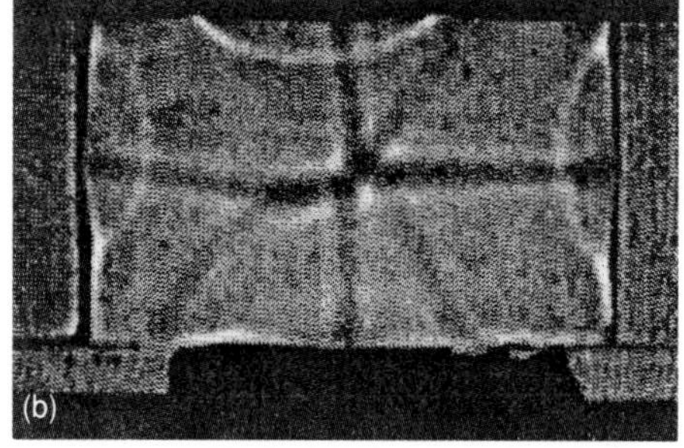


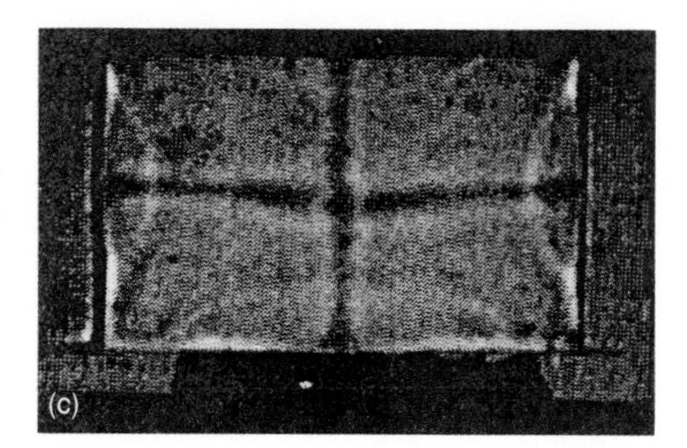


**AN AVERAGE CONVECTION PATTERN**

**Fig. 16. The first two of these Rayleigh-Bernard convection patterns (a and b) are snapshots of the flow in the silicone oil experiment, whereas (c) is a sum of ten such instantaneous pictures. (Photos courtesy of Pierre Bergé, Commissariat a L'Énergie Atomique, France.)**

more. Although one does not have a detailed understanding of the phenomenon, clearly boundary effects are causing the system to "pin" itself to these more stable configurations. Figure 16 demonstrates that a *mean structure*, or average pattern, can exist in such flows. The first two images (Figs. 16a and b) are snapshots of the flow, whereas the third (Fig. 16c) is a sum of ten such instantaneous pictures and clearly indicates the presence of a highly symmetrical average configuration.

Fluid dynamics abounds with other examples of complex configurations and pattern formation. Particularly relevant in technological applications is *shear instability*, which occurs when a fluid moves rapidly past a fixed boundary or when two fluids move past each other at different velocities. The performance and fuel-efficiency of aircraft, for example, are strongly affected by the turbulent boundary layer formed as a consequence of shear instabilities.

Figure 17 is a sequence of images of the "Kelvin-Helmholtz" shear instability simulated using the two-dimensional Euler equations that model compressible but inviscid fluid flow. (Strictly speaking, because the study does not resolve the thin turbulent boundary layer, it is technically a "slip-surface" instability.) The study reveals an incredible wealth of information, only some of which we will discuss here. Two streams of identical fluid flow past each other, both moving at the speed of sound. Initially, a small sinusoidal perturbation is given to the vertical velocity component of the flow at the boundary between the layers, and the resulting entrainment and roll-up phenomena that lead to the mixing of the two fluids is followed.

Shortly after the simulation starts, the roll-up of the boundary has already begun to generate coherent structures (Fig. 17a). These grow (actually, in a self-similar manner) until the periodic boundary conditions in the $x$ direction cause the structures to interact (note the four vortex-like structures in Fig. 17b). In addition, sudden jumps in the intensity of the colors in the top and bottom regions reveal the presence of shock waves. The four vortices merge into two (Fig.17c) and thereafter entrain, forming a *bound* vortex pair (Fig. 17d). The roll-up phenomenon creates incredibly complex

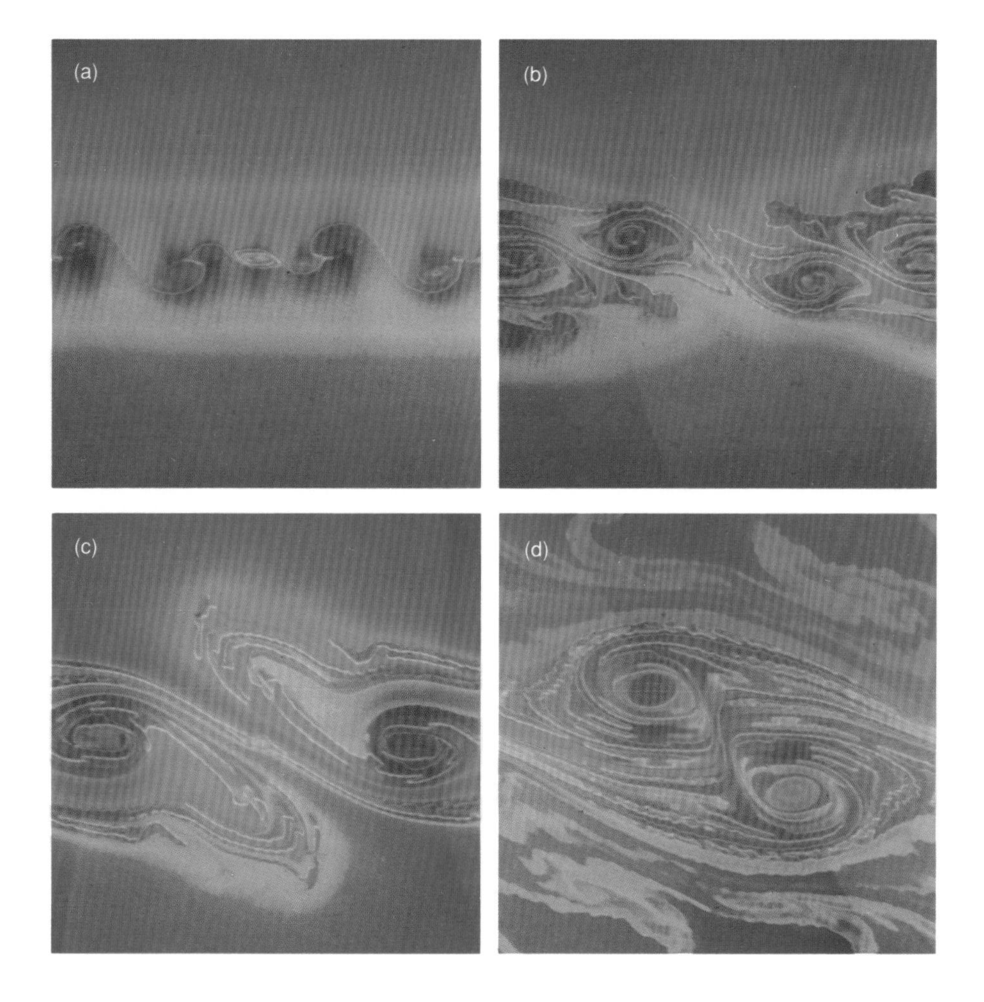


**SHEAR INSTABILITY**

**Fig. 17. Two streams of identical fluid flow past each other with the top stream (colored green to blue) moving to the right at Mach 1 (the speed of sound) and the bottom stream (colored red to purple) moving to the left also at Mach 1. The boundary between layers is a yellow line, and, initially, a small vertical sinusoidal velocity perturbation is applied at this boundary with the colors indicating the initial *y* value of a given bit of fluid. The series show the roll-up of the boundary (a) and the development of coherent structures in the form of vortices. By (d), a bound vortex pair has formed. (Figures made at Lawrence Livermore National Laboratory by Paul Woodward, University of Minnesota, and David Helder.)**

(fractal) structure from the initially smooth boundary. Thus, we see in Fig. 17 precisely the interplay between large-scale coherent structures and chaotic, fractal dynamics that typifies complex configurations in extended nonlinear systems. Further, although different in detail, Fig. 17d clearly resembles in outline the more familiar shape of Fig. 3a; art—in this case, computer art—is indeed imitating Nature.

One feature common to all our previous examples is the limited number of coherent structures that participate in the observed patterns of the system. In each case, this limitation arises from the small size (relative to the scale of the coherent structures themselves) of the "box"—be it computational or physical—in which the pattern-forming system is contained.

An example in which this constraint is relaxed is a numerical simulation, carried out at Los Alamos a decade ago by Fred Tappert, of the self-focusing instability that arises in the interaction of an intense laser beam with a plasma (Fig. 18). The instability is closely related conceptually to the mechanism by which solitons are formed in optical fibers and reflects an important difficulty in attempts to develop inertial confinement fusion. On a much different scale, this phenomenon leads to significant electromagnetic disturbances in the ionosphere.

The particular equation used in the simulation is a two-space-dimension variant of the nonlinear Schrödinger equation (Eq. 12). Here the equation has the specific form

$$i\frac{\partial E}{\partial t}+\frac{\partial^2 E}{\partial x^2}+\frac{\partial^2 E}{\partial y^2}+\left(1-e^{-|E|^2}\right)E=0, \tag{30}$$

where $E=E(x,y,t)$ is the electric field envelope function. For small $|E|^2$, the equation contains an effective cubic nonlinearity and thus becomes the direct two-dimensional

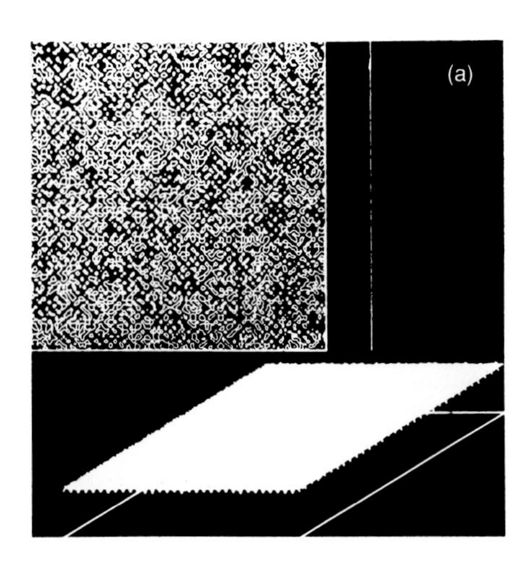


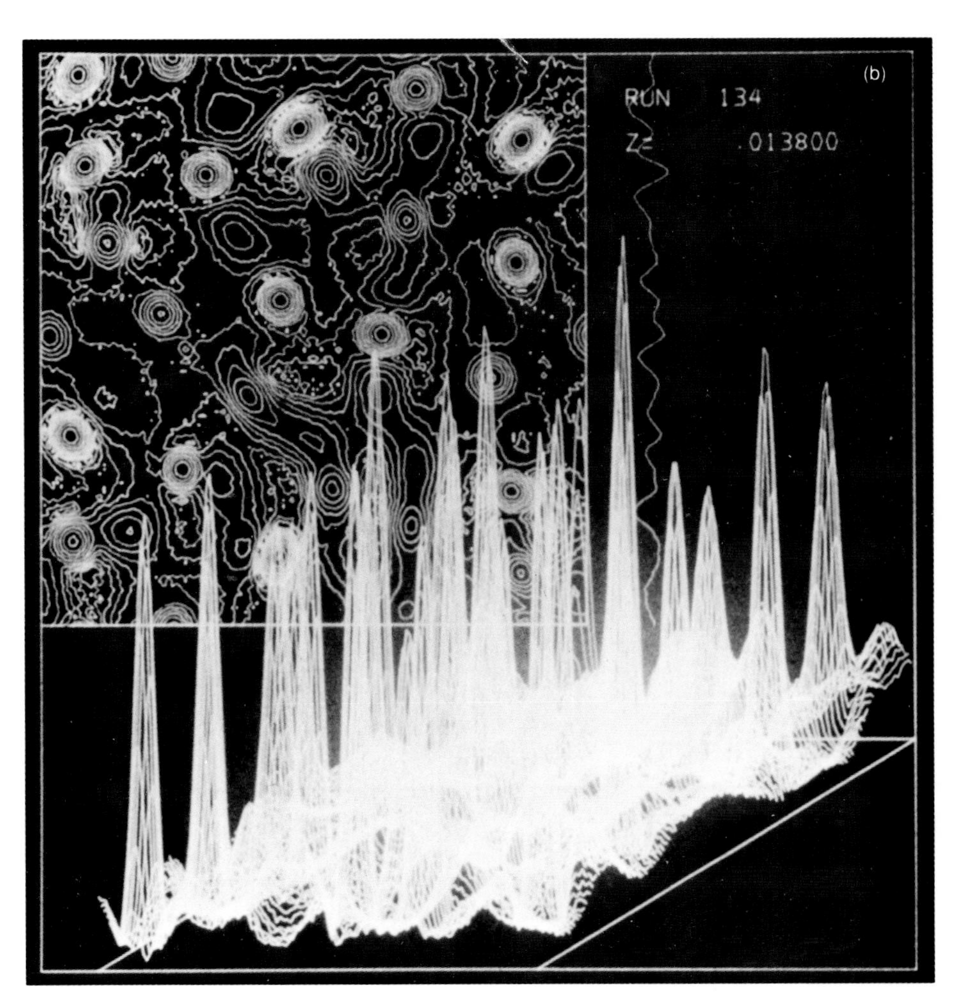


## CAVITONS: SELF-FOCUSING INSTABILITIES

**Fig. 18. The development of a self-focusing instability in a laser beam passing through a plasma. These frames, taken from a computer-generated movie, show both a contour plot (upper left) and a projected plot of the laser intensity across the profile of the beam. (a) Initially, the beam is essentially uniform with a small amount of random spatial "noise," but as it moves into the plasma, the self-focusing instability generates filaments of high intensity that (b) grow dramatically as the beam progresses further. (Photos made at Los Alamos by Fred Tappert, University of Miami.)**

generalization of Eq. 12. As $E$ approaches infinity, the nonlinearity saturates, and Eq. 30 becomes effectively linear.

From a random initial condition of spatial white noise (Fig. 18a), a complex configuration involving a large number of coherent structures develops (Fig. 18b). Having observed these complex patterns involving many coherent structures, Tappert went on to isolate the individual coherent structures—now known as *cavitons*—and to study their interactions numerically. Since the dynamics can not be properly appreciated without showing the time evolution, I will not attempt to describe it here; however, this study is an excellent example of using experimental mathematics to unravel the role that analytically inaccessible coherent structures play in the formation of complex configurations.

**Analytic Developments.** Our brief pictorial survey of numerical and experimental studies of pattern formation should make clear the daunting nature of the general problem. Thus it is hardly surprising that current analytic approaches focus on special and isolated instances of pattern formation that reduce the problem to a more tractable form. Although much remains to be done, this "divide and conquer" philosophy has recently demonstrated such substantial promise that, in the next decade, we should witness a revolution for patterns comparable to those for solitons and chaos.

One line of analytic attack is to start with a system that has exact solitons—the one-dimensional sine-Gordon or nonlinear Schrödinger equations, for example. One then perturbs the system with driving and damping forces and studies the patterns that emerge from the evolution of the analytically known coherent structures under the influence of the chaotic dynamics. This approach has been used extensively in the case of the damped, driven sine-Gordon equation, and a very rich phenomenology has developed. However, detailed quantitative understanding, even in the case of a perturbed integrable system, can only be produced at present if the purely analytic approach is guided and supplemented by numerical simulations.

To describe other semi-analytic approaches, let me focus on pattern formation in fluid flows. I shall discuss three related techniques that derive approximate or effective equations appropriate to specific situations called the *amplitude-*, *phase-*, and *prototype-equation* techniques.

As previously observed, a nonlinear system often exhibits bifurcations or sharp transitions in the qualitative behavior of its solutions as a function of one of its parameters. The Rayleigh-Benard instability in a fluid heated from below is one such case (Figs. 15 and 16). When the rate of heating is less than a certain threshold, the fluid simply conducts the heat from the hot bottom to the cooler top, in effect acting like a solid object. At a critical value of the heating rate, this conducting state becomes unstable and *convection*—the familiar rolling motion that can be seen in boiling water—sets in. This transition is the nonequilibrium analog of a phase transition. We can model the temporal and spatial structure of the transition with a phenomenological equation written in terms of a parameter that describes the amplitude of the convecting state. This convection *order parameter* will be zero below threshold and nonzero above. A variety of near-threshold phenomena have been treated successfully using such amplitude equations.

Now consider a nonlinear system already in a state with an overall, regular pattern: for example, a sequence of straight convection rolls in a large box or the rectangular arrangement of convection cells in Fig. 16c. Let this pattern be described by a dominant wave vector (or vectors) that we call $k_0$. Many patterns close to the initial one can be studied by considering slow spatial and temporal modulation of $k_0$. The resulting phase equations can be viewed as the nonequilibrium analogs of hydrodynamics since they apply to low-frequency, long-wavelength motions near a given state. Again, such phase equations have been used to analyze many specific fluid flows.

Prototype equations, although perhaps motivated by specific fluid motions, are not necessarily strictly derivable from the fundamental Navier-Stokes equations but rather are intended to capture the essence of certain nonlinear effects. More precisely, prototype equations often serve as a means of gaining insight into competing nonlinear effects and are thus extremely important in developing analytic understanding. The Korteweg-deVries equation (Eq. 10), which played a central role in the discovery of solitons, can be viewed as an example of a prototype equation. That it is also derivable for surface waves in shallow, narrow channels is an added bonus. Similarly, the Kuramoto-Sivashinsky (KS) equation (Eq. 28)—is another prototype equation.

Very recently, pattern formation in convecting fluid flows in large containers has been studied using the Kolmogorov-Spiegel-Sivashinsky, or KSS, equation—a generalization of the original KS equation. Because some very interesting results about the interplay of coherent structures and chaos have come from these studies, I will use the KSS equation to illustrate the prototype-equation technique.

The specific form of the equation is

$$\frac{\partial \phi}{\partial t} + \beta\phi + \gamma\left(\frac{\partial \phi}{\partial x}\right)^2 + \frac{\partial}{\partial x}\left[\left(\alpha - \delta\left(\frac{\partial \phi}{\partial x}\right)^2\right)\frac{\partial \phi}{\partial x}\right] + \frac{\partial^4 \phi}{\partial x^4} = 0, \tag{31}$$

where $\alpha$, $\beta$, $\gamma$, and $\delta$ are adjustable parameters. This equation models large-scale unidirectonal flow. An example is the Kolmogorov flow in which an effectively two-dimensional viscous fluid is subjected to a unidirectional external force field periodic in one of the spatial directions. Such a flow can be realized in the laboratory using thin layers of electrolytic fluids moving in a periodic magnetic field.

In Eq. 31 $\phi = \phi(x, t)$ is the fluctuating part of the stream function (that is, the stream function minus the mean periodic field component), $\partial\phi/\partial t$ is the familiar local derivative for the fluid motion, $\beta\phi$ represents the classical linear damping of the fluctuations, $\gamma(\partial\phi/\partial x)^2$ is the convective derivative for the fluid motion in an unfamiliar form, and the last terms describe viscosity effects.

One can see the role of the local and convective derivative terms more directly by differentiating Eq. 31 with respect to $x$ and considering the gradient of the stream function: $u(x, t) = \partial\phi/\partial x$, which is related to the velocity. In the differentiated equation, the two terms assume the form $\partial u/\partial t + 2\gamma u \partial u/\partial x$, familiar from, for example, the Korteweg-deVries equation (Eq. 10). Note that the convective term in Eq. 31 increases rapidly when $\phi$ is varying rapidly in $x$ (that is, for large wavenumbers $k$), correctly suggesting that this term leads to a flow of energy from large to small spatial scales.

A careful examination of the viscosity effects—given by the final three terms in Eq. 31—reveals the interplay and competition essential to the pattern-forming properties of this model. The lowest-order diffusion term, $\alpha\partial^2\phi/\partial x^2$, has, since $\alpha > 0$, the wrong sign for stable diffusion and hence leads to an exponential growth of the solution for all wave numbers $k$. The higher order linear diffusion term $\partial^4\phi/\partial x^4$, controls the growth for large $k$. As a consequence, only a certain range of wave numbers ($0 < k < k_c$) exhibits the linear instability that leads to rapid growth. This *negative viscosity* region causes a flow of energy from small to larger spatial scales and thus creates the coherent structures observed in the equations. In turn, the growth of these structures is limited by the nonlinear terms—in particular, by the convective derivative terms—and the resulting competition between the negative viscosity and convective terms provides a mechanism for energy to cycle back and forth between small and large scales. Finally, the nonlinear viscosity term, $\delta\frac{\partial}{\partial x}(\partial\phi/\partial x)^3$, provides an important *local* variation in viscosity; in regions where $\partial\phi/\partial x$ is small, the effective local viscosity is negative, but as $\partial\phi/\partial x$ grows, the nonlinearity leads to a locally positive viscosity and to normal dissipation of energy at small scales.

For the KSS equation, recent analytic studies have shown that the full partial differential equation is strictly equivalent to a set of coupled ordinary differential equations corresponding to a finite-dimensional dynamical system. Further, the phase-space dimension of this dynamical system is proportional to the number of linearly unstable modes and hence increases linearly with the length of the system $L$. In addition, the finite dynamical system possesses a universal strange attractor with fractal dimension also proportional to $L$. These results are concrete examples of the mode-reduction program, and their attainment in an equation involving *local* negative viscosity effects marks a step forward in analytic understanding of turbulence. However, as in the case of the KS equation, the results are not of themselves sufficient to identify the natural coherent structures that arise in Eq. 31 nor to relate them directly to the reduced modes.

To search for the coherent structures, extensive numerical simulations of the KSS equation are currently in progress. Figure 19 depicts one solution (for $\alpha = 2$, $\beta = 0.15$, $\gamma = 1$, and $\delta = 0.58$) with a system size such that there are fifty unstable modes. The black cross-hatched structures are regions of (spatially homogeneous) chaos. Note that, with the horizontal axis representing time and the vertical axis representing position, these regions often propagate through the system, as indicated by the diagonal "motion" of the cross-hatched areas.

The most striking features in Fig. 19 are the orange horizontal bands, which intermittently appear and disappear at various locations and times within the system.

These are relatively quiescent, large-scale spatial subdomains and are the candidates for the coherent structures. Notice that the propagating chaotic regions do not penetrate these structures. However, as suggested in particular by the region around the long-lived coherent structure in the lower right corner, one may be able to describe interactions of the propagating chaotic regions with the coherent structures. Such interactions may involve phase shifts (as in the case of solitons) as well as creation and annihilation of both the propagating chaos and the coherent structures. At present, these and related questions are under active investigation.

**THE KSS SYSTEM**

**Fig. 19. This solution of the KSS equation (for $\alpha$ = 2, $\beta$ = 0.15, $\gamma$ = 1, and $\delta$ = 0.58 in Eq. 31) has both regions of chaos (crosshatched) and regions of relatively quiescent behavior (orange). Since time is represented by the horizontal axis and the spatial variable by the vertical axis, the diagonal "motion" of crosshatched areas represents propagation of these chaotic regions through the system. (Photo courtesy of Basil Nicolaenko and Hughes Chaté, Los Alamos National Laboratory.)**

From our discussion it is clear that, although exciting results are beginning to appear, development of the paradigm of complex configurations and pattern formation will occupy researchers in nonlinear science for years to come. It is perhaps of interest to suggest a few of the broad questions that must be addressed.

- To what extent can the complex structures and patterns be thought of as superpositions of coherent structures, and in what "space" can these structures be superposed? In this regard, we know that for weakly perturbed soliton-bearing systems, the appropriate space in which approximate superposition holds is the inverse scattering transform space. Further, some recent studies suggest that certain turbulent flows can usefully be decomposed as sums of terms, each having vorticity parallel to the velocity.
- What is the dynamics of competition among patterns? How does this competition depend on the nature of the interactions among individual coherent structures? For what systems can one view the different patterns as local minima in a "pattern accessibility" space? What can one say about the basins of attraction in this space?
- In systems with constrained geometry—such as the fluid experiment of Figs. 15 and 16—can one understand quantitatively the observed selection of more symmetric patterns over less symmetric ones? Here the analogy to pinning phenomena in solid state systems may be useful.
- For what pattern-forming systems can one construct a hierarchy of equations in which successive levels of approximation lead downward from the Navier-Stokes equations through an approximate partial differential equation to a finite set of coupled ordinary differential equations? How can one match the solutions across various levels of this hierarchy? Such matching will be essential, in particular to understand what happens when the effective equations lower in the hierarchy break down.

• What can one say about pattern formation in fully-developed, three-dimensional turbulence? For the full Navier-Stokes equations, can anything analogous to the competing mechanisms in the KSS equation be identified, so that both a cycle involving a flow of energy from large to small spatial scales and the re-emergence of large-scale coherent structures can exist? If so, this cycle could form the basis for a "turbulence engine," which would explain at least the major features of transport in turbulent flows.

Since most of our remarks have focused on problems in fluid dynamics, it is important to re-emphasize the broad impact of our last paradigm. The complex fractal structures observed in ceramic cracks and in oil recovery problems, although treated for convenience in our discussion of fractals, are, in fact, more accurately viewed as examples of patterns. Similarly, dendritic growth is a solidification process critically dependent on a pattern selection mechanism that is itself exquisitely sensitive to anisotropy and extrinsic noise. The development of mesoscale textures—that is, patterns larger than the atomic scale but yet not macroscopic—remains an important issue for metallurgy.

In fact, in the microscopic theories of solid state materials, the mechanism underlying pattern dynamics is a question not yet fully resolved. Here, in distinction to the case of fluids, one does not have a fundamental model such as the Navier-Stokes equations to rely on, so one cannot naively assume diffusive coupling among the patterns. Instead, a variety of possible mechanisms must be looked at closely.

In biology, pattern formation and selection is ubiquitous, with applications from the cellular to the whole organism level. And in ecology, nonlinear reaction-diffusion equations suggest spatial patterns in predator-prey distributions and in the spread of epidemics.

To conclude this section, I will look at an intriguing feature of nonlinear pattern-forming systems—the property of pattern *self reproduction*—using a *cellular automaton.* Cellular automata are nonlinear dynamical systems that are discrete in both space and time and, importantly, have, at each site, a finite number of state values (allowed values of the dependent variable). Such systems were invented, and first explored, by John von Neumann and Stan Ulam. Currently, they are being studied both for their fascinating intrinsic properties and for a number of applications, including pattern recognition. They are also being used as novel computational algorithms for solving continuum partial differential equations (see "Discrete Fluids" for the example of lattice-gas hydrodynamics).

Figure 20 shows four stages in the growth of a self-reproducing pattern found in a cellular atuomaton with eight possible states per site. At each step in time, the new state of a given cell is determined by a small set of rules based on the current state of the cell and the state of its nearest four neighbors on a square lattice. The particular pattern shown generates copies of itself, forming a colony. On an infinite lattice the colony would continue to grow forever. Despite its simplicity and the rigidity of its predetermined rules, the self-reproduction of this automaton is intriguingly reminiscent of the development of real organisms, such as coral, that grow in large colonies.

## The Future of Nonlinear Science

From the many open questions posed in the previous sections, it should be clear that nonlinear science has a bright and challenging future. At a fundamental level issues such as the scaling structure of multifractal strange sets, the basis for the ergodic hypothesis, and the hierarchy of equations in pattern-forming systems remain unresolved. On the practical side, deeper understanding of the role of complex configurations in turbulent boundary layers, advanced oil recovery, and high-performance ceramics would provide insight valuable to many forefront technologies. And emerging solutions to problems such as prediction in deterministically chaotic systems or modeling fully developed turbulence have both basic and applied consequences. Further, the nonlinear revolution promises to spread to many other disciplines, including economics, social sciences, and perhaps even international relations.

If, however, one had to choose just one area of clearest future opportunity, one would do well to heed another of Stan Ulam's well-known *bons mots*:

> "Ask not what mathematics can do for biology,
> Ask what biology can do for mathematics."

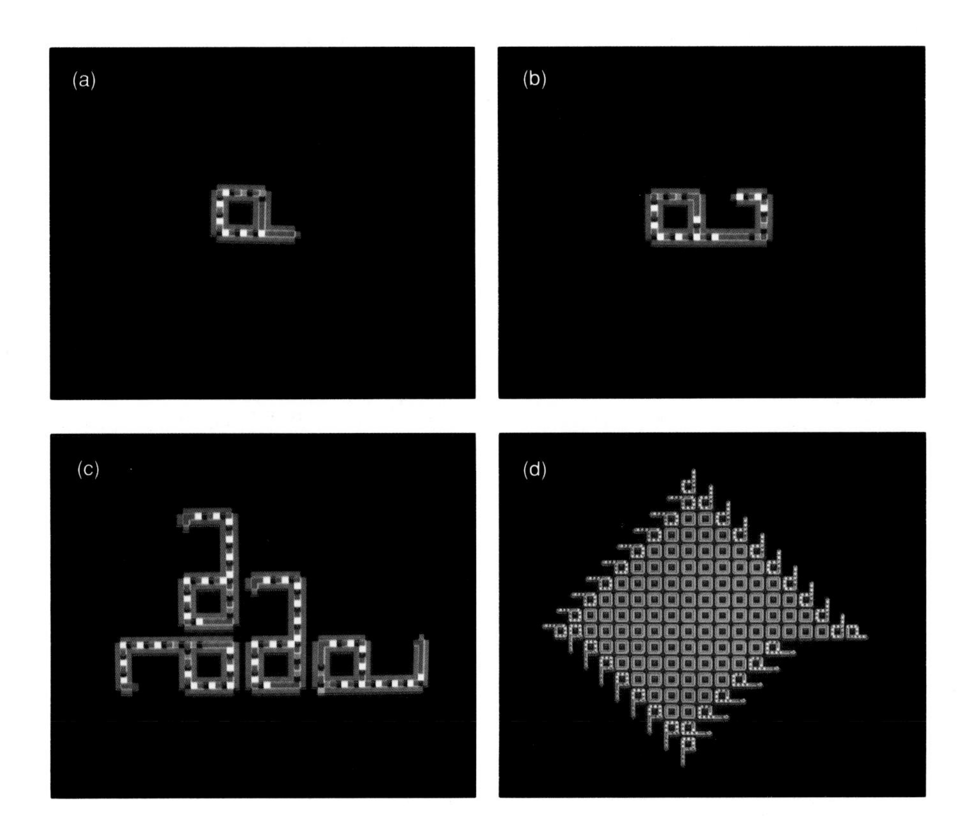


## CELLULAR AUTOMATON

**Fig. 20. This cellular automaton consists of a grid of square cells with each cell able to take on any of eight possible states (indicated by different colors). (a) The basic building block of a repeating pattern for this automaton is a hollow square occupying an area of 10 by 15 cells with a tail that develops (b) until it produces a second hollow square. (c) The pattern continues to grow in time until (d) it has produced a large colony of the original pattern. (Figures courtesy of Chris Langton, Los Alamos National Laboratory.)**

If we replace "mathematics" with "nonlinear science," Stan's comment becomes even more appropriate to the present situation. We have already seen the beginnings of an understanding of many aspects of morphology, from fractal structures in ferns to nonlinear pattern-selection models for human digits. Similarly, the role of chaos in biological cycles, from heartbeats to cell densities, is rapidly being clarified. And the basic observation that incredibly complex behavior—including both pattern formation and self-reproduction—can emerge in systems governed by very simple rules has obvious implications for modeling biological phenomena.

But the greatest challenge is clearly to understand adaptation, learning, and evolution. *Adaptive* complex systems will have features familiar from conventional dynamical systems, including hierarchical structures, multiple basins of attraction, and competition among many metastable configurations. In addition, they must also have a mechanism for responding to, and taking advantage of, changes in their environment.

One approach to adaptation is to construct an explicit temporal hierarchy: one scale describes the actual dynamics and a second, slower time scale allows for changes in the nonlinear equations themselves. Models for the human immune system and for autocatalytic protein networks are among the prospective initial applications for this concept.

A second approach to adaptation, sometimes termed *connectionism*, is based on the idea that many simple structures exhibit complex collective behavior because of connections between the structures. Recent specific instances of this approach include mathematical models called *neural networks*. Although only loosely patterned after true neurological systems, such networks show remarkable promise of being able to learn from experience. A related set of adaptive models, called *classifier systems*, show an ability to self-generate a hierarchy of behavioral rules: that is, the hierarchy is not placed a priori into the system but develops naturally on the basis of the system's experience. In general, connectionist models suggest a resolution of the long-standing issue of building a reliable computer from unreliable parts.

In all these future developments, the tripartite methodology incorporating experimental mathematics, real experiments, and novel analytic approaches will continue to play a critical role. One very exciting prospect involves the use of ultraspeed interactive graphics, in which enormous data sets can be displayed visually and interactively at rates approaching the limits of human perception. By using color and temporal evolution, these techniques can reveal novel and unexpected phenomena in complicated systems.

To insure the long-term success of nonlinear science, it is crucial to train young researchers in the paradigms of nonlinearity. Also, interdisciplinary networks must be fostered that consist of scholars who are firmly based in individual disciplines but are aware of, and eager to understand, developments in other fields.

In all these respects, nonlinear science represents a singularly appropriate intellectual legacy for Stan Ulam: broadly interdisciplinary, intellectually unfettered and demanding, and—very importantly—fun. ■

**David Campbell**, the Laboratory's first J. Robert Oppenheimer Fellow (from 1974 to 1977), is currently the Director of the Laboratory's Center for Nonlinear Science. He received his B.A. in chemistry and physics from Harvard College in 1966 and his Ph.D. in theoretical physics from Cambridge University in 1970. David has extended his activities in physics and nonlinear science to the international level, having been a National Academy of Sciences Exchange Scientist to the Soviet Union in 1977, a Visiting Professor at the University of Dijon, Dijon, France in 1984 and 1985, and a Ministry of Education Exchange Scientist to the People's Republic of China in 1986. He and his wife, Ulrike, have two children, Jean-Pierre and Michael.

## Acknowledgments

I am grateful to the many colleagues who, over the years, have shared their insights and helped shape my perspective on nonlinear science. In the preparation of this article I have benefited greatly from the comments and assistance of Jim Crutchfield, Roger Eckhardt, Doyne Farmer, Mitchell Feigenbaum, Jim Glimm, Erica Jen, and Gottfried Mayer-Kress. I wish also to thank my generous coworkers in nonlinear science who permitted the use of the figures and pictures so essential to this article.

# Further Reading

**General Discussions of Nonlinear Science and Background**

H. H. Goldstein and J. von Neumann. 1961. On the principles of large scale computing machines. In *John von Neumann: Collected Works*, Volume V, edited by A. H. Taub, pp. 1–32. New York: Pergamon Press.

E. Fermi, J. Pasta, and S. Ulam. 1965. Studies of nonlinear problems. In *Enrico Fermi: Collected Papers*, Volume II, pp. 978–988. Chicago: University of Chicago Press.

David Campbell, Jim Crutchfield, Doyne Farmer, and Erica Jen. 1985. Experimental mathematics: the role of computation in nonlinear science. *Communications of the ACM* 28:374–384.

**Solitons and Coherent Structures**

D. J. Korteweg and G. DeVries. 1895. On the change of form of long waves advancing in a rectangular canal and on a new type of long stationary waves. *Philosophical Magazine* 39:422–443.

Alfred Segger, Hans Donth, and Albert Kochendörfer. 1953. Theorie der Versetzungen in eindimensionalen Atomreihen. III. Versetzungen, Eigenbewegungen und ihre Weshselwirkung. *Zeitschrift für Physik.* 134:171–193.

J. K. Perring and T. H. R. Skryme. 1962. A model unified field equation. *Nuclear Physics* 31:550–555.

N. Zabusky and M. Kruskal. 1965. Interaction of "solitons" in a collisionless plasma and the recurrence of initial states. *Physical Review Letters* 15:240–243.

Alwyn C. Scott, F. Y. F. Chu, and David W. McLaughlin. 1973. The soliton: a new concept in applied science. *Proceedings of the IEEE* 61:1443–1483.

A. R. Osborne and T. L. Burch. 1980. Internal solitons in the Andaman Sea. *Science* 208:451–460.

C. H. Tze. 1982. Among the first texts to explain the "soliton revolution." *Physics Today*, June 1982, 55–56. (This article is a review of *Elements of Soliton Theory* by G. L. Lamb, Jr., *Solitons: Mathematical Methods for Physicists* by G. Eilenberger, and *Solitons and the Inverse Scattering Transform* by M. Ablowitz and H. Segur.)

Akira Hasegawa. 1984. Numerical study of optical soliton transmission amplified periodically by the stimulated Raman process. *Applied Optics* 23:3302–3309.

David K. Campbell, Alan C. Newell, Robert J. Schrieffer, and Harvey Segur, editors. 1986. *Solitons and Coherent Structures: Proceedings of the Conference on Solitons and Coherent Structures held at Santa Barbara.* Amsterdam: North-Holland Publishing Co. (reprinted from *Physica D* 18:Nos. 1–3).

C. G. Slough, W. W. McNairy, R. V. Coleman, B. Drake, and P. K. Hansma. 1986. Charge-density waves studied with the use of a scanning tunneling microscope. *Physical Review B* 34:994–1005.

L. F. Mollenauer, J. P. Gordon, and M. N. Islam. 1986. Soliton propagation in long fibers with periodically compensated loss. *IEEE Journal of Quantum Electronics* QE-22:157–173.

**Deterministic Chaos and Fractals**

Henri Poincaré. 1952. *Science and Method*, translated by Francis Maitland. New York: Dover Publications, Inc.

W. A. Bentley and W. J. Humphreys. 1962. *Snow Crystals*. New York: Dover Publications, Inc.

M. V. Berry. 1978. Regular and irregular motion. In *Topics in Nonlinear Dynamics: A Tribute to Sir Edward Bullard*, edited by S. Jorna, A.I.P. Conference Proceedings, No. 46, pp. 16–120. New York: American Institute of Physics.

Edward N. Lorenz. 1979. On the prevalence of aperiodicity in simple systems. In *Global Analysis: Proceedings of the Biennial Seminar of the Canadian Mathematical Congress*, edited by M. Grmela and J. E. Marsden, pp. 53–75. New York: Springer Verlag.

Mitchell J. Feigenbaum. 1980. Universal behavior in nonlinear systems. *Los Alamos Science* 1 (Summer 1980):4–27 (reprinted in *Physica D* 7:16–39, 1983).

Robert H. G. Helleman. 1980. Self-generated chaotic behavior in nonlinear mechanics. In *Fundamental Problems in Statistical Mechanics*, edited by E. G. D. Cohen, pp. 165-233. Amsterdam: North-Holland Publishing Co.

B. A. Huberman, J. P. Crutchfield, and N. H. Packard. 1980. Noise phenomena in Josephson junctions. *Applied Physics Letters* 37:750–753.

J.-P. Eckmann. 1981. Roads to turbulence in dissipative dynamical systems. *Reviews of Modern Physics* 53:643–654.

G. Mayer-Kress and H. Haken. 1981. The influence of noise on the logistic model. *Journal of Statistical Physics* 26:149–171.

Benoit Mandelbrot. 1983. *The Fractal Geometry of Nature*. New York: W. H. Freeman and Company.

David K. Umberger and J. Doyne Farmer. 1985. Fat fractals on the energy surface. *Physical Review Letters* 55:661–664.

Gérard Daccord, Johann Nittmann, and H. Eugene Stanley. 1986. Radial viscous fingers and diffusion-limited aggregation: fractal dimension and growth sites. *Physical Review Letters* 56:336–339.

Johann Nittmann and H. Eugene Stanley. 1986. Tip splitting without interfacial tension and dendritic growth patterns arising from molecular anisotropy. *Nature* 321:663–668.

James Gleick. 1987. *Chaos: Making a New Science*. New York: Viking Penquin, Inc.

David K. Campbell. 1987. Chaos: chto delat? To be published in *Nuclear Physics B*, Proceedings of Chaos, 1987.

**Complex Configurations and Pattern Formation**

J. D. Farmer, T. Toffoli, and S. Wolfram, editors. 1984. *Cellular Automata: Proceedings of an Interdisciplinary Workshop, Los Alamos*. Amsterdam: North-Holland Publishing Co. (reprinted from *Physica D* 10:Nos. 1–2).

Alan R. Bishop, Laurence J. Campbell, and Paul J. Channell, editors. 1984. *Fronts, Interfaces, and Patterns: Proceedings of the Third Annual International Conference of the Center for Nonlinear Studies, Los Alamos*. Amsterdam: North-Holland Publishing Co. (reprinted from *Physica D* 12:1–436).

Basil Nicolaenko (Nichols). 1987. Large scale spatial structures in two-dimensional turbulent flows. To be published in *Nuclear Physics B*, Proceedings of Chaos, 1987.

Pierre Bergé. 1987. From temporal chaos towards spatial effects. To be published in *Nuclear Physics B*, Proceedings of Chaos, 1987.

Paul R. Woodward, David H. Porter, Marc Ondrechen, Jeffrey Pedelty, Karl-Heinz Winkler, Jay W. Chalmers, Stephen W. Hodson, and Norman J. Zabusky. 1987. Simulations of unstable flow using the piecewise-parabolic method (PPM). In *Science and Engineering on Cray Supercomputers: Proceedings of the Third International Symposium*, pp. 557–585. Minneapolis: Cray Research, Inc.

Alan C. Newell. The dynamics of patterns: a survey. In *Propagation in Nonequilibrium Systems*, edited by J. E. Wesfried. New York: Springer Verlag. To be published.

**Adaptive Nonlinear Systems**

Doyne Farmer, Alan Lapedes, Norman Packard, and Burton Wendroff, editors. 1986. *Evolution, games, and learning: models for adaptation in machines and nature: Proceedings of the Fifth Annual International Conference of the Center for Nonlinear Studies, Los Alamos*. Amsterdam: North-Holland Publishing Co. (reprinted from *Physica D* 22:Nos. 1-3).

# THE ERGODIC HYPOTHESIS

## A Complicated Problem in Mathematics and Physics

*by Adrian Patrascioiu*

There are a few problems in physics that stir deep emotions every time they are discussed. Since physicists are not generally speaking an emotional group of people, the existence of these sensitive issues must be considered a strong indication that something is amiss. One such issue is the interpretation of quantum mechanics. I will take a moment to discuss that problem because it bears directly on the main topic of this article.

In quantum mechanics, if the question asked is a technical one, say how to compute the energy spectrum of a given atom or molecule, there is universal agreement among physicists even though the problem may be analytically intractable. If on the other hand the question asked pertains to the theory of measurement in quantum mechanics, that is, the interpretation of certain experimental observations performed on a microscopic system, it is virtually impossible to find two physicists who agree. What is even more interesting is that usually these controversies are void of any physical predictions and are entirely of an epistemological character. They reflect our difficulty in bridging the gap between the quantum mechanical treatment of the microscopic system being observed and the classical treatment of the macroscopic apparatus with which the measurement is performed. It is usually argued that we, physicists, have difficulty comprehending the formalism of quantum mechanics because our intuition is macroscopic, hence classical, in nature. Now if that were the case, we should have as much difficulty with special relativity, since we are hardly used to speeds comparable to that of light. Yet, strange as it seems at first, I have never heard physicists argue about the "twin paradox," the classic example of an unexpected prediction of Einstein's relativity. So there must be something about quantum mechanics that "rubs us" the wrong way. The question is what?

Perhaps the best way in which the strange predictions of quantum mechanics can be quantified is a certain inequality first formulated by Bell (Bell 1965). For illustration, consider a positronium atom, with total angular momentum zero, that decays into an electron and a positron. Suppose we let the electron and the positron drift apart and then measure their spin components along two axes by passing them through two magnetic fields. Now in quantum mechanics the state of the positronium atom is a linear superposition of spin-up and spin-down states: $(|\uparrow\rangle_+ |\downarrow\rangle_- - |\downarrow\rangle_+ |\uparrow\rangle_-)/\sqrt{2}$. We could therefore ask ourselves whether in each passage through the apparatus the electron and the positron have a well-defined spin (up or down), albeit unknown to us. Some elementary probabilistic reasoning shows immediately that if that were the case, the probabilities for observing up or down spins along given axes would have to obey Bell's inequality. The experimentally measured probabilities violate this inequality, in agreement with the predictions of quantum mechanics. So the uncertainties in quantum mechanics are not due to incomplete knowledge of some local hidden variables. What is even stranger is that in a refinement of the experiment in which the axes of the magnetic fields are changed in an apparently random fashion (Aspect, Grangier, and Roger 1982), the violation of Bell's inequality persists, indicating correlations between space-like events (that is, events that could be causally connected only by signals traveling faster than the speed of light). While in this experiment no information is being transmitted by such superluminal signals, and hence no conflict with special relativity exists, the implication of space-like correlations hardly alleviates the physicist's uneasiness about the correct interpretation of quantum mechanics. Of course this uneasiness is not felt by all physicists. Particle physicists, for instance, take the validity of quantum mechanics for granted. To wit, anybody who reads *Time* knows that they, having "successfully" unified weak, electromagnetic, and strong interactions within the framework of quantum field theory, are presently subduing the last obstacle, quantizing gravity by unifying all interactions into a quantum field theory of strings. And they are doing so in spite of the fact that the existence of classical gravitational radiation, let alone that of the quantized version (gravitons), has not been established experimentally.

An even older controversy, which in the opinion of some physicists has long ceased to be an interesting problem, concerns the ergodic hypothesis, the subject of this discussion. I will try to elaborate on this topic as fully as my knowledge will allow, but, by way of introduction, let me just say that the ergodic hypothesis is an attempt to provide a dynamical basis for statistical mechanics. It states that the time-average value of an observable—which of course is determined by the dynamics—is equivalent to an ensemble average, that is, an average at one time over a large number of systems all of which have identical thermodynamic properties but are not identical on the molecular level. This hypothesis was advanced over one hundred years ago by Boltzmann and Maxwell while they laid the foundations of statistical mechanics (Boltzmann 1868, 1872 and Maxwell 1860, 1867). The general consensus is that the hypothesis, still mathematically unproven, is probably true yet irrelevant for physics. The purpose of this article is to review briefly the status of the ergodic hypothesis from mathematical and physical points of view and to argue that the hypothesis is of interest not only for statistical mechanics but for physics as a whole. Indeed the mystery of quantum mechanics itself may possibly be unraveled by a deeper understanding of the ergodic hypothesis. This last remark should come as no surprise. After all, the birth of quantum mechanics was brought about by the well-known difficulties of classical statistical mechanics in explaining the specific heats of diatomic gases and the blackbody radiation law. I shall elaborate on the possible connection between the ergodic hypothesis and the resolution of these major puzzles in the last part of this article.

## The Mathematics of the Ergodic Hypothesis

I shall begin my presentation with the easier part of the problem, the mathematical formulation of the ergodic hypothesis. Consider some physical system with $N$ degrees of freedom and let $q_1, \ldots, q_N$ be its positions and $p_1, \ldots, p_N$ its momenta. We shall assume that the specification of the set of initial positions $\{q_0\}$ and momenta $\{p_0\}$ at time $t = 0$ uniquely specifies the state of the system at any other time $t$ via the equations of motion:

$$\frac{\partial q_i(t)}{\partial t} = \frac{\partial q_i\left(\{q(t)\},\{p(t)\}\right)}{\partial t}$$

and

$$\frac{\partial p_i(t)}{\partial t} = \frac{\partial p_i\left(\{q(t)\},\{p(t)\}\right)}{\partial t}. \tag{1}$$

The time evolution of the system can be represented as a path, or trajectory, through phase space, the region of allowed states in the space defined by the $2N$ independent coordinates $\{q\}$ and $\{p\}$. An observable of this system $O$ is an arbitrary function of $\{q\}$ and $\{p\}$, $O(\{q\},\{p\})$. The time-average value of some observable $O(\{q\},\{p\})$ along the phase-space trajectory starting at $t = 0$ at $\{q_0\},\{p_0\}$ is defined as

$$\overline{O}_T(\{q_0\},\{p_0\}) \equiv \frac{1}{T}\int_0^T dt\, O(\{q(t)\},\{p(t)\}). \tag{2}$$

Obviously the integral in Eq. 2 makes sense only for suitable functions of $\{q\}$ and $\{p\}$, which are the only ones we shall consider. In fact we shall further restrict the class of observables to those for which $\lim_{T\to\infty}\overline{O}_T$ exists. (This is not a severe restriction; for instance, if $O(\{q(t)\},\{p(t)\})$ is bounded along the trajectory, the limit clearly exists.) The notation in Eq. 2 makes clear that, a priori, time-average values depend upon the initial conditions $\{q_0\}$ and $\{p_0\}$.

As time passes, the trajectory of the system winds through the phase space. If the motion takes place in a bounded domain, one might expect that as $T \to \infty$ the average values of most observables settle down to some sort of equilibrium values (time-independent behavior). What would the phase-space trajectory look like if the system approached dynamical equilibrium? One could characterize it by saying that the frequency with which different neighborhoods of the phase space are visited converges to some limiting value $\mu(\{q\},\{p\})$ at each point in phase space. That such limiting frequencies exist under quite general circumstances was shown in 1927 by Birkhoff (see Birkhoff 1966) and constitutes the first step towards bridging the gap between dynamics and statistics. Indeed, Birkhoff's theorem allows one to replace time averages by ensemble averages, defined as follows. Let the state of the system be specified by the sets $\{q\}$ and $\{p\}$, and postulate that the probability for the system to be in the neighborhood of the state $(\{q\}, \{p\})$ is

$$\left(\prod_{i=1}^{N} dq_i \, dp_i\right) \mu(\{q\}, \{p\}). \tag{3}$$

That is, the general form of the probability measure is the time-independent frequency $\mu$ times the volume element of the phase space. A particular probability measure specifies completely a particular ensemble of representative systems; that is, it gives the fraction of systems in the ensemble that are in the state $(\{q\}, \{p\})$. In keeping with usual probabilistic notions, I shall assume that the probability measure has been normalized so that the integral of the probability measure for all possible states $(\{q\}, \{p\})$ is unity,

$$\int \left(\prod_{i=1}^{N} dp_i \, dq_i\right) \mu(\{q\}, \{p\}) = 1. \tag{4}$$

The ensemble average of the observable $O(\{q\}, \{p\})$ is defined as

$$\langle O \rangle_\mu \equiv \left(\prod_{i=1}^{N} \int dq_i \, dp_i\right) \mu(\{q\}, \{p\}) \, O(\{q\}, \{p\}). \tag{5}$$

Birkhoff's theorem states that, if the motion is restricted to a bounded domain, then for many initial conditions there exists an ensemble (probability measure) such that the time-average value of the observable equals an ensemble average:

$$\lim_{T \to \infty} \overline{O}_T(\{q_0\}, \{p_0\}) = \langle O \rangle_\mu. \tag{6}$$

Please note that Eq. 6 indicates that the time-average value of $O(\{q\}, \{p\})$ becomes independent of the initial conditions $\{q_0\}$ and $\{p_0\}$ as $T \to \infty$. As already mentioned above, this is true for many, but generally not all, initial conditions. If Eq. 6 is true for almost all initial conditions (for all points in the allowed phase space except for a set of measure zero), the flow through phase space described by Eqs. 1 must be fully ergodic; that is, for almost all initial conditions $\{q_0\}, \{p_0\}$ and with probability 1, the flow passes arbitrarily close to any point $\{q\}, \{p\}$ in phase space at some later time. The assumption in statistical mechanics that time averages of macroscopic variables can be replaced by ensemble averages (that is, that Eq. 6 holds) is therefore called the ergodic hypothesis.

In general, however, the flow through the phase space defined by the equations of motion may not cover the whole of the allowed phase space for almost all initial conditions. Instead the allowed phase space is divided into several "ergodic" components, that is, subregions $\Omega_i$ of the phase space such that if the flow starts in subregion $\Omega_i$, then there exists a time $t$ at which the flow will touch any given neighborhood in the set of neighborhoods covering $\Omega_i$. Moreover the flow remains in $\Omega_i$ for all time. Consequently, time-average values do depend on knowing in which "ergodic component" the system was started.

**The Ergodic Hypothesis and the Equipartition of Energy**. In statistical mechanics the ergodic hypothesis, which proposes a connection between dynamics and statistics, is sometimes regarded as unnecessary, and attention is placed instead on the assumption that all allowed states are equally probable. In this paper I emphasize that when time averaging is relevant to a problem, the assumption of equal a priori probabilities is essentially equivalent to the ergodic hypothesis (Eq. 6). To see this I will restate the general problem and gradually narrow it down to the context of classical statistical mechanics.

In general, given a phase space $\Omega$ and a probability density $\mu(\{q\}, \{p\})$, one has defined an ensemble. Furthermore one can consider a map of the phase space onto itself. (An example is provided by Eqs. 1, which are really a set of maps indexed by the continuous parameter $t$). A natural question to ask is whether the probability measure

$$\left(\prod_{i=1}^{N} dq_i \, dp_i\right) \mu(\{q\}, \{p\})$$

is invariant under this map. As we have said, Birkhoff's theorem states that under many circumstances such invariant measures exist and allow the replacement of time averages by ensemble averages. Thus the existence and construction of all the invariant measures for a certain flow is the first of two mathematical problems related to the ergodic hypothesis.

As stated so far this problem is much more general than the one of interest to Boltzmann and Maxwell in connection with the foundations of statistical mechanics. Indeed, the existence of a probability measure left invariant by a given set of maps can be investigated whether or not the sets $\{q\}$ and $\{p\}$ defining the maps are canonically conjugate variables derivable from a Hamiltonian, whether the set of maps is discrete or continuous, etc. At present the construction of such invariant measures is being actively pursued by researchers studying dynamical systems, especially dissipative ones such as those relevant to the investigation of turbulence (for example, systems described by the Navier-Stokes equations). (See the section Geometry, Invariant Measures, and Dynamical Systems in "Probability and Nonlinear Systems.")

Of particular interest in statistical mechanics, especially in connection with the ergodic hypothesis, is the invariant measure appropriate for describing physically isolated systems. The ensemble specified by this measure is traditionally called the microcanonical ensemble. The systems of interest are characterized by nonlinear interactions among the constituents and by a very large number of degrees of freedom. Generically, certain observables of a physically isolated system, such as the total energy and electric charge, are conserved; that is, they remain constant at their initial values. So let $\{I_i(\{q\}, \{p\})\}$, $i = 1, \ldots, M$ be the complete set of independent, conserved observables of a system with $N$ degrees of freedom. Obviously $M \leq 2N$. Since the flow in Eqs. 1 obeys all these conservation laws, it is clear that any invariant measure of the flow must be compatible with all the conservation laws. Consequently the probability measure must contain a delta function for each conserved quantity so that the probability is nonzero only when the conservation law is satisfied. (A delta function $\delta(x - x_0)$ can be thought of as having the value $\frac{1}{2\epsilon}$ for $x$ values between $x_0 - \epsilon$ and $x_0 + \epsilon$ for any $\epsilon$, no matter how small, and the value 0 everywhere else. The integral of a delta function is thus equal to unity.)

The fundamental hypothesis in statistical mechanics is that for isolated systems of physical interest (complicated nonlinear systems with many degrees of freedom), the measure

$$\left(\prod_{i=1}^{N} dq_i\, dp_i\right) \prod_{j=1}^{M} \delta\big(I_j(\{q\},\{p\}) - I_j(\{q_0\},\{p_0\})\big) \tag{7}$$

is left invariant by the equations of motion and is the only such measure. In other words, the hypothesis states that the microcanonical ensemble is defined by the measure in Eq. 7. Note that the probablity density in Eq. 7 is flat; that is, all regions of phase space consistent with the conservation laws are equally probable.

To understand why this assumption of equal a priori probabilities is, in effect, a restatement of the ergodic hypothesis, one must realize that the only systems under consideration in classical statistical mechanics are Hamiltonian systems (systems for which the equations of motion can be derived from a Hamiltonian principle). The existence of a Hamiltonian function $H(\{q\},\{p\})$ means that the equations describing the flow through phase space, Eqs. 1, can be written in the form

$$\dot{q}_i = [q_i, H]$$

and

$$\dot{p}_i = [p_i, H] \tag{8}$$

Here $[f, g]$ denotes the Poisson bracket:

$$[f, g] = \sum_{i=1}^{N} \left( \frac{\partial f}{\partial q_i}\frac{\partial g}{\partial p_i} - \frac{\partial f}{\partial p_i}\frac{\partial g}{\partial q_i} \right). \tag{9}$$

The existence of a simplectic structure (the Poisson bracket) is a very restrictive condition on the flow, much more so than the mere conservation of the energy. Indeed, through Liouville's theorem, it guarantees the conservation of the phase-space volume element

$$\left(\prod_{i=1}^{N} dq_i\, dp_i\right), \tag{10}$$

and thus it proves that the measure in Eq. 7 is invariant under Hamiltonian flows. Thus the first mathematical problem of constructing an invariant measure is solved for Hamiltonian systems. Consequently the ergodic hypothesis (Eq. 6) is automatically satisfied *provided that the flow is fully ergodic*. Proving that the flow is fully ergodic is the second mathematical problem related to the ergodic hypothesis and is the one that remains to be solved for Hamiltonian systems. If in fact the flow is not ergodic, then the assumption of equal a priori probabilities would not describe the time-average behavior of the system, at least not for all possible observables.

Note that if the flow is fully ergodic and all allowed states are equally probable, then we have an equipartition of energy; that is, the energy of the system is divided equally among the $N$ degrees of freedom. Indeed, let us consider for simplicity the case of a Hamiltonian system in which only the total energy is conserved. The microcanonical measure then is simply

$$\left(\prod_{i=1}^{N} dq_i\, dp_i\right) \delta\big(H(\{q\},\{p\}) - E\big). \tag{11}$$

Quite often the Hamiltonian has the form

$$H = \sum_{i=1}^{N} \frac{p_i^2}{2m_i} + V(\{q\}), \tag{12}$$

where the $m_i$'s are the particle masses. Because of the symmetry of the measure defined by Eqs. 11 and 12 under the interchange of the $p_i$'s, one can easily show that the average kinetic energy $\langle p_i^2/2m_i\rangle$ is independent of $i$. Usually one uses that fact to define a temperature $T$ via $\langle p^2/2m\rangle = kT/2$ (where $k$ is the Boltzmann constant). Such considerations can be extended to the normal modes of a lattice, which will be discussed later, and are generically referred to as the equipartition of energy.

**Mathematical Results.** Having formulated the mathematical problem, it may be of interest to state briefly what rigorous results have been obtained so far about the circumstances under which a flow is fully ergodic.

i) Oxtoby and Ulam proved in 1941 that in a bounded phase space the continuous ergodic transformations are everywhere dense in the space of all continuous measure-preserving transformations. In other words, a topology can be chosen such that ergodic transformations form the "bulk" of the whole space of continuous measure-preserving maps. This theorem says nothing about the measure of the ergodic transformations, which may even be vanishing. (See page 110 in "Learning from Ulam.") A corresponding theorem stating an analogous property of a real dynamical system with a finite number of degrees of freedom does not exist, and in fact the KAM theorem proves the contrary (see below). It is also known that Hamiltonian flows are quite rare among measure-preserving maps, and therefore the Oxtoby and Ulam result guarantees nothing about the density of ergodic Hamiltonian flows in the space of all Hamiltonian flows.

ii) For finite $N$ the Kolmogorov-Arnold-Moser (KAM) theorem (see Arnold and Avez 1968) guarantees that the ergodic hypothesis is violated for a certain class of systems. The theorem considers a completely integrable system ($M = N$ in Eq. 7) and its response to an arbitrary, weak nonlinear perturbation. By a canonical transformation one can show that a completely integrable system with $N$ degrees of freedom is equivalent to $N$ decoupled harmonic oscillators; hence it is a linear system, and its motion in phase space occurs on hypertori rather than on the whole phase space. The KAM theorem states that in the phase space of a weakly nonintegrable (weakly nonlinear) Hamiltonian, some motions still are restricted to tori, and these tori occupy a nonzero measure of the phase space. (Figure 1 shows a typical structure of the

**PHASE SPACE OF A WEAKLY NONINTEGRABLE HAMILTONIAN SYSTEM**

**Fig. 1. The system has four degrees of freedom, but conservation of energy allows us to display the phase space in three dimensions, which represent the variables $x, y$, and $\frac{\partial y}{\partial t}$. The phase space contains nested invariant tori on which motion is quasiperiodic so that a single orbit covers a torus densely. The gaps between the tori are chaotic regions in which the orbits appear as random as the toss of a coin. Since the nested tori have a finite measure in the phase space, this Hamiltonian system violates the ergodic hypothesis.**

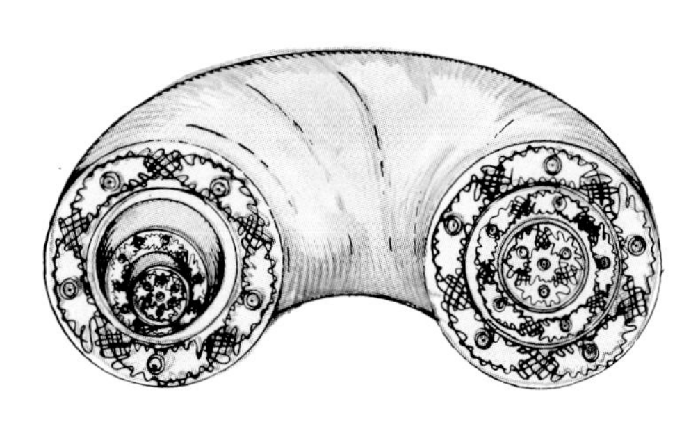

phase space for a weakly nonintegrable Hamiltonian.) Thus these systems have several ergodic components.

iii) In 1963 Sinai proved the ergodic hypothesis for certain billiard systems (Hamiltonian systems in which hard spheres bounce elastically off each other and the container walls). The geometry of the boundary turns out to be a crucial factor in proving that the flow is ergodic.

iv) It has not been possible to prove the ergodic hypothesis even for a gas of hard spheres, although it is generally believed to be true in this case.

v) For a long time the general belief was that the KAM theorem poses no problem for the ergodic hypothesis once the thermodynamic limit (the limit as $N \rightarrow \infty$ at fixed density) is taken. Counterexamples to this claim have recently been constructed (Bellissard and Vittot 1985), but it is premature to judge their generality.

vi) There exists no satisfactory formulation of the ergodic hypothesis for continuous media (field theory), since it is not known how to generalize the microcanonical measure to systems with an infinite number of degrees of freedom, especially when the total energy of the system is finite. It is interesting that while appropriate ensemble averages have not been defined, the existence of global solutions (in time), and therefore the existence of time averages, for several interesting field theories (such as classical electrodynamics and Yang-Mills theories) has been established (Eardley and Moncrief 1982).

In conclusion, from a mathematical point of view, the ergodic hypothesis has proved to be one of the most difficult problems in the last hundred years or so. Only two flows, both billiards, have been proven to be ergodic. Perhaps today's computers will speed up the rate of analytical progress by helping our intuition about the nature of the flow.

## The Physics of the Ergodic Hypothesis

Next I wish to analyze the ergodic hypothesis from a physical point of view. Undoubtedly, a dynamical approach to a physical system with many degrees of freedom, such as a gas, is impossible, and a statistical one must be developed. In doing so one must endeavor to capture the right physics. If the attempt has been really successful, the theory will withstand experimental scrutiny. But what should be done if the predictions go astray, as did the predictions of classical statistical mechanics for blackbody radiation? A sensible approach is to go back and examine what fundamental assumptions were made, which is what I shall do now.

The first question that must be settled is what should be considered "the system." Indeed the instruction in statistical mechanics is to integrate over all canonical positions and momenta with a certain measure. However, one must decide which degrees of freedom to include. For instance, take the case of the diatomic gas. Each molecule has two atoms, each atom has its own electrons and nucleus, and the latter in turn is made of quarks and gluons, say. Moreover, since the constituents are charged, they are coupled to the electromagnetic field inside the container (and also to the gravitational field). Probably most readers will think that this is not a serious question: at a certain temperature only certain degrees of freedom are excited, and these are the only ones to be integrated over. Hidden within this superficially sensible-sounding answer is one of two extremely important assumptions:

i) The ergodic hypothesis is strictly false, so that certain degrees of freedom, although dynamically coupled, never get excited and act as spectators to the thermal equilibrium that sets in for the remaining degrees of freedom.

ii) Or, the system dynamically develops largely different time scales, and the number of degrees of freedom that are more or less in equilibrium keeps increasing with time.

In either case the use of statistical mechanics becomes more subtle, since only by gaining a good grasp of the underlying dynamics can one decide what degrees of freedom are relevant in certain circumstances. In particular, there is no a priori reason to believe that the contributions to the specific heat of the vibrations and the rotations of a diatomic gas ought to be equal at all temperatures and during a typical time of observation, as was assumed in the classical predictions of statistical mechanics. Neither is there any reason to predict the Rayleigh-Jeans distribution (Fig. 2) for blackbody radiation (which assumes the equipartition of energy between all modes of the electromagnetic field), since some modes of the cavity may be effectively decoupled (case i above) or so weakly coupled that they haven't had time to thermalize (case ii). Thus the standard examples for the breakdown of classical statistical mechanics may reflect an inappropriate application of the ergodic hypothesis rather than a need for quantization, as is usually argued in physics textbooks.

The second important question that must be addressed in deciding the relevance of the ergodic hypothesis for physics is why we are using a statistical description in a given physical situation. Consider, for instance, the measurement of the specific heat of a diatomic gas. Typically one lets the gas "reach equilibrium" with a reservoir at a given temperature and then makes a certain macroscopic measurement during a certain time interval. To obtain reasonable statistics, the measurement is repeated several times. Clearly the process just described involves three types of averaging at the molecular dynamics level:

i) over initial conditions (each repetition of the measurement involves a different set of initial conditions);

ii) over time (each measurement extends over a certain time, during which the gas evolves as a dynamical system); and

iii) over microscopic degrees of freedom (this type of averaging is inherent in the measurement of macroscopic variables).

Before analyzing in detail the likely statistical relevance of each of these averaging operations, let me hasten to say that clearly only the averaging over time has anything to do with the ergodic hypothesis. Those physicists who believe that the ergodic hypothesis is not important for the foundations of statistical mechanics dismiss the

**BLACKBODY RADIATION AT 1600 K**

**Fig. 2. Theoretical predictions and experimental data for the power radiated by a blackbody at 1600 K. The classical Rayleigh-Jeans law, $u(\nu, T) = (8\pi/c^3)\nu^2 kT$, is based on equipartition of energy among all the modes of an electromagnetic field. The total (kinetic plus potential) energy in each mode is $kT$, and the number of modes in the frequency interval $(\nu, \nu+d\nu)$ is $8\pi\nu^2/c^3$, which is proportional to $\lambda^{-2}$. The quantum Planck law, in agreement with experiment, yields a peaked distribution that decreases rapidly with wavelength. The Planck law is based on the assumption that the energy in each mode is quantized; that is $E = nh\nu$, where $n$ is an integer and $h$ is Planck's constant.**

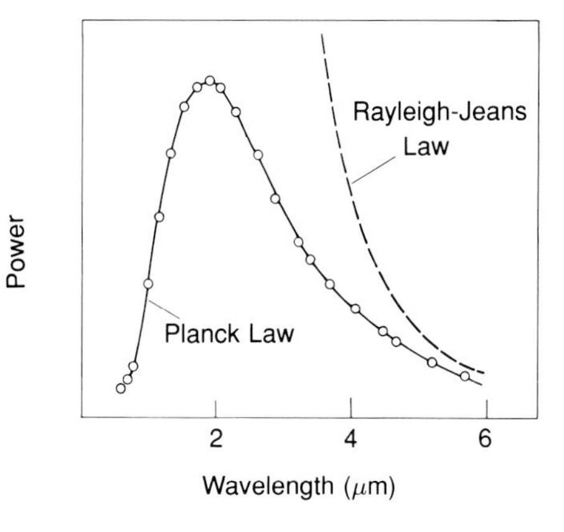

# The FPU Problem

*Excerpts from "Studies of Nonlinear Problems" by Fermi, Pasta, and Ulam*

This report is intended to be the first one of a series dealing with the behavior of certain nonlinear physical systems where the nonlinearity is introduced as a perturbation to a primarily linear problem. The behavior of the systems is to be studied for times which are long compared to the characteristic periods of the corresponding linear problems.

The problems in question do not seem to admit of analytic solutions in closed form, and heuristic work was performed numerically on a fast electronic computing machine (MANIAC I at Los Alamos).* The ergodic behavior of such systems was studied with the primary aim of establishing, experimentally, the rate of approach to the equipartition of energy among the various degrees of freedom of the system. Several problems will be considered in order of increasing complexity. This paper is devoted to the first one only.

We imagine a one-dimensional continuum with the ends kept fixed and with forces acting on the elements of this string. In addition to the usual linear term expressing the dependence of the force on the displacement of the element, this force contains higher order terms. For the purposes of numerical work this continuum is replaced by a finite number of points (at most 64 in our actual computation) so that the partial differential equation defining the motion of this string is replaced by a finite number of total differential equations. ...

The solution to the corresponding linear problem is a periodic vibration of the string. If the initial position of the string is, say, a single sine wave, the string will oscillate in this mode indefinitely. Starting with the string in a simple configuration, for example in the first mode (or in other problems, starting with a combination of a few low modes), the purpose of our computations was to see how, due to nonlinear forces perturbing the periodic linear solution, the string would assume more and more complicated shapes, and, for $t$ tending to infinity, would get into states where all the Fourier modes acquire increasing importance. In order to see this, the shape of the string, that is to say ... [its displacement,] and the kinetic energy ... were analyzed periodically in Fourier series. ...

Let us say here that the results of our computations show features which were, from the beginning, surprising to us. Instead of a gradual, continuous flow of energy from the first mode to the higher modes, all of the problems show an entirely different behavior. Starting in one problem with a quadratic force and a pure sine wave as the initial position of the string, we indeed observe initially [see figures on next page] a gradual increase of energy in the higher modes as predicted (e.g., by Rayleigh in an infinitesimal analysis). Mode 2 starts increasing first, followed by mode 3, and so on. Later on, however, this gradual sharing of energy among successive modes ceases. Instead, it is one or the other mode that predominates. For example, mode 2 decides, as it were, to increase rather rapidly at the cost of all other modes and becomes predominant. At one time, it has more energy than all the others put together! Then mode 3 undertakes this role. It is only the first few modes which exchange energy among themselves and they do this in a rather regular fashion. Finally, at a later time mode 1 comes back to within one per cent of its initial value so that the system seems to be almost periodic. All our problems have at least this one feature in common. Instead of gradual increase of all the higher modes, the energy is exchanged, essentially, among only a certain few. It is, therefore, very hard to observe the rate of "thermalization" or mixing in our problem, and this was the initial purpose of the calculation.

If one should look at the problem from the point of view of statistical mechanics, the situation could be described as follows: the phase space of a point representing our entire system has a great number of dimensions. Only a very small part of its volume is represented by the regions where only one or a few out of all possible Fourier modes have divided among themselves almost all the available energy. If our system with nonlinear forces acting between the neighboring points should serve as a good example of a transformation of the phase space which is ergodic or metrically transitive, then the trajectory of almost every point should be everywhere dense in the whole phase space. With overwhelming probability this should also be true of the point which at time $t = 0$ represents our initial configuration, and this point should spend most of its time in regions corresponding to the equipartition of energy among various degrees of freedom. As will be seen from the results this seems hardly the case. ...

In a linear problem the tendency of the system to approach a fixed "state" amounts, mathematically, to convergence of iterates of a transformation in accordance with an algebraic theorem due to Frobenius and Perron. ... Such behavior is in a sense diametrically opposite to an ergodic motion and is due to a very special character, linearity of the transformations of the phase space. The results of our calculation on the nonlinear vibrating string suggest that in the case of transformations which are approximately linear, differing from linear ones by terms which are very simple in the algebraic sense (quadratic or cubic in our case), something analogous to the convergence to eigenstates may obtain. ... ■

*Editor's note: The interpretation of the unexpected recurrences is now different. See David Campbell's discussion on page 244.*

*We thank Miss Mary Tsingou for efficient coding of the problems and for running the computations on the Los Alamos MANIAC machine.

statistical relevance of time averaging for macroscopic observables.

The averaging over initial conditions should not be of much consequence statistically. Indeed, even if one assumes that the gas is simply a collection of hard spheres (with no internal structure), the gas still constitutes a dynamical system with somewhere on the order of $10^{23}$ degrees of freedom. Unless the initial state is very special or the time of observation very short, repeating an experiment ten or a hundred times should not have important consequences. In fact, a typical measurement lasts at least a few minutes; during such a time interval each molecule undergoes, at room temperature and normal pressure, about $10^7$ collisions. Hence the number of states through which the gas passes dynamically (in time) is much larger than that due to the repetition of the experiment. Of course, as one lowers the temperature or the pressure, the collisions become more rare, so the time of observation must be increased to avoid large fluctuations in individual measurements.

Perhaps the most important averaging is the "coarse graining" involved in obtaining macroscopic variables. Two large numbers are involved in a typical measurement: the total number of degrees of freedom of the system and the number of degrees of freedom that are averaged together to obtain a macroscopic variable. The second number appears naturally in a system containing a large number of indistinguishable constituents. For instance, in determining the local density in a gas, one does not care about the trajectory of any single particle but rather about the average number of trajectories crossing a macroscopic volume at any time. Use of the laws of large numbers (see "A Tutorial on Probability, Measure, and the Laws of Large Numbers") in this context guarantees that, in spite of the fact that the underlying dynamics may be time-reversal invariant, macroscopic variables (almost) always tend to relax to their equilibrium values. In other words, because of the large numbers involved in specifying macroscopic variables, the macroscopically specified state of the system has overwhelming probability to evolve towards the equilibrium state, even if the microscopic dynamics is time-reversal invariant. Hence, an arrow of time exists at the macroscopic level even if it does not at the microscopic level. This frequently stated paradox of statistical mechanics is a straightforward consequence of the laws of large numbers.

## Confronting the Ergodic Hypothesis with Experiment

Having discussed the types of averaging involved in a real experiment, let us reconsider the experimental circumstances under which classical statistical mechanics could be expected to work. Historically, statistical mechanics appeared in connection with the endeavors to study, for example, very nearly ideal gases. (In an ideal gas the molecules are free except for occasional elastic collisions with each other or with the walls of the container.) Its foundations were statistical (predictions were based on considering an ensemble of systems, primarily the microcanonical or the canonical ensemble), in spite of the efforts of Boltzmann and Maxwell to give it a dynamical basis by invoking the ergodic hypothesis.

The fundamental assumption of statistical mechanics for an isolated system is the equal a priori probability on the hypersurface (in phase space) determined by all the conservation laws (Eq. 7). This probability measure defines the microcanonical ensemble. If the underlying dynamics is derivable from a Hamiltonian, by Liouville's

### FIGURES FROM THE FERMI-PASTA-ULAM PAPER

**Fig. 1. The quantity plotted is the energy (kinetic and potential in each of the first five modes). The units for energy are arbitrary. $N$ [the number of parts into which the string is divided] = 32; $\alpha$ [the coefficient of the quadratic term in the force equation] = $\frac{1}{4}$; $\delta t^2$ = $\frac{1}{8}$ [$\delta t$ is the length of the computational cycle]. The initial form of the string was a single sine wave. The higher modes never exceeded in energy 20 of our units. About 30,000 computation cycles were calculated.**

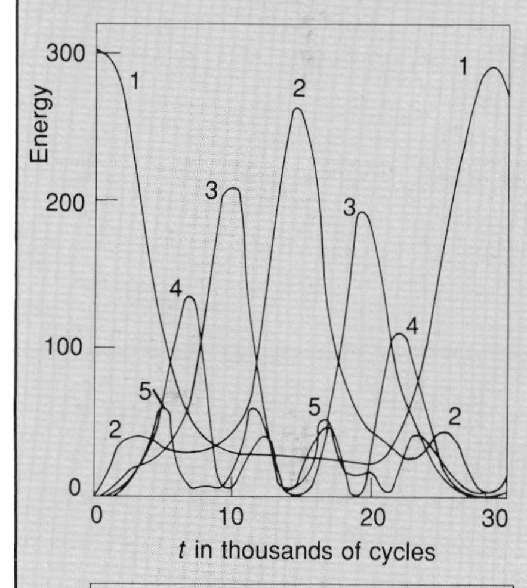


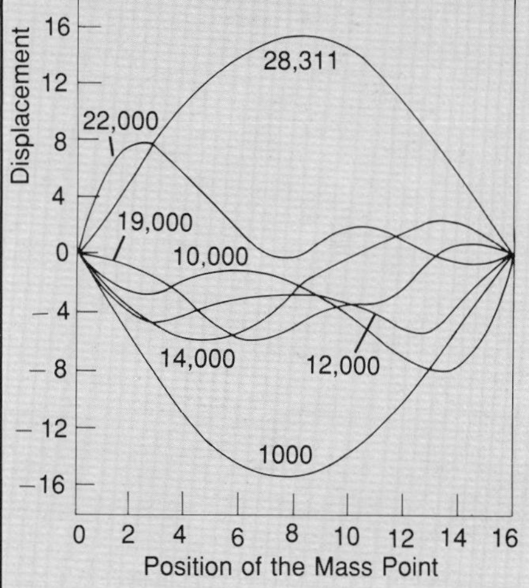


**Fig. 8. This drawing shows not the energy but the actual *shapes*, i.e., the displacement of the string at various times (in cycles) indicated on each curve. The problem is that of Fig. 1.**

theorem such a probability measure is invariant in time. Thus the only reason time averages could be different from ensemble averages would be a lack of ergodicity in the flow. In the case of a system consisting of only one species of indistinguishable particles, this potential difficulty is suppressed first by averaging over many initial conditions (so that even if the flow is not ergodic, the starting points may fall in different "ergodic" subregions) and second by measuring time-average values of macroscopic, not microscopic, variables. The chances that under these circumstances one would observe a difference between the predictions of statistical mechanics and experiment are very slim (recall the laws of large numbers), and indeed under these experimental conditions the predictions of classical statistical mechanics enjoyed great success. This explains the utter confidence of most physicists in the predictive power of statistical mechanics and their dismissal of the ergodic hypothesis as a technical, probably irrelevant detail.

On the other hand, suppose one uses the theory to make predictions about a diatomic gas, which even under the most simplifying assumptions has at least two species of indistinguishable degrees of freedom, say vibrations and translations. Without invoking the ergodic hypothesis, I can think of no a priori reason for the contributions to the specific heat of these two types of motions being found equal in typical measurements. In fact, even if the ergodic hypothesis is true, it is possible that the coupling of these two types of motions is so weak that during typical times of observation they do not reach equilibrium with each other. Yet it was the assumption that the two types of motion are in equilibrium that led to the discrepancy between classical statistical mechanics and experiment. Therefore I feel that it is unjustified to rely upon the many successes of statistical mechanics to dismiss questions regarding its foundations. On the contrary, an understanding of the ergodic hypothesis and especially of the times involved for exciting certain degrees of freedom should be equally challenging for the mathematician and the physicist.

**Quantum Mechanics: A Case of Mistaken Identity?** I would like to close this brief review of these complicated and long-standing problems with some speculations about a possible connection between the ergodic hypothesis and the necessity of using quantum mechanics at the microscopic level. First a few words about the blackbody radiation law. I have tried to emphasize the importance of measuring macroscopic averages, as well as that of particle indistinguishability, in obtaining agreement between the predictions of statistical mechanics and experiment. I think the case of the blackbody falls outside this realm. Consider a cubic lattice in $D$ dimensions. At each site let there be a particle sitting in some anharmonic potential, attached through harmonic springs to its $2D$ nearest neighbors. If the boundary conditions are periodic, the system consists of identical yet distinguishable (by site coordinates) particles. We could form macroscopic quantities by averaging over the positions or velocities of all the particles in a cube of macroscopic size and expect reasonable agreement with the predictions of statistical mechanics. Alternatively we could describe the system in terms of its normal modes and attempt to verify the classical prediction, namely, the Rayleigh-Jeans energy distribution shown in Fig. 2 (that is, the equipartition of the energy among all the normal modes). Many such studies have been performed numerically, the first being the celebrated 1955 work of Fermi, Pasta, and Ulam (see "The Fermi-Pasta-Ulam Problem"). It is always found that at sufficiently low energy density, the distribution of energy among the modes of the lattice differs drastically from the statistical prediction and in fact depends upon the initial conditions. Obviously either these systems are not ergodic, or at least the times of thermalizing the different modes are much longer than a typical time of numerical integration. And no macroscopic averaging is available to save the day! It is also known that leaving the energy density fixed and refining the lattice (taking the continuum limit) increases the discrepancy (Patrascioiu, Seiler, and Stamatescu 1985). Although such results have been accumulating for over thirty years now, they are not yet understood. Some say the systems are so close to being integrable that KAM tori or very slow diffusion rates occur in the phase space. Others claim that statistical mechanics should hold only in the thermodynamic limit (which is clearly not attainable numerically). Most physicists dismiss the whole story, since they "know" that statistical mechanics works in real life. I think this is a very narrow point of view: the problem being discussed is very much like that of the blackbody radiation law, and that was one of the failures of classical statistical mechanics. Is there a good theoretical (dynamical) basis for predicting the Rayleigh-Jeans distribution in classical physics, as the standard textbooks claim? Or are we pushing the statistical predictions in a domain for which there is no reason to expect them to hold? In "Does Equipartition of Energy Occur in Nonlinear Continuous Systems?" I describe some numerical experiments I have performed to test the validity of the statistical-mechanics predictions for a one-dimensional version of the blackbody problem and for the specific heats of systems with more than one species of degrees of freedom. Notably I found that, over the times of observation available in computer experiments, the systems failed to fulfill the ordinary expectations of an equipartition of energy. The same discrepancy has been found in many other numerical experiments.

It is well known that the resolution of the above-mentioned experimental difficulties of statistical mechanics (specific heats and blackbody radiation) was found in abandoning the classical approach to physics in favor of the quantum one. As mentioned in the introduction, this revolution has had an unqualified experimental success, although it has raised serious epistemological questions, which continue to haunt us more than sixty years after the advent of the quantum theory. I would like to give a brief outline of a heresy that I have advocated for a few years now (Patrascioiu 1983), one directly connected to the ergodic hypothesis. As I mentioned earlier, if one contemplates a dynamical basis for statistical mechanics, one is faced with a real dilemma. The accepted formulation of the electromagnetic and the gravitational interactions demands that, in essence, everything in the universe interact with everything else. (This is so because of the long-range nature of these interactions.) In fact, the notion of an isolated object (or even system) is clearly an abstraction without any a priori physical basis, since ultimately everything is coupled to everything else through the electromagnetic and gravitational fields. All we can hope is either that the ergodic hypothesis is strictly false or that the times needed to excite certain degrees of freedom are so large that we can ignore them under some circumstances. In either case certain prejudices that have been passed from generation to generation should be abandoned and their bases be opened for investigation. For instance, in the absence of a dynamical calculation, there is no basis to claim that Planck's distribution for blackbody radiation is irreconcilable with classical electromagnetism. (In fact, the distribution found numerically and shown in Fig. 2 of the sidebar very much resembles Planck's law.)

*continued on page 278*

# Does Equipartition of Energy Occur in Nonlinear Continuous Systems?

The celebrated work of Fermi, Pasta, and Ulam was the first of numerous attempts to study the distribution of energy in nonlinear continuous media. These attempts have all been indirect in that the systems are simulated by lattices of particles interacting through nonlinear potentials. The results have consistently failed to support the classical point of view regarding equipartition of energy—and yet they have stirred little excitement in the physics community. Perhaps this is so for two reasons: (i) the systems analyzed may be subject to an infinite number of conservation laws (and thus may be effectively linear), so that the individual degrees of freedom are not coupled and equipartition of energy cannot occur; (ii) the results may simply be artifacts of the lattice simulations.

Here I present some results from two of my own studies, the first of a one-dimensional model of the blackbody problem (Adrian Patrascioiu, *Physical Review Letters* 50(1983): 1879) and the second of a three-dimensional system that may give insight into the specific heats of systems with two species of degrees of freedom, such as the rotations and vibrations of diatomic molecules (K. R. S. Devi and A. Patrascioiu, *Physica D* 11(1984): 359).

In the case of blackbody radiation, the continuous medium (the electromagnetic field) is linear. Nonlinearity is introduced into the problem through the interaction of the field with the atoms in the walls of the cavity. Let us investigate a one-dimensional version of this problem, two nonlinear oscillators (particles and nonlinear springs) interacting through a linear string (Fig. 1). The string represents the electromagnetic field, and the oscillators represent the atoms. This model has the advantage that the string can be treated exactly so that no spatial lattice is needed.

The string and the particles move in the $z$ direction only. The equation of motion for the string is

$$\frac{\partial^2 z(x,t)}{\partial t^2} - \frac{\partial^2 z(x,t)}{\partial x^2} = 0, \quad \text{for } x \neq \pm 1, \tag{1}$$

and the equations of motion for the particles on the left and right, respectively, are

$$m \left.\frac{\partial^2 z(x,t)}{\partial t^2}\right|_{x=-1} = \mu \left.\frac{\partial z(x,t)}{\partial x}\right|_{x=-1} + F\big(z(-1,t)\big) \tag{2}$$

and

$$m \left.\frac{\partial^2 z(x,t)}{\partial t^2}\right|_{x=1} = -\mu \left.\frac{\partial z(x,t)}{\partial x}\right|_{x=1} + F\big(z(1,t)\big). \tag{3}$$

Here $m$ is the mass of each particle, $\mu$ is the string tension, and the nonlinear spring force $F(z)$ is defined by

$$F(z) = -\frac{dV}{dz},$$

where

$$V = k\frac{z^2}{2} + \lambda\frac{z^4}{4} + c\,|z|.$$

These equations are written in units such that the length of the string is 2 and the speed of sound is 1. The most general form for the solution of Eq. 1 is $z(x,t) = f(t+x) + g(t-x)$. Substituting this general solution into Eqs. 2 and 3 yields a system

### A ONE-DIMENSIONAL MODEL OF THE BLACKBODY PROBLEM

**Fig. 1. The blackbody problem was modeled as the interaction of a linear string (which represents the electromagnetic field) and two nonlinear oscillators (which represent atoms in the walls of the cavity). Motion of both the string and the oscillators is restricted to the *z* direction and is described by the function *z*(*x*, *t*).**

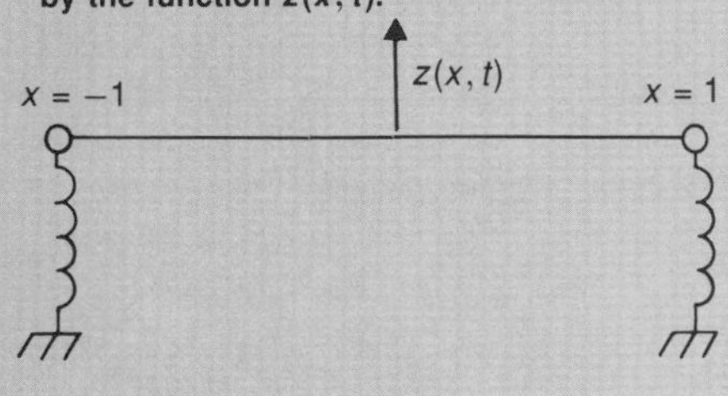

of two coupled ordinary differential equations for the functions $f$ and $g$.

The excitation of the string at $t = 0$ was specified by setting $f(x) = a \sin(\omega x + \pi/2)$ and $g(x) = 0$. The differential equations were integrated numerically, and conservation of energy was used to verify the accuracy of the calculations.

I would like to emphasize what outcome one would predict by following the same line of thought used to derive the Rayleigh-Jeans formula. The system, being nonlinear and (probably?) sufficiently complicated, will wander with equal probability throughout its phase space of given total energy. Let us choose initial conditions such that the total energy is finite. If ensemble averages and time averages are equal for this microcanonical ensemble, that is, if

$$\langle A \rangle_\mu = \lim_{T \to \infty} \frac{1}{T} \int_0^T dt\, A(t),$$

then the time-average kinetic energy of either particle should tend to zero for any initial conditions since the number of degrees of freedom is infinite. Over my times of observation, this did not seem to be the case! Under the assumption that the times of observation were sufficiently long, this result indicates that the microcanonial measure (Eq. 7 in the main text) is not applicable. We are left with two possibilities: (i) the motion of the system is quasiperiodic, or (ii) the phase space is broken into an infinite number of ergodic cells of finite size.

I also investigated the distribution of energy among the normal modes of the string. Figure 2 shows typical results for the time-average values of the fraction of the string energy in the $n$th normal mode. In all the runs performed the distribution of energy of the string among its normal modes is highly peaked (like the Planck distribution) and shows no tendency to become flat. Its shape does depend on the values of the various parameters in the problem and on the initial conditions. If all the parameters are kept fixed and the total energy is increased, the peak broadens. The shape of the distribution also varies with the frequency chosen for the initial excitation of the string, remaining constant over some range of $\omega$ and then jumping to a new shape.

The results of this study raised naturally several questions: (i) Was the observed unequal partition of energy among the normal modes of the string (the continuous medium) related to the one-dimensional nature of the medium? (ii) The unequal partition of energy reflected in the specific heats of diatomic gases results from motions of particles (rather than motions of a field, as in the case of blackbody radiation). Can this phenomenon be reproduced in a classical dynamical system?

To help answer these questions, Devi and I performed a study of a three-dimensional version of the system shown in Fig. 1. This system included four particles and six strings (Fig. 3). Our results exhibited several notable features over the times of observation: (i) time averages of, for example, total energies of particles and strings seemed to reach their asymptotic values; (ii) unequal partition of energy among the normal modes of the strings persisted, and the distributions obtained were reminiscent of that given by Planck's law; and (iii) for a variety of initial conditions, the four particles did not achieve the same average kinetic energy, a situation similar to the unequal partition of energy between the vibrational and the rotational degrees of freedom of diatomic gases. The fact that we obtained these types of results using several nonlinear (spring) potentials suggests their generality. ■

## UNEQUAL PARTITION OF ENERGY

**Fig. 2. Typical results for the distribution of energy among the normal modes of the string in the one-dimensional model of the blackbody problem (see Fig. 1). The exact shape of the energy distribution depends on the values assigned to various parameters, but in all cases the distribution was similar to a Planck distribution (see Fig. 2 of the main text) and was never flat, as it would be if the energy were partitioned equally among all the normal modes.**

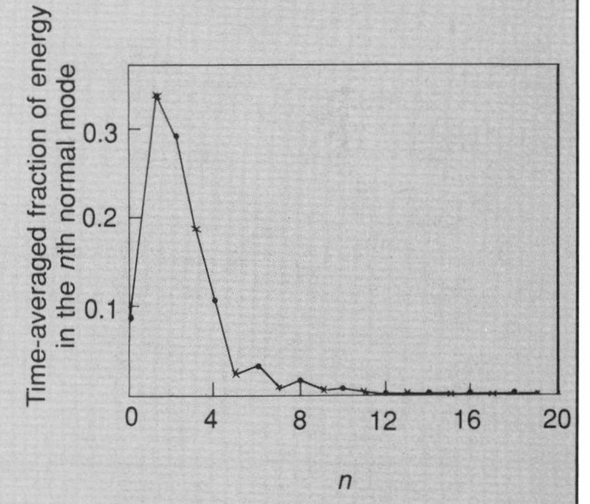


## A THREE-DIMENSIONAL LATTICE

**Fig. 3. The lattice below was used to investigate the dynamical behavior of a three-dimensional nonlinear system of particles and fields. Each of the four particles at the vertices of the regular tetrahedron is coupled to a nonlinear spring, and the particles are coupled to each other through linear strings.**

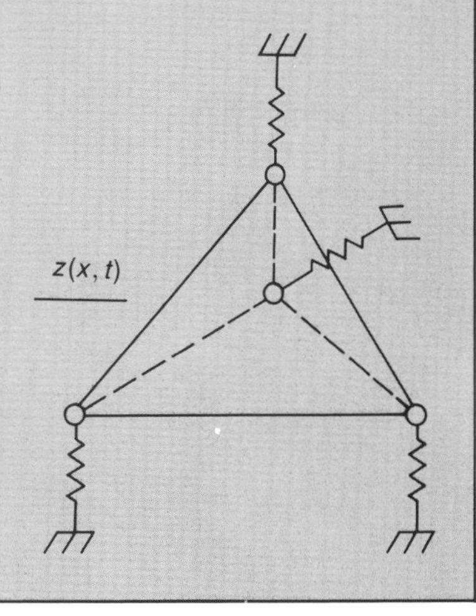

*continued from page 275*

Nor is there any basis to the claim that the classical atom is inevitably unstable because of the "ultraviolet catastrophe" (escape of all of the energy into the ultraviolet modes of the electromagnetic field, as required by the equipartition-of-energy principle of classical statistical mechanics). After all, maybe classical electromagnetism leads to a nonergodic flow (if the notion of ergodicity makes sense at all for a continuous medium) or maybe the diffusion of energy to the high modes is so slow that it has not occurred appreciably in the twelve to eighteen billion years since the big bang. That such slow diffusion is not a far-fetched supposition follows from some results obtained in the last few years. Since point charges have infinite self-energies, let us spread them by introducing a charged scalar (zero-spin) field. It has been shown rigorously that, in a certain gauge (axial), the system of coupled nonlinear equations describing the interaction of the classical electromagnetic field with this classical charged field has finite-energy-density solutions for all times. Moreover, these solutions retain their initial smoothness (number of derivatives). Using this latter property one can show that after an arbitrarily long time of evolution, an infinite number of normal modes of these fields are arbitrarily close to their initial energies (Patrascioiu 1984). Whereas there is no guarantee that this model captures the true physics in the universe, it seems hard to imagine a field whose modes thermalize in a finite amount of time.

So perhaps quantum mechanics is nothing more than classical statistical mechanics done the right way in a universe filled with particles interacting primarily via electromagnetic and gravitational forces. If so, its mysteries should be understandable once the complicated Brownian process produced by particles constantly absorbing and emitting radiation is mastered. While this scenario may seem far-fetched to many, I think it arises inescapably from contemplating the foundations of statistical mechanics. It does not contradict the experimentally observed violation of Bell's inequality unless the latter persists for truly space-like settings of the magnets. It has epistemological value and would, for example, allow the computation of the fine-structure constant and its variation with temperature (Patrascioiu 1981).

In conclusion, I think neither physicists nor mathematicians should close the book on the venerable problem of the ergodic hypothesis, and I guess some big surprises may be in store once the problem is better understood. ■

**Adrian Patrascioiu,** a native of Rumania, received his Ph.D. in theoretical physics from the Massachusetts Institute of Technology in 1973. He then spent two years as a member of the Institute for Advanced Study and two more years as a research associate at the University of California, San Diego. Since 1978 he has been a member of the Physics Department at the University of Arizona, where he is now a Professor. His honors include a Sloan Fellowship and a Los Alamos National Laboratory Stanislaw M. Ulam Scholarship.

## Further Reading

J. S. Bell. 1964. On the Einstein-Podolsky-Rosen paradox. *Physics* 1: 195. A simple discussion of Bell's inequality and its experimental verification can be found in "The Quantum Theory and Reality" by Bernard d'Espagnat. *Scientific American*, November 1979, p. 128.

Alain Aspect, Philippe Grangier, and Gerard Roger. 1982. Experimental realization of Einstein-Podolsky-Rosen-Bohm *Gedankenexperiment*: A new violation of Bell's inequalities. *Physical Review Letters* 49: 91.

L. Boltzmann. 1868. Studien über des Gleichgewicht der lebendigen Kraft zwichen bewegten materiellen Punkten. *Wiener Berichte* 58: 517.

L. Boltzmann. 1872. Weitere Studien über das Wärmegleichgewicht unter Gasmolekülen. *Wiener Berichte* 66: 275.

J. C. Maxwell. 1860. Illustrations of the dynamical theory of gases. Part I: On the motions and collisions of perfectly elastic spheres. *Philosophical Magazine and Journal of Science*, Fourth Series 19: 19.

J. C. Maxwell. 1860. Illustrations of the dynamical theory of gases. Part II. On the process of diffusion of two or more kinds of moving particles among one another. *Philosophical Magazine and Journal of Science*, Fourth Series 20: 21.

J. C. Maxwell. 1860. Illustrations of the dynamical theory of gases. Part III. On the collision of perfectly elastic bodies of any form. *Philosophical Magazine and Journal of Science*, Fourth Series 20: 33.

J. Clerk Maxwell. 1867. On the dynamical theory of gases. *Philosophical Transactions of the Royal Society of London* 157: 49.

George D. Birkhoff. 1966. *Dynamical Systems*. American Mathematical Society Colloquium Publications, Volume 9. Providence, Rhode Island: American Mathematical Society. See also *Lectures on Ergodic Theory*, by Paul R. Halmos. Tokyo: Mathematical Society of Japan, 1956.

J. C. Oxtoby and S. M. Ulam. 1941. Measure-preserving homeomorphisms and metrical transitivity. *Annals of Mathematics* 42: 874.

V. I. Arnold and A. Avez. 1968. *Ergodic Problems of Classical Mechanics*. New York: W. A. Benjamin, Inc.

Ja. G. Sinai. 1963. On the foundations of the ergodic hypothesis for a dynamical system of statistical mechanics. *Soviet Mathematics—Doklady* 4: 1818.

Ja. G. Sinai. 1967. Ergodicity of Boltzmann's gas model. In *Statistical Mechanics: Foundations and Applications (Proceedings of the I. U. P. A. P. Meeting, Copenhagen, 1966)*, edited by Thor A. Bak. New York: W. A. Benjamin, Inc.

J. Bellissard and M. Vittot. 1985. Invariant tori for an infinite lattice or coupled classical rotators. Universite de Aix-Marseille preprint CPT–85 P1796.

Douglas M. Eardley and Vincent Moncrief. 1982. The global existence of Yang-Mills-Higgs fields in 4-dimensional Minkowski space. *Communications in Mathematical Physics* 82: 193.

E. Fermi, J. Pasta, and S. Ulam. 1955. Studies of nonlinear problems. Los Alamos Scientific Laboratory report LA–1940. Also in *Stanislaw Ulam: Sets, Numbers, and Universes*, edited by W. A. Beyer, J. Mycielski, and G.-C. Rota. Cambridge, Massachusetts: The MIT Press, 1974.

A. Patrascioiu, E. Seiler, and I. O. Stamatescu. 1985. Non-ergodicity in classical electrodynamics. *Physical Review A* 31: 1906.

Adrian Patrascioiu. 1983. Beyond the mystery of quantum mechanics. In *Asymptotic Realms of Physics: Essays in Honor of Francis E. Low*, edited by Alan H. Guth, Kerson Huang, and Robert L. Jaffe. Cambridge, Massachusetts: The MIT Press.

Adrian Patrascioiu. 1984. Are there any ergodic local relativistic field theories? *Physics Letters* 104A: 87.

A. Patrascioiu. 1981. On the nature of quantum mechanics. Institute for Advanced Study report.

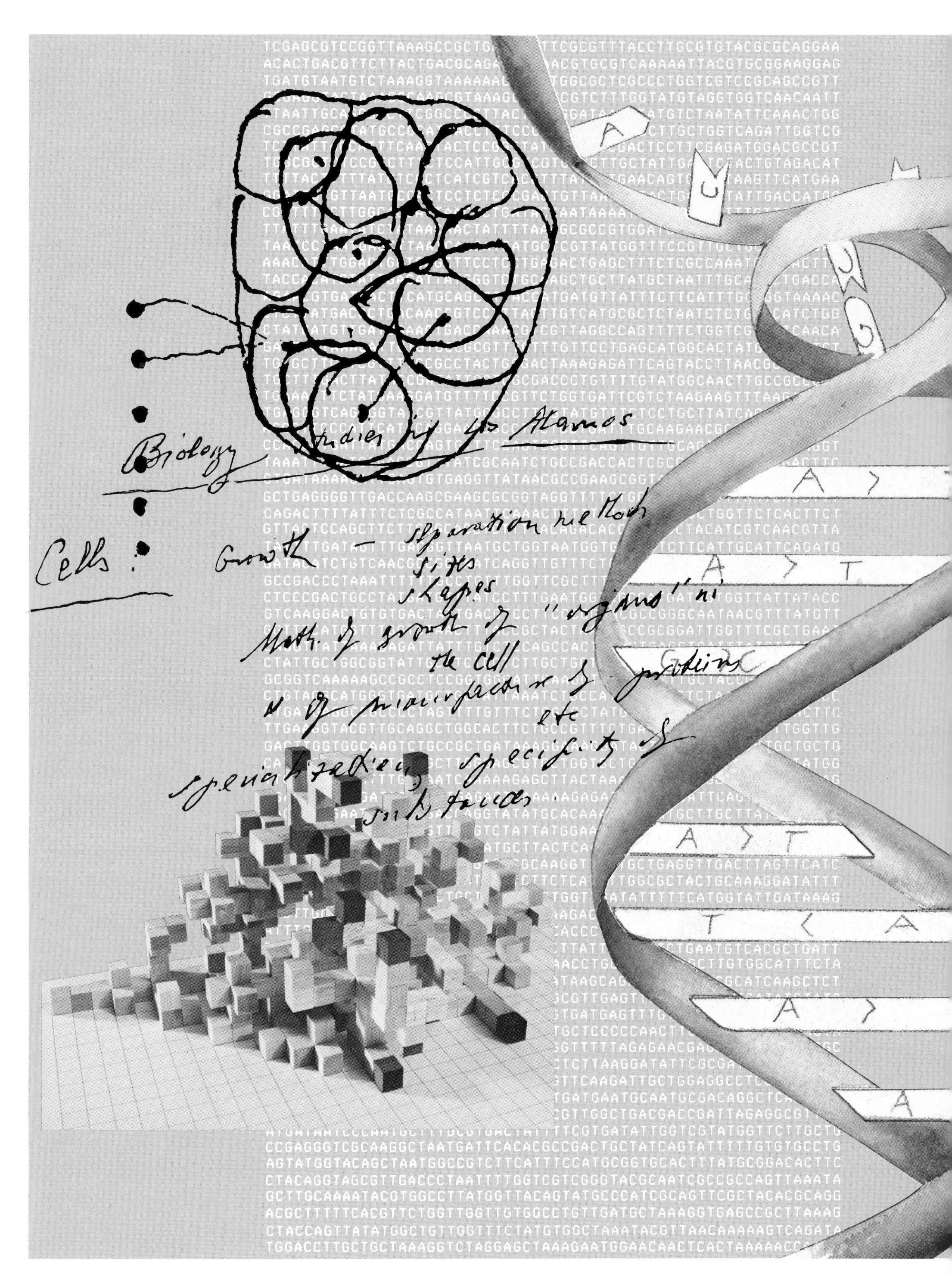
Biology
Cells :
Growth

# BIOLOGY

Ulam's genius for addressing basic questions in biology through simple mathematical models has already been encountered. In the mathematics section Hawkins and Mycielski introduced Ulam's notion of "genealogical" distance, a measure of shared ancestry. This was one among several extensions of the theory of branching processes invented by Ulam to answer questions in population dynamics and evolution.

Another area that has had even more impact is Ulam's early work on cellular automata in which he and Schrandt demonstrated how complex patterns can evolve from simple initial conditions by applying a few simple recursion rules over and over again. The idea behind these computer studies hinges on a basic question in developmental biology: How does a human being develop from a single cell, a single fertilized egg? Certainly, with $10^{12}$ cells and only $10^{6}$ genes, there are not nearly enough genes for each cell to have its own gene. Stan proposed that genes encode not just simple rules, but rules of a higher logical type that change the simple instructions, in other words "rules for the change of rules" that become operative in response to events outside the cell but in contact with it. Stan began investigating this idea by making what he called kindergarten rules and applying them to a small number of cells. In one set of rules, cells multiply along a straight line until they meet another cell at which point their line of propagation rotates by 45 degrees. Rules of this sort were run on the computer to produce patterns resembling those found in nature, such as the vein distribution in a leaf or the pattern of a capillary bed. (Several figures from these studies with Schrandt are shown on the next page.) Recent work by Gerald Edelman of Rockefeller University on morphogenesis lends credence to Ulam's basic idea of "rules for the change of rules." Edelman's work suggests that form arises through an interaction involving adhesion molecules on the cell surface that alter the primary processes of cell development. "In this case, the modifications of the rules correspond to the developmental process of induction. For during induction, as a result of associations between adjacent groups of cells, particular cells undergo alterations in their properties through the process of cell differentiation, and these alterations, in turn, modify their subsequent interactions." (This quote is from an article by Leif Finkel and Edelman in volume 10, numbers 2 and 3 of *Letters in Mathematical Physics*, a special volume in memory of Stan Ulam.)

Ulam's cellular automaton models of growth patterns were just a start. Now cellular automata of various kinds are being used to model the complex networks associated with food webs and kin selection, and even neurons in our brain.

Stan was deeply interested in the organization and function of the human brain and the structure of memory. In the Gamow Memorial Lecture of October 5, 1982 (which is published here for the first time), he outlined some speculations about the mechanism

that allows us to recognize the alikeness of members of a class (for example, chairs) and the difference between two classes of objects (chairs and tables). In the same article referred to above, Finkel and Edelman comment on those ideas:

"Ulam was concerned with the problem of categorization—how to group objects on the basis of similarity and dissimilarity. Together with the related problem of generalization—how to define a class given only a few exemplars—this constitutes perhaps the most profound problem in biology. Ulam put forward the idea of generalizing the Hausdorff metric to deal with classes of objects ... He had proven in an early paper that various compositions of two nonlinear transformations can, with some degree of accuracy, deform any plane figure into any other. If such a set of transformations are applied separately to two objects, two classes of related objects are generated. Ulam defined a generalized Hausdorff metric between these two classes that characterizes the 'similiarity' of the two original objects. He pointed out that such a recognition system would yield a substantial saving in memory since, given the transformation mechanism, only one generating member of each class need actually be stored. Applying the transformations to new inputs and/or to stored memories would then allow matching between the two based on the metric. Inputs which did not match any of the stored memories might be stored as new memories ...

The process of generating a class from a single object is used as a mechanism in the immune system, the biological recognition system concerned with recognizing foreign substances in the body. In this case, the transformations are effected by several mechanisms, including somatic recombination in the genes coding for the antibody molecules. [Edelman] has ... demonstrated ... that introducing these transformations in a repertoire of recognizing elements can actually improve the recognizing ability of the system.

With regard to pattern recognition by the brain, there are no currently known mechanisms to generate such transformations in the central nervous system. However, one of the outstanding problems of psychophysics is the mechanism responsible for so-called visual constancies. These are the invariant properties that allow us to recognize objects regardless of their spatial location or orientation (i.e., after arbitrary translations, rotations, zooming, etc.). Such a transformation generating mechanism, if present at some low level of the central nervous system, might account for these phenomena."

The question of similarity and dissimilarity, so important to recognition, comes up in a different form in the context of DNA. Here Ulam invented a new kind of metric space to measure the "distances" between the sequences of nucleotides that make up DNA segments. Walter Goad, one of Stan's "influencees" at Los Alamos, describes this invention and discusses its importance for the human genome project and for tracing the evolution of life. Summing up his experience working with Stan on both biological and weapons-oriented problems, Walter comments, "Stan habitually turned things to view from a variety of directions, much as he would see an algebraic structure topologically, and vice versa, and often supplied the connection that dispelled a gathering fog."

*A Gamow Memorial Lecture delivered at the University of Colorado, Boulder, on October 5, 1982*

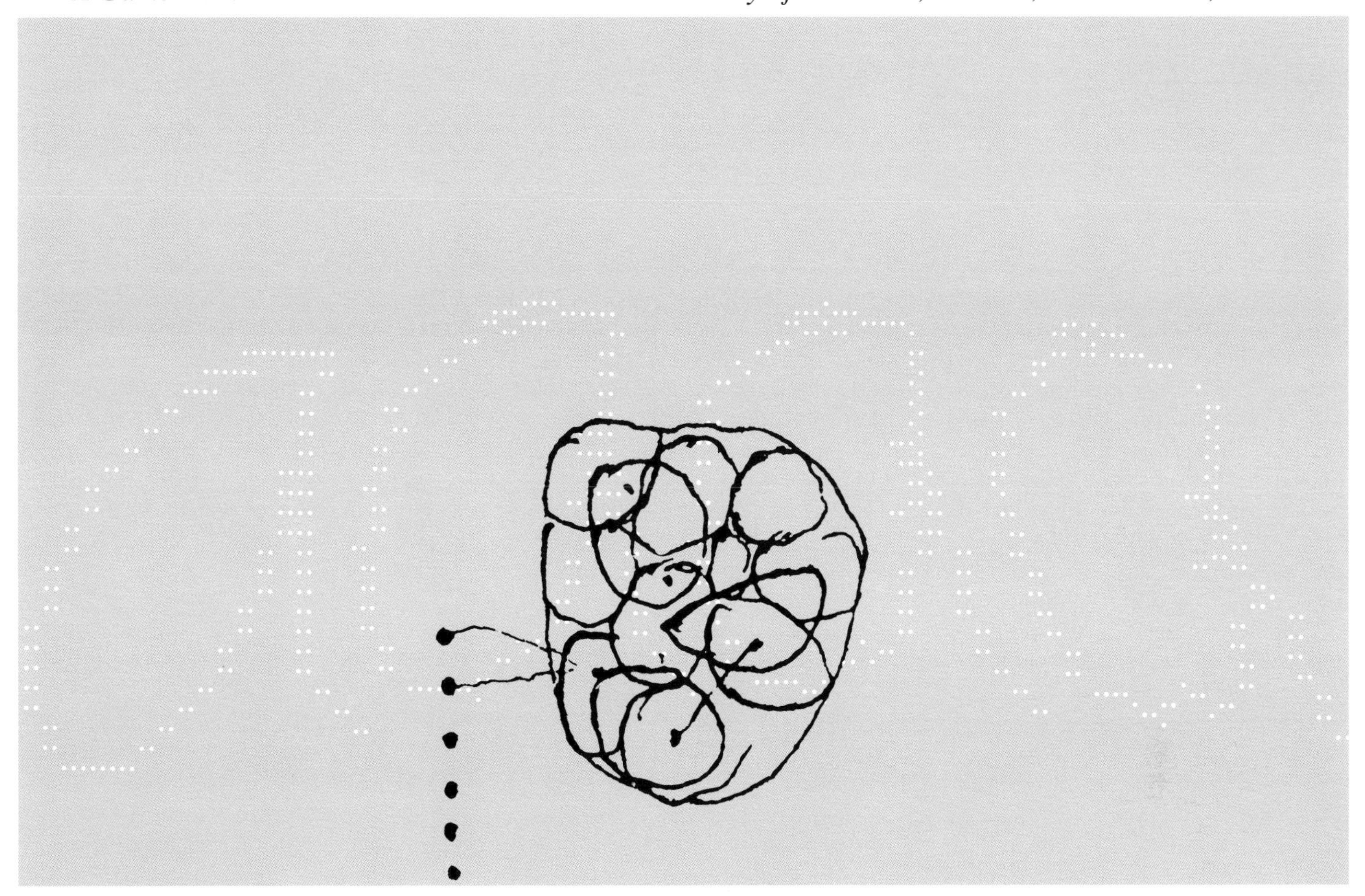

# *Reflections on the Brain's Attempts to Understand Itself*

**by S. M. Ulam**

My choice of subject for this talk may seem strange, since I am not a psychologist, a physiologist, or a neurologist, merely a mathematician and an amateur, a dilettante, in the workings of the brain. However, it is fitting that I give such a talk in memory of the late George Gamow, a friend of mine. Though by training a physicist, he was able to make famous contributions in other sciences, such as astronomy and biology, that interested him toward the end of his life. He was, like me, an amateur, a dilettante, in biology. Nevertheless one of the most important discoveries of recent times in that field is due to him. It was Gamow who first pointed out that ordered arrangements of four chemical units—four "letters"—along the DNA double helix, or chain, as he called it, might be codes for many biological processes, and that the codes for the manufacture of proteins might consist of three- or four-letter "words."

What I want to do today is talk about several of my own speculations, with some mathematical symbolism, concerning the operation of the brain. I believe that discoveries and breakthroughs within the next twenty years will lead to a better understanding of the mechanisms of the brain, of the processes of thought. It will not be a complete understanding—that would be too much to hope for—but it will give us some ideas of how the nervous system operates in lower animals and in humans.

Mathematicians may help in reaching this understanding, although for the time being I think that 99 percent of the progress will come from physiological and anatomical experiments. However, mathematics can be useful, for it is clear that the similarities between electronic computers and the nervous system are of great importance.

Another friend of mine, the late John von Neumann, was one of the pioneers in the planning and building of electronic computers. His book *The Computer and the Brain*, which was published posthumously in 1957, is still one of the most elegant and understandable general introductions to the subject. I remember the discussions we had on how the advent of computers would enlarge the scope of experimentation in mathematical and physical sciences and about his specific interest in the partial analogies between computers, as they were planned in the early forties, and the processes of deductive thinking. We saw each other frequently at the time, either in Los Alamos or in

Princeton, and we would marvel at the few physiological facts then known about the brain, such as the number of neurons it contains. That number was of the order of ten billion, and their interconnections in the human cortex were known to be still more numerous. He would say: "Not only are there ten billion computing elements, but each is connected to many others, one hundred maybe! And maybe even to one thousand in the central part of the brain!" Well now, forty years later, the number of interconnections has been shown to be of the order of thousands, up to one hundred thousand in the central part of the brain. And the total number of connections, of axons and synapses, is of order $10^{14}$. So you see, in the recent past the purely anatomical and physiological knowledge has vastly changed. The locations of certain centers in the brain and the differences between its right and left halves are also better known. And today more information is being gathered through studies of the electromagnetic signals being emitted constantly by the brain.

However, I do not believe that now, or even in the near or distant future, it will be possible to gain what might be called a complete understanding of the brain's operation. My belief rests on very important and strange results in pure mathematics. These results, which date from 1930, are associated mainly with the name of Gödel, a mathematician who worked at the Institute for Advanced Study in Princeton. Gödel proved a theorem that says, roughly speaking, that in any mathematical system, any logical system, there exist statements that have sense but cannot be proved or disproved. So in every mathematical discipline one can conceive of at present, there are undecidable propositions, finite statements that, starting from axioms, one cannot demonstrate or show to be false.

Mathematics has a store of problems, some very old, whose solutions are not known. But it was assumed that, ultimately, yes or no solutions would be found. That was the belief of Hilbert, one of the greatest mathematicians of the last hundred years. Then Gödel came along and showed that such a belief is no longer valid, that there are statements that are undecidable. This fact is of great philosophical significance. And beyond that, it could be a sort of consolation for our inability to attain a complete knowledge of various real phenomena.

So it is possible that some of the still unresolved mathematical problems are *in principle* undecidable on the basis of our present system of axioms. Many such problems are technically complicated, but let me give you one that is simple to state and understand.

A prime number is an integer that is not divisible by any number except itself. The numbers 2, 3, 5, 7, ..., 41, 43, 47, et cetera, are all prime. The Greeks knew that there are infinitely many prime numbers. That is one of the oldest, greatest, and most beautiful discoveries in mathematics. Now certain pairs of prime numbers, such as 5 and 7, 11 and 13, 17 and 19, are called twins because they differ by only 2. The question is: How many twin primes are there, a fixed finite number or an infinity? Nobody knows the answer to this question, and it may be undecidable. I asked Professor Schmidt, a very famous number theorist, if he knew who first proposed this very old problem and whether he thought it might be undecidable. He did not know the answer to the former, and to the latter he answered, "One might not be able to decide whether it is undecidable!"

I mention Gödel's theorem to show the limitations of man's program to try to understand everything, even in a restricted domain. Perhaps the scope of the human brain is finite, or conversely, perhaps the growth of humanity, of its collection of brains, will, in terms of evolution, continue indefinitely and may reveal new points of view.

To continue the speculation on what the role of mathematics might be in the study of the brain, the time is not yet ripe to say its operation can be understood with abstract theories alone. But Gamow, who was perhaps the last great amateur in science, has shown us that it is possible to speculate—fruitfully, given some luck—on the great mysteries of nature. A Greek philosopher said that many are the wonders of the universe, but the greatest of all is the human mind. And Spinoza said that it is better to begin with small and modest truths. Starting from these premises, I want to give you now a few examples of biological questions that I think mathematics has already proved somewhat useful in answering, and how similar attempts and schematizations might possibly be of some use in partially understanding the nature of human perception.

One such question concerns the mechanism of recognition of external stimuli, say sights or sounds, and ultimately of ideas. Before recognition, there is perhaps discernment, discrimination. A priori it seems easier to see the difference between two objects than their similarity or analogy. We need to map the tremendous web of connections in the human brain into overlapping classes. But before we do this, here is an example of a mathematizable biological idea, one concerning the codes for the manufacture of proteins.

Gamow's suggestion about the existence of three- or four-letter codes for the constituent amino acids of proteins was almost correct. Many of the characteristics of living organisms are coded in very long sequences of four chemical units, which biologists call by the letters A, C, G, and T. Words are short strings of these letters. Finite sentences of several hundred words are codes for proteins, such as hemoglobins of various kinds. Today tens of thousands of these codes for proteins are known, and in some cases even the spatial forms of the proteins are known. A "reader" molecule goes along the DNA

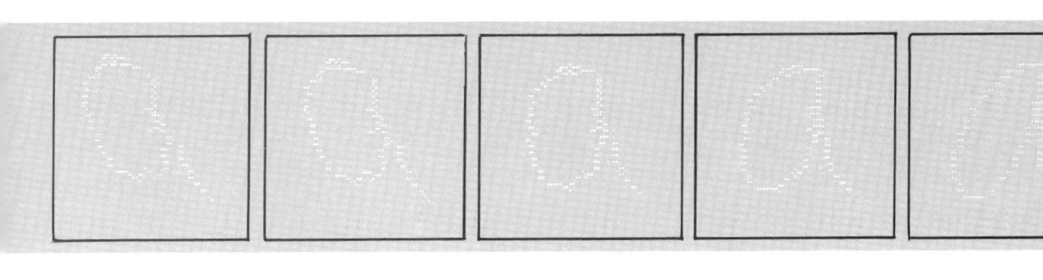



"tape," reads the code, and deposits the information in other parts of the cell, in the ribosomes. This much is now understood. The functions of other parts of the long sequences, such as those called introns, are not yet understood, but they are not codes for proteins.

Some biologists are beginning to speculate on the importance of small differences that have been found to exist between the codes for a given protein in different species. For example, cytochrome $c$, which is important for the transmission of electrical impulses in nerves, differs slightly from one species to another but remains the same within a species. The biologist Emanuel Margoliash has tried to establish an evolutionary tree based on the quantitative differences in cytochrome $c$ codes, on the gradations among them.

Mathematicians have studied in general the idea of comparing two elements $a$ and $b$—two points in some space—by expressing the degree of their difference with a quantity called a distance. This distance, which is usually denoted by $\rho(a,b)$, should have the following properties. It should be positive definite: $\rho(a,b) > 0$ if $a \neq b$ and $\rho(a,a) = 0$. It should be symmetric: $\rho(a,b) = \rho(b,a)$. And it should satisfy the triangle inequality: $\rho(a,c) \leq \rho(a,b) + \rho(b,c)$. This last property means that to go from $a$ to $c$ is no more difficult than to go from $a$ to $b$ and then from $b$ to $c$. If such a distance exists for all pairs of points in a set $S$, then $S$ is called a metric space.

I have said that the elements of the genetic code are sequences of symbols for four chemical units. For simplicity's sake and without changing any essentials, let us consider sequences of just two symbols, 0 and 1. For example, one such sequence $x$ could be 0110101 and another sequence $y$ could be 1000110. To get an idea of how much they differ, we want a distance $\rho(x,y)$ between $x$ and $y$. Let $x_i$ be the $i$th symbol in $x$ and $y_i$ be the $i$th symbol in $y$, where $i$ = 1, 2, 3, and so on. One distance we might consider is the sum of the absolute values of the differences between $x_i$ and $y_i$:

$$\rho(x,y) = \sum |x_i - y_i|.$$

Suppose $x$ and $y$ are both of length $N$ and $x = 010101\ldots0$ and $y = 101010\ldots1$. Then $\sum |x_i - y_i| = N$ since they differ in every place. This is one distance used by mathematicians. Another is the so-called Euclidean distance, $\sqrt{\sum(x_i - y_i)^2}$.

But our contention is that these distances are not suitable for biological objects. They are suitable for fixed objects, for sequences of symbols that are, so to say, rigid points of geometrical spaces, et cetera. But they are not well suited for flexible objects, such as strings of codes. To see this, consider the previous example of the two long sequences that differed in every place. They are in one sense almost identical since by erasing one symbol in each sequence they become the same. Two changes make the sequences identical! But according to the previous definitions of distance, the distance between them is $N$ or $\sqrt{N}$ instead of just 2.

Let us try another definition of a distance. For example, we could define the new $\rho$ as the minimum number of allowed changes that must be effected on one or the other sequence to make them identical. What could these allowed changes be? One might be the substitution of a 0 for a 1, or vice versa. Another could be the erasure, or the intercalation, of a 0 or a 1 at any place in the sequence. One can prove that this $\rho$ has all the properties that a distance should.

A quantitative formulation of distance can be tried not only for the sequences of symbols in the genetic code but for a great variety of other objects. For example, one can try to define a distance between two sequences of musical notes, of acoustic signals, or between two drawings or sculptures, sets of points in two or three dimensions.

It is my speculation that in the brain, or more generally in the nervous system, there must be a mechanism that, perhaps in a qualitative way only, determines a distance between a perception stored in the memory and a newly presented perception. Recognition of the newly presented perception as known or unknown might mean that this distance is below or above a certain threshold. A perception insufficiently close to any of those already in the memory would be stored as a new perception.

I want to talk about this sort of approach to the recognition of visual perceptions. Let us take, for example, the case of recognition of two-dimensional pictures. My conjecture is that the brain uses several different distances to compare such pictures after they are registered on the retina, recorded or recoded on several layers behind the retina, and deposited in the brain.

What distance might be appropriate for comparing two two-dimensional pictures, that is, two sets of points in a plane? Let the two sets be $A$ and $B$. We are interested in some possible $\rho(A,B)$. Distances between sets have been studied by mathematicians. One of these, the Hausdorff distance $\rho_{\mathrm{H}}(A,B)$ is defined as follows. Let $\rho_{\mathrm{E}}$ be the ordinary distance between two points. Given a point $x$ in $A$, find the point $y$ in $B$ for which $\rho_{\mathrm{E}}(x,y)$ is a minimum; that is, find $\min_{y\in B} \rho_{\mathrm{E}}(x,y)$. Do this for all $x$ in $A$ and then find the maximum of these minima, $\max_{x\in A} \min_{y\in B} \rho_{\mathrm{E}}(x,y)$. Now find $\min_{x\in A} \rho_{\mathrm{E}}(x,y)$ for a given $y$ in $B$ and $\max_{y\in B} \min_{x\in A} \rho_{\mathrm{E}}(x,y)$. Then

$$\rho_{\mathrm{H}}(A,B) = \max_{x\in A} \min_{y\in B} \rho_{\mathrm{E}}(x,y) + \max_{y\in B} \min_{x\in A} \rho_{\mathrm{E}}(x,y).$$

But this Hausdorff distance, like some of the distances mentioned in connection with one-dimensional sequences, can be objected to in biological applications. Obviously, $\rho_{\mathrm{H}}$, as defined, depends on aspects of $A$ and $B$ that are of little consequence to recognition. For example, $B$ may be just a magnified version of $A$ or congruent to $A$ but rotated or translated. In these cases the meaningful distance should be very small.

By repeating, or iterating, the idea of Hausdorff as follows one can arrive at a more satisfactory distance. For a given set, or picture, $A$, let us consider the class of $A$'s that "look like" $A$, that, for example, are replications of $A$ in various sizes or are obtained from $A$ by some rotation or translation. Call this class of sets an impression of $A$ and denote it by $\mathcal{A}$. We proceed analogously for $B$ and obtain a class of sets $\mathcal{B}$. Now we may define a distance between the impressions of $A$ and $B$ as follows:

$$\rho(\mathcal{A},\mathcal{B}) = \max_{A\in\mathcal{A}} \min_{B\in\mathcal{B}} \rho_{\mathrm{H}}(A,B) + \max_{B\in\mathcal{B}} \min_{A\in\mathcal{A}} \rho_{\mathrm{H}}(A,B).$$

This is a more satisfactory measure of the difference between $A$ and $B$. Needless to say, distances between three-dimensional objects can be defined analogously.

One can define still other distances between sets of points, or signals, in two or more dimensions. It is possible, for example, to express such a measure of similarity or dissimilarity as a distance between encodings of the set points in terms of orthogonal functions, such as those used in Fourier expansions. [See "An Ulam Distance."]

I shall now describe a computer experiment Robert Schrandt and I did in the early sixties at Los Alamos. The experiment concerned the use of distances in the recognition of handwritten letters and involved the second conjecture that I want to present in this talk, one about the role of impressions, or examples, in the process of recognition.

The idea of the experiment was to provide the computer with a great many handwritten examples of the letters *a* and *b*—actually with a great many sets of coordinates of points outlining the letters—and then make the computer decide if a new example was an *a* or a *b*. It would have been prohibitively tedious to provide, say, 512 examples of each letter. (Powers of 2 are convenient when dealing with computers; hence the number 512.) Instead we used a stratagem by which the computer itself generated the examples. I remembered a proof of mine that there exist on the interval, and analogously on spaces of higher dimension, two functions $f$ and $g$ such that any continuous function can be approximated by one of their compositions—$fg$, $ffg$, $fgf$, $fggf$, $fgfg$, et cetera. So we gave the computer only one example each of *a* and *b* and also two transformations of each, which served as $f$ and $g$. By programming the computer to produce compositions up to the order of 10 of the transformations, we obtained 512 examples each of *a* and *b*. When displayed on a screen, these looked indeed like various handwritten versions of the original *a* and *b*. Some were slightly tilted, others appeared to have been written by a shaky hand, and so on. Then the computer was asked to decide whether a new handwritten sample was an *a* or a *b* by computing the Hausdorff distances between the sample and the examples it had created. The computer's decisions were correct in more than 80 percent of the cases! Of course, the same method works in the case of more than two letters or other standardized figures.

The conjecture is that in the brain, in the visual system and in the memory, perhaps only a few visual perceptions are permanently stored, and, when presented with another, the brain produces, for comparison, many deformations either of what is in the memory or of what is presented. If this is so, the storage capacity of the memory would be enormously enhanced.

At present one can only speculate about the mechanisms by which the brain might produce the deformations. Some are obvious, such as a tilt of the head or a change in size. One can also only speculate about what distances or how many are used in the decision. One may also speculate that a similar mechanism directs the recognition of objects within the body. Could it be that the antibodies produced by the immune system have an analogous way of recognizing antigens? Again, deformations might be used to produce a large number of examples for such discrimination and recognition.

The next higher stage in the operation of the brain might be a more complicated analysis of impressions. Instead of considering impressions of single objects, the brain might study a succession of two or three, even a "movie" of ten or more. Combined with recognition of the passage of time, this could lead to development of primitive logic or elementary reasoning, perhaps in the form of the statement *post hoc ergo propter hoc* (after, therefore because) or its reverse *ante hoc ergo qua hoc* (before, therefore as a reason for).

Our comprehension of less elementary learning should involve the mathematical idea of measuring complexity. In recent years quite a number of mathematicians, including Jan Mycielski and André Ehrenfeucht, both professors at this university, have done some very interesting work on this subject. With proper changes some of their results could be applied to investigating the operation of the nervous system.

It is clear that one of the most important mysteries about the brain is the organization of the memory, including the means of access. As I surmised earlier, some form of memory must exist in the visual, auditory, olfactory, and immune systems—and even in the system for differentiation itself. A mechanism for producing many examples from one would certainly seem a very efficient way of using the storage capacity of the visual and auditory memories. In the course of evolution, special devices, or tricks, must have developed to increase the scope of recognition and of the complementary process of registering perceptions as new.

Let me give an example of a trick for efficient use of a computer. Suppose we have stored in its memory a great many, say $10^6$, eight-digit numbers arranged sequentially and want the computer to decide whether a given number is among those stored. The computer can do this extremely fast by comparing in succession the digits from first to last. Suppose now that we want the computer to decide whether the given number differs from any of the stored numbers by, say, 1 in any of the eight positions. We might program the computer to do this by deciding whether any of the $10^6$ numbers in its memory is that close. That would be a very lengthy operation. There is a much better way to proceed, a way that requires only sixteen times the effort required for the computer to decide whether a single number is among those stored. We first program the computer to produce from the given number the sixteen numbers that do differ by 1 in any of the eight positions and then to decide whether any of the sixteen is among those in its memory.

This example illustrates that a mechanism for producing auxiliary perceptions for comparison with perceptions stored in the memory would be an advantageous acquisition of the nervous system. So also would a mechanism for producing variations of what is stored in the memory for comparison with external stimuli. Perhaps a physiological or anatomical arrangement might serve such functions. Clearly these are merely guesses as to special characteristics the nervous system may have acquired in the course of evolution. ■

# An Ulam Distance

***by William A. Beyer***

Stan had often referred, as he did in this lecture, to a distance between sets based on an encoding of the set points in terms of orthogonal functions. However, he had never explicitly defined such a distance. I do so now to honor the originator of so many seminal ideas.

Let $A$ and $B$ be two-dimensional finite sets enclosed in a square. Let $n_A$ and $n_B$ be the number of points in $A$ and $B$, respectively. Let $\{f_{i,j}\}$ be a complete set of orthogonal functions on the square, such as two-dimensional Fourier trigonometric functions. Define $\mu^A_{i,j}$ and $\mu^B_{i,j}$, the encodings of $A$ and $B$ mentioned above, as follows:

$$\mu^A_{i,j} = \frac{1}{n_A} \sum_{x \in A} f_{i,j}(x)$$

and

$$\mu^B_{i,j} = \frac{1}{n_B} \sum_{x \in B} f_{i,j}(x).$$

Then $\mu^A_{i,j}$ and $\mu^B_{i,j}$ are functions on the nonnegative lattice points of the plane. Finally, let $\rho(f_1,f_2)$ be some selected distance between such functions. Then $\rho(\mu^A_{i,j}, \mu^B_{i,j})$ is a distance between the sets $A$ and $B$—an alternative to the Hausdorff distance defined in the lecture. ■

# SEQUENCE ANALYSIS

## *Contributions by Ulam to Molecular Genetics*

***by Walter B. Goad***

Lord Rayleigh once introduced a key idea with "It is tolerably obvious once remarked. . . ." Yes, I think now, that is just how it was—Stan Ulam providing us with a steady stream of ideas and observations "tolerably obvious" only in retrospect, and then striking in the way they became integral to one's tangible world of evolved and evolving forms and actions. Here I would like to sketch ideas developed during the sixties and seventies as an avalanche of detail, still growing, gathered about the way sequences of nucleotide bases in DNA encode instructions for development and propagation of living organisms. Stan showed us a very general way of thinking precisely about relationships among sequences, in particular, how to devise quantitative measures of relationship that, together with the computer, are of immense help in ferreting out meaning in the very great quantities of data now pouring forth.

I met Stan soon after arriving in Los Alamos at the end of 1950. I came ostensibly to finish a thesis begun at Duke under Lothar Nordheim, who had arrived several months earlier while I stayed in Durham awaiting security clearance. At last a telegram came from Carson Mark that read, "Your clearance not available." An anxious telephone call established that the "not" had been garbled in transit from "now." I was immediately swept up in the thermonuclear program, kept busy with the rest dissecting schemes and designs, and sometimes new phenomena, usually standing around a blackboard. Introducing the right factors, right at least in order of magnitude, was both vital and enjoyably competetive, laced with humor—esoteric, malicious, or plain—and an occasional flash of ego. The key, of course, was to discern the dominant phenomenon and to estimate its role in the matter at hand. One always had a feeling, almost visceral, as to how deeply an argument was rooted in the web of our knowledge of physics and mathematics. Stan habitually turned things to view from a variety of directions, much as he would see an algebraic structure topologically, and vice versa, and often supplied the connection that dispelled a gathering fog.

Around 1960 Jim Tuck invited Leonard Lerman, who was in the thick of the gathering revolution in biology and then at the University of Colorado, to visit Los Alamos. The "phage group" gathered loosely around Max Delbruck had established a mode of analysis that is still driving the biological revolution: Changes in a single DNA molecule are amplified by biological reproduction, usually in a microorganism, to the macroscopic level; there the consequences of those changes, however ramified, can be studied with the resources of physics and chemistry. The amplification is made possible by an immensely powerful, and growing, armory of molecular tools based on enzymes that carry out specific operations on specific DNA's. As we grasped those ideas from Leonard and began to see the clarity and concreteness with which the mechanisms of life would emerge from such analysis, many of us were galvanized. We soon responded in a way typical of the culture, organizing a seminar, hungrily seeking out the many aspects of the subject. As I recall, the seminar continued through the sixties and early seventies with a varying membership but with Stan, Jim Tuck, George Bell, and me as regulars. We were frequently visited, and enormously encouraged, by Ted Puck, who has built a distinguished school of molecular and cell biology at the University of Colorado and who was, and is, exceedingly optimistic about the contribution systematic theory can make to biology.

A quick tour of systematic theory inevitably would start with Darwin's grand synthesis. For physicists a key way point would be the publication in 1944 of Erwin Schrödinger's short book *What Is Life?*, which equates that grand question with one congenial to physicists: What generates "negentropy," the high degree of order that living systems are continually creating from the environment? Ever since, theorists of all kinds have looked to the formulation of some powerful physical theory of life. Short of that, what we do know is that living systems escape from the determinism of ordinary chemistry by interposing molecular adaptors to control molecular interactions. An example is provision by the complex protein structure of hemoglobin of an effective interaction between $O_2$ molecules that is completely unrelated to their interactions as free molecules: Within a hemoglobin molecule up to four $O_2$'s bind at distinct sites and thus effectively stick together. Furthermore, three or four stick more tightly than one or two. So, where there is much oxygen, four are tightly bound; where there is little, departure of one causes the others to more easily depart. Invoking the adaptor principle, Francis Crick predicted the existence of what are now called transfer RNA's—small RNA molecules, a particular species of which adapts each three-base codon to molecules of a particular amino acid. A Zen-like consciousness of physical necessity—for the way in which electrons and nuclei, and thus atoms and molecules, do what they must—leads first to puzzlement at living systems and then to resolution: Molecular adaptors free the logic of higher levels of organization to adopt and express a logic of their own, exploiting, not circumventing, physical necessity.

Proteins and RNA's provide an array of complex and highly specific adaptors, and their structures are encoded in sequences of nucleotide bases in DNA. To a large extent the double-helical structure of DNA wraps the information-conveying part of the DNA into a protected interior and so in the main removes chemical constraints on the propagation and selection of sequences.

Working on DNA as a substrate, evolution has produced the marvelously complex web of living systems we see today. The working hypothesis, to which no exception is yet known, is that all of the information for propagation and development of individual organisms is encoded somehow in the sequence of four bases adenine (A), thymine (T), guanine (G), and cytosine (C) along the DNA molecules (or, in some cases, RNA molecules) that compose its genome. The "somehow" includes the great triumphs of the past two decades, the present frontiers of molecular biology, and, undoubtedly, a great deal that we do not now even glimpse. Less than a decade after Watson and Crick determined the structure of DNA, researchers at the laboratories of Nirenberg, Khorana, and Ochoa fully worked out the "genetic code" by which the base sequences of particular segments of DNA—genes—are translated into sequences of amino acids that fold up as particular proteins. For a few years many people felt that, in principle, DNA function was now completely understood. But in the mid seventies methods were worked out for determining sequences of bases in DNA, and it amost immediately emerged that not even the sequences that are translated into proteins are simple, continuous coding sequences. The last few years have seen the discovery of a great many distinct "signals" that control the replication of DNA and the expression of genes. However, it is not yet known how the action of those signals is coordinated, as it must be, to yield the patterns seen during reproduction and development. On the other hand, an outline is emerging of the organization within DNA of repetitive sequences, which make up a substantial fraction of the genome in higher organisms. That organization may or may not have signaling capabilities, but it is almost surely important in evolution. Perhaps most striking of all is the growing knowledge of phenomena—such as the mobility and duplication of pieces of DNA and its rearrangement—that introduce into the genome a degree of dynamism far beyond what classical genetics had led us to suspect.

Most of this was yet to come in the late sixties, when the amino-acid sequences of a few proteins were the only biological sequences known. However, it was already clear that the information on which a cell acts is encoded in sequences of bases, and the question of how to characterize relationships among sequences hundreds or thousands of bases long was at hand. With his almost visceral feeling for representation of natural phenomena by general mathematical structures, Stan immediatedly framed the question in terms of defining a distance between sequences or, more generally, of defining a usable metric space of sequences (Ulam 1972). This he did by considering certain elementary base changes by which one sequence might be transformed into a second: Replacement of one base by another and insertion or deletion of a base. (Combinations of these changes can result from errors in DNA replication, chromosomal crossover during meiosis, insertion of viral or other DNA, or the action of mutagens.) Obviously, one sequence can be transformed into another by more than one set of elementary changes, as shown in the accompanying figure. What Stan proposed was to compute a measure, a "size," for each such set and to define as the distance between the sequences the minimum value of the measure.

In simplest form the measure is a sum of weights, one for each of the elementary changes that compose a transformation set. The set corresponding to the minimum measure—the distance between the two sequences—can be interpreted as the minimal mutational path by which one sequence could have evolved from the other. In 1974 Stan, with Bill Beyer, Temple Smith, and Myron Stein applied the idea of distance to discerning evolutionary relationships among various species from variations in the amino-acid sequences of a protein they all share. Also in 1974 Peter Sellers, after hearing Stan talk at Rockefeller University, proved that such a distance can indeed satisfy the conditions of a metric, the most demanding of which is satisfaction of a triangle inequality. Without that, one's sense of what it means for some among several sequences to be close and others distant would be quite unreliable.

Finding the distance between two sequences of length $N$ by brute force, that is, by computing the measures for all the possible sets of elementary changes, requires on the order of $N!$ computer operations. An algorithm for determining the distance in $N^2$ operations was discovered by the biologists Saul Needleman and Christian Wunsch in 1970 and independently by Sellers in 1974. Essentially, the algorithm proceeds by induction: The minimal set of changes needed to transform the first $n$ bases of one sequence into the first $m$ bases of the other is found by extending already computed minimal transformations of shorter subsequences, then $n$ and $m$ are increased, and so on until the ends of the sequences have been reached.

By the end of the 1970s, it was apparent that DNA sequencing would take off, and that investigators from all areas of biology, biomedicine, and bioagriculture would increasingly apply it to their particular research problems. It was also obvious that computer manipulation and analysis of sequences, much of it flowing from Stan's idea for a metric, would play an increasingly large role in exploiting the information. Mike Waterman had joined Beyer and Smith in working on sequence analysis, and Minoru Kanehisa, a postdoc from Japan, and I made genetic sequences and their analysis our principal preoccupation from then on. In 1982 a consortium of federal agencies funded GenBank, the national genetic-sequence data bank. Los Alamos collects and organizes the sequence data, and Bolt Beranek and Newman Inc. distribute them to users. By the end of 1986, DNA sequences totaling about 15 million bases, from several hundred species, had been deposited in GenBank.

In the 1980s a series of problems in sequence comparison have been faced with varying degrees of success. One problem now solved concerns global versus local closenesss (closeness, that is, in the sense of a distance between sequences). Often of interest are sequences that are close to each other although em-

## DISTANCE BETWEEN DNA SEQUENCES

**Consider the two short DNA sequences GTTAAGGCGGGAA and GTTAGAGAGGAAA. As shown in (a), one of these can be transformed into the other by four base substitutions. If the "weight" assigned to a base substitution is $x$, then the "measure" of the set of changes in (a) is $4x$. Alternatively, as shown in (b), one sequence can be transformed into the other by two base insertions, two base deletions, and two base substitutions. Since base insertions (deletions) occur less frequently than do base substitutions, the weight $y$ assigned to an insertion (deletion) is different from that assigned to a substitution; in particular $y$ is assigned a value greater than that of $x$. The measure of the set of changes illustrated in (b) is $2x+4y$, which is greater than $4x$. The distance between the two given sequences is defined as the minimum of the measures calculated for all possible sets of elementary changes that transform one sequence into the other.**

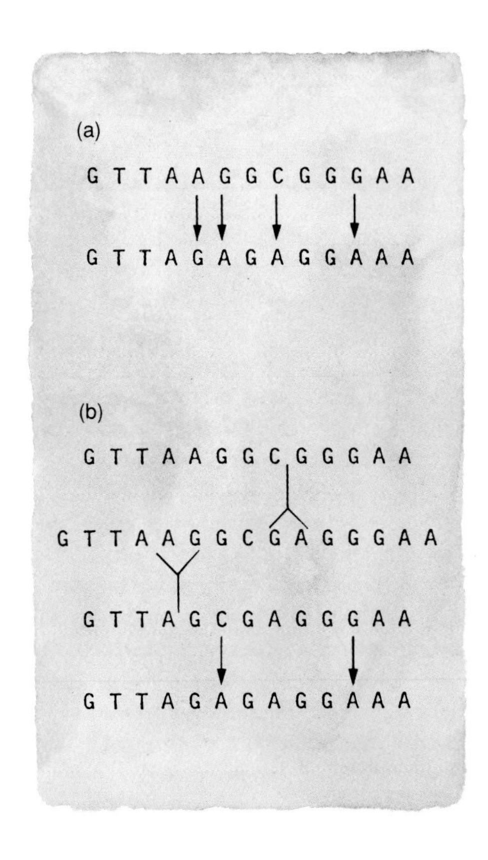

bedded in otherwise unrelated longer sequences. Peter Sellers first introduced the important distinction between local and overall closeness in 1980. A measure suited to the local problem (essentially the number of weighted changes per base, formulated so that the algorithm of Needleman, Wunsch, and Sellers can still be used) was introduced in slightly different forms by Kanehisa and me in 1982 and by Smith and Waterman in 1981. Another class of problems stems from the sheer quantity of data—examining 15 million bases, even with an $N^2$ algorithm, requires hundreds of hours on a Cray. That problem has been reasonably successfully dealt with by presecreening sequences for likely candidates for significant relationships. A table of pointers to the locations of short subsequences (a simple hash table) is created and searched for short matching sequences. At this writing the method is being implemented with new hardware features of the Cray XMP. For a general review of sequence-comparison algorithms, see Goad 1986; for a review that emphasizes mathematical aspects, see Waterman 1984.

Devising a metric appropriate to the investigation at hand is probably not a problem that can be precisely posed, much less solved. A simple metric in which each elementary change is given the same weight may well suffice when the object of study is a virus under great pressure to preserve a small genome. But such a metric may show misleading relationships when applied to segments of DNA from a more complicated organism, as Fitch and Smith found in 1983 for mammalian hemoglobins. Some relationships may depend on similarities in three-dimensional structure of DNA that are preserved through a set of sequences, as may be the case for the elements that control initiation of expression of particular genes. To discover such relationships, one needs a measure of structural similarity, expressed of course in terms of sequences. That problem is just beginning to be faced. A good sense of the problem, and of the limitations of sequence comparison, is given by analogy to another idea of Stan's. He proposed that perception, and thought itself, be considered in terms of a metric space. This frames the question: How is the distance between the visual fields corresponding to, say, two tables—which will vary greatly with circumstances—computed in our brains so that it is small compared with the distance between the visual fields corresponding to a table and a chair? Clearly the metric appropriate to a particular class of problems depends on the mechanisms one hopes to discover or illuminate.

Mathematical analysis has spread into nearly every corner of molecular genetics; its spread and development is still accelerating. In early 1986 the Department of Energy took the initiative in seriously exploring sequencing of the complete human genome, some 3 billion bases. In that project computerized management and analysis of information will play a key role.

Speaking of sequence analysis, GenBank, and all that, Stan once said, "I started all this." Yes. ■

## Further Reading


S. M. Ulam. 1972. Some ideas and prospects in biomathematics. *Annual Review of Biophysics and Bioengineering* 1: 277–291.

Willaim A. Beyer, Myron L. Stein, Temple F. Smith, and Stanislaw M. Ulam. 1974. A molecular sequence metric and evolutionary trees. *Mathematical Biosciences* 19: 9–25.

Peter H. Sellers. 1974. On the theory and computation of evolutionary distances. *SIAM Journal on Applied Mathematics* 26: 787.

Saul B. Needleman and Christian D. Wunsch. 1970. A general method applicable to the search for similarities in the amino acid sequence of two proteins. *Journal of Molecular Biology* 48: 443.

Peter H. Sellers. 1980. The theory and computation of evolutionary distances: Pattern recognition. *Journal of Algorithms* 1: 359–373.

Walter B. Goad and Minoru I. Kanehisa. 1982. Pattern recognition in nucleic acid sequences. I. A general method for finding local homologies and symmetries. *Nucleic Acids Research* 10: 247.

T. F. Smith and M. S. Waterman. 1981. Identification of common molecular subsequences. *Journal of Molecular Biology* 147: 195–197.

Walter B. Goad. 1986. Computational analysis of genetic sequences. *Annual Review of Biophysics and Biophysical Chemistry* 15: 79–95.

Michael S. Waterman. 1984. General methods of sequence comparison. *Bulletin of Mathematical Biology* 46: 473–500.

Walter M. Fitch and Temple F. Smith. 1983. Optimal sequence alignments. *Proceedings of the National Academy of Sciences of the United States of America* 80: 1382–1386.


**Walter B. Goad** received a B.S. in physics from Union College in 1945 and a Ph.D., also in physics, from Duke University in 1954. He has been a member of the staff of Los Alamos since 1950. In 1982 he received a Distinguished Performance Award from the Laboratory in recognition of his efforts at establishing GenBank, and in 1987 he was named a Fellow of the Laboratory. Until recently he directed the activities of GenBank, in which he continues to participate. His research focuses primarily on analysis of biological sequences. He is a Fellow of the American Physical Society and of the American Association for the Advancement of Science.

PART III

# The Ulam Touch

## UNPUBLISHED ITEMS

*On the lighter side, here are a few unpublished items that surfaced as we were gathering illustrations and data from Stan's files. They will accent what provocative fun Stan could be.*

# a memorable memo

In 1947 Los Alamos was in the throes of its postwar reorganization. Administration and Services, or A&S for short—a predecessor of Mail and Records—was wont to distribute a large number of mimeographed memos, in purple ink on plain white paper, to keep Laboratory members abreast of the latest developments.

One December day in a T-Division office, two individuals busily compiled a list of numbers, which they labeled "A&S Memorandum no. 10,742" and gave to a secretary to type for dissemination through the A&S channels.

Needless to say, once distributed this document caused a certain amount of stir and puzzlement, and at a Laboratory meeting an outraged member of the directorate suggested the perpetrator should be fired, which would free some much needed housing.

By and by the memo found its way to the AEC in Washington where it got a much better reception: A highly entertained Commissioner remarked that it was "The best thing to come out of Los Alamos yet!"

The culprits were J. Carson Mark (T Division Leader) and Stan Ulam (Group Leader).

Carson admitted later that their list had been hastily drawn and contained two "flaws": It was confusing to begin with "a dozen" for the number 12, and they altogether overlooked the number 10.

December 18, 1947

A & S Memorandum No. 10742

To All Concerned:

For your convenience and ready reference, we have had prepared the following list of the numbers 0-100 (inclusive) in alphabetical order.

12, 8, 18, 80, 88, 85, 84, 89, 81, 87, 86, 83, 82, 11, 15, 50, 58, 55, 54, 59, 51, 57, 56, 53, 52, 5, 40, 48, 45, 44, 49, 41, 47, 46, 43, 42, 4, 14, 9, 19, 90, 98, 95, 94, 99, 91, 97, 96, 93, 92, 1, 100, 7, 17, 70, 78, 75, 74, 79, 71, 77, 76, 73, 72, 6, 16, 60, 68, 65, 64, 69, 61, 67, 66, 63, 62, 13, 30, 38, 35, 34, 39, 31, 37, 36, 33, 32, 3, 12, 20, 28, 25, 24, 29, 21, 27, 26, 23, 22, 2, 0.

FOR THE ASSOCIATE DIRECTOR

# Sub Rosa

## A Trialogue

*After dinner one evening in 1965, Stan in a playful mood dictated to Françoise all in one breath, so to speak—without corrections or rewrites—the following "top secret" skit, which was not meant for public consumption. When asked what was to be done with it, he replied, "File it away and posterity will decide." It was filed and forgotten and no one else ever knew of its existence until it resurfaced recently. Now "posterity," in the form of the editors of this magazine, has decided to print it as a perfect example of Ulamian humor, which, built on a great sense of ridicule and classical erudition, was concise, incisive, and capturing of the essence.*

Sub Rosa, *which Stan called a play—as if he had meant to dictate a five-act opus!—was his way of making fun of the horrendous nuclear debates that had filled the councils of state, civilian and military, and the nation's press and airwaves since the advent of the atomic age. More specifically, the skit represents a few of the scientists and some of the political and scientific issues that surfaced after the frantic bomb tests of 1961 and 1962 and the Test Ban Treaty of 1963, which limited testing to underground.*

*It would not take a great reach of imagination to transpose this into the present. Twenty years later the arguments and even some of the people are still the same!*

*The footnotes are Stan's own and self-explanatory, but to heighten a younger and less familiar generation's appreciation of the satire, the following remarks and the numbered endnotes will we hope prove helpful.*

Sub rosa *(literally under the rose) refers to an alleged ancient French custom of hanging a rose over a council table to swear those present to secrecy.*

*The principal characters engaged in a three-sided conversation are Bethe, Teller, and Ulam, thinly camouflaged as Benefacius, Totilus, and Ulfilas. Benefacius is a play on the name Bonifacius, a German saint; Totilus was a belligerent king of the Ostrogoths; and Ulfilas was a bishop of the Goths. (Note the reference to chess, a favorite game of Ulam's.)*

*Rounding the cast, we have Gregarius, the geophysicist David Griggs, a great admirer of Teller and for a while chief scientist for the Air Force; Fos-terasis, John Foster, then director of Livermore, known as an outspoken hawk; and Vertihumerus (Stan's Latin for green horn), an anonymous young man. The scientist with a lapsed Q clearance is the Russian-born physicist George Gamow, whose speech was characteristically laced with misplaced articles.*

*There also appear allusions to Hermandus Canaan, read Herman Kahn of the Rand Corporation, a well-known California think-tank; Libius, Willard Libby, an AEC Commissioner; and Manilius, John Manley, secretary of the General Advisory Committee to the AEC in its early years.*

*Enjoy!*

*A Trialogue*

*Dramatis Personae*†

BENEFACIUS, a physicist

TOTILUS, a weapon scientist

ULFILAS, a chess player

GREGARIUS, a former Air Force scientist

FOS-TERASIS‡

VERTIHUMERUS, a young scientist

A SCIENTIST WITH A LAPSED Q CLEARANCE

CHORUS OF SPACE SCIENTISTS

AN ECHO OF DOD CHORUS

A CHORUS OF FEGATELLO§ SCIENTISTS

A CHORUS OF SHERWOOD[2] SCIENTISTS

AN INVERTED JACOB'S LADDER

*The trialogue takes place in Limbo, which explains the use of Latin and other classical references.*

BENEFACIUS: I have just read an interesting article in *Life*, but I think it contains some technical mistakes. For example, the data on—

TOTILUS: Many wonderful things can come from testing. Think of obtaining oil from shale, for instance.

BENEFACIUS: One mistake I noticed—

TOTILUS: Excuse me, may I say that the pressure obtained at the site of an underground explosion can produce new minerals, perhaps even diamonds. Harbors could be built in Alaska, in Greece. There is oil to be squeezed out in Texas.

BENEFACIUS: It is not correct to say—

TOTILUS: Very small, economical—I mean cheap—bombs can be tested for tactical use in small wars.

ULFILAS (*standing up, having kept quiet with some difficulty*): This article describes in dramatic tones the horribly difficult decision the President has to make—as usual in solitude—after weighing the pros and cons of testing. This comes at the end. If one reads the beginning and the middle, there seems, however, no question as to what the decision has to be. The author makes it clear what anybody in his right mind would decide. He describes the tragedy of the moratorium. Since the Russians made so much progress in testing after the end of the moratorium, it would have been much better if they and we had tested all the time.

Isn't it possible that the Russians, with their devilish cleverness, might really want us to concentrate on little improvements of warheads instead of working on the really militarily important developments, like rocketry?

CHORUS OF SPACE SCIENTISTS:
The gap exists, but it is narrow.
If we miss the moon, we go around the sun.
How can we lose!

ECHO FROM DOD CHORUS:
The credible second-strike capability is firming up.
The stable deterrent might be upgraded.

ULFILAS: Unfortunately both sides have an incredible first-strike capability.

CHORUS OF FEGATELLO SCIENTISTS:
Not all is lost,
because we've got—
or have we got?—

*With apologies to the spirit of Anatole France, who wanted to write a story so named.
†Any lack of resemblance to real persons is purely coincidental.
‡Greek for strange light.
§Italian for little liver.[1]

a neutron bomb,
and they have not.

ULFILAS (*to himself*): Produce neutrons or get off the pad!

TOTILUS: Peaceful applications are very important. We might be accused of bloodthirstiness otherwise.

There is also the possibility that if we work steadily and vigorously some of the old ideas which did not work might be revived. Something close to the proposals which were proved to be impossible can be revised to show that I was right.

A body of scientists must have something to do. *Si vis bellum para pacem.**

BENEFACIUS: There has been criticism of Los Alamos for working on peaceful aspects like nuclear rockets or the Sherwood project.

CHORUS OF SHERWOOD SCIENTISTS:
We have twelve approaches
to the problem of peaceful fusion.
Six are good and two might
even be promising.
Neutrons abound; the
instabilities are small.
A breakthrough is just beyond the
horizon.

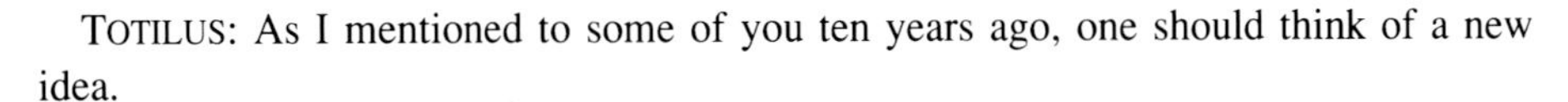

TOTILUS: As I mentioned to some of you ten years ago, one should think of a new idea.

ULFILAS: I mentioned that some twelve years ago.

TOTILUS: Production of all fissionable material must be enthusiastically increased.

ULFILAS: Why? Isn't our stockpile infinite? Although I agree that there are degrees of infinity—countable and noncountable.

It seems to me that the interesting concept of "overkill" is attacked by people on the left because it is wasteful, i.e., not economically sound, whereas people on the extreme right support it for psychological reasons—it gives them a feeling of virility, which they otherwise miss.

TOTILUS: According to the calculations of my friend Hermandus Canaan, with whom I discussed the subject in detail, if only fifty to eighty million people are killed, the country can rebound vigorously in forty to fifty years, and the forward march of the consumer economy will resume.

Shelters are important. Libius has written that one can improve one's chances of survival by a million, or maybe as much as a billion.

ULFILAS: How? If this chance after improvement should be of the order of one, then it must before have been only one in a million or one in a billion. How does that jibe with sixty million people being killed, which is one-third of the population?

*If you want war prepare peace. Really, *Si vis pacem para bellum* (If you want peace prepare war).[3]

TOTILUS: You see, that all depends. If we improve our weapons, the chances might be much greater. A toothbrush can be improved indefinitely, as, I believe, Manilius said.[4]

GREGARIUS (*scolding everybody vigorously except Totilus, to whom he makes a mild reproach for talking too much*): My committees state almost unanimously that our posture has deteriorated and needs considerable firming up both front and rear.

TOTILUS: One will never be sure without testing.

BENEFACIUS: One should really make more calculations.

TOTILUS: Perhaps you remember we discussed this fourteen years ago.

ULFILAS: A report which I wrote sixteen years ago mentions that very explicitly.

FOS-TERASIS: We have had very bad luck recently. According to the laws of probability it cannot continue indefinitely. Given enough testing, progress will be assured.

ULFILAS: I doubt it.

BENEFACIUS: That is right.

TOTILUS: My friends from the Brand* Corporation have computed that the danger of fire storms and fallout has been vastly exaggerated. Also, people who try to prevent testing exaggerate the effects of the tidal waves. One could have a million megaton explosions without the waves reaching the Rocky Mountains. I do not know where the real danger point comes, but many more explosions can safely be made in a war than people think at present.

BENEFACIUS: That is not right.

ULFILAS: Somehow these things seem to me not good. I agree that we must be strong, but it is as futile not to test as to test. Our only cleverness recently was to induce the Russians to test underground too.

VERTIHUMERUS: It is *not* futile to test, and it is *not* futile not to test. I keep my guarded pessimism.

BENEFACIUS: Why should the neutron bombs lead to such a great advantage? I agree to a possible small advantage. In the wars of the eighteenth century there were nice close formations marching with officers and drummers in front. A neutron bomb would have been useful then; it could have killed the whole group without ruining the wristwatches of the officers.

ULFILAS: I don't think there were any wristwatches in those days.

BENEFACIUS: That is right.

TOTILUS: Small bombs would enable one to have a lot of small wars. If one should exercise additional self-restraint, these might be contained. And perhaps, even in the eighteenth century a Napoleon would not have been possible.

*German for conflagration.

AN INVERTED JACOB'S LADDER *appears, and a Brand Corporation scientist starts climbing it to advocate confining and concealing nuclear explosions underground.*

CHORUS OF SPACE SCIENTISTS:
*Facilis descensus Averno*
*Sed revocare gradum superasque*
*evadere ad auras,*
*Hoc opus hic labor est.**

ULFILAS: That must mean that it is more difficult to conceal testing up in space than underground. If you are quoting Virgil, as you seem to be doing, you must know the end of the story.[6]

I must say that my main worry is not war as a result of premeditated action, but accidents. If the probability in any one year is alpha, then—

TOTILUS: Accidents, like mutations, are not always bad. You forget, Ulfilas, that there is perhaps a small probability beta—(*catching himself*) not so small, I should like to say—that something good can come out of it.

ULFILAS: It seems to me that we should all go more into space. This might be a tremendous distraction for all of us, for many reasons.

I am in favor of hyperbolic orbits and space research. A lot of spare energy can go into it. It stimulates the economy as well, and the rivalry is perhaps less dangerous there.

TOTILUS: I like parabolas myself. It is on these, you see, that you can show strength and deliver bombs.

BENEFACIUS: I like ellipses. They are useful for communication satellites and weather prediction.

TOTILUS (*angry for not having said that himself, especially since elliptical orbits can also be used for ejecting clean bombs*): I might agree to that. They might be important for the uncontaminated countries.

SCIENTIST WITH LAPSED Q CLEARANCE (*looking upon them with pity*): I too have consulted for many years. But now I am studying the astronomy and the biology. Is better. Come, Ulfilas, let's go have drink. **Finis**

*It is very easy to descend to hell, but to retrace one's steps and reach the upper regions, that's the task, that's the toil.[5]

[1] Livermore, of course.
[2] Sherwood was the name of the research program at Los Alamos on controlled fusion.
[3] A Roman proverb.
[4] John Manley actually said that a toothbrush can *not* be improved forever.
[5] From Book 6 of Virgil's *Aeneid.*
[6] The significance of this reference to "the end of the story" is not clear.

# CONVERSATIONS *with* ROTA

*The cultural affinities and intellectual differences between Stan and Rota were such that they could talk together for hours, though they were worlds apart mathematically and philosophically and never collaborated. I was fascinated by these spontaneous and informal discussions and recorded a number of those that took place in my presence, to transcribe and assemble loosely in a little collection.*

Los Alamos Science *selected the following fragments to illustrate the way Stan and Rota interacted and, more generally, the way mathematicians converse about what occupies their minds.*

Françoise Ulam

## The Mathematical Dictionary

ULAM: I think it is a very good idea to try to write a mathematical dictionary. First we must settle how many words to think about. Would you say two hundred or less?

ROTA: Two hundred! No. Ten, maybe!

ULAM: No, no. At least a hundred. They will have to be very diverse. It will be a long project. Logical words like *but* and *even* have a different character from words which have a topological or kinematical meaning like *mix, find, search*. Then there is another class of words like *involve, intuitive, imaginary*. There are many categories. I think we should have a few in each. We must say what it is for in the introduction.

ROTA: May I make an introductory sentence?

ULAM: Yes.

ROTA: Let me make an attempt to give a precise mathematical definition ...

ULAM: Not too precise.

ROTA: ... a precise mathematical definition of commonplace terms. We will take words like *but, furthermore, nevertheless,* or *crest, degenerate, skid* and describe them in terms of accepted mathematical terminology.

ULAM: And physical. Physics is almost completely mathematized now.

ROTA: I have *already,* and *perhaps,* and *pending*. They are close.

ULAM: *Already* is difficult mathematically. What about starting with *but*?

ROTA: Logicians claim *but* is the same as *and*.

ULAM: No! Its meaning is entirely different. How would you describe *but* logically? Something that leads us to a conclusion but does not? A disappointment in probability? A whole essay could be written about it. Someday there will be a tremendous theory devoted to its ramifications. It could be a germ like the word *continuous*. The study of topology is nothing else but the study of the word *continuous*.

ROTA: When I was at Princeton, Alonzo Church gave a two-hour lecture on the meaning of *but* and *and*. It is now written up in his great *Introduction to Mathematical Logic*.

ULAM: So you see! And what does he say? I never read it. I knew he was a logician but did not know he did things like that. Now let's discuss things intelligently, professor.

ROTA: O.K. Let us begin with the word *but*. Stan?

ULAM: I would say that the word *but* suggests to me the following (we'll be more precise later): an element of an algebra whose elements are uttered sentences. I can imagine it as a point in a universe of points interpreted as sentences—physical facts. I see that it won't be easy to avoid circular definitions; we must not use the word *but* in developing a theory of *but*, right? The word *but* means that an element does not belong to a given set of points that was defined *before*. But—I am just saying this on purpose now—*but* expresses that an element belongs to a set which is similar or slightly larger than the already given set. Of course, I did not really need to use the word *but* in my explanation. However—Oh! I just used the word *however*; you see how hard it is to avoid these words? By the way, this poses another interesting philosophical problem, the fact that we cannot explain a mathematical ...

ROTA: Let's not digress.

ULAM: I just want to see what is in my mind. I do not have a perfect definition right away. Do you agree that *but* is an element which does not belong to a set that was defined before?

ROTA: Yes. Now let me try my definition. We have two sets, $A$ and $B$, and a new relation between $A$ and $B$ which we will call contrast. The word *but* is the contrast between set $A$ and set $B$.

ULAM: The set $B$, in every example I know of, is usually given by the speaker. Set $A$ is mentioned; set $B$ is intended. It is not there at the beginning or maybe it is only in the mind of the second speaker. Would it be a good idea to consider it as a part of *conversation*? One person proposes something, and the other replies, "No, but ..."?

ROTA: No. I don't think it is a good idea to formalize *conversation*. It would get us too far from our purpose. If we are going to give definitions, they have to be objective.

ULAM: Whatever is done, you always stick to tempus acti, and you do not want to do something unorthodox. Why reduce it to the existing formalism? It is good to try, but it is not necessary.

ROTA: If possible, do it. Only when you have to, give up.

ULAM: O.K. I agree. Continue.

ROTA: So you have two sets and the contrast between two sets, and the word *but* is an expression of this contrast. And now I would say the word *but* is used when this contrast has to be brought out.

ULAM: Very good. But is that really always true?

ROTA: That is my story.

ULAM: We should have examples, like in dictionaries. They always give quotes from Shakespeare.

ROTA: Let me give an example: "We were going to go out touring today, but it is raining and we didn't go." Analysis: There are two situations, or sets, if you wish. One, going touring; two, assumption that the weather is fair. Then the weather turns out not to be fair, so there is a contrast between fair and unfair and the word *but* arises.

ULAM: I agree. Let me give another example: "The snail is not an insect, but it is still an animal."

ROTA: Here again you have a set. You presume the snail to belong to this set. The contrast arises because you see a further subdistinction inside this set.

ULAM: Simply, these sets are not equal; one set contains the other.

ROTA: Be that as it may, either not equal or partitioned, one contains the other.

ULAM: This is part of it but perhaps not all. We will have to have detailed discussions like that about every word, as they do at l'Académie française!

ROTA: Let me say that any definition is necessarily incomplete. It is a property of definitions to be incomplete.

ULAM: Incomplete perhaps, but still it should try to be as broad as possible.

ROTA: Then it will never end. There is a point where one says fine, adequate, even though it is not the whole story.

ULAM: Yes, I agree.

ROTA: Let's take another example. One says of a person: "He is good, but he is also careless." How would you analyze that?

ULAM: A point belongs to two sets. If you say *good*, the presumption is that everything is good about him, so you add another set.

ROTA: Suppose I replaced the word *but* by the word *and*. In your opinion how would the meaning of the sentence change?

ULAM: It would be an entirely different meaning.

ROTA: Why?

ULAM: Because the set of *carelessness* is not a set which normally is associated with the set *being good*.

ROTA: It is not a complete explanation. *But* always requires a contrast.

ULAM: True. What about a distinction?

ROTA: Distinction is too weak. *But* requires contrast and unexpectedness.

ULAM: Unexpectedness. Exactly. This is the essential thing to my mind. Namely the first set suggests something, and the second implies the suggestion does not hold. A set of properties implies a lot of others, but an exception is made. *But* suggests exception.

ROTA: No exception is involved. For example, "I was going to go out but the phone rang." That is no exception.

ULAM: That is yet a different meaning of *but*. It says that the normal pattern is being abruptly changed.

ROTA: Let me say this. A lot of examples have the following structure. You have two sets, $A$ and $B$, and you have an element $c$. You expect $c$ to belong to $A$, but then it turns out to belong to $B$. That is the typical use of *but*.

ULAM: Right. So it is not a relation of the contrast but of difference.

ROTA: We have abstracted a set-theoretic relationship for the word *but*; namely we have two sets and an element. The element may belong to $A$ but instead it belongs to $B$.

ULAM: Very good! However the two sets are somehow close. They are not too different or one contains the other because you could not say, "The pencil is long but it is black."

ROTA: Right.

ULAM: Why?

ROTA: There must be a similarity between the sets.

ULAM: Ah! Now we have caught one essential point.

ROTA: So there are these properties of sets which somehow are similar, and then there is the confusion of one element belonging to one instead of the other.

ULAM: In general the two sets are in some relation of similarity, close in the sense of a Hausdorff distance or whatnot, and not completely separate.

ROTA: Two or more sets are in turn subsets of a set of sets which is predetermined. They are members of the same family of sets.

ULAM: What does it mean, the same family?

ROTA: The family is the similarity class.

ULAM: Right, that is what one could say. Very good. We are getting somewhere.

ROTA: You see, I am becoming Ulamian. Set of sets!

ULAM: My example about the pencil was crucial. It did not make sense. Let us take something else. For example, *however*. It is not quite the same as *but*.

ROTA: Later. Let us finish with *but*.

ULAM: We have to warn our readers and ourselves that there are words that mean almost the same, with subtle shades of difference. In French there is *mais* and *cependant*. We ought to analyze that.

ROTA: What about *nevertheless* and *yet*?

ULAM: *Nevertheless* has a greater degree of something. We should analyze all these. They are all coming together.

ROTA: *In spite of*...

ULAM: I would very much like to define the word *key* or *lock*, because there is a sort of labyrinth, a maze. You have to enter a lock a certain way, which at random is difficult, and perform a sequence of operations.

ROTA: *Key* is absolutely one of the best.

ULAM: *Key, lock, labyrinth*— there is a whole topological, combinatorial meaning there. Logical too. *Key* also has an abstract meaning, a *key* to something. We are just beginning. This is a project for several months.

ROTA: We could get a grant!

ULAM: From some cultural whatnot— there are such. Philosophers do not give grants, but we are rich old men, as Erdös says. If we could meet an hour a day, we could get somewhere in one month.

FRANÇOISE: Next summer in Santa Fe.

Gainesville
January 1974

## On Teaching And Learning

ULAM: Most of what I've learned was subconscious, by osmosis. When I read, I am not aware that I am learning. I learn mainly from conversations, from people rather than from lectures, and I did not realize until a few years ago that I have a good memory.

I could start teaching mathematics with courses for college freshmen and go to junior or senior courses without any preparation, because in mathematics one thing leads to another.

Let us discuss whether teaching mathematics really makes any sense. Either the student is so good that he does not need a teacher, or else, if he needs help, he is not cut out to become a mathematician. At Harvard I had some good students with whom I could talk and feel that teaching was not merely an empty gesture.

I think I influence people more than I teach them. I influence their taste or their choices.

ROTA: I learned from you to argue in a short way, to give only ideas followed by simple examples. That is the Ulamian influence.

ULAM: I don't mind teaching, but I don't like to do it regularly. When I have to do something at a fixed hour, even if it is a pleasant dinner or cocktail, I fret. I hate not feeling completely free. But of course, being completely free immediately brings on a feeling of restlessness, of not knowing what to do!

Each of us has taught several thousand hours. If you think that a normal working year in America has about 2000 hours—an 8-hour day for about 250 days—that is quite a bit of your waking time, isn't it? But maybe it is not entirely waking time. One does it in a trance, partly asleep sometimes!

I am told I teach calculus well. It is possible, for I believe one should concentrate on the essence. One should not teach everything at a uniform level either. One should stress some important as well as some unimportant details on purpose—in a sense to follow the way I think the memory works.

When you remember a proof you remember a sequence of pleasant, unpleasant points, zeros and ones. Here comes a difficulty you try to remember, and you make an effort. Then you come to something that goes automatically and it is zero, zero, zero. Then again a special trick that has to be remembered. It is like going through a labyrinth.

ROTA: I am amazed at your labyrinth!

ULAM: I learn best from conversations. I love them, and that is how I learned physics in Los Alamos.

Some people are different in this respect. They prefer to learn slowly and methodically. How about you?

ROTA: I learn best when I am forced to do it.

ULAM: Speaking of being forced to learn, in Poland it happened several times that I announced that I would speak on a certain subject at a meeting of the mathematical society before I had a proof. I felt absolutely confident that once I had agreed to speak, I would get a proof. It could have been an embarrassment otherwise.

On the other hand, when I look at a paper of mine which has been published, I discard it after one glance, from fear that I will discover that it is wrong. There is also this tiny gnawing doubt about whether the result is new or not. Yet even in a field about which I know nothing, I can always tell whether a theorem or a point of view is good or not. This feeling comes somehow from the way the quantifiers are arranged, from the tone or music of the piece.

Do you remember what Galois wrote in his final letter before his fatal duel? He wrote that in their publications mathematicians really conceal the way they obtain their results because the process of discovery is different from what appears in print. It is important to repeat this again and again.

Gainesville
February 1974

## John von Neumann

ULAM: Hot! What is the temperature?

ROTA: 80 or so.

ULAM: Pas possible! It must be the hottest day in thirty years. Which reminds me, once flying back to Los Alamos on Carco on a hot summer day, I opened the little window and my handkerchief flew out of the plane. Behind us there was a second plane carrying Johnny and others. What do you think the probability is that my handkerchief could have gotten enmeshed in the propeller of the other plane?

ROTA: Von Neumann was older than you.

ULAM: Six, seven years.

ROTA: An older man!

ULAM: Yes. You know how it is. In the beginning the percentage was twenty or so; later it went down to ten.

ROTA: So you considered him a senior, and yet you made fun of him?

ULAM: Oh always! Of Banach too. I was always impudent.

ROTA: He did not treat you as someone younger?

ULAM: No. I don't think he knew anybody more intimately and vice versa, despite our difference in age. For a man of his stature he was curiously insecure, but his understanding, intelligence, mathematical breadth, and appreciation of what mathematics is for, historically and in the future, was unsurpassed. His immense work stands at the crossroads of the development of exact sciences. The rationalization of the idea of infinity—the life blood of its history—with its mysterious power to encode succintly and generally the properties of numbers and the patterns of geometry, received some of its definite formulations from his work. His ideas also advanced immeasurably the attempts to formalize the new, strange world of physics in the philosophically strange work of quantum theory. Fundamental ideas of how to start and proceed with the formal modes of operations and the scope of computing machines owe an immense debt to his work, though they still today give hints that are only dimly perceived about the workings of the nervous system and of the human brain itself.

Other mathematicians strike me as virtuosi who play their own special instruments. None are comparable to Johnny.

By the way, you were supposed to ask about the foods von Neumann liked.

ROTA: List the foods von Neumann liked and those he did not like!

ULAM: He was not a gourmet, but he liked to eat. He liked to go to restaurants, mainly, I think, to escape from the usual scene or routine. It was an excuse for not working, because he was a very hard worker. At home he worked at a desk, writing, a thing which irritated me a bit. When I stayed at his house and saw him suddenly leave to go upstairs and write, I, cruelly and foolishly I must say, would make fun of it. So for relaxation he liked to drive out for dinner. In Princeton we often went to a restaurant called Marot, on the highway to Trenton.

He never smoked, but he ate voluminously, which accounted for his increasing rotundity and portliness as the years went by. Sometimes when Klari, his second wife, could not finish what was on her plate, she would give it to Johnny or to me and say, "Both of you are human garbage cans!" Klari, by the way, was a very intelligent, very nervous woman who had a deep complex that people paid attention to her only because she was the wife of the great von Neumann, which was not true of course.

Johnny liked Mexican food, hot peppery stuff. I suspect it was because if he had a stomachache later, he would know what to blame it on! I always have such Machiavellian suspicions. It is probably just that he was used to Hungarian goulashes and hot paprika. He liked sweets too, but on the whole what he wanted was volume, like me, like you too. You like the volume of pasta.

He had this nervous trait, an almost automatic response. For example, whenever he saw the words *chicken mole* on a menu, he would automatically intone *Moles Hadriani*, and I would respond *Jacques de Molay*—you know, the Grand Master of the Knights Templar. It was a game of association, just like you always add *Pal* [Hungarian for Paul] when you hear the word *Erdös!*

He also had occasionally an infrequent but noticeable stutter. He would say a word and repeat it two or three times in quick succession. I wonder whether it could have been an incipient physical lesion, for he died of things affecting his brain. Actually, on second thought it could not, because his cancer started somewhere else. Sometimes I suspect that his stutter was in order to gain time while thinking over a riposte or considering quickly some other angle for a statement, like a person lighting a pipe to gain time.

ROTA: How long did you know von Neumann?

ULAM: I first met him in Warsaw in 1935, but I had already started corresponding with him the year before, and that is when he invited me to visit him at the Institute in Princeton.

ROTA: What was he working on at Los Alamos during the war?

ULAM: On everything. He was one of the originators, one of the "influencers" of implosion. By the way, you are my most eminent "influencee"; it is a relationship different from teacher-student.

He worked on the whole project, scientifically and politically, especially with the hydrogen work.

ROTA: But actual *work*?

ULAM: Of course, mostly hydrodynamics.

ROTA: Did he know much physics?

ULAM: To some extent, but he did not have the physicist's feeling for experiment. His interest was more modern than Hilbert's. His interest was in the foundations of quantum mechanics, which were mathematical. And that could be taken as an example of mathematics not really useful for real physics.

But there was no bullshit in him. That is an expression he used about certain people. He would say, "It is very rare, but there is no bullshit in so-and-so."

Of course he worked, in answer to your question. In fact he was unable to play the role of senior scientist or advisor without being actively engaged, like with computing. Even towards the end of his life, when he was chairman of the ICBM Committee, a committee established by the President after Sputnik.

ROTA: I still don't have a picture of von Neumann's personality.

ULAM: He loved jokes, though I don't think he invented many, but he remembered and repeated them, and occasionally he made original and very witty remarks or saw comparisons which were comical. Most are unprintable.

A propos of the church knowing about the atom bomb, he said, "Priests will bless the active cores." And when he noticed all the churches of Los Alamos, he was much amused when I pointed to one church and called it "San Giovanni delle Bombe"! One of the first solid non-wood buildings in Los Alamos was built for the offices of the AEC. He called it "El Palacio de la Seguridad"!

Oh! One thing about Johnny, he tended to tell people what they wanted to hear. He also used to tell me his little tactical discoveries. Once he said, "In Los Alamos it is very difficult to introduce novelty, but once introduced, it is impossible to get rid of it!"

After the war he was for a Pax Americana, and one could probably have established it, but the historical perspective, the desire to do it were not present in the country. The general population was not thinking in those terms. Although, when World War II ended, Americans were like Roman citizens during the Roman Empire. By commuting through the American bases one could go anywhere in Europe without encountering the native populations. This was really a beginning of that sort of thing, but for good or for bad—who knows—it quickly dissipated.

What else would you like to know about von Neumann?

ROTA: Always well dressed, wasn't he?

ULAM: Not really well dressed, but simple, decent, well-cut, classic city dress.

ROTA: I still don't have a picture of the man.

ULAM: He became an important government figure and very influential in ballistic missile development.

ROTA: It is strange how you like everything about him except his work in mathematics.

ULAM: Really? No, not quite so. But he was not a mathematician's mathematician. He did little in number theory, some in continuous geometry and operators and Hilbert space, and some in measure theory and group theory.

To my mind and to my taste, the most important work he did is what he did when he was getting older, which mathematicians don't appreciate, namely his speculations on automata, on the brain, and his contributions to computing and to problems in hydrodynamics.

He knew about quantum theory and some parts of theoretical physics, which few mathematicians did. He contributed to the grammar of physics, so to say. One must also mention the theory of games. What interested me less was his work in the almost-periodic functions of groups.

ROTA: Can you tell me something about how his mind worked?

ULAM: It is curious to me that in our many mathematical conversations on topics belonging to set theory and allied fields, he always seemed to think formally. Most mathematicians, when discussing problems in these fields, seem to have an intuitive framework based on geometrical or almost tactile pictures of abstract sets, transformations, and such. Johnny gave the impression of operating sequentially by formal deductions. His intuitions seemed very abstract; they involved a complementarity between the formal appearance of a collection of symbols, the games played with them, and the interpretation of their meanings. Something like the distinction between a mental picture of the physical chess board and a mental picture of a sequence of moves on it written down in algebraic notation!

The quickness of his thinking was quite remarkable. He saw immediately the possibilities of Monte Carlo. To my mind this was much more important than one hundred papers in partial differential equations! It is at least a general procedure—I would not quite call it a method—and he invented many tricks for it and specific ways to get random distributions. It was very pleasant to discuss it with him.

Too bad he did not live to see how computers have revolutionized everything and what influence they will have on science in general and even on pure mathematics. His role in their development was tremendous, and if I may say so I would say I too played a modest role in showing how to use computers!

ROTA: How would you characterize his influence?

ULAM: There used to be a time when there were mathematicians who gave specific ideas and choice of topics and directions either explicitly or by implication to the work of other mathematicians. Not to go back centuries but less than a hundred years, let us say Poincaré, Hilbert, in more recent times Herman Weyl. Hilbert had laid what was hoped would be a foundation for the final axiomatization of mathematics and beyond, of all science. Little did he know that in the thirties the unavoidable limitations of this approach would be revealed.

Von Neumann was one of these giants too in the breadth of his knowledge, especially when one remembers that now the diversity and complexity of contemporary problems enormously surpass the situation confronting Poincaré and Hilbert. Yet, he admitted to me that he felt he did not know even a third of mathematics, that he did not think it was possible nowadays for any one brain to have more than a passing knowledge of more than one-third of pure mathematics.

So, at his suggestion and for his amusement I concocted an oral doctoral examination in various fields in such a way that he would not be able to pass it. And indeed, when I thought about what problems to give him in each domain, I found one in differential geometry, one in number theory, one in algebra and a couple of others. And he agreed that he could not have answered any of the questions and the exam would have been a complete failure. Which goes to show that doctoral exams are to some extent meaningless. Of course, if one prepares for some specific topics, that is something else.

ROTA: Who was von Neumann a student of?

ULAM: He considered himself a student of Ehrardt Schmidt. It was not easy for me to get to the bottom of this. One reason, I suspect, is that Schmidt did some work in combinatorics which always interested Johnny very much.

ROTA: It was the Hilbert space. Schmidt was the only person at the time who studied nonlinear operators.

ULAM: But Johnny did not.

ROTA: That is why he admired Schmidt!

ULAM: Also I remember that he told me that Schmidt did not like to write. That surprised Johnny. I also think he secretly admired it. He said that Schmidt had told him that he felt faint whenever he saw a blank sheet of paper. Johnny was not at all like that. On the contrary, whenever he had a mathematical thought, he immediately wanted to write it down and elaborate.

ROTA: Did he have any students?

ULAM: Not really, even though at the Institute he gave several courses every year. Murray and Halperin may be considered his students.

ROTA: What about Gödel and von Neumann?

ULAM: One summer before the war when I was returning to the States, Johnny was waiting for me at the pier. His first words were that Gödel had shown that the continuum hypothesis was undecidable. This was how I heard for the first time about the existence of undecidable propositions in any formal system. So I said to him, "Oh! That is because he defines what is meant by a set." Johnny opened his eyes wide and expressed surprise that I had seen right away what was indeed the essential point. He thought I had some supernatural intuition.

I asked him whether Gödel was not a little afraid that his result was nothing but a sort of super paradox of the existing set theory, merely a diagonal method. In a sense it *is* a diagonalization. He agreed that this was probably right and that Gödel did not quite realize the importance of his discovery because of the fear that it would turn out to be merely another version of the whole series of set-theoretical paradoxes. Of course it was much more than that because he had made it all formal. The other paradoxes were special and dependent on metamathematical considerations that were not truly part of mathematics, whereas his results were. Curious how nervous people can be about their own work when it is *the* work!

ROTA: You have a higher opinion of Gödel than I have.

ULAM: Yes, I know. It was so unexpected at the time, and poor Hilbert was ...

ROTA: Not to speak of poor von Neumann.

ULAM: Johnny told me that Gödel's results made him very downcast, not quite despairing but disappointed. You must remember that his work on the axiomatization of set theory, which was way back in the twenties, constitutes to this day one of the best foundations for set-theoretical mathematics. Basically he believed in Hilbert's goal of a final and conclusive axiomatization of mathematics, and yet, in a 1925 paper, in a mysterious flash of intuition, he pointed out the limits of any axiomatic formulation of set theory. That was perhaps a vague forecast of Gödel's result. But it was left to Gödel to follow it through, and it has changed the direction of all science.

Gainesville
January 1974

## On Ethnic Minds

ROTA: What is the difference between the Slavic mind and the German mind?

ULAM: The German mind is systematic; the Slavic is not. Slavs tend to be soulful, expansive, pensive, but they are not as nebulous or as much carried away by the sound of words as Germans are. In the German language syllables and words concatenate, and they concatenate thoughts which sometimes don't go very well together.

ROTA: Whereas the Slavic?

ULAM: Slavs tend, I think, to be self-analytic, more psychological than philosophical, full of regrets, feelings of guilt, but more fundamentally optimistic than the German, and with humor, which if it is not showing, is at least not far away. German humor is based on ridicule, I don't know why. Latins are something else.

ROTA: Describe the Latin mind.

ULAM: Order. Clarity is always there. Words are separated, they don't stick together. It is like well-cooked rice compared to the sticky overcooked stuff that comes out of a German mess.

What would you say about the Jews? Would you say there is a Jewish mind?

ROTA: I don't think so. Italian Jews are Italian, German Jews are German, and so on.

ULAM: Don't you think that the Jewish mind is a little truculent, that there is a bent for contradiction? I feel I have myself this Jewish characteristic of always wanting to change what exists. It is a sort of rebelliousness, an inability to kowtow to authority. Think of the great revolutionaries—Jesus, Marx, Freud, Einstein, Cantor. Cantor by the way was only half Jewish. Most Jews are only part Jewish, you know, but the Jewishness comes through all the same.

This rebellious spirit of the Jews does not show in music, where the Jews are much less creators than performers, interpreters.

Gainesville
January 1974

## Gamow And Teller

ULAM: It is Gamow who brought Teller to George Washington University originally.

ROTA: From zero?

ULAM: From Europe, in Hitler time, when he had no job.

ROTA: How did Gamow and Teller get along?

ULAM: Gamow ruined Teller a bit. Gamow had this fantastic talent—an intuition, a lightness of touch for what is important, without doing too much work, without much mathematics, without any laborious Gründlichkeit. Teller wanted to be like that. He had other talents, complementary perhaps. Comedians always want to be tragedians and vice versa. Under Gamow's influence Teller wanted to have "ideas" at any cost.

Gainesville
January 1974

## Paul Erdös

ULAM: Plutarch compared lives, and it may have a certain sense, a certain value, to compare pairs of mathematicians. Take Erdös. Erdös and I have something in common, a tremendous facility for finding difficult combinatorial problems out of thin air.

I'd like to take Erdös, Rota, and Everett and, like in the theory of colors, see whether by mixing them one could produce all other colors!

ROTA: Your style is completely different from Erdös's. He is interested in proofs; you are not interested in proofs. You are interested in problems interesting to state and don't care very much how they are solved. Erdös cares about techniques that he uses all the time.

ULAM: Really? He likes to think from the beginning; he does not quote somebody's theorem to prove something else.

ROTA: Your typical problem can almost always be restated as follows: Develop a theory of ... along the following lines. That is what your problems are about, whereas Erdös's are never this way.

ULAM: Maybe he exaggerates by trying to put everything on paper immediately.

ROTA: There is a primitivism to Erdös.

ULAM: Yes. I have that feeling too, very much. Once you said something which if true is very flattering, namely that things I mention are germs of whole theories, whereas his on the whole are more special.

Erdös is not really narrow, but it is hard to get things out of him. I think he knows a great deal, though I don't think he has read much belles-lettres.

ROTA: He has no outside interests.

ULAM: I think he reads quickly and efficiently and gets the gist of things. I don't know how much he knows, say, of French literature, the classics, history. He does know some history because he is interested in politics. He reads about current things, progress in medicine, a little about physics. He forms impressions.

He is really very nice, never diminishes people, does not make fun of anybody, and is very much interested in young people in the sense that he is always searching for young geniuses. Wouldn't you say that in a sense he is more human than von Neumann or Fermi? Fermi was enormously aware of but not warmly interested in others.

ROTA: I really don't understand Erdös as a person. I understand him mathematically.

ULAM: He wants to be famous. He is very well known. Every mathematician knows him. He has written over 800 papers. You know the "Erdös number"—who wrote papers with him. People have a weakness for him. He has some sense of humor. Politically he is not naive at all. He is very well wishing, and really I have never heard him speak badly of anyone. Very few people are like that. You speak badly of people. I speak badly of people.

The death of his mother was a terrible blow; he still has not recovered. She was ninety-one, and he says she still could have lived another three or four years.

Erdös is interested in human destiny, in sickness, in death. He has no home. Now he refuses to go to Hungary because of their attitude towards two Israelis. Last summer, at the time of a meeting in his honor, Hungary did not let two Israeli mathematicians in. This infuriated him, and he said he would not return for several years. He is a true man of principles and in a way very courageous.

Gainesville
March 1974

*Erdös*
*Paul Erdös is the most prolific mathematician of modern times and is second only to Euler in the volume of work produced. He was a long-time friend and collaborator of Ulam. In Ulam's files are 191 letters from Erdös, mostly in longhand. Erdös collaborates the world over and has done more for collaboration in mathematics than anyone else.*

## Teaching Physics To Rota

ROTA: What are your views on classical physics versus quantum mechanics?

ULAM: Quantum mechanics uses variables of higher type. Instead of idealized points, or groups of points or little spheres or atoms or bodies, the primitive notion is a probability measure. Quite a logical leap from the classical point of view.

Nevertheless you find in quantum mechanics the strange phenomenon that a theory dealing with variables of higher type has to be imaged on variables of lower type. It is the complementarity between electron and wave.

In our minds, because of habit or historical conditions, an electron is a localized small object, whereas a wave is something diffuse. But some phenomena show a dual nature; they share properties of one and the other. I don't think there is yet a satisfactory logical or mathematical discussion of this duality. In my opinion it does not do any good to write down axioms which sanctify the usual dicta. People accept what works. Quantum theory is very successful at describing atomic phenomena, and some of its general features seem to be valid even in the subatomic nuclear and elementary-particle phenomena. But the overall success is not too striking, except perhaps in quantum electrodynamics.

To me the situation in theoretical physics seems to be the following. There are about one hundred bright young physicists in the country, all mathematically very skillful and learned—too much so for my taste! To predict or explain some of their observations, they fudge a little, which is only natural. However the next experiments at CERN or Fermilab always seem to invalidate their calculations. You would think that among so many guys making so many different predictions, at least a few would get some correct answers, but no! Whatever the prevailing beliefs or attempts, the new experiments show something else. How can this be? Nature is not that malicious. Maybe today's physicists are technically very skilled but not really imaginative or innovative enough.

ROTA: What is to your mind problem number one in physics?

ULAM: Is there a true infinity of structures going down into smaller and smaller dimensions? Is it not a precise problem, or recognized as such.

In physics there has always been an atomistic or a field point of view. If there is a field, then points are mathematical points and they are all the same. But another possibility is a very strange structure of successive stages, each stage different. The topology or the scene on which they exist, that is, space and time themselves, need not be the uniform, smooth Euclidean topology. The miracle is that physics would not be possible if protons and electrons were not very much the same. If this similarity or identity of subsets of the universe did not exist, there would be no physics. The role of physics to some extent is to divide the existing groupings—call them particles—into entities isomorphic or almost isomorphic to each other.

The great hope of physics lies in the fact that one can almost repeat the same situations. Having twenty or twenty-two bodies does not radically change a physical law. In mathematics too there are similar analogies. In physics such analogies are essential.

It may be that in reality for phenomena in the small and involving high energy, there may be an underlying true infinity that does not allow for similarities. It may be that at the present stage of evolution of the universe a sufficient number of identical situations has not yet been produced. If this is so, then physics will become fundamentally more complicated.

Who knows whether there are not fundamental complications in the nature of subparticles? Are the billions of protons that compose our bodies or this table really the same? This stability is far from guaranteed. There might be critical numbers, critical crises not only in technology but in fundamental physics itself.

Since Gödel, even in mathematics it is not simple anymore. Have I told you that van Hove asked me to give a talk on infinities in physics at CERN?

ROTA: What did you say?

ULAM: I intend to write it up in my future *Physics for Mathematicians*.

In recent years you seem to have lost your feeling of horror towards physics!

ROTA: I did not understand. I like to understand.

ULAM: Do you understand mathematics? It is easier to get accustomed quickly to a fixed symbolism, like that of mathematics. But this is largely an illusion. Mathematics has a restricted range; it has not changed since Archimedes. There are axioms, proofs, lemmas, theorems. In physics it is not clear what one really does and at what point one becomes satisfied that the formulation is correct.

Santa Fe
July and August 1974

## Miscellaneous Comments About Mathematics

**Mathematics and Games**

ULAM: A French philosopher whose name I forget said that nowhere has the human mind shown itself so inventive as in devising new games.

ROTA: Inventors of games are always anonymous. Why? What is your philosophy of the anonymity of games?

ULAM: Probably other people quickly perfected the original invention, and it is difficult to find out who thought of it first.

Are games part of combinatorics or the other way around? I claim that much of mathematics can be "paisaised," a Greek word which means to play.

Here is an example of a problem inspired by a game. Suppose $n$ is a given integer and we are to build, you and I, two permutations of $n$ letters. We construct them in turn as follows: For the first permutation I take $n_1$, you take $n_2$, I take $n_3$, and so on. Finally we get a permutation. Then we play for the second permutation. If the two permutations generate the group of all permutations, I win; if not you win. Who has a winning strategy in this game? I don't know.

If we do it at random, what is the chance [that there is a winner]? This then becomes a combination of measure, probability, and combinatorics. I talk about this racket in my book of problems. It is amusing, isn't it? It can be done in any branch of mathematics.

Paris
April 1972

**Combinatorics**

ULAM: Combinatorics is devoid of general methods. It is full of nice individual curiosities, it is Erdösian. I have nothing against it, it is amusing. But it throws no light on anything else.

ROTA: You are not being fair.

ULAM: Complex functions, the idea of entropy are broader. Ramsey's theorem, interesting as it is, is like progress in zoology when a new species of insects with one red eye and one green eye has been discovered!

ROTA: Ramsey's theorem tells more about the nature of sets than all the axioms of set theory!

ULAM: It is one of numerous properties of infinity. Why take two sets of pairs and divide them into two classes? My master's thesis already contains that sort of thing.

Some problems, big or small, are solved with a bang; they open new vistas. Others are solved with a whimper, in a way which is very specific and leaves nothing to be said or asked, regardless of whether the problems are important or interesting.

Paris
April 1972

*Ramsey's Theorem*

*One consequence of Ramsey's theorem is the following: Among a gathering of 6 people, there will be at least 3 all of whom know one another or else there will be 3 none of whom know one another. This is not true if only 5 are gathered together. In general, for each positive integer $k$ there is a positive integer $n = n(k)$ such that if $n$ people are gathered together, then there will be $k$ all of whom know one another or else none of whom know each other. To this date we know only that $n(k)$ exists but not its value for arbitrary $k$. It is known, however, that $n(2) = 2$, $n(3) = 6$, and $n(4) = 18$.*

**Cantor**

ULAM: Set theory revolutionized mathematics. It is largely the work of Cantor. What made set theory is the fact that Cantor proved that the continuum is not countable. It is hard to imagine that a field that arose from trigonometric series quickly transformed the shape and flavor of math.

Paris
April 1972

**Gödel**

ULAM: A second landmark on the scale of centuries was Gödel's undecidability theorem. Now there is a flood of results that show that our intuition of infinity is not complete. Cohen's results opened the flood gate.

Mathematics is not a finished object based on some axioms. It evolves genetically. This has not yet quite come to conscious realization.

Paris
May 1972

**A Few Unsolved Problems**

*The Continuum Hypothesis*
$2^{\aleph_0} = \aleph_1$
*or*
*if E is an uncountable subset of the interval [0,1], then there is a one-to-one correspondence between the elements of E and all the numbers between 0 and 1.*

*Twin Primes Conjecture*
*There are infinitely many primes p such that p + 2 is also a prime.*

*Goldbach Conjecture*
*Every even integer equal to or greater than 6 can be expressed as the sum of two odd primes in at least one way . For example,* $12 = 5 + 7$ *and* $16 = 3 + 13 = 11 + 5$

ROTA: Can you list ten unsolved problems in mathematics which you consider important?

ULAM: First, the continuum hypothesis. If you take the existing axioms for set theory, then it is independent.

ROTA: One!

ULAM: But the existing axioms are probably not enough to give expression to our intuitions about sets. In that sense the continuum hypothesis is not a closed story.

Two. In number theory, any problem is as good as any other. I don't know which to choose, the infiniteness of twin primes or the Goldbach conjecture. The fact that they are very difficult and so simple makes them in my opinion very important. I have to list the Riemann hypothesis because it has so many consequences, although it is not one of my favorite problems, for a reason which I cannot express.

ROTA: Would you list the Riemann hypothesis as third?

ULAM: I don't like to order them. Snobbism plays a role in the ranking of mathematical problems. By chance some so-called great mathematician mentions something. For example, out of Hilbert's marvelous twenty-three problems, several would not be considered important if it were not for the fact that it was Hilbert who proposed them! Now what would you say besides these?

It is like asking someone to please mention ten best dishes or paintings! I don't know whether any single problem is really important, except in foundations of set theory. They are mainly important for what they suggest or allude to. Think of Fermat's conjecture. It is important because it is difficult but probably also because whoever will solve it will have found some new trick or method. The important thing is that the break is simple and difficult. I came to this conclusion sort of gradually. I am being honest, which most people are not.

A great problem is: Why are some problems sometimes difficult to solve? That is metamathematical, but it may some day be mathematized. The notion of complexity is beginning to be made precise, and what I just said will become a super problem.

ROTA: Why should Goldbach's conjecture be more interesting than a Chinese puzzle?

ULAM: Because it is simple. Any child can understand it. Isn't it curious that a child can ask questions about numbers that no mathematician can answer?

Gainesville
January 1974

*Riemann Hypothesis*
*Let* $\xi(z) = 1 + 1/2^z + 1/3^z + \cdots$. *If* $\xi(z) = 0$, *then* $Re\ z = 1/2$.

*Fermat's "Last Theorem"*
*If* $n > 2$, *there do not exist positive integers* $x, y$, *and* $z$ *such that* $x^n + y^n = z^n$.

**Infinity**

ULAM: Why is it that calculus, which deals with limits, is so effective? Or why are asymptotic theorems so much simpler than finite approximations? Infinity does not correspond to the popular image. It is a guiding light, a star that draws us to finite ways of thinking, God knows why.

Santa Fe
July 1974

**The Value of Mathematics**

ROTA: What is the value of mathematics?

ULAM: Value? In what sense? In what market?

It has value because it trains the brain. Just like in any other game, practice sharpens the organ. I don't know if today mathematicians' brains are any sharper than in the time of the Greeks. Yet I think mathematics plays a genetic role. It is one of the few ways to perfect the brain, to perhaps develop new connections in the brain. It has a peculiar sharpening value. Nothing could be more important. I don't know if any other science plays the same role. Another value is the aesthetic one, which is for the practitioners.

ROTA: What is its ugliness? Could you state an ugly theorem?

ULAM: Ugliness lies in the fact that one has to be punctilious, make sure of every step. In mathematics, one cannot paint with a wide brush, one has to fill in all the details.

The same is true in chess. There are chess games which have flaws. In fact most do. Otherwise there would not be a loser.

ROTA: Compare mathematics to the classics as an educational technique.

ULAM: I would say they are complementary. Latin grammar is good training in logic, not Boolean logic, but relational logic.

Santa Fe
July 1974

**Foundations of Mathematics**

ULAM: Mycielski disagrees with me when I say there will be systems of axioms for set theory other than the Zermelo-Fraenkel point of view. He claims that everything that we can think of can be expressed in those terms. This may be true but there might someday be entirely new points of view, even about sets or classes of sets. Sets may some day be considered as "imaginary." I think that will come to pass, though at present it is not admissible.

Everything that is conceivable somehow eventually comes into existence, in what form we cannot say. Ideas which begin in a prosaic way, like the study of complexity, are the ones that go very far.

ROTA: As a phenomenologist I agree.

Santa Fe
July 1974

**The Future of Mathematics**

ROTA: What about *l'avenir des mathématiques* today?

ULAM: Mathematics will change. Instead of precise theorems, of which there are now millions, we will have, fifty years from now, general theories and vague guidelines, and the individual proofs will be worked out by graduate students or by computers.

Mathematicians fool themselves when they think that the purpose of mathematics is to prove theorems, without regard to the broader impact of mathematical results. Isn't it strange?

In the next fifty years there will be, if not axioms, at least agreements among mathematicians about assumptions of new freedoms of constructions, of thoughts. Given an undecidable proposition, there will be a preference as to whether one should assume it to be true or false. Iterated this becomes: Some statements may be undecidably undecidable. This has great philosophical interest.

ROTA: I disagree. I don't think the current work in set theory is going anywhere, and I deny that it has philosophical import. It is a bunch of technicians doing Talmudic, irrelevant exercises.

ULAM: You may not like it, but it is as relevant as Heidegger!

Set theoreticians are workers, not generals, discovering interesting facts on the behavior of axioms and how incomplete they are. To me this is of great interest. One used to assume certain ideas of infinity and suddenly, lo and behold, they are incomplete.

Santa Fe
August 1976

# Publications of Stanislaw M. Ulam

## Set Theory

Remark on the generalised Bernstein's theorem.* *Fundamenta Mathematicae* 13(1929): 281–3.

Concerning functions of sets.* *Fundamenta Mathematicae* 14(1929): 231–3.

Über gewisse Zerlegungen von Mengen.* *Fundamenta Mathematicae* 20(1933): 221–3.

On the equivalence of any set of first category to a set of measure zero (with J. C. Oxtoby).* *Fundamenta Mathematicae* 31(1938): 201–6.

On visual hulls of sets (with G. H. Meisters).* *Proceedings of the National Academy of Sciences of the United States of America* 57(1967): 1172–4.

Note on the visual hull of a set (with W. A. Beyer).* *Journal of Combinatorial Theory* 4(1968): 240–5.

On equations with sets as unknowns (with Paul Erdös).* *Proceedings of the National Academy of Sciences of the United States of America* 60(1968): 1189–95.

## Measure

Zur Masstheorie in der allgemeinen Mengenlehre.* *Fundamenta Mathematicae* 16(1930): 140–50. Also in *Mengenlehre*, edited by U. Felgner, 223–33. Darmstadt: Wissenschaftliche Gesellschaft, 1979.

Sur une propriété de la mesure de M. Lebesgue (with J. Schreier).* *Comptes Rendus de L'Académie des Sciences* 192(1931): 539–42.

Zum Massbegriff in Produkträumen.* In *Verhandlungen, Internationaler Mathematikerkongress Zürich 1932*, volume 2, 118–9. Zürich: Orell Füssli Verlag, 1932.

Sur la théorie de la mesure dans les espaces combinatoires et son application au calcul des probabilités: I. Variables indépendantes (with Z. Lomnicki).* *Fundamenta Mathematicae* 23(1934): 237–78.

On the existence of a measure invariant under a transformation (with J. C. Oxtoby).* *Annals of Mathematics, Second Series* 40(1939): 560–6.

Measure-preserving homeomorphisms and metrical transitivity (with J. C. Oxtoby).* *Annals of Mathematics, Second Series* 42(1941): 874–920.

What is measure?* *American Mathematical Monthly* 50(1943): 597–602.

## Topology

On symmetric products of topological spaces (with Karol Borsuk).* *Bulletin of the American Mathematical Society* 37(1931): 875–82.

Quelques propriétés topologiques du produit combinatoire (with C. Kuratowski).* *Fundamenta Mathematicae* 19(1932): 247–51.

Eine Bemerkung über die Gruppe der topologischen Abbildungen der Kreislinie auf sich selbst (with J. Schreier).* *Studia Mathametica* 5(1934): 155–9.

Über topologische Abbildungen der euklidischen Sphären (with J. Schreier).* *Fundamenta Mathematicae* 23(1934): 102–18.

## Transformation Theory

Sur les transformations isométriques d'espaces vectoriels normés (with S. Mazur).* *Comptes Rendus de L'Académie des Sciences* 194(1932): 946–8.

Sur les transformations continues des sphères euclidiennes (with J. Schreier).* *Comptes Rendus de L'Académie des Sciences* 197(1933): 967–8.

Sur un coefficient lié aux transformations continues d'ensembles (with C. Kuratowski).* *Fundamenta Mathematicae* 20(1933): 244–53.

Über gewisse Invarianten der $\epsilon$-Abbildungen (with Karol Borsuk).* *Mathematische Annalen* 108(1933): 311–8.

On approximate isometries (with D. H. Hyers).* *Bulletin of the American Mathematical Society* 51(1945): 288–92.

Approximate isometries of the space of continuous functions (with D. H. Hyers).* *Annals of Mathematics, Second Series* 48(1947): 285–9.

Random processes and transformations. In *Proceedings of the International Congress of Mathematicians (Cambridge, Massachusetts, August 30–September 6, 1950)*, volume 2, 264–75. Providence, Rhode Island: American Mathematical Society, 1952.

Quadratic transformations. Part I (with M. T. Menzel and P. R. Stein). Los Alamos Scientific Laboratory report LA–2305, 1959.

Some properties of certain non-linear transformations.* In *Mathematical Models in Physical Sciences: Proceedings of the Conference at the University of Notre Dame, 1962*, edited by Stefan Drobot, 85–95. Englewood Cliffs, New Jersey: Prentice- Hall, Inc., 1963.

Computer studies of some history-dependent random processes (with W. A. Beyer and R. G. Schrandt). Los Alamos Scientific Laboratory report LA–4246, 1969.

Non-linear transformation studies on electronic computers (with P. R. Stein).* *Rozprawy Matematyczne* 39(1963): 1-66. The Introduction and Part I are also in *Essays on Cellular Automata*, edited by Arthur W. Burks. Urbana, Illinois: University of Illinois Press, 1970.

Lectures in non-linear algebraic transformations (with P. R. Stein). In *Studies in Mathematical Physics*, edited by A. O. Barut, 263–314. Dordrecht, The Netherlands: D. Reidel Publishing Company, 1970.

*This publication appears in *Stanislaw Ulam: Sets, Numbers, and Universes*, edited by W. A. Beyer, J. Mycielski, and G.-C. Rota. Cambridge, Massachusetts: The MIT Press, 1974.

## Group Theory

Sur le groupe des permutations de la suite des nombres naturels (with J. Schreier).* *Comptes Rendus de L'Académie des Sciences* 197(1933): 737–8.

Über die Permutationsgruppe der natürlichen Zahlenfolge (with J. Schreier).* *Studia Mathematica* 4(1933): 134- -41.

Sur le nombre de générateurs d'un groupe semi-simple (with H. Auerbach).* *Comptes Rendus de L'Académie des Sciences* 201(1935): 117–9.

Sur le nombre des générateurs d'un groupe topologique compact et connexe (with J. Schreier).* *Fundamenta Mathematicae* 24(1935): 302–4.

Über die Automorphismen der Permutationsgruppe der natürlichen Zahlenfolge (with J. Schreier).* *Fundamenta Mathematicae* 28(1937): 258–60.

On ordered groups (with C. J. Everett).* *Transactions of the American Mathematical Society* 57(1945): 208–16.

On some possibilities of generalizing the Lorentz group in the special relativity theory (with C. J. Everett).* *Journal of Combinatorial Theory* 1(1966): 248–70.

## Miscellaneous Mathematical Topics

Problème 56. *Fundamenta Mathematicae* 20(1933): 285.

Sur une propriété caractéristique de l'ellipsoïde (with H. Auerbach and S. Mazur).* *Monatshefte für Mathematik und Physik* 42(1935): 45–8.

Problème 74. *Fundamenta Mathematicae* 30(1938): 365.

Problèmes P34; P35; P35,R1 (with S. Banach). *Colloquium Mathematicum* 1(1947): 152–3.

Approximately convex functions (with D. H. Hyers).* *Proceedings of the American Mathematical Society* 3(1952): 821–8.

A property of randomness of an arithmetical function (with N. Metropolis).* *American Mathematical Monthly* 60(1953): 252–3.

On the stability of differential expressions (with D. H. Hyers).* *Mathematics Magazine* 28(1954): 59–64.

An open problem. In *Recent Advances in Game Theory*, 223. Princeton, New Jersey: Princeton University Conference, 1962.

Some problems of a dilettante. Problems 110, 111, and 112. In *Proceedings of the 1963 Number Theory Conference, University of Colorado, Boulder, Colorado, August 5–24, 1963*, edited by S. Chowla and B. Jones, 114-5.

Problems and games in mathematics (with R. D. Mauldin). *Advances in Applied Mathematics* 8(1987): 281–344.

## Branching Processes

Theory of multiplicative processes (with D. Hawkins). Los Alamos Scientific Laboratory report LA–171, 1944.

Multiplicative systems, I (with C. J. Everett).* *Proceedings of the National Academy of Sciences of the United States of America* 34(1948): 403–5.

Multiplicative systems in several variables. Parts I, II, and III (with C. J. Everett). Los Alamos Scientific Laboratory reports LA–683, LA–690, and LA–707, 1948.

Computations on certain binary branching processes. In *Computers in Mathematical Research*, edited by R. F. Churchhouse and J.-C. Herz, 168–171. Amsterdam: North-Holland Publishing Company, 1968.

## Weapons Research

Fourteen weapons-related reports written by Ulam and his collaborators between 1944 and 1958 are still classified. These are listed in "Publications of Stanislaw M. Ulam," compiled by B. Hendry (Los Alamos Scientific Laboratory report LAMS–3923, 1968) and in *Stanislaw Ulam: Sets, Numbers, and Universes*, edited by W. A. Beyer, J. Mycielski, and G.-C. Rota (The MIT Press, 1974).

## Algebra

Projective algebra I (with C. J. Everett).* *American Journal of Mathematics* 68(1946): 77–88.

Generators for algebras of relations (with A. R. Bednarek). *Bulletin of the American Mathematical Society* 82(1976): 781–2.

Projective algebra and the calculus of relations (with A. R. Bednarek). *Journal of Symbolic Logic* 43(1978): 56–64.

## Monte Carlo Method

Statistical methods in neutron diffusion (with J. von Neumann). Report written by R. D. Richtmyer and J. von Neumann. Los Alamos Scientific Laboratory report LAMS–551, 1947. Also in *Von Neumann: Collected Works, 1903–1957*, edited by A. H. Taub, volume 5. Oxford: Pergamon Press, 1963.

The Monte Carlo method (with Nicholas Metropolis).* *Journal of the American Statistical Association* 44(1949): 335–41.

On the Monte Carlo method. In *Proceedings of the 1949 Symposium on Large-Scale Digital Calculating Machines*, 207–12. Cambridge, Massachusetts: Harvard University Press, 1951.

Monte Carlo calculations in problems of mathematical physics. In *Modern Mathematics for the Engineer, Second Series*, edited by Edwin F. Beckenbach, 95–108. New York: McGraw-Hill Book Company, Inc., 1961.

## Mathematical Physics

Heuristic studies in problems of mathematical physics on high speed computing machines (with J. Pasta). Los Alamos Scientific Laboratory report LA–1557, 1953.

Studies of nonlinear problems. Part I (with E. Fermi, M. Tsingou, and J. Pasta).* Los Alamos Scientific Laboratory report LA– 1940, 1955. Also in *Enrico Fermi: Collected Papers*, volume 2, edited by E. Amaldi, H. L. Anderson, E. Persico, E. Segrè, and A. Wattenberg. Chicago: University of Chicago Press, 1965.

Infinite models in physics.* In *Proceedings of the Seventh Symposium in Applied Mathematics (Brooklyn Polytechnic Institute, April 14–15, 1955)*. American Mathematical Society Symposia in Applied Mathematics, volume 7, 87–95. New York: McGraw-Hill Book Company, Inc., 1957.

On the possibility of extracting energy from gravitational systems by navigating space vehicles. Los Alamos Scientific Laboratory report LAMS–2219, 1958.

Heuristic numerical work in some problems of hydrodynamics (with John R. Pasta).* In *Mathematical Tables and Other Aids to Computation* 13(1959): 1–12.

Stability of many-body computations. In *Hydrodynamic Instability*, edited by Garrett Birkhoff, Richard Bellman, and C. C. Lin, 247-58. American Mathematical Society Symposia in Applied Mathematics, volume 13. Providence, Rhode Island: American Mathematical Society, 1962.

The entropy of interacting populations (with C. J. Everett). Los Alamos Scientific Laboratory report LA–4256, 1969.

Ideas of space and space-time.** *Rehovot Magazine*, Winter 1972–73, 29–33.

Infinities. In *The Heritage of Copernicus*, edited by J. Neyman, 378–92. Cambridge, Massachusetts: The MIT Press, 1974.

Physics for mathematicians.** In *Physics and Our World: A Symposium in Honor of Victor F. Weisskopf (Massachusetts Institute of Technology, 1974)*, edited by Kerson Huang, 113–21. AIP Conference Proceedings, number 28. New York: American Institute of Physics, Inc., 1976.

On the operations of pair production, transmutations, and generalized random walks. *Advances in Applied Mathematics* 1(1980): 7–21.

Further applications of mathematics in the natural sciences.** In *American Mathematical Heritage: Algebra and Applied Mathematics*, edited by J. Dalton Tarwater, 101–14. Texas Tech University Mathematics Series, volume 13. Lubbock, Texas: Texas Technological University Press, 1981.

Speculations on some possible mathematical frameworks for the foundations of certain physical theories. *Letters in Mathematical Physics* 10(1985): 101–6.

## Dynamical Systems

On the ergodic behavior of dynamical systems. In "Series of lectures on physics of ionized gases." Los Alamos Scientific Laboratory report LA–2055, 1956.

On some statistical properties of dynamical systems.* In *Proceedings of the Fourth Berkeley Symposium on Mathematical Statistics and Probability (University of California, Berkeley, June 20–July 30, 1960)*, edited by Lucian M. Le Cam, Jerzy Neyman, and Elizabeth Scott, volume 3, 315–20. Berkeley: University of California Press, 1961. Translated into Russian (1963).

On general formulations of simulation and model construction. In *Prospects for Simulation and Simulators of Dynamic Systems*, edited by George Shapiro and Milton Rogers, 3–8. New York: Spartan Books, 1967.

Transformations, iterations and mixing flows. In *Dynamical Systems II*, edited by A. R. Bednarek and L. Cesari, 419–26. New York: Academic Press, 1982.

## Space Propulsion

On a method of propulsion of projectiles by means of external nuclear explosions (with C. J. Everett). Los Alamos Scientific Laboratory report LAMS-1955, 1955.

Statement before the U.S. House of Representatives. *Hearings on Astronautics and Space Exploration.* 85th Congress, 2nd session, April 15–May 12, 1958.

**This publication appears in *Science, Computers and People: From the Tree of Mathematics*, edited by Mark C. Reynolds and Gian-Carlo Rota. Boston: Birkhaüser, 1986.

Nuclear propelled vehicle, such as a rocket (with C. J. Everett). British Patent 877,392, 1961.

The future of nuclear energy in space: A panel discussion sponsored by the Aerospace Division, American Nuclear Society, at the 1963 winter meeting in New York City, N. Y. on November 20, 1963 (with F. deLuzio, W. von Braun, M. Hunter, and I. Asimov), edited by R. F. Trapp. Also in *Nuclear News*, July 1964.

The Orion project.** *Nuclear News*, January 1965, 25–7.

## Number Theory

On certain sequences of integers defined by sieves (with Verna Gardiner, R. Lazarus, and N. Metropolis).* *Mathematics Magazine* 29(1956): 117–22.

A visual display of some properties of the distribution of primes (with M. L. Stein and M. B. Wells).* *American Mathematical Monthly* 71(1964): 516–20.

An observation on the distribution of primes (with M. Stein).* *American Mathematical Monthly* 74(1967): 43–4.

Some probabilistic remarks on Fermat's last theorem (with P. Erdös).* *Rocky Mountain Journal of Mathematics* 1(1971): 613–6.

## Combinatorics

Study of certain combinatorial problems through experiments on computing machines (with P. R. Stein). In *Proceedings of the 1955 High-Speed Computer Conference (Louisiana State University, Baton Rouge, Louisiana, February 14–16, 1955)*, 101–6.

Combinatorial analysis in infinite sets and some physical theories. *SIAM Review* 6(1964): 343–55.

Some combinatorial problems studied experimentally on computing machines. In *Applications of Number Theory to Numerical Analysis*, edited by S. K. Zaremba, 1–10. New York: Academic Press, Inc., 1972.

## Computers and Computing

Experiments in chess (with J. Kister, P. Stein, W. Walden, and M. Wells).* *Journal of the Association for Computing Machinery* 4(1957): 174–7.

Experiments in chess on electronic computing machines (with P. R. Stein).** *Chess Review*, January 1957, 13–5. Also in *Computers and Automation*, September 1957.

On some new possibilities in the organization and use of computing machines. IBM research report RC-86, 1957.

The late John von Neumann on computers and the brain.** *Scientific American*, June 1958, 127.

Electronic computers and scientific research. In *The Age of Electronics*, edited by Carl F. J. Overhage, 95–108. New York: McGraw-Hill Book Company, Inc., 1962. Also in *Computers and Automation*, August 1963 and September 1963.

Computers.** *Scientific American*, September 1964, 202–216.

La machine créatrice. In *Rencontres Internationales de Genève "Le robot, la bête et l'homme (1965)*, 31–42. Neuchatel: Éditions de la Baconnière, 1966.

Some remarks on relational composition in computational theory and practice (with A. R. Bednarek). In *Fundamentals of Computation Theory: Proceedings of the 1977 International FCT-Conference (Poznań-Kórnik, Poland, September 19–23, 1977)*, edited by Marek Karpiński, 22–32. Lecture Notes in Computer Science, volume 56. Berlin: Springer-Verlag, 1977.

On the theory of relational structures and schemata for parallel computation (with A. R. Bednarek). Los Alamos Scientific Laboratory report LAMS–6734, 1977.

An integer valued metric for patterns (with A. R. Bednarek). In *Fundamentals of Computation Theory*, 52–7. Berlin: Akademie-Verlag, 1979.

Von Neumann: The interaction of mathematics and computing.** In *A History of Computing in the Twentieth Century: A Collection of Essays*, edited by N. Metropolis, J. Howlett, and Gian-Carlo Rota, 93–9. New York: Academic Press, Inc., 1980.

A mathematical physicist looks at computing.** *Rehovot Magazine*, volume 9, number 1, 47–9.

## Biomathematics

On some mathematical problems connected with patterns of growth of figures.** *Applied Mathematics* 14(1962): 215–24. Also in *Essays on Cellular Automata*, edited by Arthur W. Burks. Urbana, Illinois: University of Illinois Press, 1970.

How to formulate mathematically problems of the rate of evolution?** In *Proceedings of the Symposium on Mathematical Challenges to the Neo-Darwinian Interpretation of Evolution, New York, April 5–8, 1961*, edited by Paul S. Moorhead and Martin M. Kaplan. Providence, Rhode Island: American Mathematical Society. Wistar Institute Monograph 5: (1967) 21–33, April 25-26, 1966 New York: A. Liss, 1985.

On recursively defined geometrical objects and patterns of growth (with R. G. Schrandt).** Los

Alamos Scientific Laboratory report LA–3762, 1967. Also in *Essays on Cellular Automata*, edited by Arthur W. Burks. Urbana, Illinois: University of Illinois Press, 1970.

On the pairing process and the notion of genealogical distance (with Jan Mycielski).* *Journal of Combinatorial Theory* 6(1969): 227–34.

Some elementary attempts at numerical modeling of problems concerning rates of evolutionary processes (with R. Schrandt). Los Alamos Scientific Laboratory report LAMS–4573, 1971.

Metrics in biology: An introduction (with W. A. Beyer, T. F. Smith, and M. L. Stein). Los Alamos Scientific Laboratory report LA–4973, 1972.

Some ideas and prospects in biomathematics.** *Annual Review of Biophysics and Bioengineering* 1(1972): 277– 91.

A molecular sequence metric and evolutionary trees (with William A. Beyer, Myron L. Stein, and Temple F. Smith). *Mathematical Biosciences* 19(1974): 9–25.

The role of abstract mathematical ideas in possible conceptual advances in natural sciences, more specifically biology. In *Proceedings of International Colloquium on the Role of Mathematical Physics in the Development of Science (Collège de France, Paris, June 13–15, 1977)*, edited by Dominique Akl, Moshe Flato, and Daniel Sternheimer, 12–25. UNESCO, 1978.

Speculations about the mechanism of recognition and discrimination. Los Alamos National Laboratory unclassified release LAUR 82–62, 1982.

Reflections on the brain's attempts to understand itself. *Los Alamos Science*, number 15, 1987, 283–7.

## Astronomy

Possibility of an accelerated process of collapse of stars in a very dense centre of a cluster or a galaxy (with W. E. Walden). *Nature* 201(1964): 1202.

Collapse of stellar systems. In *Proceedings of the 25th International Astronomical Union Symposium (Thessaloníki, Greece, August 16–22, 1964)*, 76–7. International Astronomical Union, 1966.

Numerical studies of star systems. In *Colloque sur le Problème des N Corps*, 265–7. Éditions du Centre National de la Recherche, 1968.

## Complexity

The notion of complexity (with W. A. Beyer and M. L. Stein). Los Alamos Scientific Laboratory report LA–4822, 1971.

On the notion of analogy and complexity in some constructive mathematical schemata. Los Alamos National Laboratory report LA- -9065, 1981. Also in *Probability, Statistical Mechanics, and Number Theory*, edited by Gian-Carlo Rota. Advances in Mathematics: Supplementary Studies, volume 9. New York: Academic Press, Inc., 1986.

## Graph Theory

Generalizations of product isomorphisms. In *Recent Trends in Graph Theory*, 215. Lecture Notes in Mathematics, volume 186. Berlin: Springer-Verlag, 1971.

Minimal decomposition of two graphs into pairwise isomorphic subgraphs (with F. R. K. Chung, P. Erdös, R. L. Graham, and F. F. Yao). In *Proceedings of the Tenth Southeastern Conference on Combinatorics, Graph Theory, and Computing (Florida Atlantic University, Boca Raton, Florida, April 2–6, 1979)*, volume 1, 3–18. Congressus Numerantium, volume 23. Winnipeg, Manitoba: Utilitas Mathematica Publishing Incorporated, 1979.

## Obituaries

Stefan Banach, 1892–1945.** *Bulletin of the American Mathematical Society* 52(1946): 600–3. Homage to Fermi. *Santa Fe New Mexican*, January 6, 1955.

Marian Smoluchowski and the theory of probabilities in physics.** *American Journal of Physics* 25(1957): 475–81.

John von Neumann, 1903–1957.** *Bulletin of the American Mathematical Society* 64(1958): 1–49.

Kazimierz Kuratowski, 1896–1980.** *Polish Review* 26(1981): 62–6.

Kazimierz Kuratowski, Wspomnienia. *Wiadomosci Matematyczne*, 1983. Translated into Polish by R. Engelking. (1983).

## Miscellaneous Topics

Review of *Funkcje Rzeczywiste* by Roman Sikorski. *Bulletin of the American Mathematical Society* 65(1959): 305–6.

Statement before the Joint Committee on Atomic Energy. In *Frontiers in Atomic Energy Research: Hearings before the Subcommittee on Research and Development of the Joint Committee on Atomic Energy, Eighty-sixth Congress, Second Session, March 22–25, 1960*, 282–5. Washington, D.C.: U.S. Government Printing Office, 1960.

Communication to the U.S. Senate Committee on Foreign Relations. In *Nuclear Test Ban Treaty: Hearings before the Committee on Foreign Relations, United States Senate, Eighty-Eighth Congress, First Session, on Executive M*, 505–6 and 993. Washington, D.C.: U.S. Goverment Printing Office, 1963.

Thermonuclear devices.** In *Perspectives in Modern Physics: Essays in Honor of Hans A. Bethe*, edited by R. E. Marshak and J. Warren Blaker, 593–601. New York: Interscience Publishers, 1966.

An education in applied math. In *Proceedings of May 24-27, 1966 SIAM Conference (Aspen, Colorado)*, edited by James Ortega, Paul I. Richards, and Frank W. Sinden. *SIAM Review* 9(1967): 343–4.

Philosophical implications of some recent scientific discoveries.** In *Science, Philosophy and Religion.* Proceeding Symposium Kirtland Air Force Laboratory, Albuquerque, New Mexico 44–48. (1968).

Wspomnienia Kawiarni Szkockiej (Reminiscences of the Scottish Café). *Wiadomosci Matematyczne* 12(1969): 49–58. Available in English only in manuscript form.

The applicability of mathematics.** In *The Mathematical Sciences: A Collection of Essays*, edited by the National Research Council's Committee on Support of Research in the Mathematical Sciences, 1–6. Cambridge, Massachusetts: The M.I.T. Press, 1969.

Foreword to *My World Line: An Informal Autobiography* by G. Gamow. New York: Viking Press, 1970.

Testimony before the United States District Court, District of Minnesota, Minneapolis, Minnesota, September 17, 1971, in the case of Honeywell Incorporated versus Sperry Rand Corporation, 7342–438.

Gamow—and mathematics.** In *Cosmology, Fusion & Other Matters: George Gamow Memorial Volume*, edited by Frederick Reines, 272–9. Boulder, Colorado: Colorado Associated University Press, 1972.

New rules and old games. *Outlook*, Spring 1974, 32–3.

Arthur Koestler et le défi du hazard: Entretien avec Stan Ulam de Pierre Debray-Ritzen. In *Arthur Koestler*, 428–32. Cahiers de l'Herne. Paris: Èdition de L'Herene, 1975.

Przygody matematyka. *Kultura* 9(30 Lipca 1978). Translated into Polish by Jerzy Jaruzelski.

Banach i inni. *Kultura* 10(6 Sierpnia 1978). Translated into Polish by Jerzy Jaruzelski.

Narodziny "Ksigi Szkockiej." *Kultura* 10(13 Sierpnia 1978). Translated into Polish by Jerzy Jaruzelski.

Preface to *A Half Century of Polish Mathematics: Remembrances and Reflections* by Kazimierz Kuratowski. International Series in Pure and Applied Mathematics, Volume 108. Oxford: Pergamon Press Ltd., 1980.

An anecdotal history of the Scottish Book. In *The Scottish Book: Mathematics from the Scottish Café*, edited by R. Daniel Mauldin. Boston: Birkhäuser, 1982.

Reflections of the Polish masters: An interview with Stan Ulam and Mark Kac. *Los Alamos Science*, volume 3, number 3, 1982, 54–65.

Introduction** to *Selected Studies: Physics-Astrophysics, Mathematics, History of Science. A Volume Dedicated to the Memory of Albert Einstein*, edited by Themistocles M. Rassias and George M. Rassias. Amsterdam: North- Holland Publishing Company, 1982.

## Books

*The Scottish Book: A Collection of Problems*. An edited translation of a notebook kept at the Scottish Café for the Lwów Section of the Société Polonaise de Mathématiques. Privately mimeographed and distributed by S. M. Ulam in 1957. Reprinted as Los Alamos Scientific Laboratory report LA–6832, 1967.

*Mathematics and Logic: Retrospect and Prospects* (with Mark Kac). New York: Frederick A. Praeger, Inc., 1968. The text of this book first appeared as the article entitled "Mathematics and logic: Retrospect and prospects" in Britannica Perspectives, volume 1, 557–732. Chicago: Encyclopædia Britannica, Inc., 1968. Translated into French (1973), into Serbo-Croatian (1977), and into Spanish (1979).

*Stanislaw Ulam: Sets, Numbers, and Universes*, edited by W. A. Beyer, J. Mycielski, and G.-C. Rota. Cambridge, Massachusetts: The MIT Press, 1974.

*Adventures of a Mathematician*. New York: Charles Scribner's Sons, 1976. Paperback editions published in 1977 and 1983. Translated into Japanese (1979).

*A Collection of Mathematical Problems*.* New York: Interscience Publishers, 1960. Reprinted as *Problems in Modern Mathematics*. John Wiley & Sons, Inc., 1964. Translated into Russian (1964).

*Science, Computers, and People: From the Tree of Mathematics*, edited by Mark C. Reynolds and Gian-Carlo Rota. Boston: Birkhaüser, 1986.

*Mathematics at Los Alamos: The Collected Los Alamos Reports of S. Ulam and Collaborators, 1944–1984*, edited by D. Sharp and M. Simmons. To be published by University of California Press.

# Index





On the cover: Stanislaw Ulam. This portrait and the color art on pages 5, 35, and 293 are the work of Jeff Segler.

*One day when little Claire Ulam was watching some children playing ball with their father, a friend asked whether her father ever played like that with her. The answer was an emphatic "No! No! All my father does is think, think, think! Nothing but think!"*

Published by the Press Syndicate of the University of Cambridge
The Pitt Building, Trumpington Street, Cambridge CB2 1RP
32 East 57th Street, New York, NY 10022, USA
10 Stamford Road, Oakleigh, Melbourne 3166, Australia



First published 1987 as *Los Alamos Science* Special Issue
First published by Cambridge University Press 1989

Printed in the United States of America

*Library of Congress Cataloging-in-Publication Data*

From cardinals to chaos : reflections on the life and legacy of
Stanislaw Ulam / edited by N.G. Cooper.
p. cm.
"First published 1987 as Los Alamos science, special issue" – T. p.
verso.
Bibliography: p.
ISBN 0-521-36494-9. ISBN 0-521-36734-4 (pbk.)
1. Ulam, Stanislaw M. 2. Mathematicians – United States –
Biography. 3. Probabilities. 4. Nonlinear theories.
I. Cooper, Necia Grant.
QA29.U4F76 1988
510'.92'4—dc19 [B] 88-23367 CIP

*British Library Cataloguing in Publication Data*

From cardinals to chaos : reflections on the life
and legacy of Stanislaw Ulam.
1. Mathematics. Ulam, Stanislaw M. (Stanislaw Marcin)
I. Cooper, Necia Grant II. Los Alamos science
510'.92'4

ISBN 0-521-36494-9 hard covers
ISBN 0-521-36734-4 paperback

Cover illustration and illustrations on pages 4–5, 34–5, 292–3 by Jeff Segler & Associates, Inc., Los Alamos, New Mexico.

The quotations on pages 281 and 282 are reprinted with permission of D. Reidel Publishing Company.